ACS SYMPOSIUM SERIES 697

Lignin and Lignan Biosynthesis

Norman G. Lewis, EDITOR
Washington State University

Simo Sarkanen, EDITOR
University of Minnesota

Developed from a symposium sponsored by the Division
of Cellulose, Paper, and Textile
at the 211th National Meeting
of the American Chemical Society,
New Orleans, LA,
March 24–29, 1996

American Chemical Society, Washington, DC

Library of Congress Cataloging-in-Publication Data

Lignin and lignan biosynthesis / Norman G. Lewis, editor, Simo Sarkanen, editor

p. cm.—(ACS symposium series, ISSN 0097-6156; 697)

"Developed from a symposium sponsored by the Division of Cellulose, Paper, and Textile at the 211th National Meeting of the American Chemical Society, New Orleans, LA, March 24–29, 1996."

Includes bibliographical references and indexes.

ISBN 0-8412-3566-X

1. Lignin—Congresses.

I. Lewis, Norman G., 1949– . II. Sarkanen, Simo, 1946– . III. American Chemical Society. Division of Cellulose, Paper, and Textile. IV. American Chemical Society. Meeting (211th : 1996 : New Orleans, La.)

QK898.L5L54 1998
572′.56682—dc21 98-6367
CIP

This book is printed on acid-free paper.

Distributed by Oxford University Press

PRINTED IN THE UNITED STATES OF AMERICA

Advisory Board

ACS Symposium Series

Foreword

THE ACS SYMPOSIUM SERIES was first published in 1974 to provide a mechanism for publishing symposia quickly in book form. The purpose of the series is to publish timely, comprehensive books developed from ACS sponsored symposia based on current scientific research. Occasionally, books are developed from symposia sponsored by other organizations when the topic is of keen interest to the chemistry audience.

Before agreeing to publish a book, the proposed table of contents is reviewed for appropriate and comprehensive coverage and for interest to the audience. Some papers may be excluded in order to better focus the book; others may be added to provide comprehensiveness. When appropriate, overview or introductory chapters are added. Drafts of chapters are peer-reviewed prior to final acceptance or rejection, and manuscripts are prepared in camera-ready format.

As a rule, only original research papers and original review papers are included in the volumes. Verbatim reproductions of previously published papers are not accepted.

ACS BOOKS DEPARTMENT

Contents

Preface

This ACS Symposium Series volume, which together constitutes a definitive account of the current status of the field of lignin and lignan biosynthesis, has been a singularly most satisfying task for the editors. This field of science is arguably at its most exciting point ever, and it is therefore timely that this volume is the very first compilation that simultaneously addresses the biosynthesis of both lignins and lignans: the assembly mechanisms in the two cases do, indeed, appear to be closely related.

Lignins and lignans roughly account for 30% of all organic carbon circulating in the biosphere, and constitute the most abundant products of phenylpropanoid metabolism in vascular plants. Both the polymeric lignins and the (oligomeric) lignans are related structurally through some of their monomeric precursors and some of the basic skeleta that link the monomer units together. Yet, before the tutorial, which resulted in the present ACS Symposium Series volume, little had fundamentally changed over a span of four decades in terms of the common perceptions about how these molecular species are assembled in vivo.

It had been originally asserted, despite telling evidence to the contrary which has grown in magnitude ever since, that faithful facsimiles of lignin biopolymers can be produced in homogeneous solution in vitro without the need for any guiding molecular framework. Consequently, plant lignins have traditionally been characterized as possessing random configurations and being formed by a random assembly mechanism. However, this assertion could not explain why the closely related lignans exhibit absolute regio- and stereospecificity in the configurations of their inter-unit linkages, whereas the lignins apparently do not.

By 1996, the viability of an alternative working hypothesis for their biosynthesis could no longer be denied. It seemed that a dramatic paradigm shift was then about to occur in the field, particularly as regards control of monolignol coupling and modification of the biochemical steps leading to their formation. In anticipation of this turning point, a far-ranging tutorial about the subject was convened at the 211th American Chemical Society National Meeting in New Orleans on March 24, 1996, as a prelude to the Anselme Payen Award Symposium honoring Josef S. Gratzl of North Carolina State University. The symposium itself was the largest of its kind, while the tutorial was the first of its kind, in the history of the ACS Cellulose, Paper and Textile Division. This, in turn, led to this volume, and the editors believe that the chapters within the book bring together the flavor and the substance of a field undergoing a profound change.

Besides the content of the volume, the lignin and lignan biochemical pathways serve much more than simply as an immense storage reservoir of organic carbon in living systems. They were pivotal in both the successful evolution of terrestrial plants, and in their subsequent colonization of dry land. Indeed, our continued survival, as well as that of all terrestrial animal forms is—in one way or another—dependent upon vascular plants, and by extrapolation, on the lignin and lignan biochemical pathways. They also play a decisive role in carbon-dioxide turnover in the biosphere, since it is the lignins and lignans that represent the rate-limiting step in the organic carbon cycle, an area of growing interest as global warming issues come to the fore.

They also have a number of other important roles and/or functions, which include: (1) in forestry and in pulp–paper processing, since they help confer some of the most important and most desirable qualities to wood (color, durability, etc.), while at the same time these substances need to be removed during pulp–paper manufacture and biotechnological solutions are being sought to reduce their contents for this purpose; (2) in agriculture, as a means to help enable plants to withstand and respond to environmental changes and pathogen attack; (3) in the food and forage industry, whether as antioxidants for stabilizing processed foodstuffs, or in the modification of forage to render it more suitable for livestock consumption; (4) in the area of nutriceuticals, where the lignans (in particular) are of growing importance in health protection (e.g., by reducing markedly the incidence rates of breast and prostate cancers); and (5) in pharmacy, where many lignans have important pharmaceutical uses, such as podophyllotoxin (as etoposide or teniposide), which is one of only a handful of plant-derived substances used in the treatment of cancers.

Acknowledgments

The editors are particularly indebted to the generous financial support for both the ACS Tutorial and the Anselme Payen Award Symposium from the following contributors:

Dr. Leon Semke
Champion International Corporation
FMC Corporation
Federal Paperboard Company, Inc.
International Science Foundation of Russia
IMPCO-VOEST Alpine Company
Lenzing AG

Svenska Cellulosa Aktiebolaget SCA
Westvaco Corporation-Bleacher Board Division
Westvaco Corporation, New York
Weyerhaeuser Company

The success of the present ACS Symposium Series volume in no small measure rests upon their generosity.

The organizers also wish to express their gratitude to the various U.S. Federal agencies which made the support of these research endeavors possible (U.S. Department of Energy, the National Science Foundation, the U.S. Department of Agriculture and the U.S. National Aeronautic and Space Administration), as well as the other governments, agencies, and industries that supported many of the foreign scientists who participated and attended this highly successful meeting.

Kindest thanks are also extended to Hiroko Hayashi who superbly typed the edited versions of the papers, to Laurence B. Davin for the unstinting help in proof-reading and to Pamm Caldwell-Thomas for outstanding secretarial support. The editors, of course, assume responsibility for any small errors that may have gone undetected.

It is to be hoped that the reader will enjoy becoming acquainted with these new vistas in the lignin and lignan biochemical pathways, as much as the editors have enjoyed helping to compile the volume.

NORMAN G. LEWIS
Institute of Biological Chemistry
Washington State University
Pullman, WA 99164–6340

SIMO SARKANEN
Department of Wood and Paper Science
University of Minnesota
St. Paul, MN 55108–6128

Chapter 1

Lignin and Lignan Biosynthesis: Distinctions and Reconciliations

Norman G. Lewis[1], Laurence B. Davin[1], and Simo Sarkanen[2]

[1]Institute of Biological Chemistry, Washington State University, Pullman, WA 99164–6340
[2]Department of Wood and Paper Science, University of Minnesota, St. Paul, MN 55108–6128

Before 1996, the framework within which lignin biosynthesis was understood at the molecular level had not fundamentally changed for 4 decades. During the same period nothing at all had been explicitly proposed about the mechanistic basis for lignan formation *in vivo*. The associated deficit in plant biochemistry was not minor: lignins and lignans together account for roughly 30% of the organic carbon in vascular plants. On the other hand, the biochemical transformations in phenylpropanoid metabolism leading, *via* the shikimate-chorismate pathway through phenylalanine or tyrosine, to the so-called monolignols (namely, the monomeric lignin/lignan precursors) came to be reasonably well documented. Indeed, attention has more recently been drawn to the identification of *de facto* rate-limiting steps in the various metabolic segments of the pathway as potential control-points for biotechnological manipulation. The most curious characteristic usually attributed to the lignin/lignan biosynthetic pathway is that the monolignol-derived radical coupling processes leading to lignans are regio- and stereospecific, whereas those resulting in lignin macromolecules ostensibly are not. Now that the molecular basis for the dehydrogenative dimerization of monolignols to lignans has been unraveled, however, it appears likely that an analogous mechanism may be operative in the dehydrogenative polymerization of monolignols to lignins. The investigation of this possibility has indeed become a central concern in the field of lignin biosynthesis.

It has been almost three decades (*1*) since a comprehensive attempt had been made to summarize and evaluate contending views about lignin biosynthesis. Accordingly, the present volume draws together the current, yet conceptually vastly differing, ideas about how lignin composition and structure are established in living plants. It differs from all prior contributions, not only in introducing fresh evidence to support a compelling new paradigm that seeks to understand how lignin structure is established *in vivo*, but also in including lignan biosynthesis for comparative purposes. Indeed it appears that the first committed step in lignan biosynthesis bears an important relationship to the manner in which the configurations of lignins are specified *in vivo*.

Collectively, lignins and lignans are *the* major metabolic products of phenylpropanoid metabolism in vascular plants. For woody plants, they typically account for more than 20% of the weight of angiosperms and over 25% of that of the gymnosperms. Together, they constitute some of the metabolically most expensive products generated by plants (*2*), and are derived from the shikimate-chorismate pathway (Figure 1) (*3, 4*) which produces the aromatic amino acids, phenylalanine and tyrosine (Figure 2) (*5*). As shown in Figure 3, extension of the phenylpropanoid pathway in vascular plants, from phenylalanine onwards, ultimately leads to both the polymeric lignins as well as the dimeric/oligomeric lignans. The lignins fulfill essential functions by providing structural reinforcement to (woody) plant tissues, thereby allowing all vascular plants to stand upright. Moreover, this cell-wall reinforcing process provides both the corresponding vasculature which is necessary for water conduction, as well as in assisting the means for re-orientating stems and branches in response to changes in mechanical stresses and light levels. The lignans, on the other hand, are a ubiquitous group of closely related *non-structural* phenolic metabolites, which are primarily dimeric although higher oligomers exist. They play substantive roles in plant defense, through their potent biocidal (*6-9*) and antioxidant (*6, 10-13*) properties, and many also have important functions in medicine (*14-16*) and health maintenance (*17-22*).

From an evolutionary perspective, perhaps the most significant aspect of the lignins and lignans lies in the fact that, in their absence, vascular plants would *not* readily survive. Indeed, the continued existence of *all* terrestrial animal forms is in one way or another dependent on vascular plants and, hence, on the lignin/lignan biosynthetic pathway. Moreover, it is the differential expression of this pathway that is largely responsible for much of plant biodiversity: the variable deposition of these substances, in terms of their amount and specific composition, can dramatically alter the woody textures of plants, as well as affect other properties, such as heartwood color, durability and rot resistance, and even their (aromatic) fragrance.

Recognition of these factors has prompted the study of the lignin and lignan biosynthetic pathways, in order to develop biotechnological strategies directed to the rational re-engineering of woody and non-woody plants with respect to their lignin/lignan contents and composition. The goals of such studies are manifold and include: enhancing the quality (texture, color, durability, *etc.*) of specific woody plants for lumber and fine furniture applications; either lowering the lignin contents of specific woody plants, or rendering them more susceptible to chemical/biochemical delignification protocols for pulp and paper manufacture; lowering lignin contents in domestic livestock feedstocks, in order to reduce waste disposal difficulties; re-engineering more rigid (structurally reinforced) plants for agricultural purposes that would be better able to survive in harsher climates; providing intermediate chemicals for further processing (*e.g.* into new 'bio'polymers); improving health protection by increasing levels of cancer-preventing lignans in staple dietary foodstuffs; expanding sources of important (lignan-derived) pharmaceuticals (*e.g.* podophyllotoxin); exploiting the antioxidant properties of lignans; *etc.*

This overview is intended as a brief commentary upon the salient accomplishments in the field over the last six decades. It begins with the confusion attending the ascertainment of the fundamental constitution of the lignins and lignans, and then proceeds towards a clarification of the characteristic biochemical transformations that are responsible for their biosynthesis.

Brief Historical Development of Ideas

Phenolic substances account for *ca.* 30-40% of all organic carbon (*23*) in vascular plants, of which the lignins are the predominant metabolites. Yet a determination of the actual mechanisms for the formation of lignin polymers *in vitro* and *in vivo* has, for several decades, been thwarted by the persistence of a profoundly misguided paradigm. In the early 1930's, a tremendous controversy materialized over the

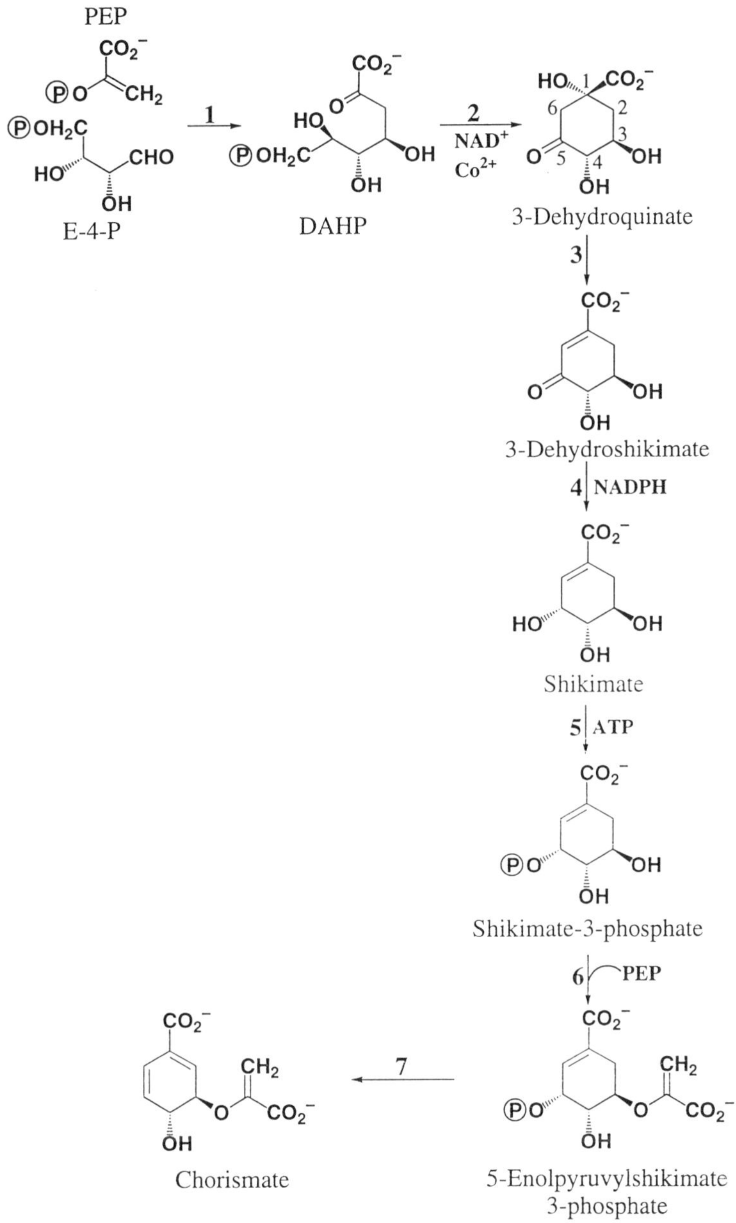

Figure 1. The shikimate-chorismate pathway: **1,** 3-deoxy-D-arabinose heptulosonic acid-7-phosphate (DAHP) synthase; **2,** dehydroquinate synthase; **3,** 3-dehydroquinate dehydratase; **4,** 3-dehydroshikimate reductase; **5,** shikimate kinase; **6,** 5-enolpyruvylshikimate-3-phosphate (EPSP) synthase; **7,** chorismate synthase.

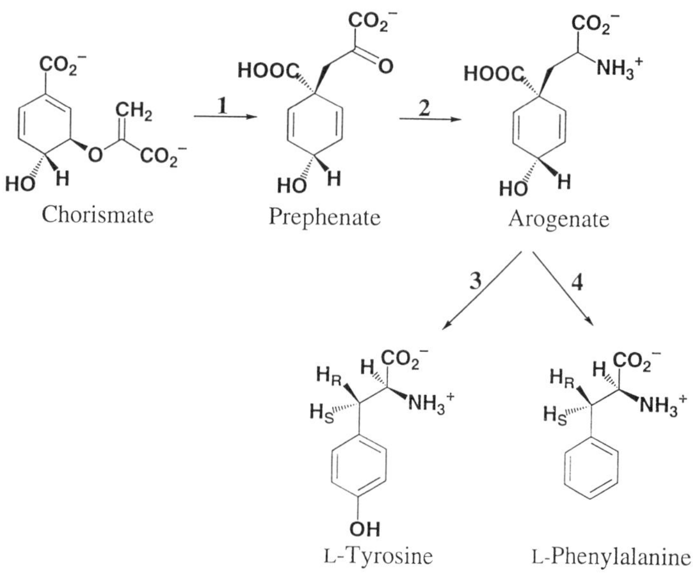

Figure 2. Biosynthetic pathway from chorismate to phenylalanine and tyrosine via arogenate: **1,** chorismate mutase; **2,** prephenate amino transferase; **3,** arogenate dehydrogenase; **4,** arogenate dehydratase.

chemical nature of lignins: various proponents described it as either a polysaccharide-derived polymer, arising from degradation of hemicelluloses, cellulose or pectins (*24*), and others proposed that even the terpenoids were products of altered lignin metabolism (*25*). Yet, much earlier, Ferdinand Tiemann (*26*) and Peter Klason (*27*) had speculated that lignin was derived from *E*-coniferyl alcohol **1** (Figure 4), and in 1933 Holgar Erdtman suggested that the monolignol (a term coined later) was converted into lignin *via* a dehydrogenative polymerization process (*28*). It was not until the 1950's that radiotracer experiments had clearly established *E*-coniferyl alcohol **1** and related monolignols to be, in fact, lignin precursors (*29*). Further studies revealed the details of the biosynthetic pathway between phenylalanine and the monolignols to involve the sequence of transformations shown in Figure 3 (see ref. (*30*) for a review).

Karl Freudenberg, Takayoshi Higuchi and others then attempted to characterize the enzymology involved in the phenolic coupling reactions that give rise to lignification (*31-35*). Curiously, they studied the dehydrogenative polymerization of *E*-coniferyl alcohol **1** using crude mushroom (laccase) (*32*) and horseradish (peroxidase) (*31, 35*) extracts, even though the enzymes employed did not originate from lignifying tissues! Thus, the enzymology responsible for the formation of an important natural product—Nature's second most abundant biopolymer, no less—relied upon enzymes from sources that had no connection with the biosynthetic pathway itself. Evidence for the involvement of comparable processes *in vivo* was thought to be provided by the observation that these crude enzyme preparations readily converted coniferyl alcohol, in the presence of O_2 and H_2O_2 as respective co-substrates, into products which *at that time* were assumed more or less to represent lignin. That is, according to these investigators, the only enzymatic control of lignin assembly involved free-radical generation from the monolignols with subsequent coupling occurring non-enzymatically. This supposition represented a departure from *all* other known biochemical processes, since no explicit control of the final configuration of the product was envisaged.

The (unresolved) problem is that, *in vitro*, the initial dehydrogenative coupling products from *E*-coniferyl alcohol **1** are primarily racemic (±)-dehydrodiconiferyl alcohols **2**, (±)-pinoresinols **3** and (±)-*erythro/threo* 8–*O*–4′-coniferyl alcohol ethers **4** in ratios of approximately 6:3:2 (see Figure 4 for structures). These products are formed through non-specific bimolecular coupling between free radical species generated by one-electron oxidation of the monolignol. On the other hand, lignin biopolymers possess frequencies of interunit linkages which differ markedly from the proportions of the different dimers produced *in vitro*. That is, natural lignins mainly embody 8–*O*–4′ (≥50%) and dehydrodiconiferyl alcohol (~10%) substructures, together with a variety of other linkages present in relatively low abundance. This has been established not only through degradative analyses (*36*), but also by specific *in situ* carbon-13 labeling, which was initially applied in order to study lignin biosynthesis (*37, 38*). There is, therefore, a marked contrast between lignin structures *in vivo* and those of the so-called monolignol dehydropolymerisates produced *in vitro*.

The early 1930's also saw the classification of an abundant group of dimeric phenylpropanoid compounds, linked through 8–8′ bonds, as lignan(e)s, a term which was used by R. D. Haworth in 1936 (*25*). Otto Gottlieb subsequently introduced the term neolignan to encompass all non 8–8′ linkages (*39, 40*), and further modified this definition to encompass the products from allylphenol coupling (*41*). However, throughout the present volume the term lignan is used to describe *all* possible coupling products as long as the linkage type is specified (8–8′, 8–5′, 8–*O*–4′ *etc.*, Figure 4). Importantly, lignans are often found in plants in optically pure form, although the particular antipode present can and does vary with the species (*42-45*). This contrasts with the racemic linkages which are believed to be incorporated into lignin biopolymers.

Erdtman (*46*) and various other investigators also studied monolignol coupling, but with the goal of attempting to determine how regiospecific or stereoselective control could be engendered to give optically active 8–8′ linked lignan products.

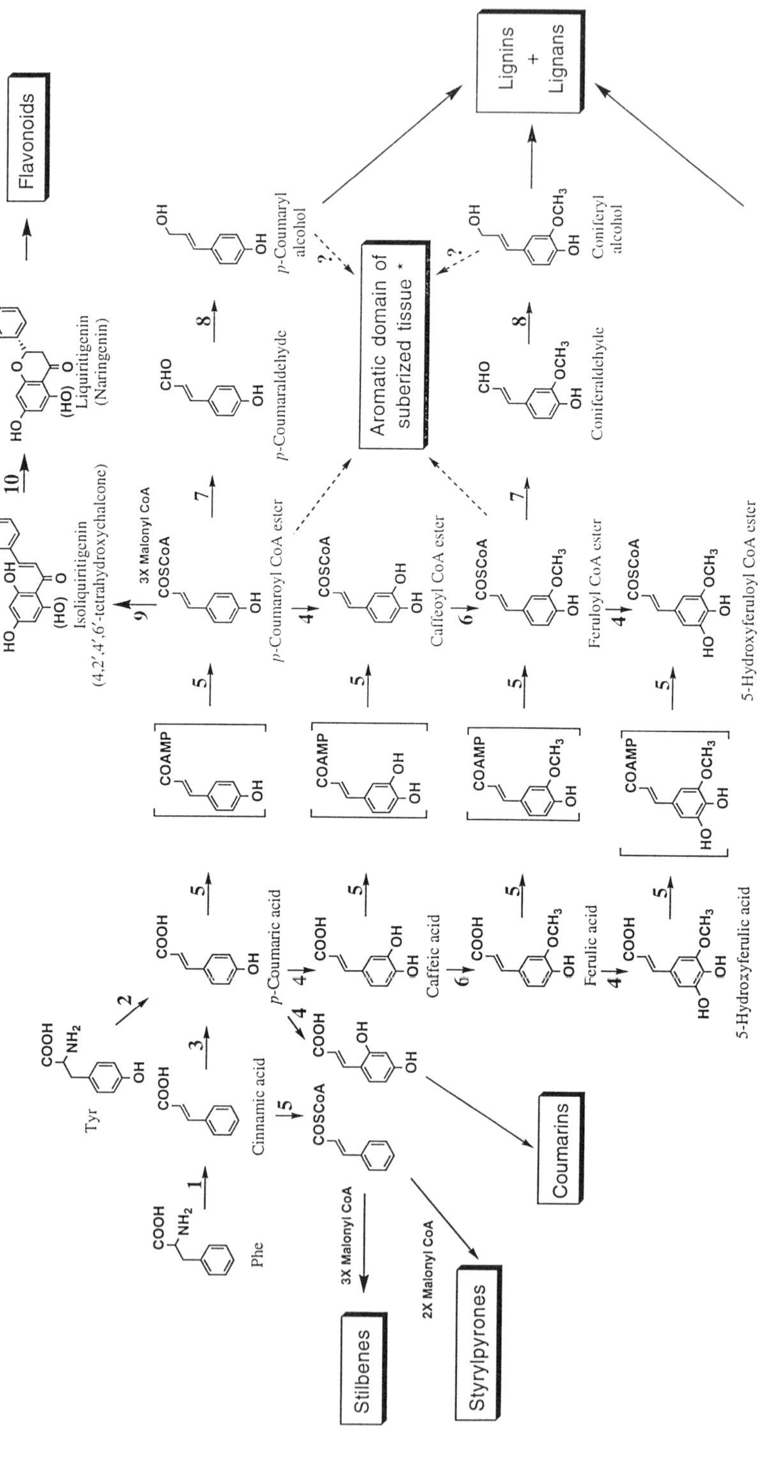
Flavonoids
Lignins + Lignans
Aromatic domain of suberized tissue *
Coumarins
Stilbenes
Styrylpyrones
Phe
Tyr
Cinnamic acid
p-Coumaric acid
Caffeic acid
Ferulic acid
5-Hydroxyferulic acid
p-Coumaroyl CoA ester
Caffeoyl CoA ester
Feruloyl CoA ester
5-Hydroxyferuloyl CoA ester
p-Coumaraldehyde
Coniferaldehyde
p-Coumaryl alcohol
Coniferyl alcohol
Isoliquiritigenin (4,2',4',6'-tetrahydroxychalcone)
Liquiritigenin (Naringenin)
3X Malonyl CoA
2X Malonyl CoA

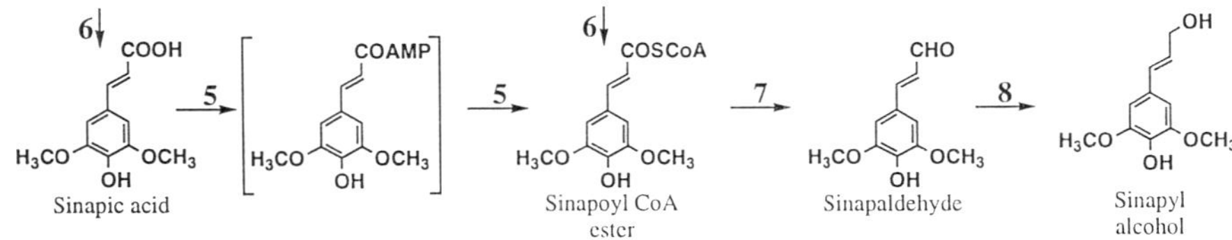

Figure 3. Main elements of the phenylpropanoid pathway: lignins and lignans are monolignol derived. **1,** phenylalanine ammonia-lyase; **2,** tyrosine ammonia-lyase (mainly in grasses); **3,** cinnamate-4-hydroxylase; **4,** hydroxylases; **5,** CoA ligases involving AMP and CoA ligation, respectively; **6,** *O*-methyltransferases; **7,** cinnamoyl-CoA:NADP oxidoreductases; **8,** cinnamyl alcohol dehydrogenases; **9,** chalcone synthase; **10,** chalcone isomerase. [Note: Conversions from *p*-coumaric acid to sinapic acid and corresponding CoA esters are boxed since dual pathways *appear* to be in effect; *: may also involve *p*-coumaryl and feruloyl tyramines, as well as *small* amounts of monolignols].

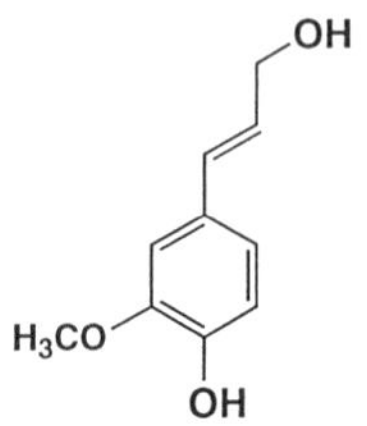

1: Coniferyl alcohol

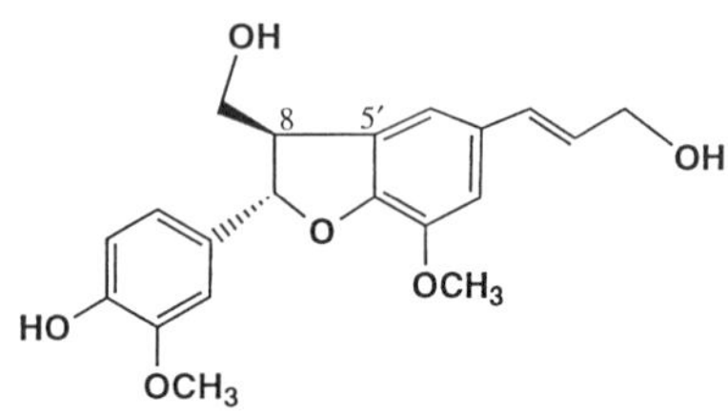

2: 8,5′-Coupling

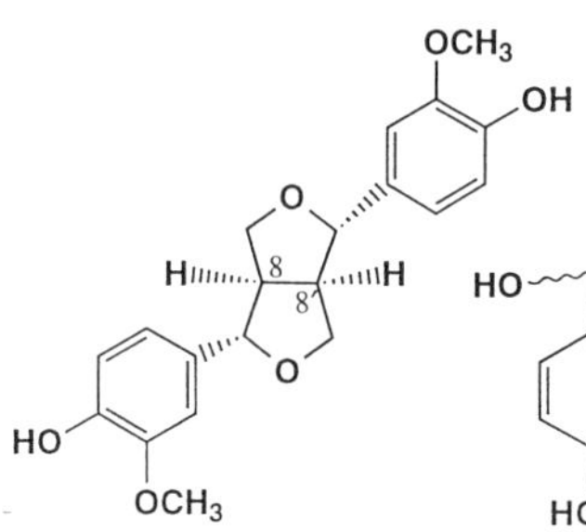

3: 8,8′-Coupling

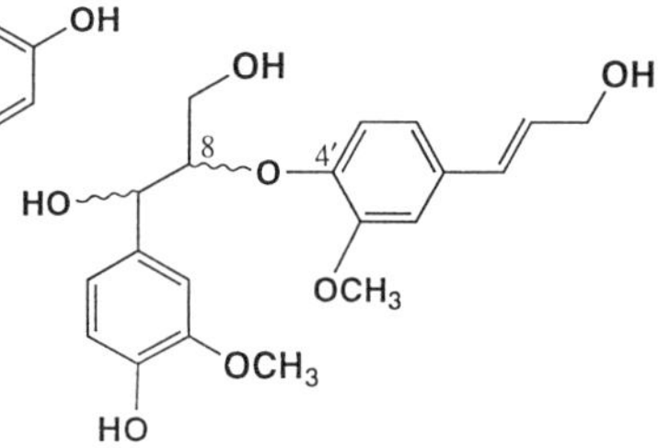

4: 8-*O*-4′-Coupling

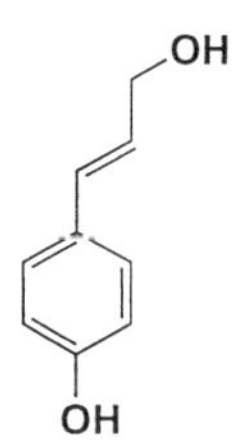

5: *p*-Coumaryl alcohol

Figure 4. *E*-Coniferyl alcohol **1** and the major racemic products (**2-4**) obtained by non-specific free-radical coupling, and *E*-*p*-coumaryl alcohol **5**.

Such studies were unsuccessful, since all attempts again engendered formation of the various well-known racemic products resulting from linkages between the different possible coupling sites present in the substrate.

Accordingly, none of the enzyme-catalyzed phenoxy radical coupling reactions *in vitro* leading to either lignans or lignins has satisfactorily duplicated those encountered *in vivo*. Given this fact, it is unclear why the non-specific dehydrogenative coupling hypothesis, which has *no* counterpart elsewhere in biochemistry, has not been substantially challenged, or at least seriously reconsidered. From a broader perspective, however, recent years have witnessed a growing interest in not only delineating how the coupling reactions leading to lignins and lignans actually occur, but also how the configurations with respect to lignin monomer composition are determined, and how the carbon (*i.e.* monomeric substrate) is allocated to the pathway. Each of these questions is now surveyed in some detail below.

Phenylpropanoid Metabolic Flux: Its Modulation and Control

In 'simple' aquatic plants, such as algae, biosynthesis of the aromatic amino acids, phenylalanine and tyrosine, is primarily directed to protein formation. The shikimate-chorismate-derived pathway to these essential proteinaceous building blocks is shown in Figures 1 and 2, and forges the link between glycolysis, the pentose phosphate pathway and aromatic amino acid metabolism. Evolution of vascular plants, on the other hand, significantly exploited the carbon flux into phenylalanine/tyrosine biosynthesis, *via* the shikimate-chorismate pathway, but not for protein biosynthesis. Instead, the phenylalanine (tyrosine) so formed was conscripted into phenylpropanoid metabolism, ultimately affording lignins, lignans and related metabolites. Altogether, some 30-40% of all organic carbon in vascular plants is 'stored' in this manner. Yet, curiously, as in the enzymatic control of phenylpropanoid phenoxy radical coupling, how carbon flux is *differentially* targeted into phenylpropanoid metabolism and its distinct biochemical branches has received little detailed attention.

For example, the shikimate-chorismate 'upstream' biosynthetic segment of phenylpropanoid metabolism has been almost completely ignored, whether in how it serves (tissue-specifically) to regulate the overall phenylpropanoid pathway or functions as a means for the storage of metabolites ultimately targeted to the phenylpropanoid pathway. Accordingly, it is appropriate that the second chapter (by Carol Bonner and Roy Jensen) in the present volume be devoted to a summary of current knowledge about the regulation and control of this metabolic segment. Significantly, however, metabolic control analysis (*47-50*) has not yet been carried out on the pathway between erythrose-4-phosphate and phosphoenol pyruvate to prephenic acid *in planta*, an essential matter if regulation of this segment of the pathway is to be fully understood.

Beyond prephenic acid, the next metabolic segment involves formation of the aromatic amino acids, phenylalanine (Phe) and tyrosine (Tyr), and their subsequent deamination to afford cinnamic and *p*-coumaric acids. This is treated as a distinct segment since it is here where both carbon and nitrogen metabolism directly interface. Thus, as can be seen from Figure 2, nitrogen is introduced at the point of prephenic acid transamination to afford arogenic (Agn) acid, which can then be converted into either phenylalanine (or, to a much lesser extent, tyrosine) in tissues undergoing active phenylpropanoid metabolism. Originally, it was thought that prephenic acid was the precursor of (*p*-hydroxy)phenylpyruvic acid(s), with the latter being transaminated to generate phenylalanine (tyrosine). However, pioneering work by Jensen *et al.* (*5*) identified arogenate as the pivotal precursor of the two aromatic amino acids. Indeed, all subsequent detailed enzymological studies to date have identified the arogenate pathway alone as leading to the aromatic amino acids in vascular plants; thus, the (*p*-hydroxy)phenylpyruvate pathway has not been validated.

Given that nitrogen is only introduced into the pathway just prior to the formation of Agn, Phe and Tyr, it is remarkable that, when phenylalanine and tyrosine are conscripted for phenylpropanoid metabolism rather than for protein synthesis, the nitrogen is immediately removed (Figure 5). The enzymes involved, phenylalanine and tyrosine ammonia lyases, were discovered by Jane Koukol and Eric Conn (*51*) and Arthur Neish (*52*) in 1961, respectively, and the former is probably the most extensively investigated in all 'secondary' metabolism. Interestingly, whether the phenylpropanoid pathway is induced or not, none of the three aromatic amino acids (arogenate, phenylalanine or tyrosine) build up to any appreciable extent, suggesting that one of the preceding steps may be 'rate-limiting' and thus a plausible candidate for a control point in the pathway. Additionally, since an equimolar amount of ammonium ion (and hence nitrogen) is liberated for every mole of phenylpropanoid product formed, then an efficient nitrogen recycling process has to be in effect or otherwise active phenylpropanoid metabolism would be nitrogen limited. In view of the 1989 assertion that such a nitrogen recycling process must therefore be operative (*2*), some considerable satisfaction attended the demonstration several years later (*23, 53, 54*) that this was achieved *via* ammonium ion reassimilation through the glutamine synthase/glutamine 2-oxoglutarate amino transferase pathway (Figure 5). In this way, regeneration of the amino donor, L-glutamate, was able to occur, the latter undergoing subsequent transamination to afford arogenate and ensuring that the overall flux of carbon into the phenylpropanoid pathway continued without any further demand on nitrogen from the plant. Clearly, any disruption of this cycle would disrupt phenylpropanoid metabolism. However, regulation of this point in the pathway would only be useful if the modulation of carbon flux (Figure 3) could be *selectively* achieved into the lignin and/or lignan pathways, respectively.

Beyond the deamination step, the cinnamic (*p*-coumaric) acid(s) formed can be metabolized *in planta* into the monolignols (and hence lignins and lignans), as well as a plethora of phenylpropanoid (acetate) derived-products, such as the suberins, flavonoids, coumarins, and styrylpyrones. It is important to note, however, that formation of a particular metabolite normally can only occur in either a tissue-specific, temporally specific, or, in some instances, species-specific manner. For example, administration of the phenylalanine ammonia lyase inhibitor, AOPP, to developing *Zinnia elegans* tracheids in cell culture resulted in the complete inhibition of lignin biosynthesis but apparently did not affect any other branchpoint pathways (*55*). Importantly, in this case, the overall architecture of the *Zinnia* tracheids was fully established during the deposition of cellulose, hemicellulose (including pectins) and structural proteins. Clearly, therefore, any *selective* targeting of this part of the pathway in a lignin specific manner (*i.e.* by targeting the appropriate tissue) could be envisaged to have a major impact on the carbon flux into *lignin* formation.

The steps beyond cinnamic and *p*-coumaric acids leading to the monolignols have received most attention in regard to their being potential regulatory steps, even though no metabolic control analysis has yet been described. Instead, the molecular approach has primarily been devoted to attempts at determining the effects of differentially expressing genes encoding putative regulatory enzymes in the *lignin* biosynthetic pathway (Figure 3) (*56-62*). As can be seen, there are essentially only four types of transformations beyond cinnamic acid, namely aromatic hydroxylations (*63-65*), *O*-methylations (*66-68*), CoA ligations and (consecutive) NADPH dependent reductions (*69*). It cannot be overemphasized, however, that the pathway as shown is deceptively simple. It has many branching points, such as to the lignins and lignans, which bifurcate at *only* the monolignol level or, in the case of the flavonoids, where the 'point of departure' from the pathway is formation of the *p*-coumaryl CoA esters. To restate what was said earlier, vascular plant species not only display various segments of the pathway to different extents, but particular tissues/cell types are only involved in *specific segments* of the pathway (*e.g.* flavonoids). Accordingly, it is bewildering that so many multifarious contributions to the literature have variously referred to each of the specific enzymatic steps between phenylalanine and the monolignols as being rate limiting or potentially regulatory for lignin. Put another way, all gene expression/activation studies to date have focused on putatively

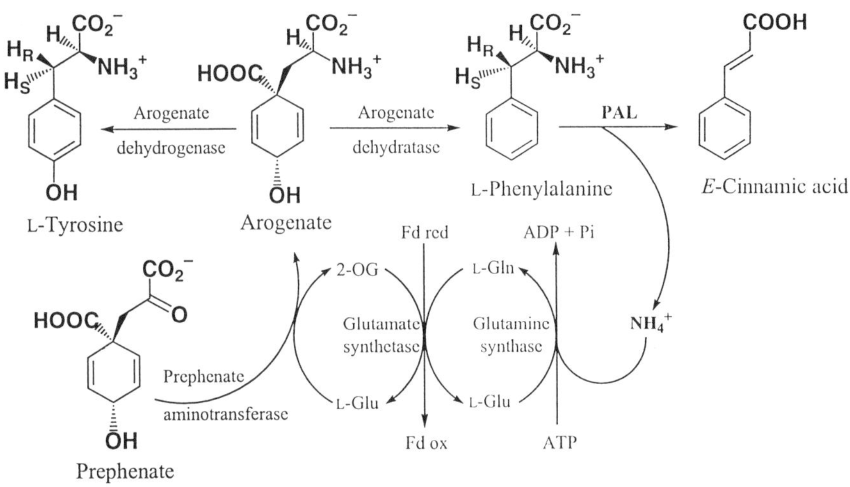

Figure 5. Proposed nitrogen cycling metabolism during active phenylpropanoid metabolism.

regulatory steps, but without any definitive data establishing that they are, indeed, rate-limiting or inevitably lignin-specific.

Considering the pathway itself, the first aromatic hydroxylation step leading to *p*-coumaric acid is catalyzed by a cytochrome P-450 NADPH-dependent oxidase (*63, 64*), as is the later conversion of ferulic acid to 5-hydroxyferulic acid (*65*). On the other hand, the step converting *p*-coumaric acid into caffeic acid is still uncertain. Originally described as a non-specific phenolase [see refs. (*70-72*)], it is now considered but not yet proven to be a distinct kind of oxidase, perhaps of a cytochrome P-450 type (*73*). Beyond the hydroxylation steps, the order in which the corresponding CoA esters are formed and *O*-methylation reactions occur is currently somewhat unclear. This is because a growing body of evidence indicates that different *O*-methyltransferases are capable of preferentially methylating, in some cases, the free hydroxycinnamic acids (*74-75*) and, in others, the corresponding CoA esters (*76-78*). The physiological significance of this differential *O*-methylation capability is as yet unknown, except that it may point either to redundancy in the pathway, or to distinct metabolite targeting of specific cell types and tissues, or to the involvement of selective metabolic pathway targeting (*e.g.* to lignins or lignans). Beyond the *O*-methylation and CoA ligation steps, the remaining transformations involve consecutive NADPH-dependent reductions, catalyzed by cinnamoyl CoA reductase (CCR) and cinnamyl alcohol dehydrogenase (CAD) (*69*), respectively. Curiously, both steps have often been described as lignin-specific *and* rate-limiting, since they catalyze the last two transformations leading to monolignols (*59, 79*). However, this may be an oversimplification, given that monolignols, such as coniferyl alcohol, can be deployed for both lignin and lignan biosynthesis. Moreover, it is as yet unclear why these processes are so often described as rate-limiting, since lignifying cells undergo the coordinated induction of the entire shikimate/chorismate and phenylpropanoid pathways in a tissue-specific manner.

For the foregoing reasons, there has long been a need to characterize the pool levels of different metabolites in the phenylpropanoid pathway under substrate saturating conditions, in order to attempt to identify what, if any, are the rate-limiting steps. In this regard, the only work of this type carried out to date has employed *Pinus taeda* cell suspension cultures, which can be induced to undergo active phenylpropanoid metabolism when grown in a solution containing 8% sucrose. Over a 96 h period, these cells respond by generating an extra-cellular monolignol-derived dehydropolymerisate, whose formation can be inhibited by addition of 20 mM KI, an H_2O_2 scavenger (*80*). Under these conditions, however, no extracellular lignin-like material is formed, but instead essentially only the monolignols, *p*-coumaryl and coniferyl alcohols, are secreted into the medium. Time course analyses for the build-up of each possible metabolic product from phenylalanine onward to the monolignols has been recently determined (A. M. Anterola *et al.*, Washington State University, unpublished obervations, 1997). These experiments, however, revealed that only *p*-coumaric, cinnamic and caffeic acids build-up during the induction of the pathway, *i.e.* none of the other intermediates preceding the steps that had been claimed to be rate-limiting in the pathway (*i.e.* the CoA esters, aldehydes or monolignols) were observed to accumulate significantly beyond the pre-existing pool levels. This indicates that *none* of the latter steps function in a rate-controlling manner. Moreover, given that the monolignols can have more than one metabolic fate, it is important that care be taken in interpreting the results of any attempts to enhance or suppress CAD or CCR enzyme levels in the pathway.

Monolignol Transport into the Cell Wall

The next issue that also remains unresolved is the nature of the intermediates being transported from the cytoplasm into the lignifying cell wall. Lignification occurs in a spatially and temporally well-defined manner (*81*), whereby, for example, *p*-coumaryl and coniferyl alcohols are *differentially* targeted to specific (initiation) sites in the lignifying cell wall (see Chapter 22). However, there is still some controversy as to whether it is the free-monolignols, or their glucosides, that are transported across the plasma membrane into the cell wall. The formation of monolignol glucosides in the cytoplasm is said to involve the action of the corresponding glucosyltransferase(s), and it is proposed that these metabolites are then transported into the cell wall where a β-glucosidase catalyzes the regeneration of the monolignols (*1*). Stefan Marcinowski and Hans Grisebach attempted sometime later to determine whether monolignol glucosides were obligatory intermediates for lignification (*82*). However, their results suggested that lignin was predominantly formed by polymerization of the monolignols being transported directly rather than as their glucoside derivatives. These observations could be interpreted as indicating either that the monolignol glucosides are storage metabolites, or that they are involved in some distinct (*e.g.* tissue-specific or lignan-specific) metabolic subsector. Indeed, it is puzzling that monolignol glucosides are apparently only found in very few plant species. Thus, the question has not yet been resolved as to whether the monolignol glucosides are required for transportation into lignifying plant cell walls or not, but ongoing work in the Ellis and Savidge laboratories is beginning to bring a substantial measure of clarification to this important subject (*83, 84*).

Oxidases and Peroxidases Involved in Lignification

For almost five decades, profound uncertainty has enveloped the enzymology of *tissue specific* bimolecular phenoxy radical coupling leading to the lignins, lignans, suberins and other phenolic natural products. Using lignin formation as an example, peroxidases (*32, 80, 85, 86*), peroxidases and laccases (*32, 87, 88*), laccases (*31, 89-91*), (poly)phenol oxidases (*92*), coniferyl alcohol oxidase (*93*), and even cytochrome oxidases (*94*) have all been implicated. This has occurred primarily because of the large variety of oxidative enzymes with broad substrate specificity that exists in plants, to which the assignment of physiological function has unfortunately all too often been arbitrary. Some of the confusion may have stemmed from the fact that all of these enzymes are present in numerous isoenzyme forms, and all are capable of converting lignin precursors (monolignols) such as *E*-coniferyl alcohol **1** into corresponding dimers like (±)-dehydrodiconiferyl alcohols, (±)-pinoresinols and (±)-guaiacylglycerol 8–*O*–4′-coniferyl alcohol ethers *via* bimolecular radical coupling processes in open solution (Figure 4). As mentioned earlier, further oxidative coupling then gives rise to the lignin polymers. It is doubtful, however, that the formation of such an important and ubiquitous class of macromolecules would be catalyzed in a haphazard manner by such a wide range of enzymes *in vivo*.

Perhaps in response to this uncertainty about the actual enzyme(s) involved, more recent attempts to establish a precise physiological function for particular peroxidase or laccase isoenzymes have taken a different approach. Transgenic plants were obtained in which genes encoding supposed lignin-forming peroxidase (*95-97*) and laccase (*98*) isoenzymes had been introduced in either sense and/or antisense orientations. Although each study was carefully executed, the results obtained, for example, in tobacco (*Nicotiana tabacum*) and yellow poplar (*Liriodendron tulipifera*), by targeting peroxidase and laccase, respectively, were equivocal. With the peroxidase sense/antisense experiments, although interesting changes in growth and development were noted, the effect on lignin formation was relatively minor. Fritig *et al.* have since suggested that these 'lignin-specific' peroxidases may only facilitate hypersensitive plant responses but not lignification proper (*99*). Somewhat comparable (*i.e.* small) effects on lignin synthesis in yellow poplar were obtained with attempts to suppress laccase gene expression. In a similar manner, studies with antisense transgenic tomato plants, targeting a putative suberin-specific peroxidase isoenzyme, had little effect on suberization (*100*).

Thus, no unequivocal proof has ever been obtained showing that any particular laccase or peroxidase isoenzyme has either regulated dehydrogenative monolignol coupling or had an exclusive role in lignification or suberization. This should not be taken to imply, however, that peroxidase(s) do not play a critical role in lignin formation; indeed, compelling evidence for the role of an H_2O_2-dependent peroxidase in 'lignin' biosynthesis has been achieved using *Pinus taeda* cell suspension cultures (*80*). Thus, when the H_2O_2 scavenger, KI, was added to the *P. taeda* culture medium, lignification in the *Pinus taeda* cell walls was *totally* suppressed, as was the formation of an extracellular 'lignin-like' precipitate. However, even though no lignin synthesis occurred, the cells were still able to biosynthesize the monolignols, *E*-coniferyl **1** and *E-p*-coumaryl **5** alcohols, *de novo* (see Figure 4 for structures). [This was shown to be the case by the secretion of both monolignols into the culture medium in radiolabeled form, following [U-^{14}C]-Phe administration and its subsequent metabolism.] This observation was important from two perspectives: first, these findings gave strong support in favor of a role for peroxidase in lignin biosynthesis and, second, a system was *at last* available to study monolignol formation *in vivo* without subsequent polymerization. This study did not, however, yield any insight into how the phenoxy radical coupling processes themselves are controlled *in vivo*.

Dehydrogenative Polymerization of Monolignols to Lignins

As already mentioned, it has long been believed that lignins are assembled by the coupling of radicals produced through the single-electron oxidation of monolignols and corresponding phenolic monomer residues in the oligomeric and polymeric components which result. In native lignin macromolecules, roughly half of the linkages are of the 8–*O*–4′ alkyl aryl ether type in lignified plant cell walls (*36*). Since dehydrogenative polymerization of monolignols *in vitro* does not yield the same relative proportions of interunit linkages as observed in native biopolymers, certain facets of the macromolecular assembly processes *in vivo* differ significantly from the dehydrogenative polymerization of monolignols in open solution.

It has not been possible accurately to account for the ratios of interunit linkages among dehydrodimers formed *in vitro* simply on the basis of unpaired electron densities on the atomic centers of the interacting monolignol radicals. In ESR spectra, the hyperfine coupling constants to adjacent protons suggest that the unpaired electron spin densities at C5 and C8 in coniferyl alcohol-derived radicals are very similar to one another (*101*). Moreover, molecular orbital calculations tend to indicate that the unpaired electron density on the (ESR-silent) 4–*O* is appreciably lower than on C5 and C8 (T. J. Elder, Auburn University, personal communication, 1997). These findings are in accord with the observation that 8–5′ and 8–8′ linked dilignols predominate over 8–*O*–4′ linked dimers in product mixtures resulting from the enzyme-catalyzed dehydrodimerization of coniferyl alcohol *in vitro* (*102-104*). In contrast, the heats of formation deduced from molecular orbital calculations indicate that the stabilities of the σ-complexes directly resulting from 8–5′, 8–8′ and 8–*O*–4′ coupling of coniferyl alcohol-derived radicals are all very similar (*105*). On the other hand, the calculated energies of the σ-complexes formed directly from 5–5′ and 4–*O*–5′ coupling are substantially higher (*105*).

The literature provides no relevant basis for understanding what occurs during dehydrogenative coupling between dilignols and monolignols as the first step towards the formation of macromolecular lignin chains. It has been reported that 4–*O*–8″ linkages predominate during the enzyme-catalyzed dehydrogenative coupling of coniferyl alcohol with 8–5′, 8–8′ and 8–*O*–4′ linked dilignols *in vitro* (*106, 107*). Unfortunately, no attempt has been made to determine unpaired electron densities on atomic centers in guaiacyl radicals where there is no conjugation with a C7–C8 double bond. These values, and the estimated heats of formation for the σ-complexes directly arising from the various possible coupling modes, could shed some light upon the unexpected absence of 4–*O*–5″ linkages in the trimers dehydrogenatively formed from coniferyl alcohol and the 8–5′, 8–8′ and 8–*O*–4′ linked dilignols.

Noncovalent interactions play an important role in certain phenoxy radical coupling processes investigated *in vitro*, and presumably contribute to the analogous events which occur in lignifying plant cell walls. For example, as far as the 8–8′ coupling mode is concerned, the dehydrodimerization of *E*- and *Z*-isoeugenol and

E-2,6-dimethoxy-4-propenylphenol has led exclusively to *threo* products (*108*). Such effects are likely to be much more generally significant than hitherto recognized; thus the frequencies of the different interunit linkages formed between (mono-, oligo- and poly-) lignol radicals *in vitro* will presumably depend to varying extents upon (i) the unpaired electron densities on the atomic centers about to become coupled, (ii) noncovalent interactions between the approaching radicals, and (iii) the energies of the σ-complexes directly resulting from each coupling mode.

Apart from those operative *in vitro*, a further factor will play an important, and perhaps decisive, role in establishing the modes of coupling between lignol radicals in lignifying plant cell walls. This effect arises from pronounced noncovalent interactions between the radicals and pre-existent macromolecules occupying particular locations within the plant cell wall matrix. When such effects become dominant, complete regio- and sterospecific phenoxy-radical coupling occurs, as in the combination of two coniferyl alcohol-derived free radicals to give (+)-pinoresinol exclusively under the control of a dirigent protein (*109*). One of the particular challenges with which the field of lignin biosynthesis must now contend involves the question of whether macromolecular lignin configurations are established in a comparably specific way.

Determinants of Lignin Configuration

Monolignols appear to be incorporated into native lignins in the same order as they are formed in their biosynthetic pathway (*110*), *viz.* *p*-coumaryl, coniferyl and sinapyl alcohol. Since these monomeric precursors differ according to whether methoxyl groups are present at C3 and C5 in the aromatic ring, the frequencies of interunit linkages to these positions will decrease in this order also. The effect is reinforced by accompanying changes in the unpaired electron density on the atomic centers in the corresponding series of monolignol-derived radicals, which at the 4-*O* more than doubles according to Austin Model 1 semi-empirical molecular orbital calculations (*101*).

The frequencies of interunit linkages in dehydropolymerisates formed from coniferyl alcohol *in vitro* vary substantially according to the conditions employed. Thus, when the monolignol is added all at once ('Zulaufverfahren' or mixing method) to the enzyme-containing solution, initial coupling to dimers is followed by further polymerization which progressively lowers the reactivity of the radical species; termination eventually occurs before the (so-called 'bulk') dehydropolymerisate components have had the opportunity of attaining high molecular weights. On the other hand, when the monolignol is added gradually ('Zutropfverfahren' or drop method) to the same dehydropolymerizing medium, dimers are likewise produced initially, but they quickly undergo further polymerization into higher molecular weight components; these then couple preferentially with the radicals being formed in low concentration from the monolignol that continues to be slowly introduced into the system. The (so-called 'end-wise') dehydropolymerisate species resulting from Zutropfverfahren embody a much higher proportion of 8–*O*–4′ linkages than those produced by Zulaufverfahren, as would be expected from the differences observed between monomer-monomer (*102-104*) and monomer-dimer (*106, 107*) dehydrogenative coupling frequencies. It has been suggested that lignins *in vivo* may be composites of 'bulk' and 'end-wise' dehydropolymerisate domains (*111*), the configurations of which represent the consequences of local fluctuations in monolignol-derived radical concentrations during lignification.

The primary structures of macromolecular lignin chains (*viz.* the sequences of interunit linkages and monomer residues), given the variability in composition among different morphological regions of plant cell walls, have been traditionally held to be random. Yet the secondary structures of lignins appear to be quite well-defined by comparison. From Raman spectral analyses with polarized incident laser beams, the aromatic rings of the lignin observed in *Picea marina* tracheid cross-sections appear to be parallel to the cell-wall lumen surface (*112*). It has been proposed that this is an outcome of the plant cell-wall matrix becoming gradually more hydrophobic during the course of lignification. As a consequence, water is claimed to be progressively displaced from what was originally a swollen polysaccharide gel, resulting in anisotropic shrinkage which tends to orient the lignin aromatic rings into directions parallel to the cell-wall surface because the effect is ostensibly greater in the radial direction (*110*). Disregarding the question of whether such a sequence of events could be consistent with the thermodynamics of lignification processes, it is worth pointing out that, when the mechanical integrity of the cell wall is preserved, shrinkage is necessarily greater in the tangential than radial direction.

Partial clarification of this matter could have been forthcoming from computer simulations of lignin structures that use simplified space-filling structures to depict monomer residues interconnected by the six most commonly occurring linkages in softwood lignins (*113, 114*). When the structures were built inside thin lamellar boundaries representing domains in the polysaccharide matrix of the secondary wall, a degree of spontaneous alignment was obtained with respect to the microfibrillar direction owing to the elongated shapes of the spaces within which the lignin representation was assembled. Analogous simulation of lignification in less restricted spaces representing the middle lamella region engendered an entirely random orientation of the monomer residues and, interestingly, a more highly crosslinked structure than inside the lamellar boundaries used to invoke the secondary wall (*113, 114*).

The fact that the monomolecular film thicknesses of lignin derivatives from woody tissues are independent of sample molecular weight has become a truism of almost classical proportions in lignin chemistry (*115*). This observation together with the molecular dimensions apparent in electron micrographs (*116*) indicates that lignin derivatives consist of disk-like macromolecular components cleaved from lamellar parent structures which are about 2 nm thick. Potassium permanganate staining of lignifying *Pinus radiata* cell walls has revealed a plausible cause for this feature of macromolecular lignin configuration (*81*). Lignin deposition in the secondary wall takes place much more rapidly in directions that are aligned with, rather than perpendicular to, the cellulose microfibrillar axes. A particularly important aspect of the process lies in the observation that the lignin domains maintain a more or less uniform density as they expand. This is most clearly evident in the middle lamella region where only after neighboring lignin domains make contact do the intervening spaces become filled in (*81*). Obviously there must be strong nonbonded attractive interactions between the polymeric lignin chains because a crosslink density of 0.052 (*117*) involving tetrafunctional branch points (*118*) in the macromolecular structure cannot account for such effects.

Each of the individual lignin domains has developed from an initiation site which was presumably incorporated into the cell wall before the onset of lignification (*81*). The nature of these initiation sites largely determines the mechanism of macromolecular lignin assembly *in vivo*, and the literature bears witness to more than one conceivable alternative in this regard. The 8–8′ linkage has been detected in ryegrass between lignin monomer residues and ferulate moieties that are ester-linked to α-L-arabinofuranose units in arabinoxylan (*119*). The possibility thus presented itself that ferulate-polysaccharide esters could act as initiation sites for lignification (*119*). On the other hand, with polyclonal antibodies, proline-rich protein epitopes have been spatially and temporally correlated with lignin formation in developing cell walls of the maize coleoptile (*120*) and in secondary walls of differentiating protoxylem elements in the soybean hypocotyl (*121*). It has accordingly been suggested that proline-rich proteins may act as a scaffold for initiating lignification at their tyrosine residues (*121*). There is no doubt that the identity of the initiation sites and their function for lignification in plant cell walls has emerged as a subject of central importance in the field of lignin biosynthesis.

Biosynthesis of Lignans

The lignans constitute a very widespread group of phenylpropanoid natural products, being found in all plant parts, including (woody) stems, rhizomes, roots, seeds, oils, exuded resins, flowers, leaves and bark tissues (*122*); their amounts differ between tissues and species. Around the turn of the last century, investigations were initiated to determine the structures of some of the most common lignans (originally called lignanes) (*25*). It was quickly concluded that they were a series of *dimeric* substances linked through 8–8′ bonds. This initial classification, unfortunately, failed to recognize that other lignan skeletal types (*e.g.* 8–5′; 8–3′; 8–1′; 5–5′; *etc.*) were also present in many plant species/tissues (*39, 40*), and that lignan(e)s can also have much higher molecular weights. It is now known that lignans encompass a wide range of structural motifs (and molecular sizes) (*39-41, 123*), rather than being restricted to the simple 8–8′ linked dilignols as previously thought. A few examples are given for illustrative purposes in Figure 6.

Frequently, although not always, lignans are found in optically active form, where the particular antipode observed can vary with plant species. For example, (+)-pinoresinol is present in *Forsythia* species (*45*), whereas the (–)-antipode occurs in *Daphne tangutica* (*44*) and (+)-sesamin occurs in *Calocedrus formosana* (*42*) with the (–)-form accumulating in a *Zanthoxylum* species (*43*). In most cases, however, lignan structures cannot arise solely from phenolic coupling, since other *post-coupling* modifications are frequently evident, such as oxidations, reductions, skeletal rearrangements, carbon-carbon bond cleavage reactions, demethylations, oligomeric assemblies and so on. In addition to these structural permutations, the deposition of lignans can also often be very extensive, as exemplified by cases such as western red cedar (*Thuja plicata*) whose heartwood can consist of up to 20% (w/w) of these components (*124*). In such instances, both the lignins and lignans coexist in large amounts in the mature tissues.

The distinction between lignans (dimers and oligomers) and macromolecular lignins has long been a contentious issue as far as their biosynthesis is concerned. This is partly because there has been no clear demarcation between the biochemical pathways, and partly because of their structural similarities. Accordingly, several years ago establishing the precise difference between lignan and lignin biosynthesis appeared to be a particularly worthwhile goal.

The initial focus addressed formation of the most common lignan types, namely those embodying the 8–8′ linkage. For experimental convenience, attention was directed towards *Forsythia* species where the lignan biosynthetic pathway has now been established to follow the scheme shown in Figure 7 (*103, 109, 125-135*). As can be seen, the entry point involves stereoselective coupling of two *E*-coniferyl alcohol molecules to afford (+)-pinoresinol (*109*). In an analogous manner, coupling of two *E*-coniferyl alcohol molecules also occurs in developing flax (*Linum usitatissimum*) seed, but here the corresponding (–)-antipode is formed (J. Ford *et al.*, Washington State University, unpublished results, 1997). Thus, the prevalent coupling mode is, not unexpectedly, dependent upon the plant species. The pinoresinol formed serves as the entry point into complex biochemical pathways leading to lignans such as sesaminol (*Sesamun indicum*) (*136*), plicatic acid (*Thuja plicata*) and podophyllotoxin (Z. Q. Xia *et al.*, Washington State University, unpublished results, 1997).

In the case of *Forsythia* species, the protein responsible for this stereoselective process, the discovery of which incidentally marked the first example of regio- and stereospecific control of phenoxy radical coupling, has been purified to apparent homogeneity (*109*), its gene cloned and the recombinant protein expressed in a functional recombinant form as described in Chapter 22. Interestingly, the dirigent protein which controls this transformation lacks oxidative catalytic capacity by itself, but in the presence of an appropriate oxidase, such as laccase, is able to confer absolute specificity to the coupling reaction. The mechanism envisaged involves the coupling of free-radicals, derived from two molecules of *E*-coniferyl alcohol, where

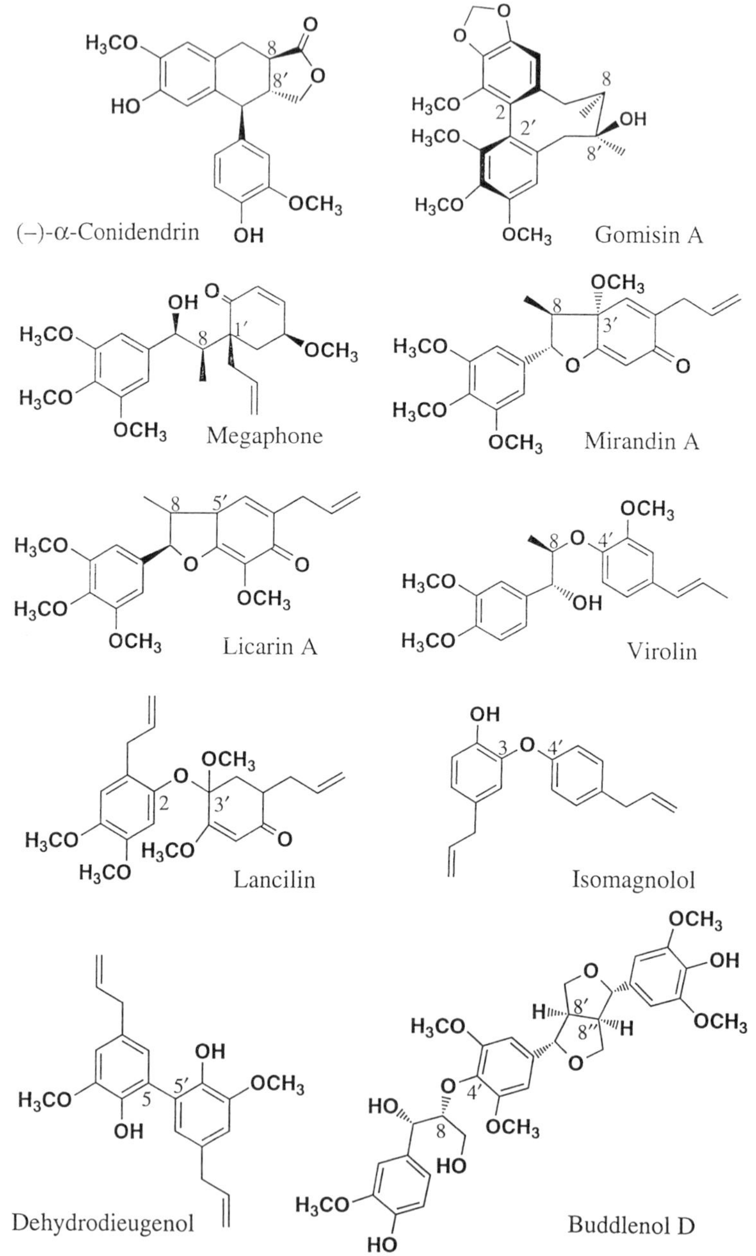

Figure 6. Examples of various lignans encompassing differing structural motifs (interunit linkages).

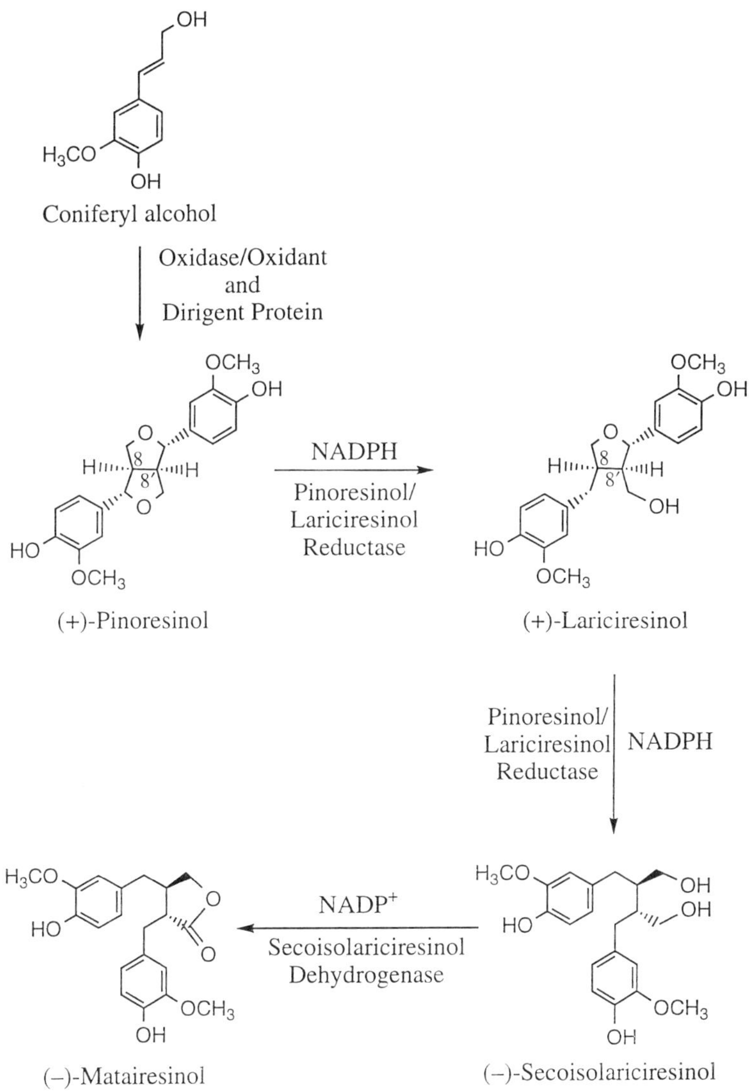

Figure 7. Lignan biosynthetic pathway in *Forsythia intermedia*.

the oxidase catalyzes the single electron oxidation step and the dirigent protein specifies the mode of coupling.

In *F. intermedia*, beyond (+)-pinoresinol, two sequential highly enantiospecific NADPH dependent reductions are catalyzed by a single reductase, namely, (+)-pinoresinol/(+)-lariciresinol reductase, to afford (–)-secoisolariciresinol (*via* (+)-lariciresinol) (*132, 135*). In this plant species, the (–)-secoisolariciresinol so formed can then undergo enantiospecific dehydrogenation to afford (–)-matairesinol (*125, 127*).

Given that other plants may possess the opposite antipodes, it is worth emphasizing that, in every case examined thus far, all biochemical processes involving coupling and post-coupling modifications in lignans are under explicit control: the biochemical processes associated with 8–8′ linked lignan formation involve highly controlled stereoselective and/or enantiospecific transformations. Furthermore, it is now evident that a new class of proteins exists that is capable of engendering the *distinct* specific coupling modes that account for the different lignan skeletal forms present in various plant species.

Another abundant lignan type is the one containing the 8–5′ linkage motif, as illustrated by dehydrodiconiferyl alcohol, its dihydro derivatives, and other metabolic variants. These substances are formed, as reported previously (*137*), by reduction of the allylic side chain of dehydrodiconiferyl alcohol and/or demethylation reactions, at least in *Pinus taeda* (Pinaceae). In other woody plants such as *Cryptomeria japonica* (Taxodiaceae) (*138*), the lignans present apparently also undergo further transformation, *i.e.* they contain dehydrodiconiferyl alcohol analogs which have not only undergone reduction of the allylic side-chain and the phenylcoumaran ring, but are also modified by acylation. A tentative biosynthetic pathway, which is being studied by Lewis *et al.*, is shown in Figure 8.

Thus, in summary, the various biochemical transformations associated with the lignan pathways are now yielding to systematic inquiry, and are revealing that the phenoxy radical coupling reactions and subsequent post-coupling modifications are under explicit biochemical control.

Relationship between Lignan and Lignin Biosynthetic Pathways

No discussion about the products of dehydrogenative monolignol coupling and any subsequent transformations would be complete without some attention being given to the contrasting distinctions between the lignan and lignin biosynthetic pathways. It is becoming increasingly evident that both pathways are *fully* independent, and that there is considerable subtlety involved in the temporal, spatial and tissue-specific expression of each. Accordingly, although many of the details await fuller documentation at the enzymatic and gene (expression) level, it is timely to compare and contrast what is now known about each metabolic system.

The lignans are initially typically formed as optically active dimers that can then undergo various modifications including the formation of higher molecular weight oligomers, which are in danger of being mistaken for lignin macromolecules. An evaluation of gene sequence(s) for the dirigent (monolignol coupling) protein from *Forsythia* revealed that it has no homology to any other protein (gene) of known function (*139*). Just as interestingly, the NADPH-dependent pinoresinol/lariciresinol reductase displays considerable sequence homology to that of isoflavonoid reductases (~62% similarity and ~42% identity), the isoflavonoids being a group of compounds involved in plant defense (as phytoalexins). Indeed, it was pointed out earlier (*140*) that the gene(s) encoding lignan reductases presumably predate the isoflavonoid reductases, given that the latter are only observed in few plant species (*e.g.* legumes) whereas the lignan reductases (based on chemotaxonomic considerations) are widespread throughout the ferns, gymnosperms and angiosperms.

In accordance with gene sequence comparisons and the biological properties of specific lignans (*e.g.* as antioxidants, fungicides, cytotoxins, antivirals, antifeedants, *etc.*), it seems evident that the *primary* role of the lignans, irrespective of their molecular size, is in plant defense, *i.e.* as part of the arsenal of compounds used to

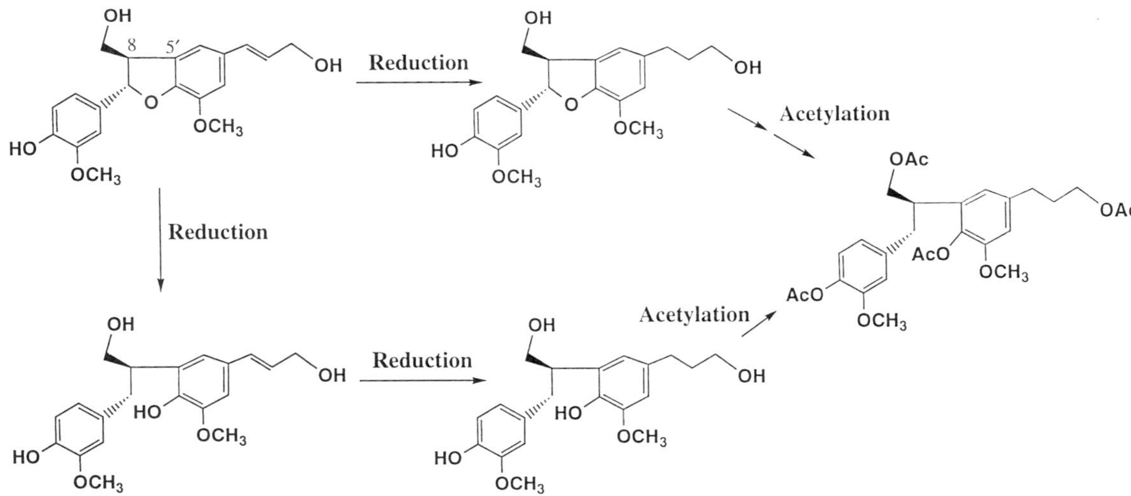

Figure 8. Proposed biosynthetic pathway to lignans present in *Cryptomeria japonica*.

ensure plant survival over extended periods. The lignans can be biosynthesized constitutively in variable amounts ranging from relatively low levels in flower petals to extensive deposition in heartwood tissues, such as cedar and redwood. In the latter case, deposition of the majority of the lignans occurs only after lignification is complete *via* extrusion into the sapwood from specialized conducting cells, such as the ray parenchyma. This deposition can continue in massive quantities, as exemplified by the *ca.* 20% (w/w) lignans found in western red cedar. Moreover, since these substances not only are present as simple dimers, but also can exist as insoluble oligomers, they have frequently been misidentified as 'secondary' or 'abnormal' lignins, even though they are formed through quite distinct biochemical pathways (*141*). Indeed, it cannot be emphasized enough that, even though their solubilization may require conditions normally associated with lignin removal, they are only present as non-structural infusions. Thus a vista is beginning to emerge where the lignan biosynthetic pathway, involving characteristic coupling processes and precise post-coupling modifications, is entirely distinct from the lignin-generating machinery.

It is, therefore, pertinent next to compare and contrast current understanding about lignin biosynthesis with what is now being observed in lignan formation. As described earlier, the prevailing dogma has insisted that lignin biosynthesis occurs in a manner whereby, following monolignol-derived free-radical generation, no direct control of macromolecular assembly is exercised at the enzymatic level. Moreover, it is most unfortunate that the initial studies which spawned this idea employed crude enzyme preparations (*e.g.* mushroom laccase) which were not even involved in lignin biosynthesis in the first place. This early working hypothesis did not, however, explain the preponderance of 8–*O*–4′ linkages in lignin biopolymers. Nor did it readily account for the heterogeneity of lignins within plant cell walls (*vide supra*).

The most important aspect of the entire lignification process lies in the fact that the individual monolignols are transported into the cell wall after deposition of all of the structural carbohydrates and proteins, *i.e.* when the overall architecture has already been established. Lignin formation itself then appears to be initiated at distinct (initiation) sites in the lignifying cell wall (*81*). These sites are temporally and spatially correlated with the deposition of particular proline-rich proteins (*120, 121*), the primary stuctures of which could conceivably encode in some explicit manner (yet undemonstrated) *lignin* primary structure. Indeed, some preliminary analysis suggests that there could possibly be a correlation between the dirigent proteins controlling lignan assembly and the proline-rich proteins associated with lignification. In the latter case, it would appear that macromolecular lignin assembly occurs *via* an end-wise dehydrogenative polymerization process where the monolignol-derived free-radicals are coupled to the growing polymer chain. If correct, the involvement of the putative proteinaceous template would provide a mechanism for the positioning of the individual monomers in such a way as to confer a specific sequence of interunit linkages upon the first macromolecular lignin chain to be synthesized, which would thereafter undergo direct replication as biosynthesis continues [(*142*) and Chapters 15 and 22].

Indeed, this *new* paradigm, together with the ideas clarifying the relationship with the lignan biosynthetic pathways, brings order into what was once believed to be a 'chaotic' biochemical series of events. It can account for the preponderance of the lignans in their various forms, lignan deposition in various tissues and cell-types, and the 'infusion' process deploying *non-structural* oligomeric lignans which have often been erroneously described as 'Brauns native lignins' or 'secondary lignins' in heartwood. Moreover, these ideas persuasively explain why heartwood had been thought to contain oligomeric components ostensibly formed *via* acidolytic degradation of lignin macromolecules in the aging tissues (*111*). It has now become more likely that they are distinct biochemical entities. Lastly, and perhaps most importantly, the new working hypothesis provides a plausible means for controlling lignin biopolymer assembly in sapwood in a biologically well-defined way. This is because the involvement of the proline-rich proteins *or other compatible polypeptides* affords a plausible mechanism for controlling macromolecular lignin assembly during monolignol transport from the plasma membrane into the lignifying cell wall.

Accordingly, a recent claim that macromolecular lignin assembly will utilize other precursors, if the proper monolignols are not available, was quite unexpected (*143*). This was based upon an analysis of a supposed mutant of loblolly pine (*Pinus taeda*) whose 'lignin' was purported to contain 30% (w/w) dihydroconiferyl alcohol substructures (*143*). It was proposed that this plant was unable to biosynthesize coniferyl alcohol fully, but instead generated dihydroconiferyl alcohol *via* some unknown biosynthetic route. Careful evaluation of the reported data revealed that only about 17% of this 'lignin' had been extracted from the tissues for analysis, and that the total contribution to the overall plant lignin from the dihydroconiferyl alcohol unit was < 5-6%. These data are much more in accord with the consequences of the infusion process resulting in the deployment of insoluble lignan oligomers, in which are incorporated the reduction products of dehydrodiconiferyl alcohol, as previously reported (*80, 137,* and Chapter 25). Therefore, there appears to be no biochemical justification for identifying these substances as 'abnormal' lignins. This, in the absence of substantive biochemical data supporting such an assertion, does little to advance the field. On the contrary, it returns to a primitive working hypothesis that far predates the Freudenberg era!

Acknowledgments

The authors thank the United States Department of Agriculture (9603622), the National Science Foundation (MCB-9631980), the U.S. Department of Energy (DE-FG03-97ER20259), the U.S. National Aeronautics and Space Administration (NAG100164), McIntire-Stennis for generous support of these studies, as well as the Lewis B. and Dorothy Cullman and G. Thomas Hargrove Center for Land Plant Adaptation Studies. Acknowledgment for support of this work is also made to the Vincent Johnson Lignin Research Fund at the University of Minnesota, and the Minnesota Agricultural Experiment Station. [Paper No. 974436805 of the Scientific Journal Series of the Minnesota Agricultural Experiment Station, funded through Minnesota Agricultural Experiment Station project No. 43-68, supported by Hatch Funds.]

Literature Cited

1. *Constitution and Biosynthesis of Lignin*; Freudenberg, K.; Neish, A. C., Eds.; Springer-Verlag: New York, NY, 1968.
2. Lewis, N. G.; Yamamoto, E. In *Chemistry and Significance of Condensed Tannins*; Hemingway, R. W., Karchesy, J. J., Eds.; Plenum Press: New York, NY, 1989; pp 23-47.
3. Campbell, M. M.; Sainsbury, M.; Searle, P. A. *Synthesis* **1993**, 179-193.
4. Dewick, P. M. *Nat. Prod. Rep.* **1995**, *12*, 579-607.
5. Jensen, R. A. In *Recent Advances in Phytochemistry*; Conn, E. E., Ed.; Plenum Press: New York, NY, 1986; Vol. 20, pp 57-81.
6. Belmares, H.; Barrera, A.; Castillo, E.; Ramos, L. F.; Hernandez, F.; Hernandez, V. *Ind. Eng. Chem. Prod. Res. Dev.* **1979**, *18*, 220-226.
7. Figgitt, D. P.; Denyer, S. P.; Dewick, P. M.; Jackson, D. E.; Williams, P. *Biochem. Biophys. Res. Commun.* **1989**, *160*, 257-262.
8. Medarde, M.; Peláez-Lamamié de Clairac, R.; López, J. L.; Grávalos, D. G.; San Feliciano, A. *Arch. Pharm. (Weinheim)* **1995**, *328*, 640-644.
9. Nakatani, N.; Ikeda, K.; Kikuzaki, H.; Kido, M.; Yamaguchi, Y. *Phytochemistry* **1988**, *27*, 3127-3129.
10. Fauré, M.; Lissi, E.; Torres, R.; Videla, L. A. *Phytochemistry* **1990**, *29*, 3773-3775.
11. Oliveto, E. P. *Chem. Ind.* **1972**, 677-679.

12. Osawa, T.; Nagata, M.; Namiki, M.; Fukuda, Y. *Agric. Biol. Chem.* **1985**, *49*, 3351-3352.
13. Xue, J. Y.; Liu, G. T.; Wei, H. L.; Pan, Y. *Free Radical Biology & Medicine* **1992**, *12*, 127-135.
14. Arnold, A. M.; Whitehouse, J. M. A. *The Lancet* **1981**, 912-915.
15. Issell, B. F. *Cancer Chemother. Pharmacol.* **1982**, *7*, 73-80.
16. Froelich-Ammon, S. J.; Osheroff, N. *J. Biol. Chem.* **1995**, *270*, 21429-21432.
17. Thompson, L. U.; Orcheson, L.; Rickard, S.; Jenab, M.; Serraino, M.; Seidl, M. M.; Cheung, F. In *Natural Antioxidants and Food Quality in Atherosclerosis and Cancer Prevention*; Kumpulainen, J. T., Salonen, J. K., Eds.; The Royal Society of Chemistry: Cambridge, England, 1996; pp 356-364.
18. Thompson, L. U.; Rickard, S. E.; Orcheson, L. J.; Seidl, M. M. *Carcinogenesis* **1996**, *17*, 1373-1376.
19. Thompson, L. U.; Seidl, M. M.; Rickard, S. E.; Orcheson, L. J.; Fong, H. H. S. *Nutr. Cancer* **1996**, *26*, 159-165.
20. Adlercreutz, H. *Gastroenterology* **1984**, *86*, 761-764.
21. Adlercreutz, H.; Höckerstedt, K.; Bannwart, C.; Bloigu, S.; Hämäläinen, E.; Fotsis, T.; Ollus, A. *J. Steroid Biochem.* **1987**, *27*, 1135-1144.
22. Adlercreutz, H. *Environmental Health Perspectives* **1995**, *103 (S7)*, 103-112.
23. van Heerden, P. S.; Towers, G. H. N.; Lewis, N. G. *J. Biol. Chem.* **1996**, *271*, 12350-12355.
24. Phillips, M. *Chem. Rev.* **1934**, *14*, 103-170.
25. Haworth, R. D. *Annu. Rept. Prog. Chem.* **1937**, *33*, 266-279.
26. Tiemann, F. *Ber. deutsch. chem. Ges.* **1875**, *8*, 1127-1136.
27. Klason, P. *Arkiv. Kemi., Mineral. Geol.* **1908**, *3*, 17.
28. Erdtman, H. *Liebigs Ann. d. Chem.* **1933**, *503*, 283-294.
29. Freudenberg, K.; Reznik, H.; Fuchs, W.; Reichert, M. *Naturwiss.* **1955**, *42*, 29-35.
30. Davin, L. B.; Lewis, N. G. In *Recent Advances in Phytochemistry*; Stafford, H. A., Ibrahim, R. K., Eds.; Plenum Press: New York, NY. (and references therein), 1992; Vol. 26, pp 325-375.
31. Freudenberg, K.; Harkin, J. M.; Reichert, M.; Fukuzumi, T. *Chem. Ber.* **1958**, *91*, 581-590.
32. Freudenberg, K. *Nature* **1959**, *183*, 1152-1155.
33. Higuchi, T. *Physiol. Plant.* **1957**, *10*, 356-372.
34. Higuchi, T.; Ito, Y. *J. Biochem.* **1958**, *45*, 575-579.
35. Higuchi, T. *J. Biochem.* **1958**, *45*, 515-528.
36. Adler, E. *Wood Sci. Technol.* **1977**, *11*, 169-218.
37. Lewis, N. G.; Razal, R. A.; Dhara, K. P.; Yamamoto, E.; Bokelman, G. H.; Wooten, J. B. *J. Chem. Soc., Chem. Commun.* **1988**, 1626-1628.
38. Lewis, N. G.; Razal, R. A.; Yamamoto, E.; Bokelman, G. H.; Wooten, J. B. In *Plant Cell Wall Polymers—Biogenesis and Biodegradation*; Lewis, N. G., Paice, M. G., Eds.; ACS Symposium Series: Washington, DC, 1989; Vol. 399, pp 169-181.
39. Gottlieb, O. R. *Phytochemistry* **1972**, *11*, 1537-1570.
40. Gottlieb, O. R. *Rev. Latinoamer. Quim.* **1974**, *5*, 1-10.
41. Gottlieb, O. R.; Yoshida, M. In *Natural Products of Woody Plants—Chemicals Extraneous to the Lignocellulosic Cell Wall*; Rowe, J. W., Kirk, C. H., Eds.; Springer Verlag: Berlin, 1989; pp 439-511.
42. Fang, J.-M.; Hsu, K.-C.; Cheng, Y.-S. *Phytochemistry* **1989**, *28*, 3553-3555.
43. Vaquette, J.; Cavé, A.; Waterman, P. G. *Planta Medica* **1979**, *35*, 42-47.
44. Zhuang, L.-g.; Seligmann, O.; Jurcic, K.; Wagner, H. *Planta Medica* **1982**, *45*, 172-176.
45. Kitagawa, S.; Nishibe, S.; Benecke, R.; Thieme, H. *Chem. Pharm. Bull.* **1988**, *36*, 3667-3670.
46. Erdtman, H. *Svensk Papperstidn.* **1939**, *42*, 115-122.
47. Fell, D. A. *Biochem. J.* **1992**, *286*, 313-330.

48. Liao, J. C.; Delgado, J. *Biotechnol. Prog.* **1993**, *9*, 221-233.
49. ap Rees, T.; Hill, S. A. *Plant Cell Environ.* **1994**, *17*, 587-599.
50. Brand, M. D. *J. theor. Biol.* **1996**, *182*, 351-360.
51. Koukol, J.; Conn, E. E. *J. Biol. Chem.* **1961**, *236*, 2692-2698.
52. Neish, A. C. *Phytochemistry* **1961**, *1*, 1-24.
53. Razal, R. A.; Ellis, S.; Lewis, N. G.; Towers, G. H. N. *Phytochemistry* **1996**, *41*, 31-35.
54. Singh, S.; Lewis, N. G.; Towers, G. H. N. *Polyphénols Actualités* **1997**, *15/16*, 16-19.
55. Ingold, E.; Sugiyama, M.; Komamine, A. *Physiol. Plant.* **1990**, *78*, 67-74.
56. Atanassova, R.; Favet, N.; Martz, F.; Chabbert, B.; Tollier, M.-T.; Monties, B.; Fritig, B.; Legrand, M. *Plant J.* **1995**, *8*, 465-477.
57. Van Doorsselaere, J.; Baucher, M.; Chognot, E.; Chabbert, B.; Tollier, M.-T.; Petit-Conil, M.; Leplé, J.-C.; Pilate, G.; Cornu, D.; Monties, B.; Van Montagu, M.; Inzé, D.; Boerjan, W.; Jouanin, L. *Plant J.* **1995**, *8*, 855-864.
58. Ni, W.; Paiva, N. L.; Dixon, R. A. *Trangenic Research* **1994**, *3*, 120-126.
59. Baucher, M.; Chabbert, B.; Pilate, G.; Van Doorsselaere, J.; Tollier, M.-T.; Petit-Conil, M.; Cornu, D.; Monties, B.; Van Montagu, M.; Inzé, D.; Jouanin, L.; Boerjan, W. *Plant Physiol.* **1996**, *112*, 1479-1490.
60. Halpin, C.; Knight, M. E.; Foxon, G. A.; Campbell, M. M.; Boudet, A. M.; Boon, J. J.; Chabbert, B.; Tollier, M.-T.; Schuch, W. *Plant J.* **1994**, *6*, 339-350.
61. Higuchi, T.; Ito, T.; Umezawa, T.; Hibino, T.; Shibata, D. *J. Biotechnol.* **1994**, *37*, 151-158.
62. Stewart, D.; Yahiaoui, N.; McDougall, G. J.; Myton, K.; Marque, C.; Boudet, A. M.; Haigh, J. *Planta* **1997**, *201*, 311-318.
63. Russell, D. W.; Conn, E. E. *Arch. Biochem. Biophys.* **1967**, *122*, 256-258.
64. Russell, D. W. *J. Biol. Chem.* **1971**, *246*, 3870-3878.
65. Grand, C. *FEBS Lett.* **1984**, *169*, 7-11.
66. Byerrum, R. U.; Flokstra, J. H.; Dewey, L. J.; Ball, C. D. *J. Biol. Chem.* **1954**, *210*, 633-643.
67. Finkle, B. J.; Nelson, R. F. *Biochim. Biophys. Acta* **1963**, *78*, 747-749.
68. Finkle, B. J.; Masri, M. S. *Biochim. Biophys. Acta* **1964**, *85*, 167-169.
69. Gross, G. G.; Stöckigt, J.; Mansell, R. L.; Zenk, M. H. *FEBS Lett.* **1973**, *31*, 283-286.
70. Kamsteeg, J.; Van Brederode, J.; Verschuren, P. M.; Van Nigtevecht, G. *Z. Pflanzenphysiol.* **1981**, *102*, 435-442.
71. Boniwell, J. M.; Butt, V. S. *Z. Naturforsch.* **1986**, *41c*, 56-60.
72. Duke, S. O.; Vaughn, K. C. *Physiol. Plant.* **1982**, *54*, 381-385.
73. Kojima, M.; Takeuchi, W. *J. Biochem.* **1989**, *105*, 265-270.
74. Kuroda, H.; Shimada, M.; Higuchi, T. *Phytochemistry* **1975**, *14*, 1759-1763.
75. Bugos, R. C.; Chiang, V. L. C.; Campbell, W. H. *Plant Mol. Biol.* **1991**, *17*, 1203-1215.
76. Schmitt, D.; Pakusch, A.-E.; Matern, U. *J. Biol. Chem.* **1991**, *266*, 17416-17423.
77. Ye, Z.-H.; Kneusel, R. E.; Matern, U.; Varner, J. E. *Plant Cell* **1994**, *6*, 1427-1439.
78. Ye, Z.-H.; Varner, J. E. *Plant Physiol.* **1995**, *108*, 459-467.
79. Lacombe, E.; Hawkins, S.; Van Doorsselaere, J.; Piquemal, J.; Goffner, D.; Poeydomenge, O.; Boudet, A.-M.; Grima-Pettenati, J. *Plant J.* **1997**, *11*, 429-441.
80. Nose, M.; Bernards, M. A.; Furlan, M.; Zajicek, J.; Eberhardt, T. L.; Lewis, N. G. *Phytochemistry* **1995**, *39*, 71-79.
81. Donaldson, L. A. *Wood Sci. Technol.* **1994**, *28*, 111-118 (and references therein).
82. Marcinowski, S.; Grisebach, H. *Phytochemistry* **1977**, *16*, 1665-1667.
83. Dharmawardhana, D. P.; Ellis, B. E.; Carlson, J. E. *Plant Physiol.* **1995**, *107*, 331-339.

84. Leinhos, V.; Udagama-Randeniya, P. V.; Savidge, R. A. *Phytochemistry* **1994**, *37*, 311-315.
85. Harkin, J. M.; Obst, J. R. *Science* **1973**, *180*, 296-297.
86. Barcelo, A. R. *Protoplasma* **1995**, *186*, 41-44.
87. Freudenberg, K. *Science* **1965**, *148*, 595-600.
88. Sterjiades, R.; Dean, J. F. D.; Gamble, G.; Himmelsbach, D. S.; Eriksson, K.-E. L. *Planta* **1993**, *190*, 75-87.
89. Driouich, A.; Lainé, A.-C.; Vian, B.; Faye, L. *Plant J.* **1992**, *2*, 13-24.
90. Sterjiades, R.; Dean, J. F. D.; Eriksson, K.-E. L. *Plant Physiol.* **1992**, *99*, 1162-1168.
91. Bao, W.; O'Malley, D. M.; Whetten, R.; Sederoff, R. R. *Science* **1993**, *260*, 672-674.
92. Mason, H. S.; Cronyn, M. *J. Amer. Chem. Soc.* **1955**, *77*, 491.
93. Udagama-Randeniya, P.; Savidge, R. *Electrophoresis* **1994**, *15*, 1072-1077.
94. Koblitz, H.; Koblitz, D. *Nature* **1964**, *204*, 199-200.
95. Lagrimini, L. M.; Bradford, S.; Rothstein, S. *Plant Cell* **1990**, *2*, 7-18.
96. Lagrimini, L. M. *Plant Physiol.* **1991**, *96*, 577-583.
97. Lagrimini, L. M.; Vaughn, J.; Erb, W. A.; Miller, S. A. *HortScience* **1993**, *28*, 218-221.
98. Dean, J. F. D. In *Lignin and Lignan Biosynthesis*; Lewis, N. G., Sarkanen, S., Eds.; ACS Symposium Series: Washington, D C, (this volume).
99. Fritig, B. Keystone Symposium on the Extracellular Matrix of Plants; Tamarron, CO; 1996.
100. Sherf, B. A.; Bajar, A. M.; Kolattukudy, P. E. *Plant Physiol.* **1993**, *101*, 201-208.
101. Russell, W. R.; Forrester, A. R.; Chesson, A.; Burkitt, M. *Arch. Biochem. Biophys.* **1996**, *332*, 357-366.
102. Freudenberg, K.; Schlüter, H. *Chem. Ber.* **1955**, *88*, 617-625.
103. Davin, L. B.; Lewis, N. G. *An. Acad. bras. Ci.* **1995**, *67 (Supl. 3)*, 363-378.
104. Okusa, K.; Miyakoshi, T.; Chen, C.-L. *Holzforschung* **1996**, *50*, 15-23.
105. Elder, T. J.; Ede, R. M. *Proc. 8th Internat. Symp. Wood Pulp. Chem.* **1995**, *1*, 115-122.
106. Freudenberg, K.; Nimz, H. *Chem. Ber.* **1962**, *95*, 2057-2062.
107. Freudenberg, K.; Tausend, H. *Chem. Ber.* **1963**, *96*, 2081-2085.
108. Sarkanen, K. V.; Wallis, A. F. A. *J. Chem. Soc., Perkin I* **1973**, 1869-1878.
109. Davin, L. B.; Wang, H.-B.; Crowell, A. L.; Bedgar, D. L.; Martin, D. M.; Sarkanen, S.; Lewis, N. G. *Science* **1997**, *275*, 362-366.
110. Terashima, N.; Fukushima, K.; He, L.-f.; Takabe, K. In *Forage Cell Wall Structure and Digestibility*; Jung, H. G., Buxton, D. R., Hatfield, R. D., Ralph, J., Eds.; American Society of Agronomy, Crop Science Society of America, Soil Science Society of America: Madison, WI, 1993; pp 247-270.
111. Sarkanen, K. V. In *Lignins—Occurrence, Formation, Structure and Reactions*; Sarkanen, K. V., Ludwig, C. H., Eds.; Wiley Interscience: New York, NY, 1971; pp 95-163.
112. Atalla, R. H.; Agarwal, U. P. *Science* **1985**, *227*, 636-638.
113. Jurasek, L. *J. Pulp Paper Sci.* **1995**, *21*, J274-279.
114. Jurasek, L. *J. Pulp Paper Sci.* **1996**, *22*, J376-380.
115. Goring, D. A. I. *ACS Symposium Series* **1977**, *48*, 273-277.
116. Goring, D. A. I.; Vuong, R.; Gancet, C.; Chanzy, H. *J. Appl. Polym. Sci.* **1979**, *24*, 931-936.
117. Yan, J. F.; Pla, F.; Kondo, R.; Dolk, M.; McCarthy, J. L. *Macromolecules* **1984**, *17*, 2137-2142.
118. Pla, F.; Yan, J. F. *J. Wood Chem. Technol.* **1984**, *4*, 285-299.
119. Ralph, J.; Grabber, J. H.; Hatfield, R. D. *Carbohydr. Res.* **1995**, *275*, 167-178.
120. Müsel, G.; Schindler, T.; Bergfeld, R.; Ruel, K.; Jacquet, G.; Lapierre, C.; Speth, V.; Schopfer, P. *Planta* **1997**, *201*, 146-159.

121. Ryser, U.; Schorderet, M.; Zhao, G.-F.; Studer, D.; Ruel, K.; Hauf, G.; Keller, B. *Plant J.* **1997**, *12*, 97-111.
122. Ayres, D. C.; Loike, J. D. *Chemistry and Pharmacology of Natural Products. Lignans: Chemical, Biological and Clinical Properties*; Cambridge University Press: Cambridge, England, 1990; pp 402.
123. Gottlieb, O. R. *Progr. Chem. Org. Nat. Prod.* **1978**, *35*, 1-72.
124. MacLean, H.; Gardner, J. A. F. *For. Prod. J.* **1956**, *6*, 510-516.
125. Umezawa, T.; Davin, L. B.; Yamamoto, E.; Kingston, D. G. I.; Lewis, N. G. *J. Chem. Soc., Chem. Commun.* **1990**, 1405-1408.
126. Umezawa, T.; Davin, L. B.; Lewis, N. G. *Biochem. Biophys. Res. Commun.* **1990**, *171*, 1008-1014.
127. Umezawa, T.; Davin, L. B.; Lewis, N. G. *J. Biol. Chem.* **1991**, *266*, 10210-10217.
128. Katayama, T.; Davin, L. B.; Lewis, N. G. *Phytochemistry* **1992**, *31*, 3875-3881.
129. Katayama, T.; Davin, L. B.; Chu, A.; Lewis, N. G. *Phytochemistry* **1993**, *32*, 581-591.
130. Lewis, N. G.; Davin, L. B.; Katayama, T.; Bedgar, D. L. *Bull. Soc. Groupe Polyphénols* **1992**, *16*, 98-103.
131. Davin, L. B.; Bedgar, D. L.; Katayama, T.; Lewis, N. G. *Phytochemistry* **1992**, *31*, 3869-3874.
132. Chu, A.; Dinkova, A.; Davin, L. B.; Bedgar, D. L.; Lewis, N. G. *J. Biol. Chem.* **1993**, *268*, 27026-27033.
133. Ozawa, S.; Davin, L. B.; Lewis, N. G. *Phytochemistry* **1993**, *32*, 643-652.
134. Paré, P. W.; Wang, H.-B.; Davin, L. B.; Lewis, N. G. *Tetrahedron Lett.* **1994**, *35*, 4731-4734.
135. Dinkova-Kostova, A. T.; Gang, D. R.; Davin, L. B.; Bedgar, D. L.; Chu, A.; Lewis, N. G. *J. Biol. Chem.* **1996**, *271*, 29473-29482.
136. Kato, M. J.; Davin, L. B.; Lewis, N. G. *Phytochemistry* **1997**, *46*, 583-591.
137. Davin, L. B.; Gang, D. R.; Fujita, M.; Anterola, A.; Lewis, N. G. *Proc. 9th Internat. Symp. Wood Pulp. Chem.* **1997**, *1*, H3, 1-4.
138. Su, W.-C.; Fang, J.-M.; Cheng, Y.-S. *Phytochemistry* **1995**, *40*, 563-566.
139. Davin, L. B.; Wang, H.-B.; Dinkova-Kostova, A. T.; Gang, D. R.; Fujita, M.; Sarkanen, S.; Lewis, N. G. U.S. Patent Application Serial No. 60/033,298.
140. Gang, D. R.; Dinkova-Kostova, A. T.; Davin, L. B.; Lewis, N. G. In *Phytochemical Pest Control Agents*; Heddin, P., Ed.; ACS Symposium Series: Washington, DC, 1997; Vol. 658, pp 58-89.
141. Hergert, H. L. In *Cellulose Chemistry and Technology*; Arthur, J. C., Jr., Ed.; ACS Symposium Series: Washington, DC, 1977; Vol. 48, pp 227-243.
142. Guan, S.-y.; Mlynár, J.; Sarkanen, S. *Phytochemistry* **1997**, *45*, 911-918.
143. Ralph, J.; MacKay, J.; Hatfield, R.; O'Malley, D.; Whetten, R.; Sederoff, R. *Science* **1997**, *277*, 235-239.

Chapter 2

Upstream Metabolic Segments That Support Lignin Biosynthesis

Carol A. Bonner and Roy A. Jensen

Department of Microbiology and Cell Science, University of Florida, Building 981, P.O. Box 110700, Gainesville, FL 32611

The synthesis of lignin in differentiating xylem cells places a quantitatively great demand upon the supply of L-phenylalanine. The mechanisms which dictate L-phenylalanine availability are largely unknown. Current knowledge of the upstream metabolic segments that ultimately form L-phenylalanine precursor molecules is reviewed, beginning with the generation of erythrose-4-phosphate and phosphoenol pyruvate from carbohydrate metabolism. Of particular interest is the nature of quinate as a carbon-reserve molecule which seems to have a special precursor relationship with lignin. Possible biochemical interactions between intracellular compartments and different cell types are considered.

The synthesis of lignin occurs in association with the normal developmental process of xylem differentiation; it can also be induced in other tissues in response to the environmental cues of mechanical wounding or infection with pathogens (*1*). The monolignol biosynthetic pathway is a prominent branch of plant phenylpropanoid metabolism, a biochemical network which utilizes L-phenylalanine as a starting substrate (*2*). Up to 60% of the dry weight of higher plants consists of molecules whose precursors once traversed the shikimate pathway to form phenylalanine (*3*), and up to 36% of the dry weight of wood is lignin (*4*). Since the metabolic flux to lignin in xylem-forming cells is vast, the question arises as to how phenylalanine can become selectively available in the correct subcellular spatial locale of one differentiated cell type in response to the demand for lignin biosynthesis? The availability of phenylalanine has been reported to be limiting for phenylpropanoid metabolism (*5, 6*). Phenylalanine is among the most biochemically expensive of the amino acids synthesized by living cells, and the flux to phenylalanine can be expected to challenge cellular reserves of phosphoenol pyruvate (PEP), erythrose-4-P (E4P) and adenosine triphosphate (ATP).

The metabolic *gestalt* (Figure 1) that generates lignin begins with two segments of carbohydrate metabolism which produce PEP and E4P, the starting precursors of aromatic compound biosynthesis. The reserve carbon that initiates carbohydrate metabolism is present as starch in plastids and sucrose in the cytosol. The potential input shown from the Calvin cycle represents a unique capability of photosynthetic chloroplasts. Phenylalanine is one product of the multi-branched, divergent, pathway

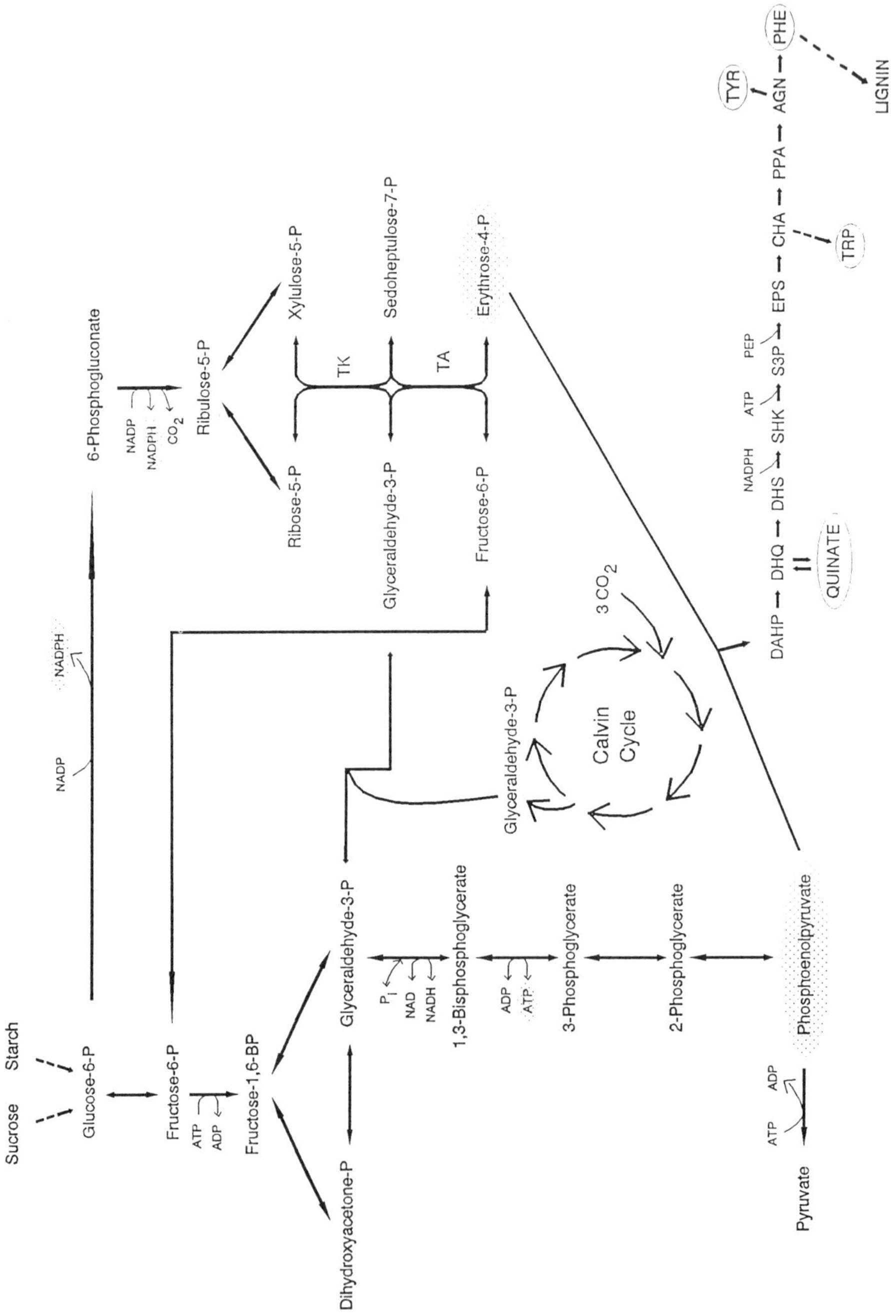

Sucrose
Starch
Glucose-6-P
Fructose-6-P
ATP
ADP
Fructose-1,6-BP
Dihydroxyacetone-P
Glyceraldehyde-3-P
P
NAD
NADH
1,3-Bisphosphoglycerate
ADP
ATP
3-Phosphoglycerate
2-Phosphoglycerate
Phosphoenolpyruvate
ADP
ATP
Pyruvate
NADP
NADPH
6-Phosphogluconate
NADP
NADPH
CO_2
Ribulose-5-P
Ribose-5-P
Xylulose-5-P
TK
TA
Glyceraldehyde-3-P
Sedoheptulose-7-P
Fructose-6-P
Erythrose-4-P
Glyceraldehyde-3-P
Calvin Cycle
3 CO_2
DAHP
DHQ
QUINATE
DHS
NADPH
SHK
ATP
S3P
PEP
EPS
CHA
TRP
PPA
AGN
TYR
PHE
LIGNIN

Figure 1. Relationships of central carbohydrate metabolism to phenylalanine and lignin biosynthesis. Glycolysis, shown at the far left, depicts the pathway present in plastids (see ref. *11* for differences in cytosolic glycolysis whereby various steps can be bypassed under conditions of P_i depletion), while the oxidative pentose phosphate pathway (OPPP) is shown at the upper right. Glyceraldehyde-3-P produced by the Calvin cycle can enter either the glycolytic pathway or the OPPP, as shown. The pathway of aromatic amino acid biosynthesis begins with the condensation of erythrose-4-P and phosphoenolpyruvate to yield DAHP. Compounds maintained within expandable pools are circled (although not circled, considerable pools of SHK may also be maintained). **Abbreviations**: DAHP; 3-deoxy-*D-arabino*-heptulosonate 7-P; DHQ, dehydroquinate; DHS, dehydroshikimate; SHK, shikimate; S3P, shikimate-3-P; EPS, enolpyruvylshikimate-3-P; CHA, chorismate; PPA, prephenate; AGN, L-arogenate; PHE, L-phenylalanine; TRP, L-tryptophan; TYR, L-tyrosine; P_i, inorganic phosphate.

of aromatic compound biosynthesis. The short phenylpropanoid pathway and the monolignol biosynthetic pathway comprise the fourth and fifth segments of the metabolic network. Contemporary reviews typically begin with the last two segments, *e.g.* the excellent review by Dixon and Paiva (*7*). Intensive ongoing research is in the process of elucidating the regulatory mechanisms which partition phenylalanine and its derivatives between lignin and the host of other products formed by competing pathways. Seldom considered has been the question of how phenylalanine availability might be an important variable. In this paper we review aspects of the connecting metabolic networks upstream of the specific pathway to lignin which might dictate phenylalanine availability.

Only a few years ago it was generally thought that the flux of carbon could not be redirected from central metabolism to peripheral metabolic pathways since many complex interactions might be perturbed. Biotechnological approaches were limited to changes that could be made in the specific pathway, *e.g.* mutations to enhance the ability of branchpoint enzymes to outperform competing enzymes, and mutations to abolish regulatory restraints. Although many interventions will undoubtedly severely unbalance metabolism to the detriment of productivity, recent successes in metabolic engineering (*8-10*) have shown that radical changes in the flux into peripheral metabolic pathways can be accomplished by manipulation of central metabolism. The elevated production of aromatic compounds in *Escherichia coli*, over and beyond that achievable by application of conventional approaches, provides one example (*8*). Using a starting background of mutations designed to abolish regulation within the pathway leading to L-tryptophan, additional changes enhancing the flux of E4P and PEP into the pathway were made. Previously unobtained increases in product yield were thus obtained. However, biotechnologists have not been the first to achieve redirection of central metabolism, and we suggest that lignin formation by higher plants may be a premier example.

Segments of Carbohydrate Metabolism

Figure 1 highlights the capacity of the intermeshed oxidative pentose phosphate pathway (OPPP) and the glycolytic pathway to cooperate in the generation of PEP and E4P, respectively. PEP and E4P are the starting precursors needed for phenylalanine (and ultimately lignin) biosynthesis. Glucose-6-P may be mobilized from carbohydrate reserves, namely from sucrose in the cytosol or from starch in plastids. Intact glycolytic pathways exist in both the cytosol and in plastids (reviewed in ref. *11*). These are spatially separate, but cross-compartmental interactions can occur since phosphorylated intermediates such as glucose-1-P, glucose-6-P, dihydroxyacetone-P, 3-phosphoglycerate, and PEP can negotiate the compartmental barrier. The OPPP is present in plastids and is generally considered to be present in the cytosol as well (but see ref. *12*).

Under photosynthetic conditions, glyceraldehyde-3-P (G3P) is potentially available directly from the Calvin cycle for input into the glycolytic and OPPP networks as a precursor for PEP and E4P molecules, as indicated in Figure 1. The ATP and NADPH consumed in the net output of triose-P are produced by the light reactions of photosynthesis.

The sufficient availability of E4P from any cell type in any cell location is a long-standing enigma (*13*). E4P is prone to dimerization because it cannot undergo intramolecular cyclization (*14*). In classical studies of plant cells, levels of E4P were below detection (Bassham, personal communication). D-Fructose-6-P and/or D-sedoheptulose-7-P have been proposed to serve as facile reservoir sources of E4P, thus allowing the rates of synthesis and utilization of E4P to be closely matched (*15*). If Delmer and co-workers are correct in their contention (ref. *16* and unpublished results) that cellulose synthesis requires sucrose as the glucose donor, it would be interesting to know what the impact would be upon E4P production since this would release substantial fructose in cells producing secondary walls.

Availability of E4P and PEP in Chloroplasts

During photosynthesis, G3P output in the chloroplast compartment fuels all subsequent carbon metabolism ($3CO_2 + 9ATP + 6NADPH + 5H_2O \rightarrow G3P + 8P_i + 9ADP + 6NADP^+$). Since G3P utilization by transaldolase (pentose phosphate pathway) and G3P dehydrogenase (glycolysis) would generate E4P and PEP, respectively, abundant resources for direct formation of E4P and PEP in the photosynthesizing plastid would seem to be present. However, uncertainty about the presence of phosphoglycerate mutase and enolase in chloroplasts has raised the question of whether chloroplasts can synthesize PEP (*17*). Furthermore, allocation of G3P to various biosynthetic pathways within the chloroplast is rigorously restrained by complex patterns of regulation for two reasons (*18*). First, five of the six molecules of triose-P formed after three turns of the Calvin cycle must be retained in order to return three 5-carbon molecules to the cycle; over-utilization of triose-P would stop the cycle. Second, export of substantial triose-P from the chloroplast to the cytosol is a mechanism which can guarantee the coupled import of P_i, which in turn is needed to maintain high rates of photosynthesis. Hence, it seems likely that a finely tuned balance is maintained in the photosynthetic chloroplast. Triose-P utilization for low-flux, essential functions in the chloroplast can probably be directly accommodated, but strong selective pressures may exist to shunt excess photosynthate in the form of triose-P to the cytosol to avoid disruption of photosynthesis. Cytosolic triose-P is appropriately mobilized to support metabolism in the various subcellular compartments throughout the plant. This includes products of triose-P metabolism which may subsequently be imported into chloroplasts.

The demand for E4P and PEP as precursors of the fraction of aromatic amino acids used for protein synthesis is quantitatively small, compared to the demand corresponding to the high-flux entry of phenylalanine into the phenylpropanoid channel. Since the aromatic biosynthetic pathway is tightly regulated by a pattern of interacting feedback loops called sequential feedback inhibition (*19*), the amount of E4P and PEP needed for aromatic amino acids that supply chloroplast protein synthesis could be directly derived from G3P photosynthate. The importance of a strictly regulated withdrawal of E4P and PEP has been demonstrated by experimental manipulations where unrestrained entry of E4P and PEP into the aromatic compound pathway of chloroplasts has been shown to disrupt photosynthesis. The normal feedback regulation of 3-deoxy-D-*arabino*-heptulosonate 7-P (DAHP) synthase-Mn (DS-Mn) by L-arogenate was eliminated by use of glyphosate which blocks the accumulation of intracellular L-arogenate (*20*). Geiger and Servaites concluded that photosynthesis was disrupted due to depletion of RuBP for carboxylation (*20*). Interference with P_i import in this case should not be a problem *per se* since the first two reactions of aromatic compound biosynthesis release P_i. However, the exit of shikimate-3-P molecules following glyphosate application may deplete P_i, as well as Calvin cycle carbon in the chloroplast. In addition the generation of P_i in the cytosol might be an important factor in maintaining balanced metabolite exchange between the mitochondrion and the cytosol. The influx of cytosolic P_i (and ADP) into the mitochondrial compartment, coupled with ATP efflux into the cytosol, is essential for continued synthesis of ATP (*21*).

Under any conditions where accumulated chloroplast starch is mobilized as an initial substrate, two molecules of glucose-6-P can enter the OPPP to yield the E4P and NADPH needed for phenylalanine biosynthesis. The fructose-6-P coproduced with E4P can be transformed via glycolysis to yield the two PEP molecules and the one ATP molecule additionally needed as phenylalanine precursors. Operation of this pathway is limited only by the availability of hexose-P and avoids the problem of 5-C depletion since the OPPP can operate freely in a $2C_6 \rightarrow 2CO_2 + C_6 + C_4$ mode, as pointed out by Schnarrenberger *et al.* (*12*).

Availability of E4P and PEP in the Cytosol

Since substantial triose-P is exported to the cytosol in exchange for P_i, this compartment is the logical spatial locale for formation of the substantial amounts of PEP and E4P that are ultimately invested in lignin, as well as in other biochemical spinoffs of phenylpropanoid metabolism. PEP formation would be accommodated by entry of G3P into the cytosolic glycolytic pathway. The presence of a cytosolic OPPP would permit E4P synthesis, and the existence of a complete cytosolic OPPP has been generally accepted for almost 30 years. However, a valid basis for doubts about the presence of all but the initial two dehydrogenases has been raised recently (*12*). This is an important question since the foregoing considerations point to a cytosolic location for the large amounts of E4P needed as precursor for lignin biosynthesis. Perhaps an unknown mechanism exists for E4P synthesis. For example, might E4P be derived via reductive carboxylation of triose-P, as appears to be the case in methanogenic bacteria (*22*)?

Is the Cytosolic Compartment Competent for Phenylalanine Synthesis?

A cytosolic species of DAHP synthase, DS-Co, is ubiquitous in plants, as is CM-2, a cytosolic isoenzyme of chorismate mutase (*23*). In contrast to the tight allosteric regulation of the plastid-localized isoenzymes (DS-Mn and CM-1), DS-Co and CM-2 are completely insensitive to allosteric control. This suggested, as the simplest possibility, the existence of a cytosolic pathway whose output is responsive to substrate availability (*24*). However, the existence of cytosolic enzymes intervening between DS-Co and CM-2 has not been demonstrated. Except for dehydroquinate synthase (step 2), genes encoding all of the common-pathway enzymes that are located in the plastid have been cloned. Use of these genes as hybridization probes has not revealed cytosolic homologs (*25*). However, possible cytosol-localized enzymes might not be homologs of plastid-localized enzymes or might have diverged substantially. Indeed, cDNAs encoding DS-Mn and CM-1 have not been successfully used as probes to obtain the established cytosolic enzymes, DS-Co or CM-2.

As one alternative possibility, we have speculated (*26*) about the existence of a different pathway (phytoshikimate pathway) in the cytosol which does not utilize E4P. This is based upon the ability of DS-Co to use G3P and PEP as co-substrates (*27*). Another possibility raised (*26*) was that DAHP formed by DS-Co in the cytosol might be translocated to the plastid where the plastid enzymes would be used for conversion to chorismate. Transport of chorismate back to the cytosol would effectively provide a mechanism where the main enzymatic machinery of the plastid can be used under conditions where the early-pathway and mid-pathway steps of plastid regulation are bypassed. Flux to phenylalanine would in this way be dictated by initial substrate availability in the cytosol.

Quinate as a Carbon Reserve

Sucrose synthesis is closely coupled with photosynthesis because the accompanying release of P_i in the cytosol is exchanged for plastidic triose-P; a constant supply of P_i is required in the chloroplast for maintenance of photosynthesis (*18*). The utilization of triose-P for starch synthesis in the chloroplast also releases P_i to drive photosynthesis. Quinate synthesis in the cytosol (*vide infra*) can be considered to be entirely comparable to sucrose synthesis in that (**i**) it is a mobile, transportable form of carbon, and (**ii**) its synthesis releases two molecules of P_i (Figure 2). In this context it is of interest that, using both *Brassica* and *Nicotiana* systems, DS-Co levels were induced under conditions of phosphate depletion (*28*). If quinate is indeed the major reserve to be used for those phenylalanine molecules entering the phenylpropanoid pathway, it is an open question whether quinate is transported back to the plastid to be converted to DHQ or SHK (Figure 2) or whether it is metabolized to phenylalanine in the cytosol by unknown steps.

Inducible Bypass Metabolism

The narrative above has indicated that it is highly unlikely that the large amounts of E4P and PEP needed as precursors for phenylalanine molecules destined to enter the phenylpropanoid pathway can be appropriated directly in the chloroplast. It follows that DS-Co in the cytosol must generate the DAHP used in the high-flux flow route. The operation of DS-Co and CM-2 in the cytosol allows various scenarios for compartmental exchange of such molecules as DAHP (or quinate) and chorismate as discussed above.

Modern biochemical advances that allow more meaningful analyses of ever-larger metabolic networks has led to an emerging realization that flux into a given metabolic branch can be modulated significantly by regulation of far-upstream enzymes which deliver substrate. The principle is well illustrated by phenazine pigment production in *Pseudomonas phenazinium* (*29*). These antibiotic agents are formed during the stationary phase of growth from anthranilate, an intermediate of L-tryptophan biosynthesis (Figure 3). During exponential growth, anthranilate is not available to the phenazine pathway because of the finely tuned feedback regulation of the tryptophan pathway. In stationary phase, a second anthranilate synthase (feedback-resistant) is induced along with other enzymes specific to the phenazine pathway. However, this in itself would not provide access to the input of chorismate needed because L-tryptophan and L-tyrosine feedback inhibit the early-pathway formation of DAHP by the two differentially controlled isoenzymes of DAHP synthase present in this group of pseudomonad bacteria. The physiologically-timed expression of a third isoenzyme, insensitive to feedback inhibition, along with the specific phenazine-pathway enzymes establishes unrestrained flux to phenazine that is limited only by substrate availability. Tight feedback control at the branchpoint leading from chorismate to L-tryptophan, L-tyrosine and phenylalanine effectively causes all chorismate formed to be channeled selectively into the phenazine pathway.

Relationship of Quinate and Lignin

The general correlation of quinate levels with lignin content was noted long ago (*30-32*). Progressively greater concentrations of quinate were measured in the comparison of herbaceous angiosperms, woody angiosperms, and gymnosperms. Tracer studies with ^{14}C-quinate demonstrated its active transformation to lignin in the wood of young pine shoots (*33*). These and other tracer studies (*34*) have shown that quinate does not accumulate as a dead-end pool, but rather forms a metabolically active pool with a rapid turnover. Ossipov and Shein showed (*33*) that an inverse relationship exists between quinate accumulation and lignification at the tissue level. Thus, quinate levels in xylem-forming cells of young shoots or in the precambial zone of trunk wood does not exceed 0.5%, which is 10-20 fold lower than in the autotrophic cells of needles. The active utilization of newly formed quinate for lignin biosynthesis accounts for the low quinate levels in xylem-forming cells. In non-lignifying tissues, quinate has alternative fates (*vide infra*). At least in conifer needles, the rate of quinate synthesis exceeds the rate of utilization (*e.g.* for protocatechuate). Quinate accumulates in vacuoles and can be mobilized for active transport throughout the plant. Quinate is extremely water soluble and is transportable in the phloem sap stream (*35*).

Novelty of Shikimate/Quinate Pools in Higher Plants

End-products of biosynthetic pathways such as amino acids, purines, pyrimidines, and vitamins are typically sequestered in metabolic pools of a size appropriate to the range of synthetic demands likely to be encountered. In contrast, intermediary metabolites are not maintained in metabolic pools and are typically present at levels that may be near the limits of detection. Indeed, it has frequently been observed that

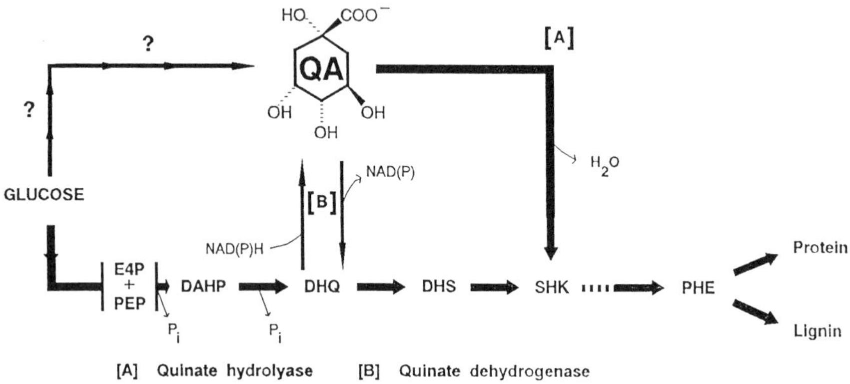

Figure 2. Enzymatic interfacing with quinate (QA). Question marks symbolize the uncertainty of a pathway to QA from glucose which does not involve DHQ. See legend of Figure 1 for abbreviations.

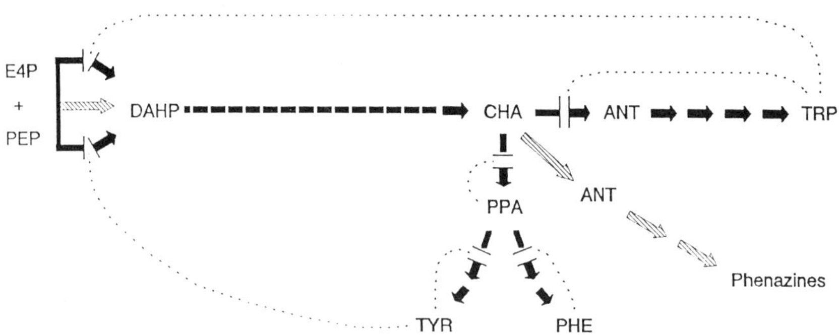

Figure 3. Inducible bypass metabolism in *Pseudomonas phenazinium*. Dotted lines indicate feedback inhibition. Solid arrows represents the enzymatic machinery in operation during primary growth, while shaded arrows identify enzymes selectively induced during stationary-phase metabolism. **Abbreviation:** ANT, anthranilate. See legend of Figure 1 for other abbreviations.

abnormally high levels of intermediates, especially phosphorylated ones, are toxic. Wild-type microorganisms, such as *Escherichia coli* and *Bacillus subtilis*, exemplify this "low-concentration-of-intermediates" rule with respect to shikimate. However, higher plants maintain distinctive pools of shikimate. In addition, the closely related compound, quinate, occupies an even more conspicuous metabolic pool. In angiosperms, shikimate and quinate have been found in most species, with herbaceous plants having low concentrations (*31*) and woody plants having high concentrations (*30*). Gymnosperms possess notably high levels of quinate (*32*), and as much as 10-14% of the dry weight of developing conifer needles is represented by quinate.

Enzymatic Ties to Quinate

Figure 2 is a composite diagram of the reactions which either use quinate as a substrate or form quinate as a product. Quinate dehydrogenase is reversible and potentially can either reduce dehydroquinate to quinate or oxidize quinate to dehydroquinate. Quinate dehydrogenase was one of the first enzymes to be studied in plant cell-suspension cultures (*36, 37*). Quinate hydrolyase, a plastid-localized enzyme, converts quinate to shikimate in corn (*38*) and in pea roots (*39*). In pea roots quinate may supplement aromatic biosynthesis in plastids at the level of shikimate to compensate for the diminution of pathway precursor in the absence of photosynthesis. Quinate hydrolyase has not been observed in conifer needles or xylem-forming cells (*40*). The source of quinate is generally assumed to be dehydroquinate. However, several studies (*41, 42*) have provided suggestive, although not compelling, evidence for the generation of quinate from glucose via a pathway separate from the aromatic biosynthetic pathway. This important possibility has not been evaluated in recent years using the more sophisticated methodology that is now available.

Quinate Metabolism in Gymnosperm Plants

Most of the data about quinate metabolism in coniferous plants comes from the research efforts of Ossipov's group in Russia. *Pinus sylvestris* and *Larix sibirica*, the main organisms studied, have yielded qualitatively similar data. Approaches taken have included rigorous analytical profiles of metabolites in different tissues at various developmental times, tracer studies with radio-labelled precursors, and a limited amount of enzymological characterization. Tracer studies have shown that quinate metabolism is dynamic, with high turnover (*43*). In the autotrophic cells of young needles as much as 10-14% of the total dry weight accumulates as quinate. An active pool of quinate is converted to hydroxybenzoic acids, mainly protocatechuate but some gallic acid as well. A significant portion of protocatechuate is catabolized to TCA cycle intermediates. Most quinate of autotrophic needle cells is maintained in a large vacuolar pool. This pool is a source of quinate for transport to other tissues, notably xylem-forming cells. In xylem-forming cells where lignin synthesis is maximal, quinate content has never been observed to exceed 0.5% of the dry weight (*33*). This correlates with a ten-fold greater specific activity of quinate dehydrogenase than in needles. Accordingly, the lower quinate content has been attributed to its accelerated metabolism by quinate dehydrogenase in order to elevate the precursor pool for lignin.

Quinate dehydrogenase has been purified about 170-fold from both the needles and developing xylem cells of *Larix sibirica* (*44*). Both enzymes were specific for $NADP^+$/NADPH as co-substrate. However, numerous differences in the two enzymes were described. Most strikingly, the enzyme from needles catalyzed an overall conversion of quinate to protocatechuate, and some gallic acid was also formed. Since the overall conversion persisted even after a purification step of affinity chromatography, either a tenacious enzyme complex or a multifunctional protein must exist. Gallic acid might be formed non-enzymatically. In contrast, quinate dehydrogenase from xylem-forming cells did not produce protocatechuate or gallic acid from quinate. Quinate dehydrogenase from needles, but not from xylem-forming cells, was activated several fold by PEP. The needle enzyme was also uniquely stimulated by Mg^{++}. The enzyme from xylem-forming cells exhibited substantially greater affinities for both quinate and $NADP^{+}$.

Role of Quinate in Bypass Metabolism?

A rough picture is emerging that is consistent with current background information about how quinate metabolism is connected with central carbon metabolism, hydroxybenzoic acid synthesis and lignin synthesis in needles and xylem-forming cells of conifer trees. In needles the high metabolic output sustained by photosynthesis generates enormous quantities of quinate, presumably from dehydroquinate formed in the cytosol. Alternative fates of quinate in the cytosol include conversion to gallic acid (a defense molecule), conversion to protocatechuate and hence to TCA cycle intermediates (perhaps as a central source of carbon during dark metabolism), and deposition within the vacuolar space. Vacuolar quinate is in a state of dynamic turnover and is a transient reservoir for reallocation to the cytosol and for transport to other tissues (notably xylem-forming cells).

Figure 4 illustrates a hypothetical scheme depicting how quinate may participate in a bypass mechanism to provide phenylalanine for lignin biosynthesis. In this scenario lignin output is directly dependent upon both quinate input and the bypassing in the plastid compartment of aromatic-pathway regulation at both the early-pathway and mid-pathway levels. In the plastids of differentiating xylem cells the feedback inhibition of DS-Mn by L-arogenate (*45*) limits DAHP formation in proportion to the demand for aromatic amino acids as substrates for protein synthesis. The metabolic capacity for production of E4P and PEP in non-photosynthetic plastids is probably quite limited. In plastids from pea roots the output of phenylalanine was enhanced by provision of intermediates such as dehydroquinate or shikimate (*46*). Quinate transported from leaf or needle tissue is imported into xylem-cell plastids where quinate dehydrogenase catalyzes the production of dehydroquinate, thus intercepting the common aromatic pathway. Excess chorismate is formed as a result of (**i**) bypassing the initial pathway step which is subject to tight allosteric control, and (**ii**) the tight allosteric control of the main enzymes which utilize chorismate (chorismate mutase-1 and anthranilate synthase). Excess chorismate thus trapped at the branchpoint in the plastid is shunted to the cytosol where an allosterically-insensitive species of chorismate mutase (CM-2) has been demonstrated in many plants (*3, 47*). Two subsequent steps (not yet proven to exist in the cytosol) generate phenylalanine as substrate for ultimate conversion to lignin molecules (or other phenylpropanoid derivatives). An increasing body of information points to the existence of channeling mechanisms that depend upon protein-protein interactions within each of the five metabolic segments considered here (reviewed in ref. *48*). Perhaps dissociable and appropriately regulated complexes exist at the interface of the terminal steps of phenylalanine biosynthesis and initial steps of phenylalanine utilization.

Conclusions

Lignin biosynthesis in differentiating xylem cells is a high-flux process which requires high phenylalanine input. This, in turn, demands a substantial commitment of E4P and PEP. It is unlikely that large quantities of E4P and PEP can be withdrawn from the chloroplast without disrupting photosynthesis. Therefore, the large amount of DAHP fated for conversion to the phenylalanine utilized by the phenylpropanoid pathway must be formed in the cytosol. Quinate is a carbon-reserve molecule that seems to have a special precursor relationship with lignin. We propose that quinate is formed in the cytosol of autotrophic cells as a source of reserve carbon, much of which is destined for lignin biosynthesis. Mobilization of quinate involves transport in the phloem to differentiating xylem cells where it is imported into plastids and converted to dehydroquinate. Entry of dehydroquinate into the shikimate pathway

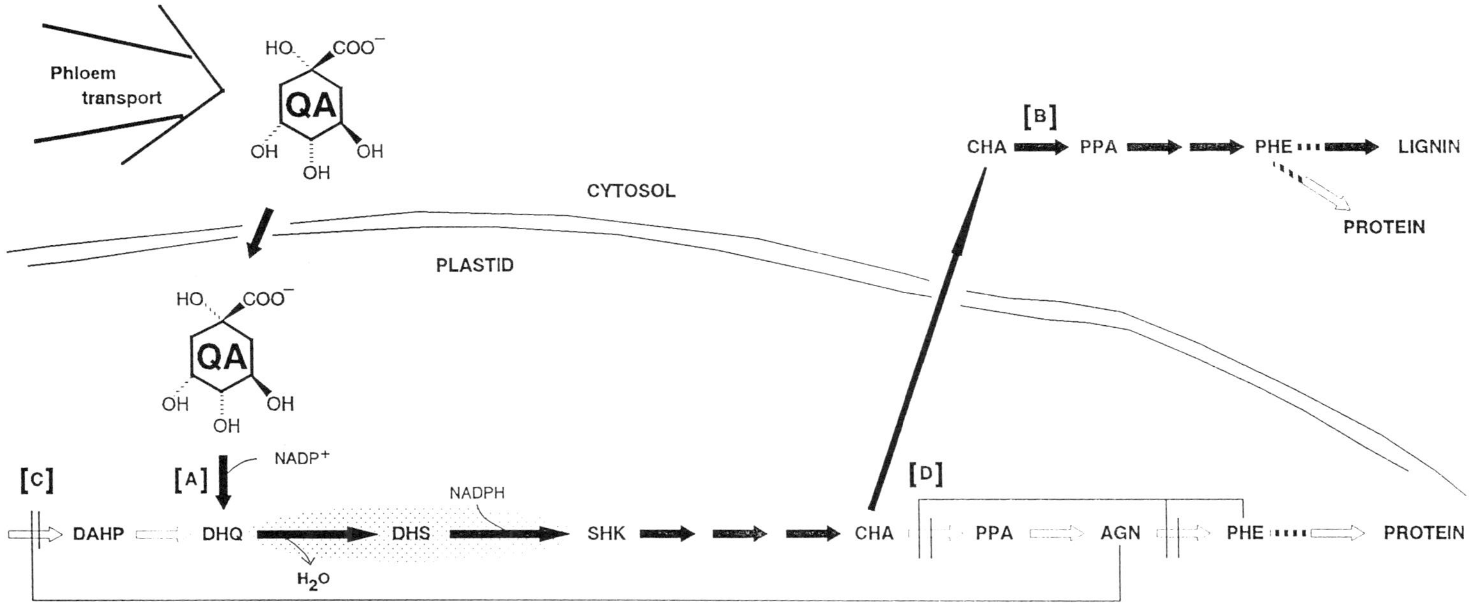

[A] Quinate dehydrogenase [B] Chorismate mutase - 2 [C] DAHP synthase - Mn [D] Chorismate mutase - 1

Figure 4. Possible mechanism for utilizing imported quinate (QA) to bypass the regulatory restraints present in the plastids of differentiating xylem cells. Solid arrows show the flow route from QA to lignin superimposed on the regulated steps of phenylalanine biosynthesis (open arrows). The two reactions marked with stippling are domains (dehydroquinase and shikimate dehydrogenase) common to a bifunctional protein.

bypasses early-pathway regulation and forms chorismate in proportion to quinate input. Due to mid-pathway regulation in the plastid, chorismate is exported to the cytosol where reactions are completed to form phenylalanine and then lignin.

Acknowledgments

Research in the principal author's laboratories has been supported by Department of Energy contract DE-FG05-86ER13581. This is Florida Agricultural Experiment Station Journal Series No. R-05329.

Literature Cited

1. Whetten, R.; Sederoff, R. *The Plant Cell* **1995**, *7*, 1001-1013.
2. Lewis, N. G.; Yamamoto, E. *Annu. Rev. Plant Physiol. Plant Mol. Biol.* **1990**, *41*, 455-496.
3. Bentley, R. *Crit. Rev. Biochem. Mol. Biol.* **1990**, *25*, 307-384.
4. Sarkanen, K. V.; Hergert, H. L. In *Lignins: Occurrence, Formation, Structure and Reactions*; Sarkanen, K. V., Ludwig, C. H., Eds.; Wiley Intersciences: New York, NY, 1971; pp 43-94.
5. Margna, U. *Phytochemistry* **1977**, *16*, 419-426.
6. Yao, K.; DeLuca, V.; Brisson, N. *The Plant Cell* **1995**, *8*, 1787-1799.
7. Dixon, R. A.; Paiva, N. L. *The Plant Cell* **1995**, *7*, 1085-1097.
8. Berry, A. *Trends Biotechnol.* **1996**, *14*, 250-256.
9. Stephanopoulos, G.; Sinskey, A. J. *Trends Biotechnol.* **1993**, *11*, 392-396.
10. Farmer, W. R.; Liao, J. C. *Curr. Opin. Biotechnol.* **1996**, *7*, 198-204.
11. Plaxton, W. C. *Annu. Rev. Plant Physiol. Plant Mol. Biol.* **1996**, *47*, 185-214.
12. Schnarrenberger, C.; Flechner, A.; Martin, W. *Plant Physiol.* **1995**, *108*, 609-614
13. Williams, J. F.; Blackmore, P. F.; Duke, C. C.; MacLeod, J. K. *Int. J. Biochem.* **1980**, *12*, 339-344.
14. Duke, C. C.; MacLeod, J. K.; Willams, J. F. *Carbohydr. Res.* **1981**, *95*, 1-26.
15. Williams, J.F.; Arora, K. K.; Longenecker, J. P. *Int. J. Biochem.* **1987**, *19*, 749-817.
16. Amor, Y; Haigler, C.H.; Johnson, S.; Wainscott, M.; Delmer, D.P. *Proc. Natl. Acad. Sci. USA* **1995**, *92*, 9353-9357.
17. Bagge, P.; Larsson, C. *Physiol. Plant.* **1986**, *68*, 641-647.
18. Macdonald, F. D.; Buchanan, B. B. In *Plant Physiology, Biochemistry and Molecular Biology*; Dennis, D. T., Turpin, D. H., Eds.; John Wiley & Sons, Inc.: New York, NY, 1990; pp 239-252.
19. Doong, R. L.; Ganson, R. J.; Jensen, R. A. *Plant, Cell and Environment* **1993**, *16*, 392-402.
20. Geiger, D. R.; Servaites, J. C. *Annu. Rev. Plant Physiol. Plant Mol. Biol.* **1994**, *45*, 235-256.
21. Douce, R.; Neuburger, M. In *Plant Physiology, Biochemistry and Molecular Biology*; Dennis, D. T., Turpin, D. H., Eds.; John Wiley & Sons, Inc.: New York, NY, 1990; pp 173-197.
22. Choquet, C. G.; Richards, J. C.; Patel, G. B.; Sprott, G. D. *Arch. Microbiol.* **1994**, *161*, 481-488.
23. Morris, P. F.; Doong, R.; Jensen, R. A. *Plant Physiol.* **1989**, *89*, 10-14.
24. Jensen, R. A. *Physiol. Plant.* **1986**, *66*, 164-168.
25. Schmid, J.; Amrhein, N. *Phytochemistry* **1995**, *39*, 737-749.
26. Jensen, R. A.; Morris, P.; Bonner, C.; Zamir, L. O. In *Plant Cell Wall Polymers: Biogenesis and Biodegradation*; Lewis, N.G., Paice, M. G., Eds.; American Chemical Society: Washington, D.C.; *ACS Symp. Ser.* 1989, Vol. 399; pp 89-107.
27. Doong, R.; Gander, J. E.; Ganson, R. J.; Jensen, R. A. *Physiol. Plant.* **1992**, *84*, 351-360.

28. Fischer, R. S.; Bonner, C. A.; Theodorou, M. E.; Plaxton, W. E.; Hrazdina, G.; Jensen, R. A. *Bioorg. Med. Chem. Lett.* **1993**, *3*, 1415-1420.
29. Pierson, L. S.; Gaffney, T.; Lam, S.; Gong, F. *FEMS Microbiol. Lett.* **1995**, *134*, 299-307.
30. Boudet, A. M. *Phytochemistry* **1973**, *12*, 363-370.
31. Yoshida, S.; Tazaki, K.; Minamikawa, T. *Phytochemistry* **1975**, *14*, 195-197.
32. Ossipov, V. *Novosibirsk, Science* **1979**, 1-111.
33. Ossipov, V.; Shein, I. *Plant Physiol.* **1990**, *37*, 518-526.
34. Boudet, A. M.; Graziana, A.; Ranjeva, R. In *Annual Proceedings of the Phytochemical Society of Europe*; Van Sumere, C. F., Lea, P. J., Eds.; Clarendon Press: Oxford, 1985, Vol. 25; p 125.
35. Ziegler, H. In *Encyclopedia of Plant Physiology*; Zimmermann, M. H., Milburn, J. A., Eds; Springer-Verlag: Berlin, 1975, Vol. 1; pp 59-100.
36. Gamborg, O. L. *Biochim. Biophys. Acta* **1966**, *128*, 483-491.
37. Gamborg, O. L. *Phytochemistry* **1967**, *6*, 1067-1073.
38. Graziana, A.; Boudet, A. In *Groupe Polyphénols*; Boudet, A., Ranjeva, R., Eds.; 1983, Vol. 11; pp 120-125.
39. Leuschner, C.; Herrmann, K. M.; Schultz, G. *Plant Physiol.* **1995**, *108*, 319-325.
40. Ossipov, V.; Shein, I. *Biochimiya* **1986**, *51*, 230-236.
41. Tazaki, K. *Plant & Cell Physiol.* **1979**, *20*, 225-231.
42. Boudet, A. M. *Plant & Cell Physiol.* **1980**, *21*, 785-792.
43. Ossipov, V. *Doctoral thesis;* A.N. Bach Institute of Biochemistry*; Moscow*, 1990.
44. Ossipov, V.; Chernov, A.; Zrazherskaga, G.; Shein, I. *Trees* **1995**, *10*, 46-51.
45. Doong, R. L.; Ganson, R. J.; Jensen, R. A. *Plant, Cell & Environ.* **1993**, *16*, 393-402.
46. Leuschner, C.; Schultz, G. *Bot. Acta* **1991**, *104*, 240-244.
47. Morris, P. F.; Doong, R. L.; Jensen, R. A. *Plant Physiol.* **1989**, *89*, 10-14.
48. Hrazdina, G.; Jensen, R. A. *Annu. Rev. Plant Physiol. Plant Mol. Biol.* **1992**, *43*, 241-267.

Chapter 3

Integrating Nitrogen and Phenylpropanoid Metabolic Pathways in Plants and Fungi

G. H. N. Towers[1], S. Singh[1], P. S. van Heerden[2], J. Zuiches[2], and Norman G. Lewis[2]

[1]Department of Botany, University of British Columbia, Vancouver, British Columbia V6T 1Z4, Canada
[2]Institute of Biological Chemistry, Washington State University, Pullman, WA 99164–6340

Phenylpropanoid and phenylpropanoid-acetate pathways are major metabolic sinks for assimilated organic carbon in vascular plants *e.g.* affording lignins, lignans, flavonoids, suberins, coumarins, and related compounds. Together, they contribute to about 30-40% of all organic plant matter. Their formation was a critical juncture in land plant adaptation, and was achieved by elaboration of complex biochemical pathways initiated by metabolism of the aromatic amino acids, phenylalanine and to a lesser extent tyrosine. Certain fungi, especially wood-decaying basidiomycetes, such as *Lentinus lepideus,* also biosynthesize phenylpropanoid derivatives in large amounts. The initial step in phenylpropanoid metabolism is catalyzed by phenylalanine and tyrosine ammonia lyases to afford the corresponding cinnamic acid derivatives and an equimolar amount of ammonium ion. In our investigations, ^{15}N-specifically labeled substrates, *i.e.* L-[^{15}N]phenylalanine, $^{15}NH_4Cl$, L-[^{15}N]glutamic acid and L-[^{15}N]glutamine were administered to potato and sweet potato tuber disks, loblolly pine cell suspension cultures and *L. lepideus* mycelia, respectively. Analyses of the resulting amino acid extracts by both ^{15}N NMR spectroscopy and GC-MS established that the ammonium ion released during active phenylpropanoid metabolism was recycled back for L-phenylalanine regeneration. The significance of this nitrogen cycling process for the evolution and adaptation of plants and fungi to land is discussed.

During evolutionary adaptation from an aquatic to a land-based habitat, plants had to contend with, among other things, different environmental effects resulting from: direct UV-B irradiation, alterations in gravitational load perception and response, metabolism in a desiccating habitat and numerous challenges/encroachments by other organisms leading to a myriad of defense-related responses (*1*). This successful adaptation to land was achieved primarily by the elaboration of new biochemical pathways leading to the formation of hitherto unknown, so-called secondary, metabolites derived from either phenylalanine, or less frequently, tyrosine (*2*). The entry point into these pathways was catalyzed by the action of phenylalanine ammonia lyase (PAL), and to a lesser extent, tyrosine ammonia lyase (TAL), *i.e.* to give cinnamic and *p*-coumaric acids, respectively. In terrestrial plants, of course, these cinnamic acids are precursors of a fantastic array of metabolic products, such as flavonoids, suberins, lignins/lignans and thousands of other phenolics. Indeed, at least 30-40% of the dry weight of all vascular plant material is derived *via* phenylpropanoid/phenylpropanoid-acetate metabolism. By contrast PAL has not yet been detected in algae with one possible exception (*Dunaliella*) (*3*). Some bacteria also employ PAL for the biosynthesis of the photoreceptor, photoactive yellow protein (PYP) (*4*), and various fungi have the ability to either constitutively or inducibly activate PAL (*5*).

In vascular plants, the flavonoid skeleton is derived from condensation of *p*-coumaryl CoA with three malonyl CoA units. Some of the resulting metabolites strongly absorb UV-B (320-400 nm) irradiation and frequently accumulate in epidermal and subepidermal cells of leaves, fruits and stems (*6*). Consequently, they are considered to be able to protect such organs from UV-B induced damage. Moreover, formation of related metabolites, such as the anthocyanins, can also be induced in response to light, this being a commonly observed phenomenon with roots or developing organs such as young leaves (*7*). While beyond the scope of this particular chapter, numerous flavonoid-derived substances have crucial roles in plant defense (*8*), pigmentation (*9*) and growth/development (*10*). Additionally, nitrogen fixation during *Rhizobium*-legume interactions (*11*) is also in large part dependent upon the employment of flavonoid signaling molecules, and the transcriptional activation of genes involved in flavonoid biosynthesis is a well-known response of seedlings and many plants (*12-14*).

Vascular plants produce suberin, a complex biopolymeric matrix which is primarily deposited in periderm tissues (*2, 15-17*). For example, a critical function of the barks of trees, as well as in many other types of plants, is to prevent uncontrolled desiccation and water loss by the whole organism and is achieved *via* formation of suberized layers. Suberized tissue formation is also often implicated in defense, since its structural matrix affords a relatively impenetrable barrier to opportunistic pathogens. The aromatic domain of suberin in potato (*Solanum tuberosum*) tuber wound-healing suberized tissue was shown to be primarily composed of hydroxycinnamate-derived entities, such as feruloyl tyramine and *p*-coumaryl tyramine, although it also apparently contains low levels of monolignol-derived substances.

The principal products of phenylpropanoid metabolism, in terms of overall carbon deployment are, however, the polymeric lignins, which account for 20-30% of all vascular plant material (*18*). They are derived mainly from hydroxycinnamyl alcohols (monolignols), although smaller amounts of hydroxycinnamates are also apparently involved in lignin deposition in grasses as well (*18*). Closely related metabolites, such as the lignans (*2, 19, 20*) have various roles such as in plant defense (*19, 21*), in helping impart color, texture and durability to certain heartwood tissues (*22*), as antioxidants (*23*), and so forth (see chapters 22 and 25).

Based on these representative examples of phenylpropanoid/phenylpropanoid acetate metabolites and their functions, it should be evident that L-phenylalanine is involved in many metabolic branchpoint pathways, some of which are illustrated in Figure 1. However, both Phe and Tyr are only present in low concentrations in plant cells being actively metabolized *via* four types of reactions. That is, they are: *predominantly* converted into *E*-cinnamate and *E*-*p*-coumarate-derived substances through the action of PAL and TAL with concomitant release of ammonium ion (*24, 25*); incorporated into proteins; decarboxylated and incorporated into aromatic alkaloids (*26*), as well as undergoing transamination followed by decarboxylation leading to the corresponding phenethyl alcohols, aldehydes and acids which are the precursors of a vast array of phytochemicals (*27*).

As shown in Scheme 1, in the pre-aromatic pathway leading to the formation of Phe and Tyr, prephenate is transaminated *via* glutamate to yield arogenate (*28*),

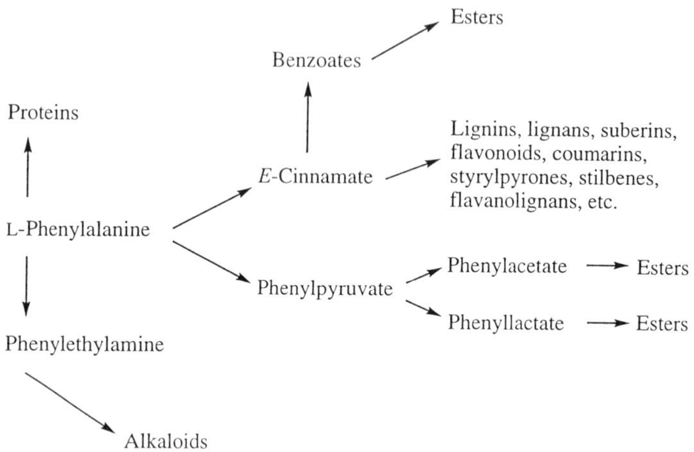

Figure 1. General metabolic fate of phenylalanine in vascular plants.

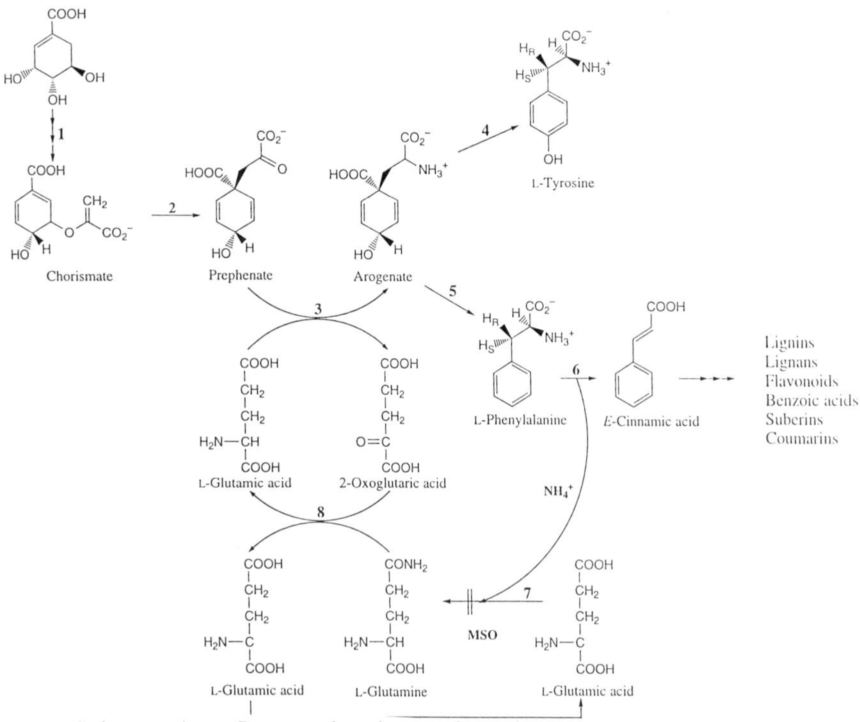

Scheme 1. Proposed scheme for nitrogen recycling during phenylpropanoid metabolism. Numbers indicate enzymes: 1, shikimate-chorismate pathway enzymes; 2, chorismate mutase; 3, prephenate:glutamate aminotransferase; 4, arogenate dehydrogenase; 5, arogenate dehydratase; 6, phenylalanine ammonia lyase (PAL); 7, glutamine synthetase; 8, glutamine 2-oxo-glutarate aminotransferase (GOGAT). (Reproduced with permission from ref. *28*.)

which in turn can then be either dehydrated to yield L-Phe or decarboxylated/dehydrated to give L-Tyr (*28, 29*) in the chloroplast.

In the deamination reactions catalyzed by PAL and TAL (Scheme 1), which have no cofactor requirements, the pro-*S* proton from C-3 of these amino acids is abstracted in an antiperiplanar fashion relative to the (leaving) amino group functionality to generate *E*-(*trans*)-cinnamate and *E*-*p*-coumarate (*24*). Although many crude enzyme preparations from plant and fungal sources display tyrosine ammonia lyase (TAL) activity (*30*), tyrosine is generally speaking not a major source of phenylpropanoid-derived compounds in higher plants and hence is not discussed further.

With an equimolar amount of ammonium ion being generated for every phenylpropanoid skeleton formed, clearly there must be a major demand for nitrogen unless some effective nitrogen recycling process is in effect in tissues and cells undergoing active phenylpropanoid metabolism. Recognition that such a mechanism for nitrogen recycling must be in effect has long been recognized (*31*); however, the actual biochemical processes involved, whereby ammonium ion is assimilated *via* the glutamine synthase (GS)/glutamine 2-oxoglutarate aminotransferase (GOGAT) system, to generate the amino donor, L-Glu, for regenerating L-Phe has only recently been established (*32, 33*).

Results and Discussion

Four experimental systems, namely tuber disks of potato (*Solanum tuberosum*) and sweet potato (*Ipomoea batatas*), loblolly pine (*Pinus taeda*) cell suspension cultures and the mycelium of a wood-decaying fungus, *Lentinus lepideus,* were selected for study. These were chosen since each displays not only highly active phenylpropanoid metabolism under our experimental conditions, but they also engender metabolism into different branches of the pathway. The results obtained for each are described individually below, and are then compared and contrasted.

Nitrogen Recycling during Phenylpropanoid Metabolism in Potato Tuber Disks. Wounding and light are known to induce phenylpropanoid metabolism in potato tuber disks. In particular, they engender *de novo* synthesis of PAL (*34*), active biosynthesis of chlorogenic acid (*35*) and other phenolic metabolites, as well as formation of suberized tissues (*15-17*). It was, therefore, instructive to examine the metabolic fate of the ammonium ion released during induction of the phenylpropanoid pathway in this species. Accordingly, aseptic potato disks were incubated with 25 μM L-[^{15}N]phenylalanine for 24 h, following which the resulting amino acid enriched extract was isolated and subjected to ^{15}N NMR spectroscopic analysis. As can be seen in Figure 2A, only a single enriched ^{15}N resonance at 90.8 ppm was observed, this presumably corresponding to the amide nitrogen of Gln. Its formation was rationalized as resulting from the assimilation of $^{15}NH_4^+$, released during PAL catalysis, and its conversion into ^{15}N-Glu (amide) by the action of glutamine synthase (GS) (*32*). This metabolic sequence of events was further confirmed by incubation of potato disks with L-[^{15}N]Phe in the presence of 5 mM methionine-S-sulphoximine (MSO), a known glutamine synthase (GS) inhibitor (*36*). Under these conditions, the amino acid enriched extract so obtained only displayed resonances corresponding to L-Phe and NH_4^+ as shown in Figure 2B. These results suggested that GS/GOGAT effectively assimilated any released NH_4^+ from the PAL reaction.

A comparable metabolic fate of NH_4^+ was observed when potato disks were next incubated with 20 mM $^{15}NH_4Cl$. Following a 24 h period incubation, and isolation of the amino acid enriched extract as before, its ^{15}N NMR spectrum revealed a resonance at 90.8 ppm again corresponding to L-Gln (amide) (Figure 3A) (*32*). As before, the addition of MSO inhibited NH_4^+ assimilation (Figure 3B). Taken together, these results indicated that ammonium ion, either supplied as $^{15}NH_4Cl$ or liberated from L-[^{15}N]Phe during PAL catalysis, was actively assimilated *via* the GS/GOGAT pathway, thereby providing the amino acid donor for regeneration of arogenate, and hence completion of the cycle. It is also worth mentioning that in these studies, no significant involvement of glutamate dehydrogenase (GDH) was noted.

Nitrogen Recycling during Phenylpropanoid Metabolism in Sweet Potato Tuber Disks. As for potato, both wounding and light exposure of aseptic sweet potato tuber disks enhanced PAL activity and chlorogenic acid accumulation. For example, after 24 h metabolism, the then freshly sliced tuber disks exposed to light now had a 5-fold higher PAL activity than those incubated in the dark (*37*).

Accordingly, a study of nitrogen metabolism in this biological system was next carried out. Incubation of aseptic sweet potato tuber disks with 20 mM L-[^{15}N]phenylalanine for 24 h ultimately gave an amino acid enriched extract, which was subjected to spectroscopic analysis as before. The ^{15}N NMR spectrum so obtained exhibited resonances corresponding to L-Gln (amide), L-Glu, L-Ala and residual L-phenylalanine, respectively (*37*). This again suggested that the GS/GOGAT pathway was involved in assimilation of the ammonium ion released during phenylpropanoid metabolism. However, in order to prove unambiguously that this metabolic pathway was in effect, comparable experiments were undertaken but now in the presence of specific inhibitors of GS, GOGAT and PAL, respectively. In this context, addition of MSO (2.5 mM), a glutamine synthase inhibitor (*38*), to sweet potato tuber disks, which had been administered 2 mM L-phenylalanine as before, now gave an amino acid enriched extract whose ^{15}N NMR spectrum was devoid of resonances due to ^{15}N-Gln-(amide-N), ^{15}N-Glu or ^{15}N-Ala. By contrast, only signals corresponding to $^{15}NH_4^+$ (0 ppm) and residual L-[^{15}N]Phe (18.4 ppm) were observed (*37*). Experiments were also carried out using the specific GOGAT inhibitor, azaserine (AZA) (0.2 mM) (*39*), and resulted in an amino acid enriched extract whose ^{15}N NMR spectrum gave ^{15}N resonances, which only corresponded to L-Gln (90.8 ppm) and residual L-[^{15}N]Phe, *i.e.* there was no metabolism into either ^{15}N-Glu or ^{15}N-Ala. Additionally, when ^{15}N-Phe was administered to aseptic sweet potato tuber disks, but now in the presence of the potent PAL inhibitor, 2-aminoinelan-2-phosphonic acid (AIP) (0.05 mM) (*40*), only a resonance corresponding to Phe was observed, *i.e.* its deamination was inhibited. Thus, as for potato, these observations clearly suggest the involvement of a GS-GOGAT system for reassimilation of PAL-generated NH_4^+ in sweet potato tuber disks. (As an aside, the incorporation of ^{15}N into Ala from L-[^{15}N]phenylalanine could be interpreted as suggesting that there may be an active transaminase involving Phe and pyruvate in these disks, or that pyruvate can directly assimilate the NH_4^+ released during the PAL reaction. However, since there is no transfer of ^{15}N nitrogen to Ala, when the PAL reaction is blocked, there is no evidence for either process occurring. Consequently, it is apparently formed *via* further metabolism of L-Glu.)

In order to ascertain if L-Glu served as nitrogen donor to the aromatic amino acids, *i.e.* arogenate, Phe and Tyr, it was important to next examine the metabolic fate of ^{15}N-Glu. Incubation of sweet potato disks with 20 mM L-[^{15}N]Glu for 24 h were next carried out, and only resulted in resonances in the amino acid extract corresponding to Ala and residual Glu. This indicated that transamination reactions were also occurring which afforded Ala during amino acid metabolism in this tissue. To prove that the proposed nitrogen cycle, *via* GS/GOGAT to arogenate/phenylalanine and tyrosine, was operative, the effect of the PAL inhibitor, AIP, was again examined. In this regard, addition of 0.05 mM AIP (*40*) to the metabolizing sweet potato tuber disks, ultimately gave an amino acid enriched extract, whose ^{15}N NMR spectrum revealed a new resonance corresponding to Phe (18.4 ppm). This again showed that Glu acts as an amino donor either directly, or indirectly through arogenate, to permit the continued biosynthesis of Phe in sweet potato tuber disks.

Comparable results were obtained when the disks were incubated with 20 mM $^{15}NH_4Cl$. After 24 h, isolation and analysis of the resulting amino acid extract revealed that Gln (amide), Glu and Ala were ^{15}N-enriched (*37*). Addition of 2.5 mM MSO as before, however, completely inhibited the assimilation of NH_4^+ into the above-mentioned amino acids, and only a resonance corresponding to NH_4^+ was

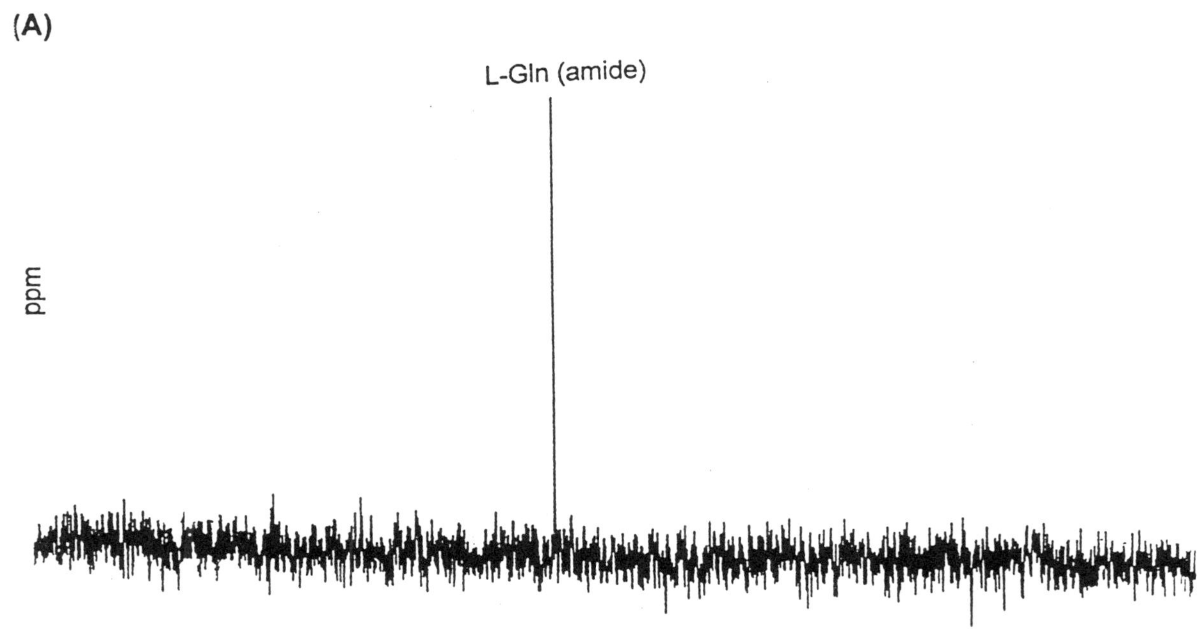

Figure 2. ^{15}N NMR spectra of amino acid extracts prepared from potato disks incubated for 24 h with (A) L-[^{15}N]Phe and (B) L-[^{15}N]Phe in the presence of MSO. (Reproduced with permission from ref. *28*.) (Copyright 1986 National Academy of Science.)

Continued on next page.

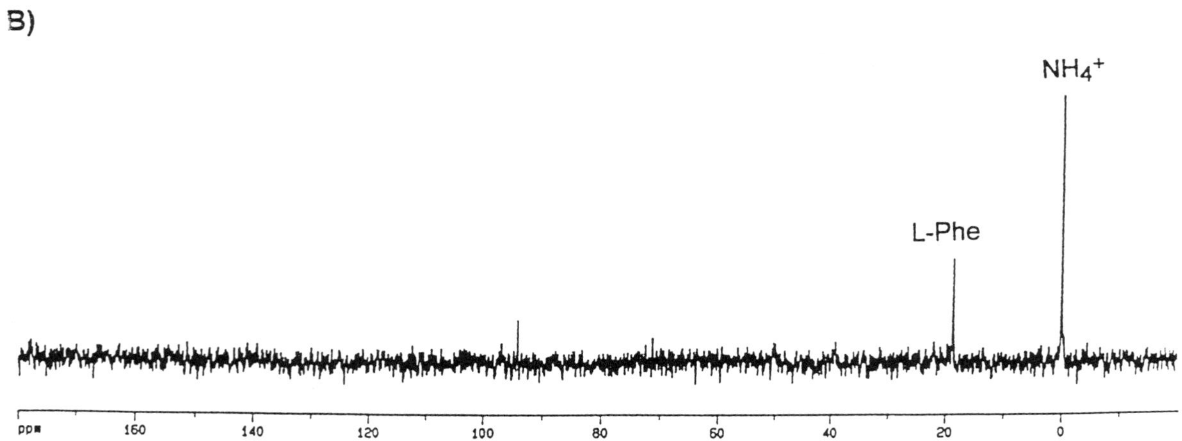

Figure 2. *Continued.*

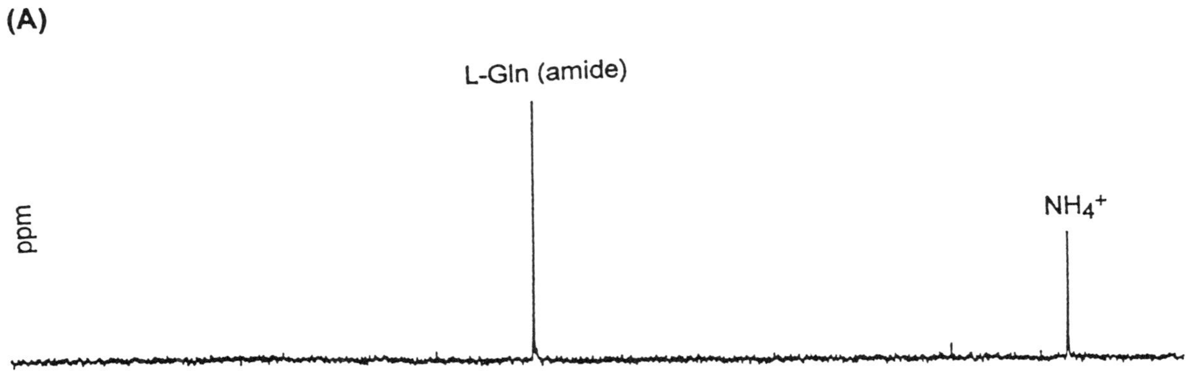

Figure 3. ^{15}N NMR spectra of extracts prepared from potato disks incubated for 24 h with (A) $^{15}NH_4Cl$ and (B) $^{15}NH_4Cl$ in presence of MSO. (Reproduced with permission from ref. *28*.)
(Copyright 1986 National Academy of Science.)

Continued on next page.

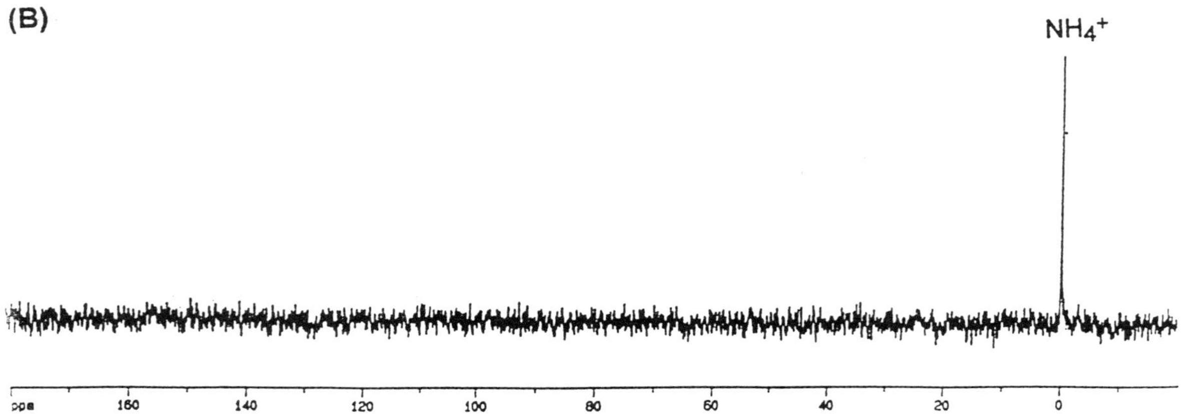

Figure 3. *Continued.*

evident. These observations are again in agreement with the hypothesis that the GS/GOGAT enzyme system is involved in reassimilation of NH_4^+ in sweet potato disks, and thereby enables active phenylpropanoid metabolism to continue.

Nitrogen Recycling during Phenylpropanoid Metabolism in Loblolly Pine Cells. Phenylpropanoid metabolism in loblolly pine (*Pinus taeda*) cell suspension cultures can be actively induced by subculturing in solutions containing high levels of sucrose (*33, 41, 42*), *i.e.* to give both lignified cell walls and an 'extracellular lignin precipitate' (see chapter 25). It was, therefore, next of interest to establish the metabolic fate of 10 mM L-[^{15}N]Phe. Thus, following a 24 h period, the resulting amino acid enriched extract so obtained was subjected to ^{15}N NMR spectroscopic analysis, and revealed ^{15}N resonances corresponding to the amide group of Gln, the α amino groups of Gln and Glu and the amino group of Phe (*33*), respectively. On the other hand, when metabolism was extended to 96 h, resonances corresponding to Ala and Ser were also evident.

Quantitative measurements of both pool size and isotopic enrichment of these amino acids were also carried out at different metabolic intervals and again clearly implicated the involvement of the GS/GOGAT pathway in the assimilation of PAL-generated NH_4^+. Moreover, when experiments were next conducted in the presence of the known PAL inhibitor, 0.1 mM L-aminooxy phenylpropionic acid (AOPP) (*43*), the dominant resonance observed was that of unmetabolized L-[^{15}N]Phe together with a very small signal presumably due to Gln and Glu amino groups, *i.e.* PAL was substantially inhibited by addition of the known inhibitor, L-AOPP. When experiments were next carried out using the GS inhibitor, MSO (5 mM), this resulted in only L-[^{15}N]Phe and NH_4^+ resonances being observed, *i.e.* no incorporation of ^{15}N into Gln, Glu, Ala and Ser was noted (*33*). Additionally, the employment of AZA (5 mM), an inhibitor of GOGAT (*39*), substantially inhibited incorporation of the ammonium ion from ^{15}N-Phe into the α amino group of Glu. Taken together, these results, using specific inhibitors, indicate that the primary metabolic fate of the nitrogen of ^{15}N-Phe is *via* its assimilation into Gln and Glu by the action of GS/GOGAT. Again, no evidence for NH_4^+ assimilation by GDH was observed.

In order to establish the further channeling of nitrogen from Glu back to Phe to complete the cycle, it was necessary to establish the metabolic fate of $^{15}NH_4^+$, L-[^{15}N]Gln and L-[^{15}N]Glu. Initial experiments were conducted using $^{15}NH_4Cl$, whose metabolism over a 96 h time period gave a ^{15}N NMR spectrum of its amino acid extract revealing resonances attributable to Gln, Glu, Ala, Arg, Pro, Lys, Orn, γ-aminobutyric acid (GABA) and NH_4^+. This suggested that the NH_4^+ derived from NH_4Cl was available for utilization in the various general pools used for amino acid and protein biosynthesis. On the other hand, the NH_4^+ ion released from Phe during phenylpropanoid metabolism was only incorporated into Gln, Glu, Ala and Ser, *i.e.* it was not apparently made available for general amino acid/protein synthesis. Additionally, incubation of loblolly pine cell cultures with 10 mM $^{15}NH_4Cl$ as before, but now in the presence of 0.1 mM of L-AOPP (*43*), gave only a resonance corresponding to ^{15}N-Phe. Similarly, when L-[^{15}N]Gln or L-[^{15}N]Glu were incubated with the induced *P. taeda* cell cultures, in the presence of L-AOPP (0.1 mM), only accumulation of ^{15}N-Phe was observed. (This was further established by quantitation of the amino acid pool sizes by HPLC and GC-MS isotopic enrichment.) Taken together, these results again prove that the ammonium ion released is transferred to L-Glu, which then functions as the amino donor to form Phe.

In summary, employing ^{15}N-substrates with and without specific inhibitors of PAL, GS and GOGAT, and with subsequent analysis of the resulting metabolic products by HPLC, ^{15}N NMR spectroscopy and GC-MS, resulted in the discovery of a novel mechanism for the recycling of nitrogen during active phenylpropanoid metabolism in loblolly pine cell cultures. The ammonium ion released is first assimilated by the GS/GOGAT pathway resulting in the synthesis of L-Glu, which then serves as the amino donor to Phe. The efficient recycling of NH_4^+, at low nitrogen levels, would allow for lignin formation, as well as that of other phenylpropanoid products, to continue under conditions where there is limited nitrogen availability in either the whole plant or in cell cultures.

Nitrogen Recycling during Phenylpropanoid Metabolism in the Basidiomycete, *Lentinus lepideus*. *Lentinus lepideus* is a wood-decaying basidiomycete which produces a 'brown' rot of wood. It is often described as a 'woody' bracket fungus, but does not produce the lignins present in higher plants. Nevertheless, it displays high PAL activity and it biosynthesizes significant amounts of methylated phenolic acids in culture medium, *e.g.* methyl *p*-methoxy cinnamate (*44*). Since it grows on wood which is a poor nitrogen substrate, it was anticipated that some mechanism for the conservation of nitrogen should also be in place in the rapidly growing mycelium.

To investigate this possibility, four-day old mycelium was incubated with ^{15}N-labeled substrates, in an analogous manner to that described for the plant specimens. Accordingly, in a time course study of L-[^{15}N]Phe metabolism, it was noted that after 2 h, the first amino acid to be detected, labeled with ^{15}N, was Gln (amide-N). After 12 h, additional ^{15}N resonances corresponding to Glu, Ala, residual Phe as well as an unidentified resonance at 91.19 ppm were also evident. Interestingly, following 24 and 48 h of incubation with L-[^{15}N]Phe, an additional resonance attributed to γ-amino-butyric acid (GABA) was also observed.

Specific inhibitors of PAL and GS were next used to investigate whether nitrogen recycling was continuing as before. Thus, addition of MSO (GS inhibitor) to the L-[^{15}N]Phe-treated mycelia completely inhibited incorporation into Gln and Glu, although resonances attributed to GABA, Ala and Phe were evident. Similar results were observed when AIP (PAL inhibitor) was used. Therefore, it appears that in *Lentinus*, apart from the GS/GOGAT pathway operating, additional mechanisms for ammonium assimilation may also be in effect.

Concluding Remarks

In summary, our results with potato and sweet potato root tuber disks, loblolly pine cells and *Lentinus lepideus* mycelia demonstrate that PAL-generated ammonium ion is actively assimilated *via* GS/GOGAT metabolism. The L-Glu formed as a consequence of GS/GOGAT activity appears to serve as a nitrogen donor for aromatic amino acid biosynthesis, leading to arogenate, Phe and Tyr. However, in the basidiomycete, *L. lepideus*, ammonium ion produced from Phe is apparently assimilated by GS/GOGAT pathway, as well as by additional mechanism(s) which need to be further clarified. No evidence supporting a role of GDH in the reassimilation of PAL-liberated ammonium ion in any of these species was found. A general mechanism for recycling the liberated ammonium ion back to Phe for further phenylpropanoid metabolism in plant and fungal species is illustrated in Scheme 1. This efficient nitrogen recycling mechanism explains why plants and fungi do not experience any obvious symptoms of nitrogen limitation under conditions of active phenylpropanoid metabolism.

Experimental

L. lepideus was subcultured on solid potato dextrose agar and then maintained on liquid culture medium containing (per L): dextrose (20 g), yeast extract (5 g), peptone (2 g), $MgSO_4.7H_2O$ (0.5 g), KH_2PO_4 (0.45 g) and K_2HPO_4 (0.5 g). Maintenance of suspension cultures of loblolly pine (*33, 41, 42*), potato (*32, 45*) and sweet potato tuber (*37*) slices were as previously reported.

The general procedure for studying the fate of the NH_4^+ ion released from the PAL reaction was as follows: ^{15}N-labeled substrates, *e.g.* L-phenylalanine, L-glutamic acid, L-glutamine (amide ^{15}N) or NH_4Cl were administered to cells or tissues under aseptic conditions. In certain experiments synthetic inhibitors of PAL, (AIP, 2-aminoindan-2-phosphonic acid and AOPP, L-aminooxy phenylpropionic acid), GS

(MSO, methionine-S-sulphoximine) and GOGAT (AZA, azaserine) were administered along with ^{15}N-labeled substrates (*32, 33, 37*).

Immediately after each incubation metabolism period, the corresponding plant or fungal material was rinsed with distilled water and frozen (liq. N_2). The amino acid enriched extracts from the soluble pool were extracted with ethanol and evaporated under reduced pressure at 30°C. The corresponding dried amino acid extracts were then resuspended in 5 ml of distilled water and extracted with an equal volume of $CHCl_3$. The aqueous phase so obtained was next centrifuged for 10 min (9,000 rpm, 4°C), with the resulting supernatants frozen (liq. N_2) and lyophilized. The lyophilized amino acid samples were dissolved in 0.1 N HCl (1 ml), containing D_2O (50 μl), and subjected to analysis by ^{15}N-nuclear magnetic resonance (^{15}N NMR) spectroscopy (*28, 29*). ^{15}N NMR spectroscopic measurements were performed on a Bruker AMX 500 spectrometer operating at 50.68 MHz at 298K, with broad band coupling, employing a Waltz-16 composite pulse sequence. Chemical shifts relative to the $^{15}NH_4^+$ resonance at 0 ppm were obtained using $^{15}NH_4Cl$ (250 mM) as external standard. Assignment of resonances in each sample was made by comparison with authentic ^{15}N-labeled amino acids.

The amino acid extracts for GC-MS analysis were prepared by the method similar to that described above for the NMR spectroscopic analyses. Amino acid extracts were derivatized, using the MTBSTFA reagent (*46*) with the resulting N-DMTBS derivatives of amino acids in the extracts analyzed by GC-MS (*32, 33*). GC-MS was performed on a Hewlett Packard 5989A GC-MS system operating in the EI-mode. The specific isotopic abundances of various amino acids (*e.g.* Phe, Gln and Glu) were determined by plotting ion current profiles and calculating ratios of the ^{14}N:^{15}N from the selected ion clusters as described elsewhere (*33*).

The total pools of amino acids in loblolly pine cell extracts were determined after derivatization with phenylisothiocyanate using the pico-tag method and analyzed by HPLC (*45*).

Acknowledgments

We thank the U.S. Department of Energy (DE-FG06-91ER20022), the National Aeronautic and Space Administration (NAG100164) and the Arthur M. and Katie Eisig-Tode Foundation (to P. S. van Heerden) for generous support of this research.

Literature Cited

1. Lewis, N. G.; Davin, L. B. In *Isoterpenoids and Other Natural Products: Evolution and Function*; Nes, W. D., Ed.; ACS Symposium Series; Washington, DC, 1994, Vol. 562, pp 202-246.
2. Davin, L. B.; Lewis, N. G. In *Recent Advances in Phytochemistry*; Stafford, H. A., Ibrahim, R. K., Eds.; Plenum Press: New York, NY, 1992; pp 325-375.
3. Löffelhardt, W.; Lugwig. B.; Kindl, H.; *Hoppe-Seylers Z. Physiol. Chem.* **1973**, *354*, 1006-1012.
4. Hoff, W. D.; Dux, P.; Hard, K.; Devreese, Nugeren-Roodzant, I. M.; Crielaard, W.; Boelens, R.; Kaptein, R.; Van Beeuman, J.; Hellingwerf, K. *J. Biochemistry* **1994**, *33*, 13959-13962.
5. Sikora, L. A.; Marzluf, G.A. *J. Bacteriol.* **1982**, *150*, 1287-1291.
6. Strack, D.; Wray, V. In *The Flavonoids. Advances in Research since 1986*; Harborne, J. B., Ed.; Chapman and Hall: London, 1993; pp 1-22.
7. Toguri, T.; Umemoto, N.; Kabayashi, O.; Ohtani, T. *Plant Mol. Biol.* **1993**, *23*, 933-946.
8. Lamb, C. J.; Lawton, M. A.; Dron, M.; Dixon, R. A. *Cell*, **1989**, *56*, 215-224.
9. Brouillard, R.; Dangles, O. In *The Flavonoids. Advances in Research since 1986*; Harborne, J. B., Ed.; Chapman and Hall: London, 1993; pp 565-588.
10. Vogt, T.; Pollack, P.; Taglyn, N.; Taylor, L. P. *Plant Cell*, **1994**, *6*, 11-23.

11. Long, S. *Cell* **1989**, *56*, 203-214.
12. Van Tunen, A. J.; Koes, R. E.; Spelt, C. E.; Van der Krol, A. R.; Mol, J. M. N. *EMBO* **1988**, *7*, 1257-1263.
13. Schmelzer, E.; Jahnen, W.; Hahlbrock, K. *Proc. Natl. Acad. Sci. USA* **1988**, *85*, 2989-2993.
14. Kubasek, W. L.; Shirley, B. W.; McKillop, A.; Goodman, H. M.; Briggs, W.; Ausubel, F. M. *Plant Cell* **1992**, *4*, 1229-1236.
15. Bernards, M. A.; Lopez, M.; Zajicek, J.; Lewis, N. G. *J. Biol. Chem.* **1995**, *270*, 7382-7386.
16. Bernards, M. A.; Lewis, N. G. *Phytochemistry* **1997** (in press).
17. Bernards, M. A.; Davin, L. B. *Polyphénols Actualités* **1996**, *14*, 4-6.
18. Lewis, N. G.; Yamamoto, E. *Annu. Rev. Plant Physiol. Plant Mol. Biol.* ***1990*** *41*, 455-497.
19. Ayres, D. C.; Loike, J. D. *Chemistry and Pharmacology of Natural Products. Lignans: Chemical, Biological and Clinical Properties*; Cambridge University Press: Cambridge, 1990; pp 402.
20. Gottlieb, O. R.; Yoshida, M. In *Natural Products of Woody Plants. Chemicals Extraneous to the Lignocellulosic Cell Wall*; Rowe, J. W., Kirk, C. H., Eds.; Springer-Verlag: Berlin, 1989; pp 439-511.
21. Markhanen, T.; Makinen, M. L; Maunuksela, E.; Himanen, P. *Drugs Exp. Clin. Res.* **1981**, *7*, 711-718.
22. Gardner, J. A. F.; MacDonald, B. F.; MacLean, H. *Can. J. Chem.* **1960**, *38*, 2387-2394.
23. Osawa, T.; Nagata, M.; Namiki, M.; Fukuda, Y. *Agric. Biol. Chem.* **1985**, *49*, 3351-3352.
24. Koukol, J.; Conn, E. E. *J. Biol. Chem.* **1961**, *236*, 2692-2698.
25. Neish, A. C. *Phytochemistry* **1961**, *1*, 1-24.
26. Kutchan, T. M. *Plant Cell* **1995**, *7*, 1059-1070.
27. Gross, G. G. In *The Biochemistry of Plants: A Comprehensive Treatise*; Conn, E. E., Ed.; Academic Press: New York, NY, 1981, pp 301-316.
28. Jung, E.; Zamir, L. O.; Jensen, R. A. *Proc. Natl. Acad. Sci. USA* **1986**, *83*, 7231-7235.
29. Jensen, R. A. *Recent Adv. Phytochem.* **1986**, *20*, 57-82.
30. Camm, E. L.; Towers, G. H. N. *Phytochemistry* **1973**, *12*, 961-973.
31. Lewis, N. G.; Yamamoto, E. In *Chemistry and Significance of Condensed Tannins*, Hemingway, R. W., Karchesy, J. J., Eds.; Plenum Press: New York, NY, 1989; pp 23-47.
32. Razal, R. A.; Ellis, S.; Singh, S.; Lewis, N. G.; Towers, G. H. N. *Phytochemistry* **1996**, *41*, 31-35.
33. van Heerden, P. S.; Towers, G. H. N.; Lewis, N. G. *J. Biol. Chem.* **1996**, *271*, 12350-12355.
34. Sacher, J. A.; Towers, G. H. N.; Davies, D. D. *Phytochemistry* **1972**, *11*, 2383-2391.
35. Levy, C. C.; Zucker, M. *J. Biol. Chem.* **1960**, *235*, 2418.
36. Lea, P. J.; Ridley, A. M. In *Herbicides and Plant Metabolism*; Dodge, A. D., Ed.; Cambridge University Press: Cambridge, 1989; pp 137-170.
37. Singh, S.; Lewis, N. G.; Towers, G. H. N. *J. Plant Physiol.* (in press).
38. Miflin, B. J.; Lea, P. J. *Biochem. J.* **1975**, *149*, 403-409.
39. Ronzio, R. A.; Rowe, W. B.; Meister, A. *Biochemistry* **1969**, *8*, 1066-1075.
40. Zon, J.; Amrhein, N. *Liebigs Ann. Chem.* **1992**, 625.
41. Eberhardt, T. L.; Bernards, M. A.; He, L.; Davin, L. B.; Wooten, J. B.; Lewis, N. G. *J. Biol. Chem.* **1993**, *268*, 21088.
42. Nose, M.; Bernards, M. A.; Furlan, M.; Zajicek, J.; Eberhardt, T. L.; Lewis, N. G. *Phytochemistry* **1995**, *39*, 71-79.
43. Amrhein, N.; Godeke, K. *Plant Sci. Lett.* **1977**, *8*, 313-317.
44. Wat, C. K.; Towers, G. H. N. *Phytochemistry* **1977**, *16*, 290.
45. Razal, R. A.; Lewis, N. G.; Towers, G. H. N. *Phytochem. Anal.* **1994**, *5*, 98-104.
46. Rhodes, D.; Rich, P. J.; Brunk, D. G. *Plant Physiol.* **1989**, *89*, 1161-1171.

Chapter 4

Biochemistry and Molecular Biology of Lignin-Specific *O*-Methyltransferases from Woody Plant Species

Huabin Meng and Wilbur H. Campbell

Phytotechnology Research Center and Department of Biological Sciences, Michigan Technological University, Houghton, MI 49931–1295

Since lignin is the second most abundant biopolymer on earth while not being very useful to humankind, there is interest in minimizing its content in woody tissues, especially those with high commercial value. To achieve this biotechnological end, we need to gain a great deal of understanding of the key enzymes involved, their genes and the regulation of these genes in woody species. Lignin-specific *O*-methyltransferases (OMTs) are clearly very central to the control of monolignol biosynthesis. This brief review provides a perspective on the history and significance of the OMTs in relation both to their role in monolignol biosynthesis and to their importance in gaining a general understanding of OMTs in higher plants.

Very early in the studies of monomeric lignin precursors or monolignols, OMTs were recognized as key enzymes since phenolic compounds in plants are often methyloxylated. Since *S*-adenosyl-L-methionine (S-Ado-Met) is a common methyl group donor in biological systems, the OMTs of lignifying tissues were found to depend on this substrate for the methylation of hydroxyl groups. In the formation of monolignols, there are two steps which are catalyzed by OMTs, namely, the methylation of the 3- and 5-hydroxyl groups on the aromatic ring. Early work by Higuchi and coworkers (*1*) demonstrated that it was likely to be the same enzyme which catalyzed both methylations in angiosperm dicotyledons. In addition, they showed that the OMT in gymnosperms was a monospecific type of enzyme with a capacity only to catalyze the methylation of caffeic acid but not 5-hydroxyferulic acid. Thus, these early studies set the stage for current investigations of OMT enzymes and the importance of their roles in the biosynthesis of monolignols. To summarize the understanding of these enzymes and their relationship to the lignins formed in woody species when we began our studies of lignin-specific OMTs about 7 years ago, basically it was thought that a single form of OMT existed in angiosperm dicots which catalyzed the methylation reactions in both the coniferyl and sinapyl alcohol pathways leading to the less condensed forms of lignin common in these species. On the other hand, in gymnosperms the OMT only acted in the more focused coniferyl alcohol pathway found in these species and thus exhibited a correspondingly more exclusive substrate specificity. The more condensed lignin of gymnosperms is more difficult to remove chemically from cellulose fibers, which increases the cost of pulping processes. Thus, the idea was enunciated that, if the two enzymes could be cloned which differ most clearly between the hardwoods and softwoods, namely the bispecific OMT and the 5-hydroxylase, then biotechnological modification of gymnosperms could be carried out to modify monolignol biosynthesis and consequently alter the type of lignin made in softwoods. Hence, we began the process by cloning the OMT of aspen developing secondary xylem, which is representative of the hardwood family of woody species (*2*).

As it turned out, much more than the simple cloning of the first monolignol-specific enzyme was accomplished when we cloned aspen xylem OMT. Thus, part of this review will illustrate the broad implications of the information gained when we cloned aspen xylem OMT, which is related to the overall importance of *O*-methyltransferase catalyzed methylation reactions in higher plants. In addition, we wish to emphasize the general approach this laboratory has adopted for studying higher plant enzymes which are difficult to obtain in large quantities from natural sources. In short, we have now demonstrated that the easiest route to obtaining detailed biochemical information on these enzymes, such as OMT and nitrate reductase, that appears at low concentrations in plant tissues, is to express them in microorganisms and study the recombinant forms (*3, 4*). This approach is leading to detailed biochemical understanding of both the structure and function of these rare enzymes, which requires large quantities of pure protein. This approach has the added advantage that key functional amino acid residues of the enzymes can be specifically mutated and the mutant enzyme forms expressed and purified, which is very difficult if not impossible to accomplish at this time in plants, especially as far as enzymes in woody species are concerned. We are, of course, not the only laboratory taking this approach to studying plant enzymes. However, our work emphasizes production of very large quantities of plant enzymes in highly purified form so that their 3-D structures can be determined.

Isolation and Cloning of Monolignol-Specific OMT

We began by isolating caffeic acid OMT in pure form from developing secondary xylem of aspen and preparing an antibody specific for the enzyme (*2, 5*). With the anti-OMT antibody, we selected a cDNA clone for the aspen xylem OMT using a library prepared from the mRNA of aspen secondary developing xylem (*2*). We sequenced the clone and also obtained enough amino acid sequence of the purified OMT to prove that the clone we had isolated was authentic. Furthermore, we used the antibody against the OMT to demonstrate that the enzyme was specifically localized in the xylem of aspen, which established our OMT clone as being for a monolignol-specific OMT (*2*). Moreover, we obtained expression in *Escherichia coli* of an enzymatically active recombinant OMT, which reacted with the antibody against the natural OMT. This incontrovertible proof of the origin and identity of our OMT cDNA, along with the availability of an antibody for it, opened up the field of study of OMTs in higher plants. Subsequent to our pioneering studies of plant OMT, many research groups have isolated OMTs from plants using the aspen OMT cDNA and the specific antibody we prepared against the natural and recombinant forms of aspen OMT. Several groups in the world are currently utilizing these valuable tools to isolate new OMT isoforms from plants.

Dimensions of the Higher Plant OMT Family

To illustrate the breadth of the higher plant OMT family, we have compared the amino acid sequences of all currently available cloned OMTs which have sequence similarity to aspen caffeic acid OMT (Figure 1). It is clear, from this comparison of OMTs involved in so many plant biochemical processes, that the aspen OMT can be considered a prototypical higher plant OMT and we have designated this family of enzymes as "Type 1" OMT. The core of the Type 1 OMT family is formed by the caffeic acid OMTs which all have very close similarity in amino acid sequence to aspen caffeic acid OMT. Another family of OMTs from higher plants, which is very different in amino acid sequence from aspen OMT and the members of the Type 1 OMT family, is the caffeoyl CoA OMT (*6*). This enzyme was first cloned in relation to disease resistance in tobacco, but has more recently been shown to exist in the xylem of woody species and is also likely to have a role in the biosynthesis of monolignols (*6-8*). We have designated this OMT family as "Type 2". In addition to significant differences in the amino acid sequences, Type 1 and 2 OMTs, which catalyze essentially the same reaction (namely the methylation of the 3- and 5-hydroxyls on the aromatic rings of monolignols), are found to be different in their molecular size and apparent functionality. Basically, Type 1 OMTs have molecular weights around 40 kDa and do not require magnesium for activity, while the Type 2 OMTs are about 25 to 30 kDa in size and do require magnesium for activity. In this latter regard, the Type 2 OMTs resemble the animal catechol OMT but differ significantly from it in sequence (*9*). The differences in the 3-D structures of the two types of plant OMT are expected to be large because of the differences in their amino acid sequences, and the comparison will be fascinating as it materializes during the next few years since these enzymes catalyze the same methylation reaction. In addition, the comparisons of their structures to that of animal catechol OMT, for which a 3-D structure has already been determined (*9*), will be quite interesting since the animal enzyme also catalyzes essentially the same reaction as the plant enzymes.

It is instructive to delve more deeply into the comparison between the amino acid sequences of Type 1 plant OMTs. We identified regions in the sequence of aspen OMT, when we first obtained it, which suggested which parts of the enzyme were involved with binding to S-Ado-Met (*2*). These were regions in the C-terminal part of the sequence with similarity to those in the OMTs from animals, fungi and bacteria. Thus, it is clear that aspen OMT is representative not only of a family of plant OMTs with differing function but also of a larger group of OMTs found in many species. Based on the more detailed comparisons which can now be made between the larger number of amino acid sequences available for plant Type 1 OMTs, we can make more detailed assignments in aspen OMT sequence regions and identify invariant residues expected to have key functional roles in the reactions catalyzed by OMT (Figure 2 and Table I). We now find that the N-terminal half of OMT exhibits regions likely to be involved in binding the phenolic substrate of the enzyme and in controlling substrate specificity. However, much work remains to be done in exactly identifying the residues that serve the enzyme in these capacities. In addition, it is now clear that the C-terminal half of OMT is likely to be involved both in binding S-Ado-Met, the methyl group donor, and in interacting with the phenolic substrate (Figure 2 and Table I). Although these assignments must be considered to be preliminary until a 3-D structure becomes available for aspen OMT, the low homology region found in the C-terminal half of OMT (designated Region VIII in Figure 2 and Table I) must be serving a function for the enzyme other than binding the common S-Ado-Met substrate. The list of invariant residues found in Type 1 OMT is still large at this stage of our understanding (Table I), but it does provide a starting point for designing site-directed mutants of OMT to investigate the functionality of these residues, as we have recently illustrated with our studies of recombinant aspen OMT (*3*). It can be expected that as more plant OMTs are cloned and sequenced, this list of invariant and highly conserved amino acid residues will become smaller and the focus of mutagenesis experiments can be narrowed.

Expression of Recombinant OMT in *E. coli*

Obtaining sufficient quantities of OMT from aspen xylem for carrying out detailed biochemical studies of its structure and function is impractical. The most logical alternative is to express the OMT as a catalytically active recombinant protein in *E. coli* or another similar well characterized system for heterologous protein expression. We have demonstrated the effectiveness of this approach to facilitating detailed

```
    1         10        20        30        40        50          60          70          80        90
A.  MGS...TG.ETQMTPT.QVSDEEAHLFAMQLASASVLPMILKTAIELDLLEIMAKAGPG..AFLSTSEIASHLPT.KNPD...APVMLDRILRLLASYSIL
B.  MGS...TGSETQMTPT.QVSDEEANLFAMQLASASVLPMVLKAAIELDLLEIMAKAGPG..AFLSPGEVAAQLPT.QNPE...APVMLDRIFRLLASYSVL
C.  MGS...TG.ETQMTPT.QVSDEEANLFAMQLASASVLPMVLKAAIELDLLEIMAKAGPG..VFLSPTDIASQLPT.KNPD...APVMLDRMLRLLASYSIL
D.  MGS...TG.ETQITPT.HISDEEANLFAMQLASASVLPMILKSALELDLLEIIAKAGPG..AQISPIEIASQLPT.TNPD...APVMLDRMLRLLACYIIL
E.  MGS...TG.ETQITPT.VVSDEEANLFAMQLASASVLPMVLKSAIELDVLEIIAKAGPG..VHLSPADIAAKLPT.TNPD...AAVMLDRMLRLLACYNIL
F.  MGS...ANPDNKNSMT.K.EEEACLSAMALASASVLPMVLKSAIELDLLELIKKSGPG..AYVSPSELAAQLPT.QNPD...APVMLDRILRLLASYSVL
G.  M.......................LFAMQLACASVLPMVLKSAIELDLLEIIR..GQD..TCMSPTEIASHLPT.TNPD...APAMVDRILRLLSCYSVV
H.  MGS...TS.ESQSNSL.THTEDEAFLFAMQLCSASVLPMVLKSAVELDLLELMAKAGPG..AAISPSELAAQLST.QNPE...APVMLDRMLYVLASYSLF
I.  ?GS...ASRKLRRSNT.KIVNEDE.LFAMQLASASVLPMVLKSAIELDLLEIIA..SQD..TCMSPTEIASHLPT.TNPH...APTMIDRILRLLSSYSIV
J.  M.......................WLLGMELGNFSCLSMAMKAAIELDVLQIIANAGNG..VQLSPRQIVAHIPT.TNPD...AAITLDRILRVLASHSVL
K.  MTT..YTNGNYTQPKT.LDKDEQLAGLAVTLANAAAFPMILKSAFELKILDIFSKAGEG..VFVSTSEIASQIGA.KNPN...APVLLDRMLRLLASHSVL
L.  MGS...TAGDVAAVV.....DEEACMYAMQLASSSILPMTLKNAIELGLLEVLQKEAGGGKAALAPEEVVARMPA.APSDPAAAAAMVDRMLRLLASYDVV
M.  MES.....STKSQIPT.QSEEERNCTYAMQLLSSSVLPFVLHSTIQLEVFEILAKSND...TKLSASQIVSQIPNCKNPD...AATMLDRMLYVLASYSLF
N.  MGNSYITKEDNQISATSEQTEDSACLSAMVLTTNLVYPAVLNAAIDLNLFEIIAKATPPG.AFMSPSEIASKLPASTQHS..DLPNRLDRMLRLLASYSVL
O.  MVTSMPLEANSNGQILQA....EAELFCHSFGYLK..SMALQSVVKLRIPDVLHRYGGA....ASLPELLSTVPI..HPN...KLPYLPRLMKMLAAAGIF
P.  MELSPNNSTDQSLLDAQL......ELWHTTFAFMK..SMALKSAIHLRIADAIHLHGGA....ASLSQILSKVHL..HPS...RVSSLRRLMRVLTTN.VF

                100       110       120         130       140         150       160       170       180
A.  TCSL.KDLPD.GKVERLYGLAPVCKFLTKNED...GVSVSPLCLMNQDKVLMESWYYL....KDAILDGGIP.FNKAYGMT.AFEYHGTDPRFNKVFNKGM
B.  TCTL.RNLPD.GKVERLYGLAPVCKFLVKNED...GVSIAALNLMNQDKILMESWYYL....KDAVLEGGIP.FNKAYGMT.AFEYHGTDPRFNKIFNRGM
C.  TYSL.RTLAD.GKVERLYGLGPVCKFLTKNEE...GVSIAPLCLMNQDKVLLESWYHL....KDAVLEGGIP.FNKAYGMT.AFEYHGTDPRFNKVFNRGM
D.  TCSV.RTQQD.GKVQRLYGLATVAKYLVKNED...GVSISALNLMNQDKVLMESWYHL....KDAVLDGGIP.FNKAYGMT.AFEYHGTDPRFNKVFNKGM
E.  NHSL.RTLPD.GKVERLYGLAPVAKFLVKNED...GVSLSALNLMNHDKVLMESWYYL....KETVLEGGIP.FNKAYGMT.AFEYHGTDPRFNKVFNKGM
F.  NCTL.KDLPD.GGIERLYSLAPVCKFLTKNED...GVSMAALLLMNQDKVLMESWYHL....KDAVLEGGIP.FNKAYGMT.AFEYHGKDPRFNKVFNQGM
G.  TCSV.RSVDD....QRVYGLAPVCKYLTKNQD...GVSIAALCLMNQDKVLMESWYHL....KDAVLDGGIP.FNKAYGMS.SFEYHGTDPRFNKVFNRGM
H.  TCSIVRTLPD.SSVERLYSLAPVC.YLTKNAD...GVSVAPLLLMNQDKVLMESWYHL....KDAVLDGGIP.ENKAYGMT.AFEYHGTDPRFMLVFMRGM
I.  TCSV.RSVDD....QRVYSPAPVCKYLTKNQD...GVSIAALCVAAQDKVLMECWYHM....KDAVLDGGIP.FNKAYGMP.IFDYFAKDLGSNKVFNKGM
J.  SCSV.TTNEN.GKAERLYGLTPLCKYLVKNQD...GVSLAPLVLMNQDKVLMESWYYL....KDAVLDGSQP.FTKAHGMN.AFEYPAMDQRFNRVFNRGM
K.  TCKL..QKGE.GGSQRVYGPAPLCNYLASNDG...QGSLGPLLVLHHDKVMMESWFHL....NDYILEGGVP.FKRAHGMI.QFDYTGTDERFNHVFNQGM
L.  RCQM..EDRD.GRYERRYSAAPVCKWLTPNED...GVSMAALALMNQDKVLMESWYYL....KDAVLDGGIP.FNKAYGMT.AFEYHGTDARFNRVFNEGM
M.  TCSIVEDEENNGGQKRVYGLSQVGKFFVRDED...GASMGPLLALLQDKVFINSWFEL....KDAVLEGGVP.FDRVHGVVHAFEYPKSDPKFNDVFNKAM
N.  TSTT.RTIED.GGAERVYGLSMVGKYLVPDES...RGYLASFTTFLCYPALLQVWMNF....KEAVVDEDIDLFKNVHGVT.KYEFMGKDKKMNQIFNKSM
O.  TAEDVPATVGDGEPTTLYHLNAVSRLLVDDASVNGGASMSPCVLLGTVPLFLGASLKLHEWLQSEEQATTETPFMLAHGGT.LYGIGGRDSEFNTVFNKAM
P.  GTQPLGGG.SDDDSEPVYTLTPVSRLLIGSQSSQLAQTPLAAMVLD..PTIVSPFSELGAWFQHELPDPCI..FKHTHGRG.IWELTKDDATFDALVNDGL

               190       200       210       220       230       240       250       260       270       280
A.  SDHSTITMK.KILETYKGFEGLTSLVDVGGGTGAVVNTIVSKYPSIKGINFDLPHVIEDAPSYPGVEHVGGDMFVSVPKADAVFMKWICHDWSDAHCLKFL
B.  SDHSTITMK.KILETYKGFEGLETVVDVGGGTGAVLSMIVAKYPSMKGINFDLPHVIEDAPPLPGVKHVGGDMFVSVPKGDAIFMKWICHDWSDDHCAKFL
C.  ADHSTITMK.KILETYKGFEGLTSVVDVGGGTGAVLNMIVSKYPSIKGINFDLPHVIEDAPQYPGVEHVGGDMFVSVPKGDAIFMKWICHDWSDEHCLKFL
D.  SDHSTITMK.KILETYTGFEGLKSLVDVGGGTGAVINTIVSKYPTIKGINFDLPHVIEDAPSYPGVEHVGGDMFVSIPKADAVFMKWICHDWSDEHCLKFL
E.  NDHTTITMK.KILEIYTGFEGLKSLVDVGGGVGAVVNMIVTKYPHIKGINFDLPHVIEDAPSYPGVEHVGGDMFVSVPKADAVFMKWICHDWSDEHCIKFL
F.  SNHSTIIMK.KILEIYQGFQGLKTVVDVGGGTGATLNMIVSKYPSIKGINFDLPHVIEDAPSYPGVDHVGGDMFVSVPKGDAIFMKWICHDWSDAHCLKFL
G.  SDHSTITMK.KVFQTYQGFQGLTSLVDVGGGTGATLTMILSKYPTIRCINFDLPHVIEDAPEYPGIEHVGGDMFVSVPKGDAIFMKWICHDWSDEHCLKLL
H.  SDHSTMSMK.KILEDYKGFEGLNSIVDVGGGTGATVNMIVSKYPSIKGINFDLPHVIGDAPTYPGVEHVGGDMFASVPKADAIFMKWICHDWSDEHCLKFI
I.  SDFSSMIIK.KILETYKGFQGLTSLVDVGGGTGATLTKILSKYPTIRGINFDLPHVIQDAPEYPGIEHVGGDMFVSVPKGDAIFMKWICHDWNEEQCLKLL
J.  SEHSTMLMN.KILDTYEGFKEVQELVDVGGGVGSTLNLIVSKYPHISGINFDMPHVVVDAPHYPAVKHVGGDMFDSVPSGQAIFMKWILHDWSDDHCLRLL
K.  AHHTILVMK.KLLDNYNGFNDVKVLVDVGGNIGVNVSMIVAKHTHIKGINYDLPHVIADAPSYPGVEHVGGNMFESIPQADAIFMKWVLHDWSDEHCVKIL
L.  KNHSVIITK.KLLDFYTGFEGVSTLVDVGGGVGATLHAITSRHPHISGVNFDLPHVISEAPPFPGVRHVGGDMFASVPAGDAILMKWILHDWSDAHCATLL
M.  INHTTVVMK.KILENYKGFENLKTLVDVGGGLGVNLKMITSKYPTIKGTNFDLPHVVQHAPSYPGVEHVGGDMFESVPEGDAIFMKWILHDWSDSHNLKLL
N.  VDVCATEMK.RMLEIYTGFEGISTLVDVGGGSGRNLELIISKYPLIKGINFDLPQVIENAPPLSGIEHVGGDMFASVPQGDAMILKAVCHNWSDEKCIEFL
O.  GASSEFVAALAVRECRDVFAGIKSLVDVAGGNGTTARTIAEAFPYVKCSVLDLPQVIQGISSHGTVEFVAGDMMEFVPPAEAVLLKYVLHNWSDQDCVKIL
P.  ASDSQLIVDVAIKQSAEVFQGISSLVDVGGGIGAAAQAISKAFPH.KCSVLDLAHVVAKAPTHTDVQFIAGDMFESIPPADAVLLKSVLHDWDHDDCVKIL
```

```
              290       300       310       320       330       340       350            365
A.  KNCYDALPE...NGKVILVECILPVAPDTSLATKGVVHVDVIMLAHNPGGKERTEKEFEGLAKGAGFQG.FEVMCCAFNTHV.IEFRKKA
B.  KNCYDALPN...IGKVIVAECVLPVYPDTSLATKNVIHIDCIMLAHNPGGKERTQKEFETLAKGAGFQG.FQVMCCAFGTHV.MEFLKTA
C.  KNCYAALPD...NGKVILGECILPVAPDSSLATKGVVHIDVIMLAHNPGGKERTEQEFQALAKGAGFQG.FNVACSAFNTYV.IEFLKKN
D.  KNCYEALPD...NGKVIVAECILPVAPDSSLATKGVVHIDVIMLAHNPGGKERTQKEFEDLAKGAGFQG.FKVHCNAFNTYI.MEFLKKV
E.  KNCYDAVPD...NGKVIVAECILPVAPDSSLATKGVVHIDVIMLAHNPGGKERTEKEFEALAKAAGFQG.FRVVCCAFNTYI.MEFLKKP
F.  KNCHEALPE...NGKVILAECLLPEAPDSTLSTQNTVHVDVIMLAHNPGGKERTEKEFEALAKGAGFRG.FIKVCCAYNSWI.MELLK
G.  KNCYDALPN...NGKVILAECILPEVPDSSLATKGVVHIDVITVAHNPGGKERTEKEFEALAKAAGFQG.FQVFCNAFNTYI.IEFSKKICN
H.  KNCYEALPA...NGKVIIAECILPEAPDTSLATKNTVHVDIVMLAHNPGGKERTEKEFEALAKGAGFTG.FARLVALTTLGS.WEFNK
I.  KNCYDALPN...NGKVIVAEYILPVVPDSSLASKLSVTADVMIVTQNSGGKERTEKEFEALAKAAGFQG.FQVFCNAFTIYI.IEFSKNISN
J.  KNCHKALPE...KGKVIVVDTILPVAAETSPYARQGFHKDLLMLAYNPGAKERTEQEFRDLAKEVGFAGGVKPVCCVNGMWV.MEFHK
K.  NKCYESLAK...GGKIILVESLIPVIPEDNLESHMVFSLDCHTLVHNQGGKERSKEDFEALASKTGFST.VDVICCAYDTWV.MELYKK
L.  KNCYDALPE...NGKVIVVECVLPVNTEATPKAQGVFHVDMIMLAHNPGGKERYEREFRELAKGAGFSG.FKATYIYANAWA.IEFIKS
M.  KNCYKALPD...NGKVIVVEAILPVKPDIDTAVVGVSQCDLIMMAQNPGGKERSEEEFRALATEAGFKG.VNLICCVCNFWV.MEFCK
N.  SNCHKALSP...NGKVIIVEFILPEEPNTSEESKLVSTLDNLMFI.TVGGRERTEKQYEKLSKLSGFSK.FQVACRAFNSLGVMEFYK
O.  TRCREAISHGEKAGKVIIIDTVVGSPSQQILESQVTMDLSMMMLF.N..GKVREEQNWHKIFLEAGFSH.YKIH.NVLGMRSLIEVQP
P.  KNCKKAIPPREAGGKVIIINMVVGAGPSDMKHKEMQAIFDVYIMFIN..GMERDEQEWSKIFSEAGYAD.YRII.PVLGVRSIIEVYP
```

Figure 1. Alignment of the Amino Acid Sequences of Plant OMTs of Type 1. The following full-length OMT sequences were used with the source identified as the GenBank accession number where it is known: A. Aspen caffeic acid OMT (*2*, M72523); B. Eucalyptus caffeic acid OMT (X74814); C. Almond caffeic acid OMT (X83217); D. Alfalfa caffeic acid OMT (M63853); E. *Stylosanthes humilis* caffeic acid OMT (L36109); F. Vanilla caffeic acid OMT (personal communication from Peter Brodelius); G. *Chrysosplenium americanum* caffeic acid OMT (U16793 with corrections provided by A. Gauthier); H. Tomato caffeic acid OMT (X74452); I. Pine OMT (personal communication from Ross Whetten); J. *Chrysosplenium americanum* 3^1 flavonoid OMT (U16794); K. Ice plant α *o*-pinene OMT (M87340); L. *Zea mays* root OMT (M73235); M. Tobacco *o*-diphenol OMT (X71430); N. Alfafla isoliquin OMT (L10211); O. Barley stress induced OMT (X77467); P. *Zea mays* developmentally induced endodermis OMT (L14063). Two sequences very closely related to aspen OMT were omitted from this comparison: hybrid poplar caffeic acid OMT (M73431) which has only 7 differences from aspen OMT and *Populus kitakamiensis* caffeic acid OMT (D49710) which has only 4 differences from aspen OMT. Several other partial sequences are available for plant OMTs, which are apparently of Type 1, and have not been included in this alignment.

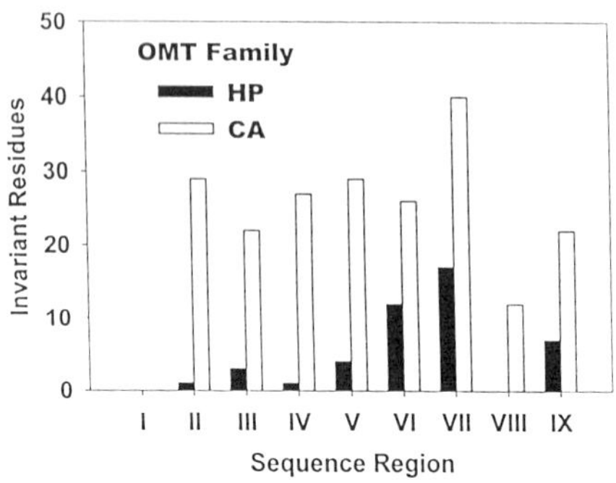

Figure 2. Invariant Amino Acid Residues in Type 1 OMTs Compared by Sequence Region. The regions of the sequence are defined in Table I. The Type 1 OMT family is divided into two parts: caffeic acid OMTs (CA), which include sequences A to H in Figure 1, and all plant OMTs of Type 1 (HP), which include sequences A to P in Figure 1.

Table I. Invariant Residues in 16 Plant Type 1 OMTs by Sequence Region.

Region	Residues	Function	Invariant Residues
I	1 to 20	?	None
II	21 to 71	Phenolic Specificity	L42
III	72 to 123	Phenolic Specificity	R79, L84, Y106
IV	124 to 161	Phenolic Substrate Binding	G159
V	162 to 201	Phenolic Substrate Binding	F163 (aromatic), D169, N177, F198
VI	202 to 236	S-Ado-Met Binding	V204, D205, V206, G208, G209, G211, I218, Y222, P223, D231, P233, V235
VII	237 to 294	S-Ado-Met Binding	H246, G249, D250, M251, F252, P257, A261, K265, H269, W271, C276, L280, C283, G291, K292, V293, I294
VIII	295 to 319	Phenolic ? Specificity ?	None
IX	320 to 365	S-Ado-Met Binding	N324 (polar), G326, G327, R330, G343, F344, E360

Sequence regions based on the residues and numbering of aspen OMT are defined by the amino acid alignment in Figure 1 and illustrated in Figure 2. S-Ado-Met binding region is defined as previously shown (*2*). Functions suggested for other regions are speculative.

biochemical studies with another rare plant enzyme, namely nitrate reductase: we have expressed catalytically active fragments of nitrate reductase in *E. coli* whereafter the 3-D structure of one of these fragments was determined (*4*). In the case of OMT, we had shown in the original study of its clone that it could be expressed in an active form in *E. coli* (*2*). More recently, we moved the OMT coding sequence to the more favorable pET vector for expression of large amounts of the active OMT in *E. coli* using the natural AUG start codon so that a natural protein was obtained (*3*). The purification of OMT from *E. coli* extracts is simpler than from xylem and the yield of protein is quite high: about 5 mg of pure OMT can be obtained from a liter of culture after induction (*3*). Using the antibody prepared against natural OMT, we have established that the recombinant protein cross-reacts and that the catalytic properties of the recombinant protein, such as its substrate specificity, are the same as those of the natural enzyme. We have prepared a large amount of the OMT and provided it to a crystallography group to attempt its crystallization.

The phenolic substrate specificity of OMT was investigated in detail using 25 different phenolic compounds (*3*). Numerous characteristics of the enzyme's active site were discovered by these means. First, the natural substrates caffeic and 5-hydroxyferulic acids were the preferred substrates, as expected. Also 5-hydroxyferulic acid was methylated about twice as fast as caffeic acid, which indicates that a specific site exists in OMT for binding the 3-methoxyl group of 5-hydroxyferulic acid as a result of which this binding energy is used to promote catalysis. Second, although aspen OMT cannot methylate the 4-hydroxyl group of substrates, it prefers substrates which have a 4-hydroxyl group, which suggests that this group is also specifically bound in the active site, perhaps by an ionic amino acid side chain of the enzyme, and that this process may be important in bringing the phenolic hydroxyl to be methylated into the proper orientation for efficient methylation. Third, large substrates such a quercetin and caffeoyl CoA are good substrates for OMT, which suggests that the active site pocket does not extend beyond the limit needed for recognizing and binding the natural substrates. Fourth, artificial substrates such as benzoic acids and benzylamines, are not as good substrates as benzaldehydes, which suggests that the charged groups on the aromatic ring interfere with substrate binding. Artificial substrates with longer side chains than those found in the natural substrates are poor ones, especially if they contain a charged residue or bulky substitutents. These data lead to a model of the phenolic substrate binding site in OMT which is discussed below.

We have also studied the kinetic mechanism of OMT in detail for both substrate/product pairs, namely caffeic/ferulic acid and 5-hydroxyferulic/sinapic acid, as well as with the methyl group donor S-Ado-Met and the corresponding product, *S*-adenosyl-L-homocysteine (S-Ado-Hcy) (*3*). The kinetic mechanism was found to be an ordered sequential process with S-Ado-Met binding first followed by introduction of the phenolic substrate. However, product inhibition patterns were not compatible with the simple ordered consecutive release of products, which led to the suggestion that the strong inhibition by S-Ado-Hcy may reflect its remaining in the active site after the methylated phenolic product has left (*3*). In addition, the phenolic product binds to the active site for the phenolic substrate suggesting that the enzyme may remain in an "activated" state for rapid binding of the phenolic substrate, which is expected to be the limiting substrate *in vivo*. In contrast to previous investigations of the kinetic mechanism of OMTs, our model suggests that OMT is probably not inhibited *in vivo* by the strong residual binding of S-Ado-Hcy, but rather that this product's presence can actually promote catalysis by preserving the OMT in a state where the phenolic substrate can bind more readily (*3*).

Site-Directed Mutagenesis of Recombinant OMT

Another important advantage of working with recombinant forms of plant enzymes in a heterologous expression system is that site-directed mutagenesis can be used to investigate the importance of invariant and highly conserved residues in the enzyme's catalytic mechanism. We have demonstrated this with the cytochrome b reductase fragment of nitrate reductase (*10*). In the case of OMT, we first focused our attention on its highly conserved Cys residues (*3*). It has been long known that OMT is inhibited by reagents that can react with thiol groups, and we showed that recombinant OMT is inhibited quite strongly by *p*-chloromercuribenzoate (*3*). So the question arises as to whether thiol groups are important to the catalytic mechanism of OMT. This could be investigated with site-directed mutagenesis by narrowing down the number of Cys residues to be mutated through the comparison between OMT amino acid sequences (Figure 1). There are 9 Cys residues in aspen OMT, but only 2 of them are highly conserved in the family of Type 1 OMTs (Table II). In this way a comparison between the sequences in a family of enzymes with very high similarity can be used to substitute for a 3-D structure in guiding site-directed mutagenesis. We generated Ser and Ala mutants for Cys residues at positions 276 and 283 of aspen OMT, both of which have been suggested to be in the region for OMT binding to S-Ado-Met (*3*). Both Ser and Ala mutants were active, as was the Ala double mutant where both Cys276 and Cys283 were replaced by Ala. This indicated that these highly conserved Cys residues are not essential for OMT activity but probably are involved with substrate binding. All the mutants were still inhibited by *p*-chloromercuribenzoate, but the Cys-276-Ala and double mutant were less sensitive (*3*). This decrease in sensitivity to the thiol inhibitor was similar to the degree of protection provided to the enzyme by S-Ado-Met and S-Ado-Hcy, which supports the suggestion that at least Cys-276 is important in binding this cosubstrate.

Conclusions and Expectations

The cloning and characterization of aspen monolignol-specific OMT has had a major impact on our understanding of lignin precursor biosynthesis in woody species. This work has also had a major impact on gaining understanding of plant OMTs in general. In addition to those aspects of its importance described here, studies of the impact of antisense constructs of OMT have also yielded important findings about the regulation of lignin composition and content in model systems (*11*). Other groups have also carried out antisense OMT experiments with transgenic plants (*12-14*). Now that these antisense studies are moving toward the alteration of lignin in woody plant species, this aspect of OMT molecular biology will grow in importance. In collateral studies, the discovery that a second form of OMT is involved in monolignol biosynthesis (*8*), namely the Type 2 OMTs or caffeoyl CoA OMT, will also bring greater understanding about the central role of the methylation reactions catalyzed by OMTs in monolignol biosynthesis in woody species. Why would a plant use two different enzymes catalyzing the same reactions in a tissue as strongly focused on metabolic functions as is xylem tissue? While we may never know the full answer to this key question, it can be expected that we will gain a greater understanding of how these two OMT forms contribute to the plant's control of monolignol biosynthesis and thus, perhaps, to how the plant cell can so precisely regulate the types of lignin deposited at different cell wall sites.

The other area where our studies of OMT are having a major impact is in advancing the understanding of the enzyme's biochemical mechanism. We expect in next few years to obtain a 3-D structure of aspen caffeic acid OMT. This will be an important step since aspen OMT not only represents a large number of Type 1 OMTs involved in many plant processes, but also is closely related to OMTs in animals, bacteria and fungi (*2*). In the meantime, we have illustrated how the comparison between amino acid sequences of all Type 1 OMTs can be used to guide the application of site-directed mutagenesis to the study of the function of Cys residues in OMT (*3*). As a summary of all our results on OMT functionality, we have created a cartoon model of the enzyme's active site (Figure 3). In the model, we illustrate the concept that the phenolic hydroxyl group to be methylated initiates an attack on the methyl group of S-Ado-Met after its conversion to an anion by a key basic residue of the enzyme; the idea comes partly from our kinetic studies of OMT showing that the

Table II. Cysteine Residues and Their Substitutions in Plant Type 1 OMTs

OMT Group	C92	C112	C127	C268	C276	C283	C297	C351	C352
8	C_6	C_6	C_3	C_8	C_8	C_8	C_8	C_7	C_4
Caffeic Acid	Y	A	N_3					V	N_2
	H		L_2						S
OMTs									A
16	C_{11}	C_{10}	C_4	C_{10}	C_{15}	C_{16}	C_9	C_{12}	C_7
Plant	Y	A_2	N_3	L_6	N		T_2	V	N_4
OMTs	H	G_2	L_4				Y	Y	S
	S	S_2	V_2				S		A
	T		A				A		I
	A		T				F		R
			M				M		P

Cys residue numbers are shown as found in aspen OMT (*2*). See Figure 1 for sequence alignments on which this set of data is based. The subscripts on the individual residues indicate their frequency in each subgroup of the OMT sequences. Single letter amino acid abbreviations are used for all residues.

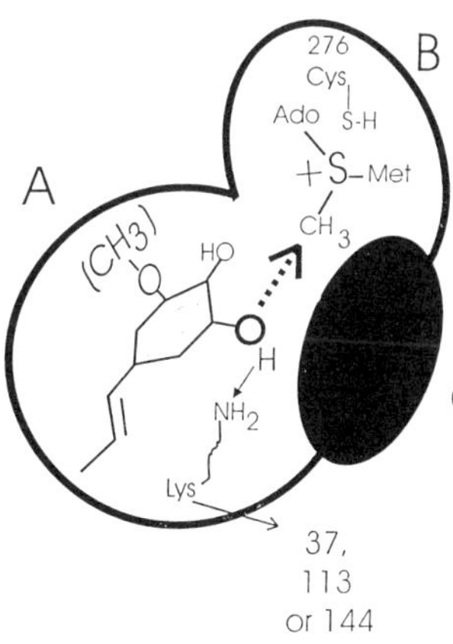

Figure 3. A Schematic Model for Aspen OMT. The sequence of aspen OMT has been divided into 3 major regions: A) phenolic substrate specificity determining and binding regions; B) S-Ado-Met binding region; and C) a region of unknown function near the C-terminus, which was defined as subregion VIII in Figure 2 and Table I with a suggestion that it is involved in phenolic substrate binding. Two key residues of OMT are shown with a Lys (possibly one of the highly conserved residues K37, K113, or K144 in aspen OMT) serving as the base to activate the phenolic hydroxyl which can then attack the S-Ado-Met being held by Cys-276. The other region emphasized in the model is the pocket for binding the 3-methoxyl group of 5-hydroxyferulic acid, which is probably filled in gymnosperm OMTs.

ternary complex of OMT with the phenolic substrate and S-Ado-Met is required for activity. This model incorporates our results from the substrate specificity of aspen OMT and site directed mutagenesis as bases for suggestions about which amino acid residues may function in specific roles for binding the phenolic substrate and S-Ado-Met. Of course, this model is highly speculative and requires confirmation from the 3-D structure of the enzyme. In the meantime, site-directed mutagenesis studies will further contribute to the identification of key functional amino acid residues in OMT. For example, our suggestion that specific residues in aspen OMT are used to bind the 3-methoxyl group of 5-hydroxyferulic acid points to a way of investigating the difference between the bispecific OMT of angiosperm dicots and the more monospecific OMT of gymnosperms. Thus, over the next few years, the structure and function of these OMT forms will become more fully understood and this knowledge may be very useful in biotechnologically modifying monolignol biosynthesis in order to alter the type of lignin formed in softwoods.

Acknowledgments

This research has been supported by grants from the U.S. Department of Agriculture and the National Science Foundation, as well as funds from the State of Michigan Research Excellence Fund provided by Michigan Technological University.

Literature Cited

1. Higuchi, T. In *Biosynthesis and Biodegradation of Wood Components*; Higuchi, T., Ed.; Academic Press: Orlando, FL, 1985; pp 141-160.
2. Bugos, R. C.; Chiang, V. L.; Campbell, W. H. *Plant Mol. Biol.* **1991**, *17*, 1203-1215.
3. Meng, H.; Campbell, W. H. *Arch. Biochem. Biophys.* **1996**, *330*, 329-341.
4. Campbell, W. H. *Plant Physiol.* **1996**, *111*, 355-361.
5. Bugos, R. C.; Chiang, V. L.; Campbell, W. H. *Phytochemistry* **1992**, *31*, 1495-1498.
6. Pakusch, A. E.; Kneusel, R. E.; Matern, U. *Arch. Biochem. Biophys.* **1989**, *271*, 488-494.
7. Ye, Z.-H.; Varner, J. E. *Plant Cell* **1994**, *6*, 1427-1439.
8. Meng, H.; Campbell, W. H. *Plant Physiol.* **1995**, *108*, 1749; Plant Gene Register PGR95-040.
9. Vidgren, J.; Svensson, L. A.; Liljas, A. *Nature* **1994**, *368*, 354-358.
10. Dwivedi, U. N.; Shiraishi, N.; Campbell, W. H. *J. Biol. Chem.* **1994**, *269*, 13785-13791.
11. Dwivedi, U. N.; Campbell, W. H.; Yu, J.; Datla, R. S. S.; Bugos, R. C.; Chiang, V. L.; Podila, G. K. *Plant Mol. Biol.* **1994**, *26*, 61-71.
12. Ni, W.; Paiva, N. L.; Dixon, R. A. *Transgenic Research* **1994**, *3*, 120-126.
13. Atanassova, R.; Favet, N.; Martz, F.; Chabbert, B.; Tollier, M.-T.; Monties, B.; Fritig, B.; Legrand, M. *Plant J.* **1995**, *8*, 465-477.
14. Doorsselaere, J. V.; Baucher, M.; Chognot, E.; Chabbert, B.; Tollier, M.-T.; Petit-Conil, M.; Leple, J.-C.; Pilate, G.; Cornu, D.; Monties, B.; Van Montagu, M.; Inze, D.; Boerjan, W.; Jouanin, L. *Plant J.* **1995**, *8*, 855-864.

Chapter 5

Genes Involved in the Final Steps of Monolignol Biosynthesis and Their Manipulation for Tailoring New Lignins

A.-M. Boudet, D. Goffner, C. Marque, and J. Grima Pettenati

UMR CNRS/UPS 5546, Centre de Biologie et Physiologie Végétales 118, route de Narbonne, 31062 Toulouse Cedex, France

The genes encoding cinnamoyl CoA reductase (CCR) and cinnamyl alcohol dehydrogenase (CAD) have been cloned and characterized in *Eucalyptus gunnii*. Their spatio-temporal expression has been investigated with a combination of techniques revealing a high level of expression associated with lignifying tissues. CCR down-regulated tobacco plants have been obtained and exhibit an unusual yellow-brown coloration of the xylem. Chemical and ultrastructural modifications including a decrease in lignin content were observed. CAD down-regulated transgenic tobacco plants have also been obtained through the use of different antisense constructs. Several chemical modifications are induced in transgenic lignins which exhibit an increased extractability, a positive attribute for the pulping industry. The results presented here confirm that lignin content and composition can be modified by directed genetic manipulation of genes specifically involved in monolignol biosynthesis.

Despite the important physiological and adaptative functions of lignins (*1*, *2*), their presence in plant tissues reduces the economic benefits to be gained from forage consumption by livestock and from trees destined for wood pulp and paper manufacture (*3*, *4*). In the latter case, lignin has to be separated from cellulose, mainly by harsh chemical treatments which are costly in both financial and environmental terms. It is therefore of obvious importance to modulate lignin content and/or composition in order to tailor characteristics in plants that would render them more suited to agroindustrial uses. It has also to be emphasized that even slight modifications in lignin content and/or composition could provide important economic benefits for the utilization of wood in the pulping industry because of the large scale involved. Since almost all of the genes of the "classical" lignin biosynthesis pathway and some genes involved in alternative pathways such as caffeoyl-coenzyme A 3-*O*-methyl transferase (CCoAOMT) have now been characterized (Figure 1), the modulation of lignin content and structure by genetic engineering has become a realistic objective in the context of overexpression and loss of function experiments.

Thus the alteration of the lignin component of the cell wall seems now feasible despite the complexity of the biosynthetic pathway leading to its formation, whereas a

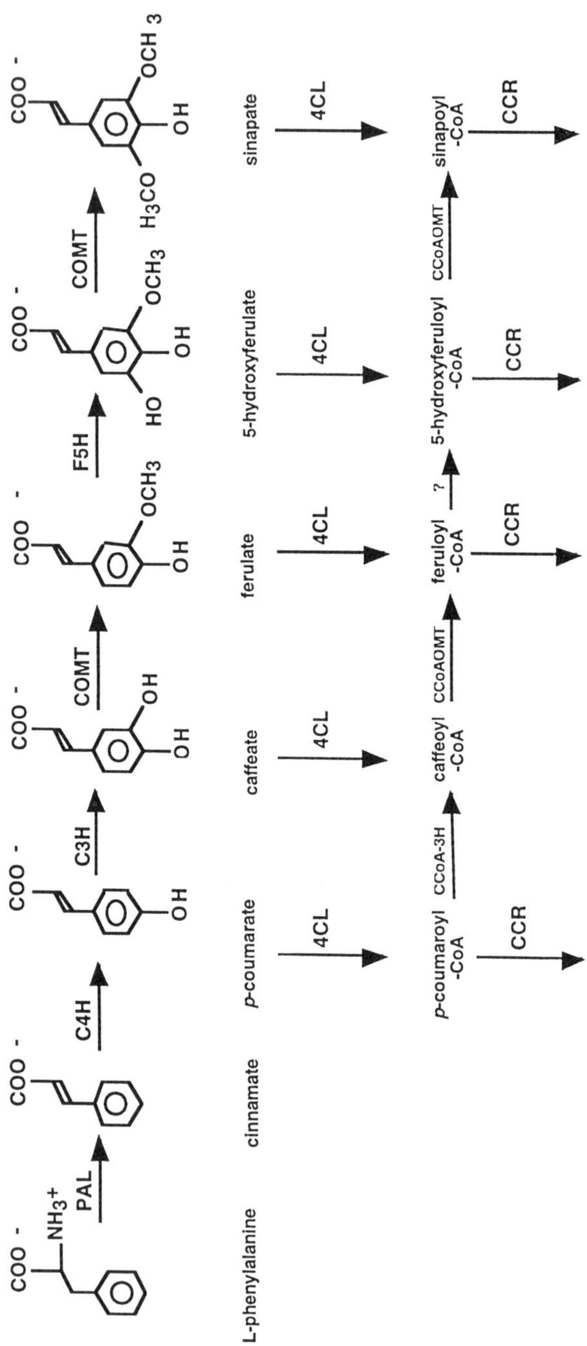
L-phenylalanine
cinnamate
p-coumarate
caffeate
ferulate
5-hydroxyferulate
sinapate
PAL
C4H
C3H
COMT
F5H
COMT
4CL
4CL
4CL
4CL
4CL
p-coumaroyl -CoA
caffeoyl -CoA
feruloyl -CoA
5-hydroxyferuloyl -CoA
sinapoyl -CoA
CCoA-3H
CCoAOMT
?
CCoAOMT
CCR
CCR
CCR
CCR

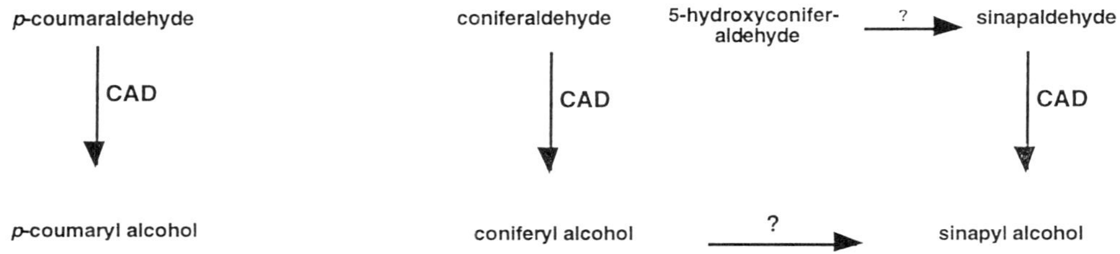

Figure 1. Potential biosynthetic pathways of monolignol synthesis. Enzymes of the common phenylpropanoid pathway include phenylalanine ammonia-lyase (PAL), cinnamate-4-hydroxylase (C4H), coumarate-3-hydroxylase (C3H), caffeic acid 3-O-methyltransferase (COMT), ferulate-5-hydroxylase (F5H) and 4-coumarate : CoA ligase (4CL). Alternatively, hydroxylation and methylation steps may occur at the cinnamoyl-CoA level via coumaroyl-coenzyme A 3-hydroxylase (CCoA-3H) and /or caffeoyl-coenzyme A O-methyltransferase (CCoAOMT).Other potential routes are suggested. The two step reduction of cinnamoyl-CoAs to cinnamyl alcohols (monolignols) catalyzed by cinnamoyl-CoA reductase (CCR) and cinnamyl alcohol dehydrogenase (CAD) respectively constitute the lignin specific pathway.

similar goal cannot presently be achieved for the polysaccharidic components of the wall due to a lack of knowledge about the corresponding enzymes and genes involved.

If one considers the potential target genes for tailoring new lignins, it appears that only a few exhibit the necessary requirements. Along the common phenylpropanoid pathway, the steps catalyzed by phenylalanine ammonia lyase (PAL), cinnamate 4-hydroxylase (C4H), 4-coumarate CoA ligase (4CL) do not seem, *a priori*, to be appropriate targets for specifically manipulating lignin since they are involved in the production of various classes of phenolic compounds. This was clearly demonstrated in plants in which PAL down-regulation led to many pleiotropic effects incompatible with normal growth and development of plants (*5*). Thus, unless lignin-specific isoforms of these enzymes sufficiently different at the sequence level from other isoforms (unassociated with lignification) are characterized, these enzymatic steps in the common phenylpropanoid pathway may not be suitable for modifying lignin synthesis. Two enzymes along the phenylpropanoid pathway may be interesting for controlling the monomer composition of lignin: *O*-methyltransferase (OMT) and ferulate 5-hydroxylase (F5H). In contrast to the previously mentioned biocatalysts, OMT and F5H do not control the overall flux of phenylpropanoid compounds, but they have an effect on the methoxylation pattern of the phenylpropanoid intermediates. Significant results have already been obtained concerning the down-regulation of OMT genes using either an antisense or sense RNA strategy in tobacco (*6-8*) and in a woody angiosperm, poplar (*9*). In the two latter studies, it has been clearly demonstrated that, through ectopic expression of an OMT antisense gene, the monomeric composition of lignins can be profoundly affected. The impact of this modification in lignin quality on paper pulp manufacture and forage digestibility has not yet been evaluated in depth.

Targeting downstream genes specifically involved in the final steps of lignin formation potentially represent a more suitable strategy for reducing lignin content. The "lignin-specific" pathway involves two reductive enzymes, namely cinnamoyl-CoA reductase (CCR, EC 1.2.1.44) and cinnamyl alcohol dehydrogenase (CAD, EC 1.1.1.195) which convert the hydroxycinnamoyl-CoA esters into monolignols (Figure 1). Monolignols are then transported to the cell wall via mechanisms which are still poorly understood. The subsequent polymerization step involves their oxidation into phenoxy radicals which spontaneously polymerize to give oligomers and finally the lignin polymer. The nature of the oxidative enzymes involved in the formation of these mesomeric free radicals is still a matter of debate. Convincing arguments suggest the potential role of peroxidases, laccases and coniferyl alcohol oxidases (*2*) in the process, but an unambiguous demonstration of the involvement of a specific class of enzyme is still lacking. Loss-of-function experiments have not hitherto allowed firm conclusions to be drawn concerning the involvement of an anionic peroxidase (*10*) and of laccase (*11*) probably because of gene redundancy or of the combined effects of different enzymes.

In this review, we will focus our attention on the enzymes and genes involved in the "monolignol-specific pathway," *i.e.* CAD and CCR, that are often considered to be specific to lignification. This assumption, however, is not strictly correct, even though lignins are without a doubt the major products derived from monolignols. Monolignols can also be used for the synthesis of lignans, such as pinoresinol and other related phenolics, *e.g.* dehydrodiconiferyl alcohol (*12*). To what extent the down-regulation of the CAD and CCR genes could also affect the synthesis of these compounds has to be demonstrated, but these genes represent, at the moment, the most suitable and readily available targets for specifically manipulating lignins through genetic engineering.

Cinnamoyl CoA Reductase

As the first committed step in the lignin branch pathway, CCR can be postulated as a potential control point regulating overall carbon flux towards lignins and, therefore, its down-regulation could affect the lignin content in the cell wall. This hypothesis could not have been tested hitherto, however, due to the lack of a molecular probe for this enzyme.

Purification of CCR and Cloning of the cDNA. In our laboratory, we have recently developed an efficient protocol for purifying CCR from *Eucalyptus gunnii* xylem (*13*). The first cloning of a cDNA encoding CCR was achieved by screening an *E. gunnii* differentiating xylem cDNA library with degenerate oligonucleotides designed from internal peptide sequences obtained from purified eucalyptus CCR (*14*). The isolated cDNA (EUCCR) of 1210 bp contained an open reading frame of 1008 nucleotides encoding a protein of 336 amino acids (calculated molecular weight 36.5 kDa). Confirmation of the identity of the isolated clone was provided by production of recombinant functional CCR in *Escherichia coli*. The K_m values and the kinetic behavior of the recombinant enzyme are in agreement with those obtained for the native *E. gunnii* CCR (*13*). Southern blots suggested the presence of only one gene in *E. gunnii*. A CCR genomic clone (gCCR1) was isolated from an *E. gunnii* genomic library containing four introns and five exons whose positions were deduced by comparison with the cDNA. Comparative sequence analysis revealed remarkable homologies between CCR and dihydroflavonol-4-reductase (DFR), the first enzyme of the anthocyanin biosynthesis pathway. Whereas the homology was found over the entire sequence, the highest similarity was in the N-terminal region and particularly in the putative NADPH binding domain. Moreover, the positions of the four introns in CCR correspond to the positions of introns I, II, III and IV in dicot DFRs, reinforcing the hypothesis of a common ancestor for CCR and DFR. It is worth noting that significant similarities were found with mammalian 3beta-hydroxysteroid dehydrogenase and bacterial UDP-galactose epimerase, suggesting that CCR shared a common ancestor with these enzymes and, therefore, can be considered to be a new member of the "mammalian 3beta-hydroxysteroid dehydrogenase/plant dihydroflavonol reductase" superfamily.

CCR Gene Expression. CCR gene expression was investigated by Northern analysis and *in situ* hybridization in eucalyptus and poplar. Localization of CCR expression to the young differentiating xylem of both stems and roots provided convincing evidence for the role of this enzyme in lignification. The spatial expression pattern was very similar to that observed for the other committed enzyme in the lignin biosynthetic pathway, CAD_2 (*15, 16*) and other enzymes involved in phenylpropanoid synthesis (*2*). *In situ* hybridization revealed that the CCR was also expressed in internal phloem fibers in eucalyptus, but no signal was associated with secondary phloem fibers or epidermal tissue. The fact that CCR is expressed to a high level in leaves is interesting since such organs are not highly lignified. Some expression is undoubtedly related to the formation of lignified vascular tissue. However it suggests that the CCR-mediated reaction products may be associated with other processes than lignification, such as formation of lignans, phytoalexin like compounds, and so forth.

Alterations of Lignin Synthesis in CCR Down-Regulated Transgenic Plants. The recent characterization of the eucalyptus CCR cDNA opened a new route for genetically engineering plants with altered cell wall properties. The first attempts to use the eucalyptus CCR cDNA to down-regulate CCR in transgenic tobacco were unsuccessful. More recently, using a tobacco cDNA isolated from Zeneca Seeds (using the eucalyptus cDNA as a probe), we obtained tobacco plants transformed with homologous antisense CCR cDNA fragments under the control of either constitutive promoters (the CaMV35S promoter or CaMV35S with a double enhancer) or a lignification-specific promoter (CAD promoter (*15*)). Preliminary results have been obtained on primary transformants showing that it is possible to down-regulate CCR activity up to 70% (*i.e.* 30% residual activity) and still obtain viable plantlets (J. Piquemal, *et al.*, unpublished data). Some of the primary transformants with very depressed CCR activity exhibit an orange-brown coloration of the xylem area, which is not observed in the control plants. These observations strongly suggested that significant chemical changes had occurred in these transformed plants. The F1 back-cross generation of one of the most depressed primary transformants gave a segregating population with azygous (control) and hemizygous (antisense) plants identified through PCR and Southern analysis. Seventy-day old plants were analyzed for lignin content through three different methods, namely those involving acetylbromide (*17*), Klason determination (*18*) and thioglycolic acid (*19*). Significant reduction of lignin content was observed with all these different techniques in CCR down-regulated plants (J. Piquemal, *et al.*, unpublished data), the most dramatic decrease of lignin content (40%) being observed with the Klason method.

In addition to these quantitative changes, qualitative changes were suggested by the unusual coloration of the xylem area already mentioned and also by a lower staining intensity with phloroglucinol-HCl. Alkaline hydrolysis released larger amounts of UV absorbing material from antisense plants than from the corresponding controls. This is of importance for the pulping industry for which a low lignin content and an improved susceptibility to alkaline hydrolysis are particularly beneficial. In the transformed plants with the greatest reduction in CCR activity, abnormal phenotypes were observed, *i.e.* with a reduction in size and collapsed xylem vessels. These observations reveal that a substantial decrease in lignin content leads to deleterious effects on normal development. As we have in hand a range of plants with variously depressed CCR activity, we are now evaluating to what extent the lignin content can be modified without affecting normal plant development. One of the advantages of the RNA antisense strategy is indeed the possibility of modulating the desired effect. Suitable transformants with maximum lignin reduction and normal development should be easily identifiable. These results underscore for the first time that it is possible to significantly reduce the lignin content of plants through the down-regulation of CCR activity, thereby confirming that CCR is a regulatory step in lignin biosynthesis.

CCR cDNA clones have also been isolated in our laboratory from maize and alfalfa (M. Pichon *et al.*, unpublished data) and by other groups in poplar (W. Boerjan, personal communication) and tall fescue (C. Grand, personal communication). These cDNAs will be useful for modifying the lignin profiles in these important crops.

Cinnamyl Alcohol Dehydrogenase

A cDNA encoding CAD was initially characterized in tobacco by Knight *et al.* in 1992 (*20*) and, subsequently, in a wide range of plant species (see ref *2* for a review). The identity of both eucalyptus and spruce cDNAs have been confirmed by functional expression in *E. coli* (*21, 22*).

CAD cDNAs throughout the plant kingdom share extensive sequence homology (80%-90% similarity at the protein level), suggesting that CAD genes have been highly conserved throughout evolution. Detailed sequence analysis of CAD indicates that it belongs to the long-chain zinc-containing alcohol dehydrogenase family. The degree of homology between eucalyptus CAD and horse liver alcohol dehydrogenase (HLADH, EC 1.1.1.1) was large enough to enable homology-based molecular modeling from the crystallographic coordinates of HLADH (*23*) and hence to identify amino acid residues of potential functional significance. As predicted by the three-dimensional structure, when a point mutation (S212D) was effected by site-directed mutagenesis and the resulting cDNA expressed in *E. coli*, the cofactor affinity towards $NADP^+$ was modified (*24*).

Recently, detailed studies have been performed on an "atypical" CAD, CAD_1 from *Eucalyptus* (*25*) which exhibits different physicochemical properties from those of CAD_2, the isoform classically associated with lignification. Using degenerate oligonucleotides derived from internal peptide sequences of purified CAD_1, the corresponding cDNA has been isolated in our laboratory (D. Goffner *et al.*, unpublished data). The identity of CAD_1 was unambiguously confirmed by functional expression in *E. coli.* Based on various criteria including the ability to convert coniferaldehyde into coniferyl alcohol and its localization in lignifying tissues, it may not be possible to rule out the hypothesis that CAD_1 may be an "alternative" enzyme in monolignol biosynthesis. Antisense experiments in tobacco are now underway in order to gain insight into the role of CAD_1.

Expression of a Fused Eucalyptus CAD_2 Promoter-Reporter Gene Construct. CAD genes have been characterized in tobacco (*26*), *Eucalyptus botryoides* (*27*), *E. gunnii* (*15*), and *Arabidopsis* (*28*). In order to study the spatial and developmental expression of the eucalyptus CAD gene, we fused the CAD promoter to the β-glucuronidase (GUS) reporter gene and transferred the construct into poplar and tobacco through *Agrobacterium tumefaciens*-mediated transformation (*15*). Quantitative fluorimetric assays showed that GUS activity was highest in roots followed by stems and leaves. Similar findings were reported in transgenic tobacco harboring a tobacco CAD promoter-GUS construct (*26*).

Histochemical analysis of GUS activity was then carried out at the cellular level. Initially, it had been demonstrated that GUS activity was predominantly localized in ray parenchyma cells in stems (*15*). However, recent results using improved localization techniques (*16*) showed that, in addition to ray parenchyma cells, a high level of CAD promoter-driven GUS activity was also associated with the cambium / differentiating xylem area in poplar stems, *i.e.* cells that are destined to become lignified. This observation supports the hypothesis that lignification may occur in two complementary stages (not necessarily chronologically separated). In the first "stage", monolignols are produced by cells undergoing lignification. Lignin deposition, cell autolysis and death may then proceed as classically described (*29*). The fact that lignin deposition continues even in dead xylem cells (*30*) suggests the occurrence of a 'second phase' in which monolignols may be produced in adjacent ray parenchyma cells and exported towards the site of polymerization. At the present time, one cannot determine to what extent mature xylem elements provide their own precursors, while they are still metabolically active, or receive precursors from neighboring cells.

CAD_2 expression was also observed in the periderm in association with lignified (and possibly suberised) cells in the phellem. These results suggest that CAD_2 is implicated in the formation of protective dermal layers in woody plants. Moreover, CAD_2 was also expressed in phloem fibers and the parenchymatic cells immediately surrounding them. In summary, our results show that the CAD gene exhibits a complex and specific expression pattern in poplar stem tissues.

Alteration of Lignins in CAD Down-Regulated Transgenic Plants. The first CAD down-regulated plants were obtained in tobacco by Halpin *et al.* (*31*) through ectopic expression of a homologous CAD cDNA antisense gene driven by the CaMV 35S promoter.

Interestingly, in line 50 plants, with the lowest residual CAD activity (7%) as well as in several primary transformants with low levels of CAD activity, the xylem was reddish-brown, reminiscent of maize and sorghum brown-midrib mutant phenotypes (*32*). In similar transgenic plants, we have shown that this phenotype appears to be developmentally regulated: the very young differentiating xylem in the apex is green, but the red coloration gradually develops in older tissues. In mature tobacco stems the red color fades to orange, suggesting structural modifications of the lignin during development. In addition, staining of the antisense tobacco stem sections with the Wiesner reagent (phloroglucinol-HCl) gave rise to an unusual reaction compared to that with control plants (leading to a red rather than purple stain).

Although the lignin contents of antisense and control plants were similar as determined by either Klason or acetyl bromide analysis, the antisense lignins were more susceptible to alkali/thioglycolic acid extraction, suggesting that changes had occurred in their chemical structure (*33*).

Detailed analysis of lignin composition was performed on antisense CAD plants by a variety of state-of-the-art techniques. Pyrolysis mass spectrometry indicated an increased incorporation of both conifer- and sinapaldehydes into the lignins of line 50 plants, thereby resulting in the production of a novel lignin. The inclusion of aldehyde monomers in place of alcohols in lignin could have a profound effect on the formation of resinols and other dimeric structures, resulting in a polymer with altered intermolecular bonding and modified chemical reactivity. Changes in the monomeric composition of lignin from CAD antisense plants were also investigated using thioacidolysis. These results indicated a depressed S/G ratio in CAD antisense plants. In addition, the proportion of the most frequently occurring interunit beta-*0*-4 linkages which are targets of lignin depolymerization processes (*33*), was reduced. However only traces of cinnamaldehydes were identified in both control and antisense plants by this method.

Additional analyses of lignin structure and chemical composition in CAD down-regulated tobacco plants have been performed more recently (*34, 35*). Large quantities of vanillin and syringaldehyde (C_6-C_1 aldehydes of G and S moieties) and significant amounts of coniferaldehyde are released from transgenic lignin following mild alkali treatment (N. Yahiaoui *et al.*, submitted).

Further evidence for the incorporation of aldehydes into lignin was provided by Higuchi *et al.* (*36*). They have shown that coniferaldehyde is converted to a dark wine-red dehydropolymerisate (DHP) in the presence of horseradish peroxidase. Their results indicate that the wine-red color of coniferaldehyde DHPs, which gradually fades with time, is related to the extended conjugation of the coniferaldehyde groups. They also suggest that the red brown coloration of trangenic plants with reduced CAD activity could be ascribed to "abnormal lignins" with increased frequencies of cinnamaldehyde groups. We have confirmed these results by a histochemical "semi *in vivo* " approach facilitating the formation of a red cell wall polymer when stem sections are incubated in the presence of coniferaldehyde and peroxidase (N. Yahiaoui *et al.*, submitted).

Finally the combined use of FT-IR and FT-Raman spectroscopies provided further evidence for the incorporation of cinnamaldehyde groups into the lignin of antisense CAD plants (*37*). Furthermore these non-destructive analyses also suggest that the CAD antisense transformed plants may contain lignin that is less condensed (cross-linked) than the wild type.

Recently Hibino *et al.* (*38*) have introduced an antisense CAD gene from *Aralia cordata* into tobacco. This study showed that this transformation with a heterologous gene was effective in reducing CAD activity. Two transgenic plants which only exhibited 55% and 20% reduction of the CAD activity did not show any change in lignin content but did exhibit an increased level of cinnamaldehydes. These results are in partial disagreement with our results and those of Halpin *et al.* (*31*) since in both cases we did not observe significant alterations in antisense plants unless a pronounced reduction (more than 80%) in CAD activity was observed. However, different methods were used to determine the aldehyde content.

The incorporation of cinnamaldehyde moieties into the transgenic lignin of some CAD down-regulated plants, apparently in place of the corresponding alcohols, is a striking example of the potential metabolic adaptation of plants and the chemical flexibility of lignins.

The effects of the down-regulation of CAD have also been investigated in antisense poplar, a readily transformable woody species (*39*). In general, the results are similar to those in tobacco in that poplar plants with strongly reduced CAD activity (up to 20% residual activity) exhibit qualitative but not quantitative changes in lignins. However, the S/G ratio was not modified in transgenic poplar trees as was the case in tobacco. Moreover, in poplar plants transformed with CAD antisense constructs the red coloration was also apparent in the xylem, but only in the outermost portion corresponding to young tissue.

The pulping characteristics of CAD down-regulated tobacco and poplar have been investigated by the Centre Technique du Papier, Grenoble, France (M. Petit-Conil, personal communication, *39*). Simulated kraft pulping experiments showed improved selective delignification in antisense plants, yielding pulp with a reduced kappa number (the kappa number reflecting the amount of residual lignin in the pulp). Antisense plants also showed an increase in pulp yield and required less chemical treatment for subsequent bleaching, without any apparent detrimental effect on paper quality. Taken together, these data indicate that chemical changes in lignin structure and composition as a result of reduced CAD activity may afford important economic benefits by exerting a marked effect on pulping characteristics.

Future Prospects

Many plant materials, and in particular wood, are essentially composed of cell walls. The structural and chemical variability of walls dictate their chemical reactivity, as well as their mechanical properties and energy content. By genetic engineering, it is now possible to exploit this variability to tailor new plants more suitable for different agro-industrial purposes. The results presented in this review confirm that lignin content and/or composition can be modified by directed genetic manipulation of genes specifically involved in monolignol biosynthesis. Moreover, it has been clearly demonstrated that genetic manipulation of genes further upstream in the lignin pathway such as those coding for OMT (*8, 9*) and F5H (C. Chapple, personal communication) has a significant impact on the monomeric composition of lignins. Downstream genes such as coniferin glucosidase or specific peroxidases and laccases may also constitute suitable targets for reducing lignin content in woody species. Genes encoding transcription factors involved in the regulation of lignin biosynthesis could also be used in transformation experiments as they become available (*40*).

In the future, experiments will be aimed at determining more precisely the chemical modifications of transgenic lignins, the stability of the transgenes in long-living woody plants, and the industrial advantages at the mill level of genetically modified plants. In addition, a fine tuning of lignin content and/or composition could be attained by simultaneous down-regulation of two or three target genes. Finally, it will be of the utmost importance to assess the effects of genetic manipulation on normal growth and development as well as in response to biotic and abiotic stresses.

The results presented here are extremely promising and allow us to predict an expansion in the development of lignin genetic engineering approaches. One of the main limitations at the moment is the lack of efficient transformation technology for many industrially important woody species, particularly in gymnosperms, but these techniques are continuously improving. The ectopic expression of different lignification gene constructs in trees used in agroforestry also provides a means for a better understanding of the control of lignification. Such new data are important to better orientate the strategies for improving tree species through more classical genetic approaches (41) that are complementary to transgenesis.

Acknowledgments

The authors thank their coworkers involved in this research and the European Community (DGXII) for financial support through projects OPLIGE "ECLAIR program" (Contract no. AGRE-CT92-0021) and TIMBER "FAIR program" (Contract no. FAIR-CT95-0424) (DGXII).

Literature Cited

1. Lewis, N. G.; Yamamoto, E. *Annu. Rev. Plant Physiol. Plant Mol. Biol.* **1990**, *41*, 455-496.
2. Boudet, A.-M.; Lapierre, C.; Grima-Pettenati, J. *New Phytol.* **1995**, *129*, 203-236.
3. Whetten, R. W.; Sederoff, R. R. *Plant Cell* **1995**, *7*, 1001-1013.
4. Dean, J. F. D.; Eriksson, K. E. L. *Holzforschung* **1992**, *46*, 135-147.
5. Elkind, V.; Edwards, R.; Mavandad, M.; Hedrick, S. A.; Ribak, O.; Dixon, R. A.; Lamb, C. J. *Proc. Natl. Acad. Sci. USA* **1990**, *87*, 9057-9061.
6. Ni, W. T.; Paiva, N. L.; Dixon, R. A. *Transgenic Res.* **1994**, *3*, 120-126.

7. Dwivedi, U. N.; Campbell, W. H.; Yu, J.; Datla, R. S. S.; Bugos, R. C.; Chiang, V. L.; Podila, G. K. *Plant Mol. Biol.* **1994**, *26*, 61-71.
8. Atanassova, R.; Favet, N.; Martz, F.; Chabbert, B.; Tollier, M. T.; Monties, B.; Fritig, B.; Legrand, M. *Plant J.* **1995**, *8*, 465-477.
9. van Doorsselaere, J.; Baucher, M.; Chognot, E.; Chabbert, B.; Tollier, M. T.; Petit-Conil, M.; Leplé, J. M.; Pilate, G.; Cornu, D.; Monties, B.; van Montagu, M.; Inzé, D.; Boerjan, W.; Jouanin, L. *Plant J.* **1995**, *8*, 855-864.
10. Chabbert, B.; Tollier, M. T.; Monties, B. *Proc. 7th Internat. Symp. Wood Pulp. Chem.* **1993**, *1*, 462-468.
11. Dean, J. F. D. *Cellulose, Paper and Textile Division, 211th American Chemical Society National Meeting,* **1996**, poster 008.
12. Lewis, N. G.; Davin, L. B. *ACS Symp. Ser.* **1994**, *562*, 202-246.
13. Goffner, D.; Campbell, M. M.; Campargue, C.; Clastre, M.; Borderies, G.; Boudet, A.; Boudet, A.-M. *Plant Physiol.* **1994**, *106*, 625-632.
14. Lacombe, E.; Hawkins, S.; van Doorsselaere, J.; Piquemal, J.; Goffner, D.; Poeydomenge, O.; Boudet, A.-M.; Grima-Pettenati, J. *Plant J.* **1997**, *11*, 425-441.
15. Feuillet, C.; Lauvergeat, V.; Deswarte, C.; Pilate, G.; Boudet, A.-M.; Grima-Pettenati, J. *Plant Mol. Biol.* **1995**, *27,* 651-667.
16. Hawkins, S.; Samaj, J.; Lauvergeat, V.; Boudet, A.; Grima-Pettenati, J. *Plant Physiol.* **1997**, *13*, 321-325.
17. Iiyama, K.; Wallis, A. F. A. *Wood Sci. Technol.* **1988**, *22*, 271-280.
18. Effland, M.J. *Tappi* **1977**, *60 No. 10*, 143-144.
19. Campbell, M. M.; Ellis, B. E. *Planta* **1992**, *186*, 409-417.
20. Knight, M. E.; Halpin, C.; Schuch, W. *Plant Mol. Biol.* **1992**, *19*, 793-801.
21. Grima-Pettenati, J.; Feuillet, C.; Goffner, D.; Borderies, G.; Boudet, A.-M. *Plant Mol. Biol.* **1993**, *21*, 1085-1095.
22. Galliano, H.; Cabané, M.; Eckerskorn, C.; Lottspeich, F.; Sandermann, H., Jr.; Ernst, D. *Plant Mol. Biol.* **1993**, *23*, 145-156.
23. MacKie, J. H.; Jouahari, R.; Douglas, K. T.; Goffner, D.; Feuillet, C.; Grima-Pettenati, J.; Boudet, A.-M.; Baltas, M.; Gorrichon, L. *Biochim. Biophys. Acta* **1993**, *1202*, 61-69.
24. Lauvergeat, V.; Kennedy, K.; Feuillet, C.; MacKie, J. H.; Gorrichon, L.; Boudet, A.-M.; Grima-Pettenati, J.; Douglas, K. T. *Biochemistry* **1995**, *34*, 12426-12434.
25. Goffner, D.; Joffroy, I.; Grima-Pettenati, J.; Halpin C.; Knight M. E.; Schuch, W.; Boudet, A.-M. *Planta* **1992**, *188*, 48-53.
26. Walter, M. H.; Schaaf, J.; Hess, D. *Acta Hort.* **1994**, *381*, 162-163.
27. Hibino, T.; Chen, J. Q.; Shibata, D.; Higuchi, T. *Plant Physiol.* **1994**, *104*, 305-306.
28. Baucher, M.; van Doorsselaere, J.; Gielen, J.; van Montagu, M.; Inzé, D.; Boerjan, W. *Plant Physiol.* **1995**, *107*, 285.
29. Esau, K. *Anatomy of Seed Plants*; Wiley: New York, NY, 1977.
30. Pickett-Heaps, G. D. *Protoplasma* **1968**, *65*, 181-205.
31. Halpin, C.; Knight, M. E.; Foxon, G. A.; Campbell, M. M.; Boudet, A.-M.; Schuch, W. *Plant J.* **1994**, *6*, 339-350.
32. Cherney, J. H.; Cherney, D. J. R.; Akin, D. E.; Axtell, J. D. *Adv. Agron* **1991**, *46*, 157-198.
33. Lapierre, C. In: *Forage Cell Wall Structure and Digestibility*; Jung, H.G., Buxton, D. R., Hatfield, R. D., Ralph, J., Eds.; ASA, CSSA and SSSA: Madison, 1993, pp 133-166.
34. Myton, K.; Stewart, D.; Yahiaoui, N.; McDougall, G.; Marque, C.; Boudet, A-M. *7th Cell Wall Meeting*: Santiago de Compostella, Spain, 1995, poster 106.
35. Tollier, M. T.; Chabbert, B.; Lapierre, C.; Monties, B.; Francesch, C.; Rolando, C.; Jouanin, L.; Pilate, G.; Cornu, D.; Baucher, M.; Inzé, D. In *Les Colloques Polyphenols '94;* INRA Ed.: Paris, 1994, *volume 69*, pp 339-340.

36. Higuchi, T.; Ito, T.; Umezawa, T.; Hibino, T.; Shibata, D. *J. Biotechnol.* **1994**, *37*, 151-158.
37. Stewart, D.; Yahiaoui, N.; McDougall, G. J.; Myton, K.; Marque, C.; Boudet, A.-M.; Haigh, J. *Planta* **1997**, *201*, 311-318.
38. Hibino, T.; Takabe, K.; Kawazu, T.; Shibata, D.; Higuchi, T. *Biosci. Biotech. Biochem.* **1995**, *59*, 929-931.
39. Boucher, M.; Chabbert, B.; Pilate, G.; Doorsselaere, J. V.; Tollier, M.-T.; Petit-Conil, M.; Cornu, D.; Monties, B.; van Montagu, M.; Inze, D.; Lise, J.; Wout, B. *Plant Physiology*, **1996**,*112*, 1479-1490.
40. Douglas, C.J. *Trends Plant Sci.* **1995**, *1 (6),* 171-178.
41. Campbell, M.M.; Sederoff, R.R. *Plant Physiol.* **1996**, *110*, 3-13.

Chapter 6

β-Glucosidases and Glucosyltransferases in Lignifying Tissues

D. Palitha Dharmawardhana and B. E. Ellis

Department of Plant Science, University of British Columbia, Vancouver, British Columbia V6T 1Z4, Canada

The biosynthesis of the lignin polymer is believed to rely upon the ability of plant cells to produce, transport, store and mobilize the precursor monolignols. These are thought to be stabilized by formation of 4-*O*-β-D-glucoside derivatives which are produced by the action of UDPG-utilizing glucosyltransferases, and hydrolyzed by monolignol-specific β-glucosidases. In conifers, the relevant enzymes have been at least partially characterized, and a cDNA encoding a pine coniferin β-glucosidase has recently been cloned from developing xylem. The properties of the pine enzyme and gene are described, as well as the possible implications that these properties may have for the organization of lignin metabolism within woody tissues.

The monolignol glucosides (*p*-hydroxycinnamyl alcohol glucoside, coniferin and syringin) have been found to accumulate in the cambial sap of all gymnosperms, and in some of the angiosperms, that have been examined (*1-3*). This 4-*O*-glycosylation of monolignols modifies their properties, making them more water soluble and much less reactive. The glucosides have been suggested to be the transport, or the storage, form of monolignols (*4*), but their exact role has yet to be established. Their importance to lignin biosynthesis is strongly suggested by the results of monolignol glucoside radiotracer experiments in both gymnosperms and angiosperms. The monolignol glucosides have been shown to be effectively incorporated into protolignin in the proper morphological location, and they were also found to be the most efficient precursors for specific labeling of lignin in the gymnosperms *Pinus* and *Ginkgo* (*5-7*), and in the angiosperms, magnolia, lilac, poplar and *Oryza* (*2, 5, 8*). These observations suggest that a glucosyltransferase/β-glucosidase system may operate in lignifying tissues for the interconversion of cinnamyl glucosides and monolignols. An UDP-glucose:coniferyl alcohol glucosyltransferase has been reported from spruce cambium (*9*) and other tissues (*10*). Enzymes capable of hydrolyzing coniferin have been reported from spruce seedlings (*11*) and cell suspension culture systems (*12, 13*). Until recently, the presence in plant xylem of β-glucosidases specific for monolignol glucosides had not been reported, even though xylem is the most active tissue for lignin biosynthesis in plants. Thus, the experimental evidence in support of the involvement of a glucosyltransferase/glucosidase system in the process of lignin formation in lignifying cells is very limited. In this article, we outline our current understanding of this portion of the lignin biosynthetic pathway, with particular emphasis on recent results concerning coniferin glucosidase.

Glucosyl Transferases

Enzyme activities capable of catalyzing the transfer of glucose to the phenolic hydroxyl groups of monolignol aglycones have only been examined in a few plant species. Interestingly, despite the rarity of monolignol glucoside accumulation outside the gymnosperm species, this activity was first reported (*14*) in an angiosperm (cell cultures of 'Paul's Scarlet' rose). A similar enzyme activity was detected in all angiosperm species examined, as well as in pteridophyte and bryophyte species, but the highest activity occurred in gymnosperms (*10*). The UDPG:coniferyl alcohol glucosyltransferase was partially purified both from rose cells and from *Forsythia* stem segments. In each case, it had a pH optimum of 7.5-8.0, a strict requirement for UDPG as a glucose donor, and the ability to glucosylate both coniferyl and sinapyl alcohols, with K_M values in the low micromolar range. The apparent molecular weight of the enzyme from both species was approximately 52 kDa, based on gel permeation chromatography. A detailed study of the enzyme recovered from *Picea abies* cambial sap yielded a homogeneous protein with properties similar to those of the angiosperm enzyme (pH optimum 7-8, molecular mass 50 kDa) (*9*). The conifer enzyme showed a strong preference for coniferyl alcohol as a substrate, relative to sinapyl alcohol. Coniferaldehyde was also an effective substrate.

A recent preliminary report described the analogous enzyme from lignifying tissue of *Pinus banksiana* and *P. strobus* (*15*). The pine enzyme had a pH optimum of 7.6, a pI of 4.9 and apparently displayed a strong substrate preference for coniferyl and sinapyl alcohols. The activity of the glucosyltransferase was developmentally regulated, and peaked during the period of maximum lignin deposition in pine branches.

These limited data suggest that the UDPG:coniferyl alcohol glucosyltransferase may be similar across the plant phyla, and show that the enzyme is amenable to full purification. It would clearly be desirable to obtain the polypeptide sequence and pursue gene isolation. The presence or absence of potential transport sequences, and comparisons with other glucosyltransferases that display different substrate specificities, would provide valuable clues concerning the subcellular location, active site and evolutionary origins of this lignification-associated enzyme.

β-Glucosidases

β-Glucosidases catalyze the hydrolysis of the glycosidic bond in aryl and alkyl β-glucosides, and in β-1,4-linked glucans. They are ubiquitous in bacteria, fungi, plants and animals. Plant β-glucosidases have been implicated in a multitude of metabolic and defense-related responses which range from the release of conjugated phytohormones (*16*) to protection against herbivores and pathogens by liberating antifeedants, cyanide and thiocyanates (*17, 18*). The diversity of glucosides and glucosidases that occur in plants led to an earlier perception that, in plants, glucosylation and de-glucosylation were relatively non-specific processes. The wide range of substrates that can be cleaved by crude glucosidase preparations such as 'almond β-glucosidase' has reinforced this view. With the development of higher resolution protein fractionation techniques, and careful attention to true substrate preferences, it has become clear that the hydrolysis of glucosides in plants is a process that can be, and generally is, remarkably specific (*19*). Our search for a glucosidase that might be responsible for liberating monolignols in conifer xylem therefore emphasized the use of coniferin, and its structurally-similar chromogenic analogue, VRA-G (5[4-(β-D-glucopyranosyloxy)-3-methoxyphenylmethylene]-2-thioxothiazolidine-4-one-3-ethanoic acid) as assay substrates.

β-Glucosidases in Differentiating Xylem of *Pinus contorta*. Anion exchange chromatography of total xylem protein extract, and assays for β-glucosidase activity using a range of synthetic substrates and coniferin, revealed three major populations of β-glucosidases (Figure 1). Only peak 3 had coniferin hydrolyzing activity, and it was also active toward the coniferin analogue, VRA-G. The β-glucosidases in peaks 1 and 2 were able to hydrolyze all the synthetic glucosides used, including VRA-G, but showed little activity toward coniferin. Based on substrate specificity and activity staining of non-denaturing gels, the β-glucosidases in peaks 2 and 3 were designated 4-methyl umbelliferyl-β-D-glucoside (MUG) β-glucosidase and 'coniferin β-glucosidase', respectively. Peak 1 appeared to be an aggregated or complexed form of the MUG β-glucosidase. The MUG β-glucosidase was able to hydrolyze a variety of synthetic aromatic glucosides efficiently but was relatively less effective toward coniferin and syringin (*20*). Its exact function has yet to be determined, although it could be involved in some aspect of cell wall polysaccharide biosynthesis or modification.

Coniferin β-Glucosidase from *P. contorta* Xylem. The coniferin β-glucosidase, when purified to apparent homogeneity, migrated as a single band on silver-stained non-denaturing gels. This protein band also demonstrated coniferin hydrolytic activity and in-gel VRA-G hydrolytic activity. The purified coniferin β-glucosidase had a high activity and affinity toward coniferin and syringin, compared to other substrates tested (Table I). The aglycone specificity demonstrated by this enzyme supports the view that plant β-glucosidases do exhibit high specificity toward their naturally occurring glycosidic substrates. In this sense, they are similar to other enzymes of intermediary metabolism which carry out a specific function in the metabolism of the cell. It is noteworthy that coniferin β-glucosidase did utilize both coniferin (K_M = 0.18 mM) and syringin (K_M = 0.29 mM) efficiently as substrates, which was also true for the β-glucosidase reported in spruce seedlings (*11*) and *Glycine max* suspension cultures (*21*). This pattern is analogous to the specificity of the angiosperm UDPG:coniferyl alcohol glucosyltransferase, which is unable to differentiate between coniferyl alcohol and sinapyl alcohol (*14*), although the *Picea* enzyme is more discriminating (*9*). Thus, it appears that the glucosidases, at least, are unlikely to play a direct role in determining the **type** of lignin (*i.e.* guaiacyl *vs.* syringyl) deposited in the wall.

The N-terminal amino acid sequence obtained from the purified coniferin β-glucosidase was used to design PCR primers that allowed amplification of a portion of the corresponding cDNA in a library constructed from differentiating pine xylem. This was then used to obtain the full sequence. The identity of the isolated cDNA was confirmed by its sequence identity to the N-terminal amino acid sequence of the purified protein, by its overall homology to other plant β-glucosidases, and by the production of an active glucosidase enzyme with appropriate substrate preferences when the cDNA was expressed in a heterologous system (*22*). Analysis of the amino acid sequence deduced from the cDNA, and comparison of the predicted physicochemical properties (M_r, pI) with those of the coniferin β-glucosidase purified from *P. contorta* xylem, showed close correspondence. Interestingly, a very different β-glucosidase capable of hydrolyzing coniferin has been isolated from cambial tissue of *P. banksiana* (*23*). This protein is reported to have a much larger molecular mass than the *P. contorta* coniferin glucosidase, and its N-terminal amino acid sequence shows no homology to known glucosidases.

Based on sequence similarity, glycosyl hydrolases have been classified into 35 distinct families (*24*). Coniferin β-glucosidase belongs to *family 1*, which includes β-glucosidases from diverse organisms. *Family 1* enzymes share common active site motifs and, probably, similar folding characteristics. Highly conserved Glu and Asp residues, which could also be tentatively identified in the coniferin β-glucosidase

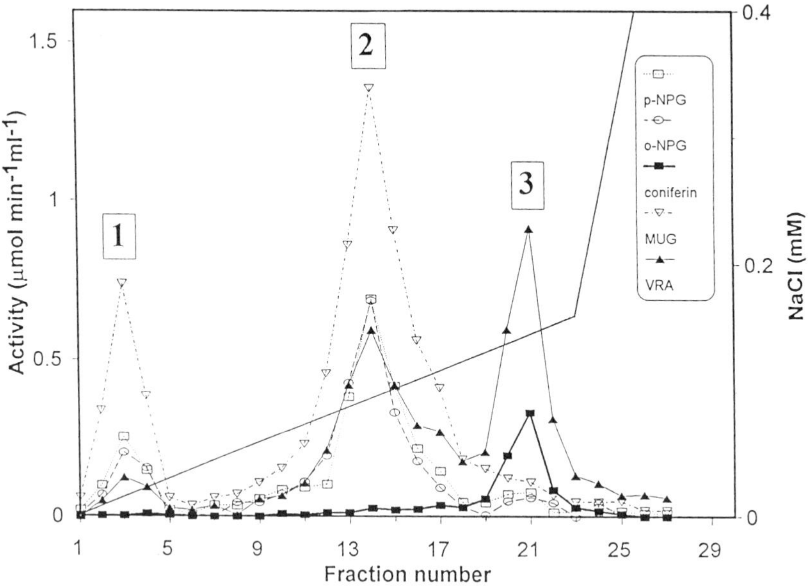

Figure 1. Fractionation of β-glucosidase activity from *P. contorta* xylem extracts by anion exchange column chromatography. Each fraction was assayed with five different β-glucosidase substrates; p-NPG (*p*-nitrophenyl-β-D-glucoside), o-NPG (*o*-nitrophenyl-β-D-glucoside), coniferin, MUG (4-methyl umbelliferyl-β-D-glucoside) and VRA-G. (*20*). Reproduced with permission from reference 20. Copyright 1995 American Society of Plant Physiologists.)

Table I. Substrate Specificity of Purified Coniferin β-Glucosidase from *Pinus contorta*.

Substrate	Relative activity (coniferin =100)	K_M (mM)
coniferin	100	0.18
syringin	51	0.29
4-methyl umbelliferyl-β-glucoside	16	2.30
2-nitrophenyl-β-glucoside	57	-
4-nitrophenyl-β-glucoside	26	1.90
VRA-G	45	-
salicin	10	-
4-methyl umbelliferyl-α-glucoside	1	-
2-nitrophenyl-β-galactoside	12	-
4-nitrophenyl-β-cellobioside	ND[a]	-

[a] Not detected.

sequence, have been shown to play the roles of acid catalyst and stabilizing nucleophile during cleavage of the glycosidic bond (*25, 26*). The coniferin β-glucosidase amino acid sequence also includes an N-terminal 23 amino acid sequence that has characteristics of eukaryotic secretory signal peptides, suggesting an extracellular location for the enzyme. This is analogous to the signal peptide found in cell wall-located pro-peroxidases (*27*) and pro-cyanogenic β-glucosidases (*28*). The presence of this signal peptide supports the earlier hypothesis that monolignol glucosides may be transported across the plasma membrane and subsequently hydrolyzed within the cell wall before dehydrogenative polymerization (*1*). However, at this stage we cannot exclude the possibility that *P. contorta* coniferin β-glucosidase is targeted to another membrane-bound compartment within the cell.

By histochemical staining, the β-glucosidases and peroxidases which are thought to be involved in the final dehydrogenative polymerization of monolignols into lignin were found to be restricted to the zone of differentiating and lignifying xylem within the actively growing pine stem (*22*). Even in these differentiating xylem cells, the enzymes were expressed only during the developmental period that coincides with deposition of the secondary wall and initiation of lignification (*20*). This same region of the xylem, a few cell layers beneath the cambium, was also found to contain the highest concentration of coniferin during spring growth (*3*). The spatially and temporally coordinated occurrence of β-glucosidases, peroxidases and coniferin in xylem cells provides support for a specific role and participation of a cinnamyl glucoside/β-glucosidase system in *Pinus* stem lignification.

Unlike gymnosperms, angiosperms (except species belonging to Magnoliaceae and Oleaceae) do not appear to accumulate monolignol glucosides in lignifying stems. This has been viewed as evidence against an essential role for a cinnamyl glucoside/β-glucosidase system in angiosperm lignification (*2*). Although our studies cannot directly counter the above view, we have used substrate-specific histochemistry and immunolocalization to show that β-glucosidases and other lignification-related enzymes are localized in actively growing poplar stems in a spatial and temporal pattern that is very similar to that seen in *Pinus* stems (*22*). This, together with the reported efficient incorporation of radiolabeled monolignol glucosides into lignin in poplar (*8*) is consistent with the general participation of β-glucosidases in both gymnosperm and angiosperm xylem lignification. We have detected an analogous temporal expression pattern of β-glucosidases and other lignification-related enzymes in the phloem fibers of poplar, which suggests that this model can be extended to most, if not all, types of lignifying cells in plants.

Mode of Coniferin β-Glucosidase Action. Although the end result of cell wall lignification has been well documented, the exact mechanism of cell wall component deposition and assembly still remains unresolved. Cellulose microfibrils are thought to be synthesized and assembled at the apoplastic face of the plasma membrane *via* multisubunit cellulose synthase complexes embedded in the membrane (*29*). In algae and plants, these complexes have been proposed to move two-dimensionally within the membrane, and thus determine the pattern of cellulose deposition. Microtubules of the cell cytoskeleton are thought to be involved in directing the movement of the complexes, possibly through direct or indirect interactions (*29, 30*). Other cell wall polysaccharides, however, are synthesized in the endomembrane system and are transported to the plasma membrane by vesicles derived from the Golgi system and rough endoplasmic reticulum (*31*). The fusion of these vesicles with the plasma membrane is again thought to be directed by the microtubules, whose orientation and localization often reflects the orientation of secondary cell wall patterning and cellulose microfibrils (*32, 33*).

Lignin accumulates within the cell wall, while most of the enzymes of phenylpropanoid metabolism appear to be cytosolic. Lignin precursors that are synthesized by these cytosolic enzymes will need to be transported from the cytosol to the apoplast (or its topological equivalent) for dehydrogenative polymerization, a reaction which must act on the free monolignols, since their glucosides are not suitable substrates. Trans-membrane movement of the lignin monomers could direct them into the cell wall proper, or perhaps into specialized vesicles. As in the transport of vesicles carrying cell wall polysaccharides, such 'monolignol vesicles' could be guided by the microtubules to regions of the cell active in secondary cell wall deposition and lignification. In this fashion, lignification would be closely linked to the previously deposited secondary cell wall patterning.

While most of the enzymes of phenylpropanoid metabolism appear to be 'soluble', the sub-cellular location of these reaction sequences is still uncertain. However, due to the membrane association of cytochrome P-450-dependent hydroxylases, there is reason to believe that most of the core reactions of phenylpropanoid metabolism would be at least loosely associated at the surface of membranes such as endoplasmic reticulum. Cinnamate 4-hydroxylase and PAL have been found to reside in the microsomal fraction in cell fractionation studies (*34*). Such behavior could also facilitate the formation of multienzyme complexes that would aid in metabolic channeling of substrates and products in the pathway, as suggested for the phenylpropanoid pathway by Hrazdina and Jensen (*35*) and Hrazdina and Wagner (*34*). A large proportion of the monolignols generated by the phenylpropanoid pathway, at least in gymnosperms, is likely to be glycosylated by UDPG:coniferyl alcohol glucosyltransferase either before, during or following transport of the monolignol into an extra-cytosolic compartment, such as membrane-bound vesicles. These monolignol-carrying vesicles could either be targeted to the large central vacuole to create a storage pool of monolignol glucosides, or be directed to the plasma membrane to provide substrates for lignification. Ultrastructural investigations using radiolabeled phenylalanine and cinnamic acid have revealed abundant vesicle traffic within lignifying tracheids, where the vesicles appeared to be fusing with the plasma membrane (*36*). Considering the high concentration of coniferin extractable from conifer xylem, a considerable amount of coniferin should be accumulating within these cells. Experiments conducted with protoplasts prepared from differentiating pine xylem have shown that most of this coniferin is located within the protoplasts, probably in the central vacuole (*37*). Angiosperms that do not show accumulation of monolignol glucosides within their developing xylem may have developed an alternative system of transport and management of monolignols that avoids large-scale accumulation of the glucosides within lignifying cells.

Since coniferin β-glucosidase is a secretory glycoprotein, synthesis on ribosomes associated with the endoplasmic reticulum may be followed by glycosylation of the polypeptide backbone, within the Golgi system. Enzyme-containing Golgi vesicles could then directly fuse with the plasma membrane to release coniferin glucosidase into the cell wall for monolignol glucoside hydrolysis. Alternatively, enzyme-containing vesicles could fuse with monolignol glucoside-loaded vesicles before the latter reach the cell wall. Takabe *et al.* (*38*) observed an increase in the proportion and size of smooth endoplasmic reticulum during active lignification and S3 layer deposition in conifer tracheids. The number of Golgi complexes and rough endoplasmic reticulum, on the other hand, had peaked during primary wall deposition and gradually decreased during secondary wall lignification. This indirectly suggests that smooth endoplasmic reticulum is associated with lignification and thus could be involved in the synthesis, storage and vesicular transport of monolignols or their glucosides to the plasma membrane.

The model discussed above would imply that coniferin β-glucosidase is ultimately deposited and active in the cell wall of lignifying cells. This would not be an unprecedented phenomenon, since the occurrence of other β-glucosidases in the apoplastic space is well documented (*30-41*). Such an arrangement would establish a spatial compartmentation of β-glucosidase in the apoplast, keeping it segregated from the substrate stored in the vacuole. The validity of this model could be tested by examining the subcellular localization of coniferin β-glucosidase using immunocytochemical techniques.

Future Prospects

The availability of a full-length coniferin β-glucosidase cDNA will permit not only opportunities for probing the structure and function of the enzyme through site-directed mutagenesis, but also for down-regulating the expression of coniferin β-glucosidase in order to definitively establish its importance. Such studies, which are currently underway, can be expected to provide valuable new insights into the regulation and metabolic control of these ultimate stages of lignin biosynthesis. The complete coniferin β-glucosidase gene sequence is also a good potential candidate for yielding a novel xylem-specific promoter whose activity is strongly linked to lignin formation. Together, these resources should provide further impetus to the quest for directed modification of lignin in plants by genetic engineering, a goal which is stimulating considerable research interest (*42*).

LITERATURE CITED

1. Freudenberg, K.; Harkin, J. M. *Phytochemistry* **1963**, *2*, 189-193.
2. Terazawa, M.; Okuyama, H.; Miyake, M. *Mokuzai Gakkaishi* **1984**, *30*, 322-328.
3. Savidge, R. A. *Can. J. Bot.* **1988**, *66*, 2009-2012.
4. Whetten, R. W.; Sederoff, R. R. *Plant Cell* **1995**, *7*, 1001-1013.
5. Terashima, N.; Fukushima, K.; Tsuchiya, S. *Wood Sci. Technol.* **1986**, *6*, 495-504.
6. Terashima, N.; Fukushima, K. *J. Wood Sci. Technol.* **1988**, *22*, 259-270.
7. Fukushima, K.; Terashima, N. *Holzforschung* **1991**, *45*, 87-89.
8. Fukushima, K.; Terashima, N. *J. Wood Chem. Technol.* **1990**, *10*, 413-433.
9. Schmid, G.; Grisebach, H. *Eur. J. Biochem.* **1982**, *123*, 363-370.
10. Ibrahim, R. K. *Z. Pflanzenphysiol.* **1977**, *85*, 253-262.
11. Marcinowski, S.; Grisebach, H. *Eur. J. Biochem.* **1978**, *87*, 37-44.
12. Hösel, W.; Fiedler-Preiss, A.; Boigmann, E. *Plant Cell Orgn. Cult.* **1982**, *1*, 137-148.
13. Campbell, M. M.; Ellis, B. E. *Planta* **1992**, *180*, 409-417.
14. Ibrahim, R. K.; Grisebach, H. *Arch. Biochem. Biophys.* **1976**, *176*, 700-708.
15. Förster, H.; Savidge, R. A. *Proc. 22nd Annu. Mtg. Plant Growth Reg. Soc. Am.* **1995**, 402-407.
16. Campos, N.; Laszlo, B.; Feldwisch, J.; Schell, J.; Palme, K. *Plant J.* **1992**, *2*, 675-684.
17. Bones, A.; Slupphaug, G. *J. Plant Physiol.* **1989**, *134*, 722-729.
18. Cuevas, L.; Niemeyer, H. M.; Jonsson, L. M. V. *Phytochemistry* **1992**, *31*, 2609-2612.
19. Hösel, W.; Conn, E. E. *Trends Biochem. Sci.* **1982**, *7*, 219-221.
20. Dharmawardhana, D. P.; Ellis, B. E.; Carlson, J. E. *Plant Physiol.* **1995**, *107*, 331-339.
21. Hösel, W.; Todenhagen, R. *Phytochemistry* **1980**, *19*, 1349-1353.
22. Dharmawardhana, D. P. *Ph.D. Dissertation*; University of British Columbia: Vancouver, 1996.
23. Leihos, V.; Udagama-Randeniya, P. V.; Savidge, R. A. *Phytochemistry* **1994**, *37*, 311-315.
24. Henrissat, B. *Biochem. J.* **1991**, *280*, 309-316.
25. Trimbur, D. E.; Warren, R. A. J.; Withers, S. G. *J. Biol. Chem.* **1992**, *267*, 10248-10251.
26. Withers, S. G.; Rupitz, K.; Trimbur, D.; Warren, R. A. J. *Biochemistry* **1992**, 31, 9979-9985.
27. Lagrimini, L. M.; Burkhart, W.; Moyer, M.; Rothstein, S. *Proc. Natl. Acad. Sci. USA* **1987**, *84*, 7542-7546.
28. Hughes, M. A.; Brown, K.; Poncoro, A.; Murrey, B. S.; Oxtoby, E.; Hughes, J. *Arch. Biochem. Biophys.* **1992**, *295*, 273-279.
29. Delmer, D. P.; Amor, Y. *Plant Cell* **1995**, *7*, 987-1000.

30. Giddings, T. H.; Staehelin, L. A. In *The Cytoskeletal Basis of Plant Growth and Development*; Lloyd, C. W., Ed.; Academic Press: London, 1990; pp 85-99.
31. Reiter, W. D. In *Arabidopsis*; Meyerowitz, E., Somerville, C. R., Eds.; Cold Spring Laboratory: Cold Spring Harbor, NY, 1994; pp 955-988.
32. Heath, I. B.; Seagull, R. W. In *The Cytoskeletal Basis of Plant Growth and Development*; Lloyd, C. W., Ed.; Academic Press: London, 1982.
33. Lloyd, C. W. *The Cytoskeletal Basis of Plant Growth and Form*; Academic Press: London, 1991.
34. Hrazdina, G.; Wagner, G. J. *Arch. Biochem. Biophys.* **1985**, *237*, 88-100.
35. Hrazdina, G.; Jensen, R. A. *Annu. Rev. Plant Physiol. Plant Mol. Biol.* **1992**, *43*, 241-267.
36. Pickett-Heaps, J. D. *Protoplasma* **1968**, *65*, 181-190.
37. Leinhos, V.; Savidge, R. A. *Can. J. For. Res.* **1993**, *23*, 343-348.
38. Takabe, K.; Fukazawa, K.; Harada, H. *ACS Symp. Ser.* **1989**, *399*, 47-66.
39. Heyn, A. N. *Arch. Biochem. Biophys.* **1969**, *132*, 442-449.
40. Kakes, P. *Planta* **1985**, *166*, 156-160.
41 Frehner, M.; Conn, E. E. *Plant Physiol.* **1987**, *84*, 1296-1300.
42. Campbell, M. M.; Sederoff, R. R. *Plant Physiol.* **1996**, *110*, 3-13.

Chapter 7

Specific Compartmentalization of Peroxidase Isoenzymes in Relation to Lignin Biosynthesis in the Plant Cell

A. Ros Barceló, M. Morales, and M. A. Pedreño

Department of Plant Biology, University of Murcia, E-30100 Murcia, Spain

The compartmentalization of peroxidase isoenzymes related to lignin biosynthesis was studied in the cell walls and the cell wall-free spaces of several woody and non-woody plant species. The results illustrate a specific compartmentalization of peroxidase isoenzymes in cell walls compared with that found in vacuoles. Cell wall peroxidase isoenzymes may be classified into three main groups: one with acidic pI (APrx), and two with a basic pI (BPrx LpI and BPrx HpI). While basic peroxidases are constitutively expressed, acidic peroxidases are normally developmentally regulated. This developmental regulation and the greater efficacy in the oxidation of coniferyl alcohol shown by acidic peroxidases suggest that this isoenzyme group plays a key role in lignin biosynthesis, although some contribution to this process by basic peroxidases cannot be ruled out.

Peroxidase (EC 1.11.1.7) is the enzyme responsible for the linking and cross-linking of monolignols during the biosynthesis of lignins in the plant cell wall. As may be expected from its specific role in lignin biosynthesis, peroxidase is mainly located in the cell wall, although other subcellular localizations have been reported. Over the last few years, considerable information has been accumulated about the subcellular localization of peroxidase isoenzymes (*1*). However, it remains unclear whether the findings concerning isoenzyme localization in a few plants can be generalized and rationalized. Combined histochemical, cytochemical and biochemical studies are necessary to rationalize some concepts which are only just beginning to be understood.

To cast light on this question, we have studied peroxidase isoenzyme localization in lupin (*Lupinus albus*) (*2-4*), grapevine (*Vitis vinifera*) (*5, 6*), pepper (*Capsicum annuum*) (*7, 8*), lettuce (*Lactuca sativa*) (*9*), catharanthus (*Catharanthus roseus*) (*10*), oats (*Avena sativa*) (Rojas, M.C.; Ros Barceló, A., unpublished data) and aleppo pine (*Pinus halepensis*) (Ros Barceló, A., unpublished data). Electron microscope cytochemistry, vacuum infiltration and subcellular fractionation, together with protoplast and vacuole isolation techniques, show that in all the studied plant materials both basic (BPrx, pI > 7.0) and acidic (APrx, pI < 7.0) peroxidase isoenzymes are located in the cell wall free spaces, probably in equilibrium with those bound to the cell walls, while basic peroxidase isoenzymes of high pI (BPrx HpI, pI > 9.2) are the only isoenzymes located in vacuoles (Table I).

Table I. Compartmentalization of Peroxidase Isoenzyme Groups in the Plant Cell

	APrx	BPrx LpI	BPrx HpI	reference
Lupinus albus	cw	nd	cw, v	*2,3,4*
Vitis vinifera	cw	cw	cw, v	*5,6*
Capsicum annuum	cw	cw	cw, v	*7,8*
Lactuca sativa	a	a	cw, v	*9*
Catharanthus roseus	a	a	cw, v	*10*
Avena sativa	cw	cw	cw, v?	np
Pinus halepensis	cw	cw	cw, v?	np

cw: cell wall; v: vacuole; nd: not determined; a: absent; np: not published; v?: vacuolar localization not studied.

The BPrx HpI Isoenzyme Group: a Distinctive Peroxidase Isoenzyme Group Located in Vacuoles and Cell Walls

The spent medium in which a plant cell culture has grown may be considered to be a large free intercellular space forming a continuum with the plant cell wall (*11*). For this reason, suspension cultured cells are useful for studying the specific compartmentalization of peroxidase isoenzymes in the cell wall once the cell wall free space that constitutes the spent medium has been isolated by non-disruptive techniques. In this simple experimental system, peroxidase isoenzymes isolated from grapevine cell cultures may be classified into three main groups: APrx (acidic peroxidases), BPrx LpI (basic peroxidases of low pI, 7.0 < pI < 9.2), and BPrx HpI (basic peroxidases of high pI, pI > 9.2).

In grapevine cell cultures, APrx and BPrx LpI are largely located in the cell walls, as can be expected from the fact that these isoenzymes are present exclusively in the spent medium in which the cells were cultured (*6, 12*), while BPrx HpI are located in both cell walls (*5, 6, 12*) and vacuoles (*5, 6, 13*). In fact, through protoplast and vacuole isolation techniques, it was found that the grapevine peroxidase isoenzyme group BPrx HpI was the only group present in protoplast and vacuole preparations. Taking α-mannosidase as a vacuolar marker, 87% of the protoplast BPrx HpI was located in the vacuolar fraction (Table II). Similarly, when using grapevine anthocyanins as vacuolar markers, the yield of vacuolar BPrx HpI was found to be 110% of the starting protoplast fraction (Table II). These results suggest that almost all, if not all, the protoplast BPrx HpI was located in the vacuolar fractions.

Table II. Comparison between Compartmentalization of the BPrx HpI Isoenzyme Group in Vacuoles and Protoplasts

	protoplast	vacuolar marker	vacuoles	reference
Lupinus albus	-	α-mannosidase	93 %	*3*
Vitis vinifera	100 %	α-mannosidase	87 %	*6*
	100 %	anthocyanins	110 %	*6*
Capsicum annuum	100 %	α-mannosidase	164 %	*7*
Lactuca sativa	100 %	α-mannosidase	28 %	*9*

These results confirm those previously obtained using *Lupinus albus* hypocotyls (*2-4*). In this case, the cell wall localization of peroxidase isoenzymes was studied by vacuum infiltration of tissue sections with buffers, while the vacuolar localization was studied by purification of vacuolar membranes through 20-40% sucrose gradients. Using this experimental approach, only the peroxidase isoenzyme group BPrx HpI was located in cell walls and vacuole fractions (Figure 1). As was the case for the grapevine peroxidase isoenzyme group BPrx HpI, lupin peroxidase isoenzyme group BPrx HpI was recovered with a yield of 93% in the vacuole fraction (Table II). Similar results (Table II) were found in *Capsicum annuum* (*7, 8*), lettuce (*Lactuca sativa*) (*9*) and catharanthus (*Catharanthus roseus*) (*10*).

Electron microscope cytochemical studies, using 3,3′-diaminobenzidine as a hydrogen donor, revealed that, in *Lupinus albus*, the vacuolar peroxidase activity, *i.e.* the peroxidase isoenzyme group (BPrx HpI), was mainly located at the internal face of the tonoplast membrane (*3*), a localization pattern similar to that found in suspension cultured grapevine cells (*5, 13*) and *Catharanthus* leaves (*14*).

Molecular Heterogeneity of the Peroxidase Isoenzyme Group BPrx HpI

The fact that the peroxidase isoenzyme group BPrx HpI is located in both the cell walls and vacuoles (Table I) suggests that this isoenzyme group follows the two secretion routes found in plant cells: (1) that where the subcellular target is the cell wall and (2) that where the subcellular target is the vacuole. The double compartmentalization of this peroxidase isoenzyme group in the cell wall and vacuoles is not surprising since similar results have been obtained with their homologous isoenzymes in other plant species, such as *Nicotiana tabacum* (*15*) and *Lupinus polyphyllus* (*16*).

The existence of two subcellular target compartments for this peroxidase isoenzyme group in plant cells raises a question about the molecular heterogeneity of this, apparently single, hemoprotein isoenzyme. The molecular heterogeneity of peroxidase isoenzymes is easily studied by non-equilibrium IEF (NEIEF), a tool which makes it possible to detect isomers differing slightly in either molecular weight or isoelectric point.

Since peroxidases are monomeric glycoproteins, this heterogeneity frequently arises from the different patterns of glycosylation, a feature common to all basic peroxidases (*17, 18*). On this basis, we may suspect that the peroxidase isoenzyme group BPrx HpI will be composed of at least two isomers, probably differing in their glycosylation patterns, as would be expected for an isoenzyme group that is secreted simultaneously into the cell wall and the vacuole.

As was anticipated above, NEIEF of the *Lupinus albus* peroxidase isoenzyme group BPrx HpI reveals the presence of two isoenzymes (de Pinto, M.C.; Ros Barceló,

A., *J. Plant Physiol.*, in press), previously known as B_3 and B_4 peroxidase isoenzymes (*3, 4*). A similar heterogeneity (*i.e.* the presence of at least two isoenzymes) for this peroxidase isoenzyme group was detected in lettuce (*Lactuca sativa*) (*19*), *Catharanthus roseus* (*10*), and strawberries (*Fragaria ananassa*) (López-Serrano, M.; Ros Barceló, A., unpublished data). Surprisingly, the two isozymes were simultaneously located in the cell walls and vacuoles in the case of *Lupinus albus* hypocotyls (*3, 4*). Target sequences (signal peptides) on the primary structure of these isoenzymes remain to be elucidated.

Physical State of Plant Peroxidase Isoenzymes in the Cell Wall

The presence of the complete set of peroxidase isoenzymes in the spent medium of cultured cells suggests that peroxidase isoenzymes move freely through the cell wall barrier, the cell wall network having pores of sufficient size to permit their free diffusion from the plasma membrane to the periplasmic free space. This is not surprising since cell wall pores admit the free diffusion of proteins as large as 60 kDa (*20*), and plant peroxidase isoenzymes have an average molecular weight of 43 kDa. Likewise, this suggests that most of the plant peroxidase isoenzymes located in the cell walls are found in the cell wall free spaces as soluble forms.

This does not preclude, however, the possibility that peroxidases may also be bound to the cell walls, linked to the polygalacturonate anionic groups of pectins particularly (*21*). In order to discriminate which peroxidase isoenzymes are bound to cell walls and which are free in the cell wall free spaces, vacuum infiltration of intact plant tissues with buffers is a useful tool. When this technique was used with oat coleoptiles, it was found that both the APrx and BPrx LpI isoenzyme groups were easily extracted from the cell wall free spaces with buffers of low ionic strength (Figure 2, lane a), suggesting that they occur unbound in the cell wall free spaces, while the BPrx HpI isoenzyme group was only extracted when $CaCl_2$ was added to the buffer (Figure 2, lane c). This is due to the fact that calcium competes with peroxidases for the anionic groups of the plant cell walls (*21*), releasing peroxidases from the bound to the soluble state.

Peroxidases from plant cell walls may be classified, according to the dependence on pH of their binding to cell walls, as being A-, N- or B-type. Type A peroxidases (generally those that are acidic in nature and therefore belong to the APrx isoenzyme group) bind to cell walls only at acidic pHs (*21*), precisely at those pH values where they are charged positively (pH values lower than their pI). Type B peroxidases (those that are basic in nature and generally belong to the BPrx HpI isoenzyme group) bind to cell walls within the entire range of physiological pH values (*4*), since they are charged positively throughout this pH range. Type N peroxidases (generally belonging to the BPrx LpI isoenzyme group) do not bind to cell walls at any pH value. Since it is known that the binding of peroxidases to cell walls is typically electrostatic in nature (*21*), N-type peroxidases must experience some steric hindrance (probably caused by the polysaccharide chains on the hemoprotein surface) that prevents binding to the cell wall surface.

Phenol Oxidase (Laccase-like) Activities of the Plant Peroxidase Isoenzyme Group BPrx HpI

It has long been known that plant peroxidases can oxidize some phenols in the absence of exogenous H_2O_2. This is due to the ability of plant peroxidases to show an 'oxidase-like activity', in which the peroxidase intermediate compound III appears to play a key role in the catalytic activity of the enzyme (*22, 23*). Compound III is formed in the presence of a suitable reducing agent (RO^-) according to the reaction scheme:

$$\text{peroxidase}(Fe^{+3}) + RO^- \rightarrow \text{peroxidase}(Fe^{+2}) + RO^\bullet \quad (1)$$

$$\text{peroxidase}(Fe^{+2}) + O_2 \rightarrow \text{compound III} \quad (2)$$

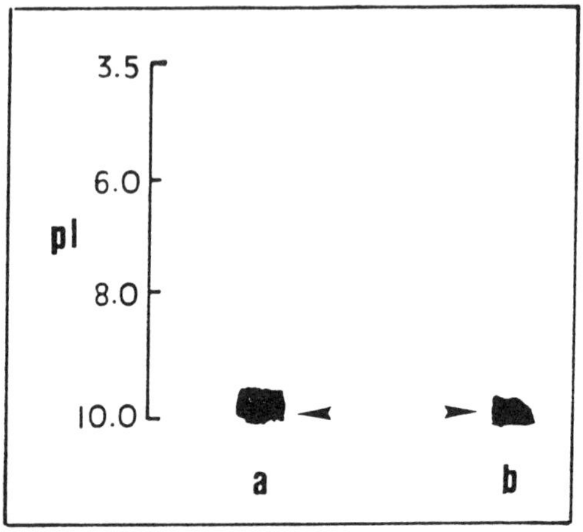

Figure 1. Isoelectric focusing in 3.5-10 pH gradients of cell wall (lane a) and vacuolar (lane b) lupin hypocotyl peroxidase isoenzymes. Peroxidase isoenzymes were stained with 1.0 mM 4-methoxy-alpha-naphthol in the presence of 0.5 mM H_2O_2. Arrowheads indicate basic peroxidase isoenzymes with high isoelectric point (BPrx HpI).

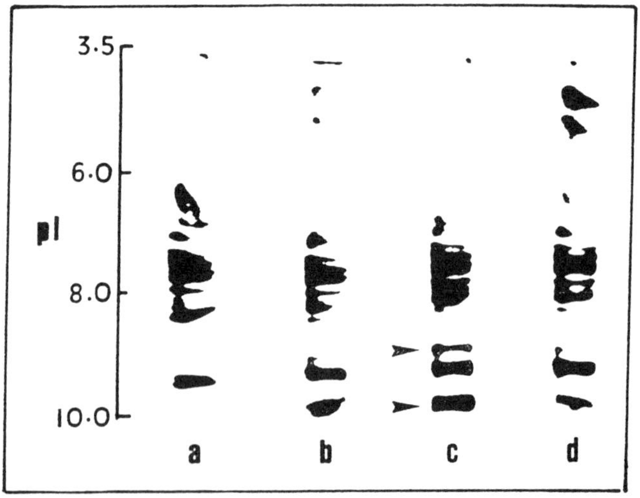

Figure 2. Isoelectric focusing in 3.5-10 pH gradients of peroxidase isoenzymes extracted from oat coleoptile cell walls by vacuum infiltration with 10 mM Tris-HCl buffer (pH 7.5), in the absence (lane a) and presence (lane c) of 25 mM $CaCl_2$. Peroxidase isoenzyme pattern in the tissues after vacuum infiltration with 10 mM Tris-HCl buffer (pH 7.5), in the absence (lane b) and presence (lane d) of 25 mM $CaCl_2$. Peroxidase isoenzymes were stained with 1.0 mM 4-methoxy-alpha-naphthol in the presence of 0.5 mM H_2O_2. Arrowheads indicate basic peroxidase isoenzymes with high isoelectric point (BPrx HpI).

Thus, in the presence of O_2 and a suitable reducing agent for Fe^{+3} (RO^-), and in the absence of hydrogen peroxide, the catalytic cycle of peroxidase is drawn towards compound III, which is generally considered to be a resonance hybrid between Fe^{+2}-O_2 and Fe^{+3}-O_2^-. The importance of compound III is that it is broken down (*22*) in the presence of a phenol (PhOH) giving:

$$\text{compound III} + \text{PhOH} \rightarrow \text{peroxidase}(Fe^{+3}) + \text{PhO}^{\bullet} + \text{H}\,O_2^- \qquad (3)$$

and thus generating the hydrogen peroxide which later may be used in the 'normal' catalytic cycle of the enzyme; the latter, of course, passes through compound I and compound II intermediates (*21*). Consequently, if a particular peroxidase isoenzyme can be reduced by a phenol (PhOH), the oxidation of the phenol by this peroxidase isoenzyme would be expected to take place with concomitant reduction of O_2.

Cell wall localized peroxidase isoenzymes may show phenol oxidase activity (*24, 25*) and, due to this behavior, they have sometimes been confused with laccase isoenzymes. The phenol oxidase activity of certain cell wall peroxidase isoenzymes is inhibited by the addition of catalase and may easily be monitored by vacuum infiltration followed by staining with a phenol in the absence of H_2O_2, once these isoenzymes have been separated, *e.g.* by isoelectric focusing.

When this was done with cell wall peroxidase isoenzymes extracted by vacuum infiltration of stems and needles (Figure 3) from aleppo pine (*P. halepensis*), it was found that only the BPrx HpI isoenzyme group showed phenol oxidase (*laccase*-like) activity. These results are not surprising since Chabanet *et al.* (*25*) found similar results with cationic peroxidase isoenzymes from *Vigna radiata* (mung bean) hypocotyls. Furthermore, this differential phenol oxidase (*laccase*-like) activity of the peroxidase isoenzyme group BPrx HpI is to be expected since it is generally accepted that basic peroxidases are potentially more easily reduced (see equation 1) than acidic peroxidases (*26*).

These results point to the special catalytic properties of the BPrx HpI isoenzyme group, and its ability to show phenol-oxidase (*laccase*-like) activity. Despite this special phenol-oxidizing behavior, when these peroxidase isoenzymes were purified from *C. roseus* leaves (Sottomayor, M.; López Serrano, M.; Di Cosmo, F.; Salema, R.; Ros Barceló, A., unpublished data), they exhibited a Soret band at 404 nm, and two visible bands at 501 nm and 633 nm, typical of high-spin ferric heme peroxidases.

Kinetic Behavior of the BPrx HpI Isoenzyme Group during the Oxidation of Coniferyl Alcohol

Because the BPrx HpI isoenzyme group is located in the cell walls and vacuoles (Table I), a study of their ability to oxidize coniferyl alcohol appears to be of special significance.

The fact that the oxidation/reduction potential of coniferyl alcohol is as high as 420 mV at pH 6.0 (*27*) suggests that it is incapable of reducing peroxidase according to equation 1. For this reason, the oxidation of coniferyl alcohol by peroxidase may only be expected to occur, in the absence of any other reducing agent, through the peroxidative cycle. In this respect, Hayashi and Yamazaki (*28*) determined redox potentials of about 950 mV for compound I/compound II and compound II/ferric couples at slightly acidic pH values, which suggests that, although coniferyl alcohol appears to be incapable of reducing ferriperoxidase to ferroperoxidase, it is a good substrate for both compound I and compound II reduction.

Oxidation of coniferyl alcohol by the basic peroxidase isoenzyme group BPrx HpI from grapevines follows Michaelis-Menten type kinetics with inhibition by substrate at higher concentrations (Figure 4). To calculate the microscopic rate coefficients (k_1 and k_3) for the oxidation of coniferyl alcohol by this basic peroxidase isoenzyme group, the rates of oxidation of coniferyl alcohol were treated, at each concentration of coniferyl alcohol and H_2O_2, according to the generally accepted mechanism for the peroxidase cycle (*29*):

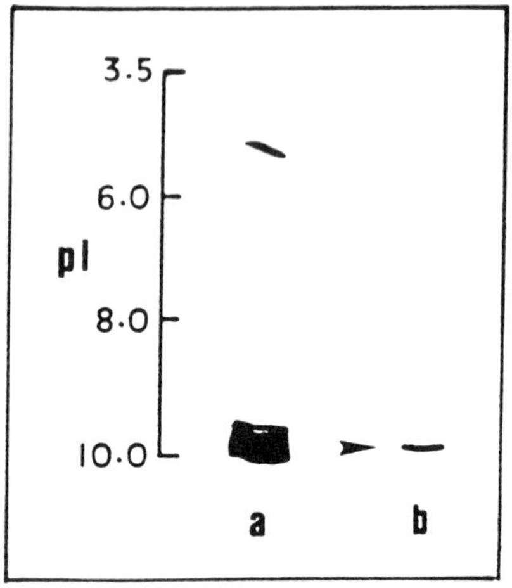

Figure 3. Isoelectric focusing in 3.5-10 pH gradients of peroxidase isoenzymes extracted from aleppo pine needle cell walls by vacuum infiltration with 0.1 M acetate buffer (pH 6.0) in the presence of 50 mM $CaCl_2$, stained with 1.0 mM 4-methoxy-alpha-naphthol in the presence (lane a) and absence (lane b) of 0.5 mM H_2O_2. Arrowheads indicate basic peroxidase isoenzymes with high isoelectric point (BPrx HpI) showing laccase-like activity.

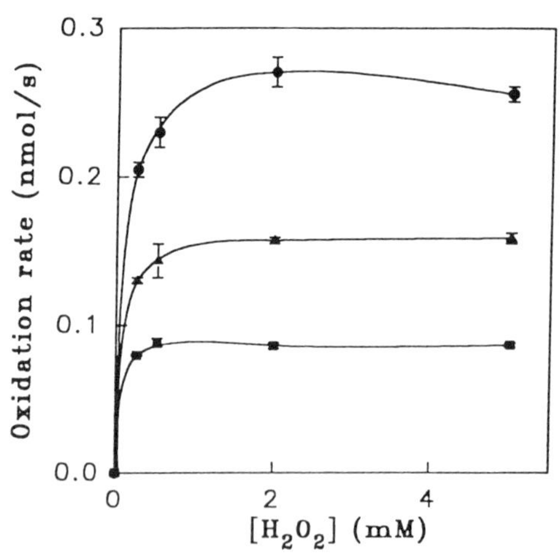

Figure 4. Dependence of the oxidation rate of coniferyl alcohol catalyzed by the basic peroxidase isoenzyme B_5 of grapevines (an isoenzyme belonging to the BPrx HpI isoenzyme group) on H_2O_2 concentration in a reaction medium containing 0.1 M Tris-acetate buffer (pH 5.0), [25 μM (■), 50 μM (▲) and 100 μM (●)] coniferyl alcohol and enzyme. Bars represent standard errors (n = 3).

$$P + H_2O_2 \xrightarrow{k_1} \text{compound I} + H_2O \tag{4}$$

$$\text{compound I} + \text{ArOH} \xrightarrow{k_2} \text{compound II} + \text{ArO}^{\bullet} \tag{5}$$

$$\text{compound II} + \text{ArOH} \xrightarrow{k_3} P + \text{ArO}^{\bullet} \tag{6}$$

where ArOH is coniferyl alcohol and compound I and compound II are the key intermediates in the peroxidase cycle.

Assuming that $k_2 >> k_3$, as is the case for most peroxidases (*29*), the steady-state rate equation may be written as:

$$v = \frac{2[E]k_3[\text{ArOH}][H_2O_2]}{(k_3/k_1)[\text{ArOH}] + [H_2O_2]} \tag{7}$$

from which the dependence of v on $[H_2O_2]$ may be written as:

$$v = \frac{A[H_2O_2]}{B + [H_2O_2]} \tag{8}$$

where A = 2 [E] k_3 [ArOH] and B = (k_3/k_1) [ArOH].

Double-reciprocal plots ($1/v$ *vs.* $1/[H_2O_2]$) (Figure 5) of the data shown in Figure 4 allow us to calculate A and B values for each coniferyl alcohol concentration. Plotting A *versus* B values gives a straight line (A = 2 [E] k_1 B, Figure 6), which confirms the proposed mechanism (Equations 4-6), and from its slope it is possible to calculate the k_1 value.

Similarly starting again from Equation 7, the dependence of v on [ArOH] may be written as:

$$v = \frac{A[\text{ArOH}]}{B + [\text{ArOH}]} \tag{9}$$

where A = 2 [E] k_1 $[H_2O_2]$ and B = (k_1/k_3) $[H_2O_2]$. Double reciprocal plots ($1/v$ *vs.* 1/[ArOH]) allow us to calculate A and B values for each H_2O_2 concentration. Plotting A *versus* B values gives a straight line (A = 2 [E] k_3 B, Figure 7), which further confirms the mechanism proposed, and from its slope it is possible to calculate the k_3 value.

From the slopes of the straight lines plotted in Figures 6 and 7, it is therefore possible to determine the values of k_1 and k_3. The values obtained from the steady-state kinetic data were k_1 (compound I formation constant) = 10.4 μM^{-1} s^{-1} and k_3 (compound II reduction constant) = 8.7 μM^{-1} s^{-1}.

Thus, the oxidation of coniferyl alcohol by peroxidases from the BPrx HpI isoenzyme group follows the accepted model for peroxidase oxidation, in which compound I and compound II appear as the main intermediates in the catalytic cycle (Equations 4-6). This was supported by the good fits to a straight line which passed through the coordinate origin when the parameters A *vs* B were plotted for varying H_2O_2 (Figure 6) and coniferyl alcohol (Figure 7) concentrations.

Catalytic Properties of Acidic and Basic Peroxidase Isoenzymes during the Oxidation of Coniferyl Alcohol

From the results shown in Table I, the plant cell wall contains both acidic and basic peroxidase isoenzymes. To date, it has been difficult to decide which of these two isoenzyme groups is responsible for lignin biosynthesis in the cell wall, and it even

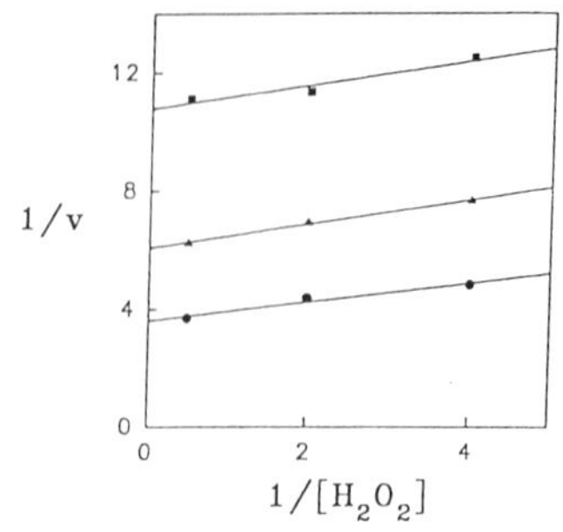

Figure 5. Double reciprocal plots of data shown in Figure 4.

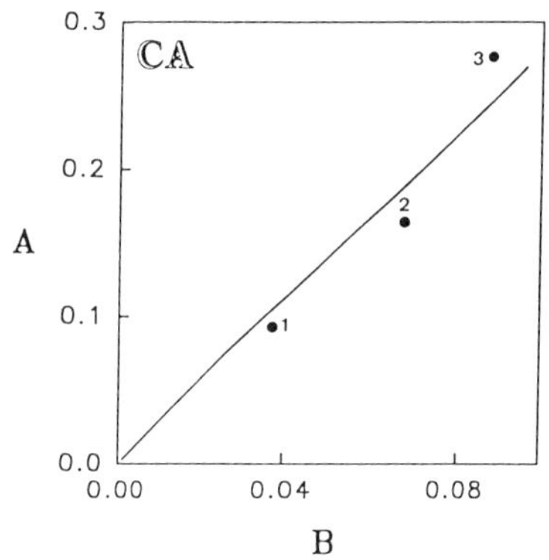

Figure 6. Plots of parameters A (nmol s^{-1}) *versus* B (mM) obtained by varying H_2O_2 at three coniferyl alcohol concentrations [25 μM (1), 50 μM (2) and 100 μM (3)]. From the slope of the straight line (2[E]k_1), k_1 was calculated to be 10.4 $\mu M^{-1}s^{-1}$.

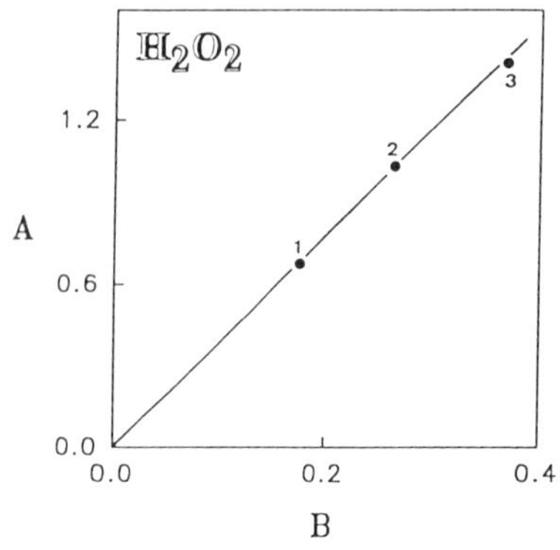

Figure 7. Plots of parameters A (nmol s^{-1}) versus B (mM) obtained by varying coniferyl alcohol at three H_2O_2 concentrations [0.25 mM (1), 0.5 mM (2) and 2 mM (3)]. From the slope of the straight line (2[E]k_3), k_3 was calculated to be = 8.7 $\mu M^{-1}s^{-1}$.

seems possible that both peroxidase isoenzyme groups may be involved. Mäder *et al.* (*30*) suggested that basic peroxidases generate the H_2O_2 which is later used by acidic peroxidases in phenolic coupling. However, both basic and acidic peroxidases are only temporary residents in the plant cell wall and, in most cases, they are not present simultaneously. Furthermore, although it is well established that acidic peroxidases play a central role in the lignification of the vascular bundles in stems (*31, 32*), this does not appear true in other cases where basic peroxidases are clearly and selectively associated with the lignification of vascular structures (*33-36*).

To cast some light on this controversy, it is interesting to compare the catalytic activities of acidic and basic peroxidase isoenzymes during the oxidation of coniferyl alcohol. An examination of the macroscopic apparent Michaelis constants (K_M^{obs}) shows that acidic peroxidase isoenzymes have a greater affinity for coniferyl alcohol (Table III), while basic peroxidases have a greater affinity for H_2O_2 (Table IV).

Table III. Observed K_M Values for Coniferyl Alcohol during Its Oxidation at pH 5.0 by Acidic and Basic Peroxidase Isoenzymes with H_2O_2

isoenzyme	K_M^{obs} μM	[H_2O_2] mM	source	reference
APrx	10	0.33	*Lupinus*	*37*
BPrx HpI	176	0.25	*Vitis*	-
BPrx HpI	266	0.50	*Vitis*	-
BPrx HpI	369	2.00	*Vitis*	-

A similar conclusion may be reached when comparing the microscopic catalytic rate coefficients for compound I formation (k_1) and compound II reduction (k_3). In this case, the reactivity of ferriperoxidase with H_2O_2 [whose magnitude is evaluated through the coefficient for compound I formation (k_1)] is greater for basic peroxidases than for acidic peroxidases (Table V), while the reactivity of compound II toward coniferyl alcohol [whose magnitude is evaluated through the coefficient for compound II reduction (k_3)] is greater for acidic peroxidases than for basic peroxidases (Table V).

Table IV. Observed K_M Values for H_2O_2 during Its Reduction at pH 5.0 by Acidic and Basic Peroxidase Isoenzymes with Coniferyl Alcohol

isoenzyme	K_M^{obs} μM	[coniferyl alcohol] mM	source	reference
APrx	160	0.100	*Lupinus*	*37*
BPrx HpI	38	0.025	*Vitis*	-
BPrx HpI	67	0.050	*Vitis*	-
BPrx HpI	88	0.100	*Vitis*	-

In any case, cell wall acidic peroxidases show greater intrinsic efficacy

$$k_{cat}/K_M^{obs} = \frac{2k_1[H_2O_2]}{(k_1 / k_3[H_2O_2]} = 2\,k_3$$

than basic peroxidases for coniferyl alcohol oxidation (see Table V), suggesting that they are evolutionarily better adapted than basic peroxidases for the *in situ* oxidation of coniferyl alcohol to lignins. Nevertheless, this redundancy in biological function shown by the different peroxidase isoenzyme groups, underlined by both their differing reactivities toward coniferyl alcohol and their specific localization in the several domains of the primary cell wall and secondary thickening domains (*32*), could explain why lignins are not uniformly deposited throughout the plant cell wall and why they are intrinsically so heterogeneous.

Table V. k_1 and k_3 Values at 25°C for the Oxidation of Coniferyl Alcohol by Acidic and Basic Peroxidase Isoenzymes

isoenzyme	k_1 $\mu M^{-1}s^{-1}$	k_3 $\mu M^{-1}s^{-1}$	pH	source	reference
APrx	1.4	16.0	4.50	*Armoracia*	*38*
BPrx LpI	12.0	2.8	4.50	*Armoracia*	*38*
BPrx LpI	6.7	2.4	3.96	*Hordeum*	*39*
BPrx HpI	10.4	8.7	5.00	*Vitis*	-

Acknowledgments

This work has been partially supported by grants from the CICYT (Spain), Projects # ALI 93/573 and ALI 95/1018. The authors thank Dr. M.C. Rojas (Departamento de Química, Universidad de Chile) for permitting us to use unpublished data.

Literature Cited

1. Pedreño, M. A.; Bernal, M. A.; Calderón, A. A.; Ferrer, M. A.; López-Serrano, M.; Merino de Cáceres, F.; Muñoz, R.; Ros Barceló, A. In *Plant Peroxidases: Biochemistry and Physiology*; Welinder, K. G., Rasmussen, S. K., Penel, C., Greppin, H., Eds.; University of Geneva: Geneva, 1993; pp 307-314.
2. Ros Barceló, A.; Muñoz, R.; Sabater, F. *Plant Sci.* **1989**, *63*, 31.
3. Ros Barceló, A.; Ferrer, M. A.; García-Florenciano, E.; Muñoz, R. *Bot. Acta* **1991**, *104*, 272.
4. Ferrer, M. A.; Pedreño, M. A.; Ros Barceló, A.; Muñoz, R. *J. Plant Physiol.* **1992**, *139*, 611.
5. García-Florenciano, E.; Calderón, A. A.; Pedreño, M. A.; Muñoz, R.; Ros Barceló, A. *Plant Growth Regul.* **1991**, *10*, 125.
6. Calderón, A. A.; García-Florenciano, E.; Pedreño, M. A.; Muñoz, R.; Ros Barceló, A. *Z. Naturforsch.* **1992**, *47c*, 215.
7. Bernal, M. A.; Pedreño, M. A.; Calderón, A. A.; Muñoz, R.; Ros Barceló, A.; Merino de Cáceres, F. *Ann. Bot.* **1993**, *72*, 415.

8. Cuenca, J.; García-Florenciano, E.; Ros Barceló, A.; Muñoz, R. *Plant Cell Rep.* **1989**, *8*, 471.
9. Gómez-Tena, M.; Pedreño, M. A.; Ros Barceló, A.; Ferrer, M. A. *J. Amer. Soc. Hort. Sci.* **1994**, *119*, 1276.
10. Sottomayor, M.; de Pinto, M. C.; Salema, R.; DiCosmo, F.; Pedreño, M. A.; Ros Barceló, A. *Plant Cell Environ.* **1996**, *19*, 761.
11. van Huystee, R.; Tam, A. S. K. *J. Plant Physiol.* **1988**, *133*, 645.
12. Calderón, A. A.; Zapata, J. M.; Ros Barceló, A. *Plant Cell Tissue Organ Cult.* **1994**, *37*, 121.
13. López-Serrano, M.; Ferrer, M. A.; Ros Barceló, A.; Pedreño, M. A., *Eur. J. Histochem.* **1995**, *39*, 69.
14. Sottomayor, M.; DiCosmo, F.; Salema, R.; Ros Barceló, A. In *Plant Peroxidases: Biochemistry and Physiology*; Obinger, C., Burner, U., Ebermann, R., Penel, C., Greppin, H., Eds; Unversity of Geneva: Geneva, 1996; pp 128-133.
15. Schloss, P.; Walter, C.; Mäder, M. *Planta* **1987**, *170*, 225.
16. Wink, M. *J. Exp. Bot.* **1993**, *44*, 231.
17. Wan, L.; Gijzen, M.; van Huystee, R. B. *Biochem. Cell Biol.* **1994**, *72*, 411.
18. Green, B. N.; Oliver, R. W. A. *Biochem. Soc. Transac.* **1991**, *19*, 929.
19. Ferrer, M. A.; Gómez-Tena, M.; Pedreño, M. A.; Ros Barceló, A. *Postharv. Biol. Technol.* **1996**, *7*, 301.
20. Tepfer, M.; Taylor, I.E.P. *Science* **1981**, *213*, 761.
21. Ros Barceló, A.; Pedreño, M. A.; Muñoz, R.; Sabater, F. *Physiol. Plant.* **1988**, *73*, 238.
22. Pedreño, M. A.; Ros Barceló, A.; García-Carmona, F.; Muñoz, R. *Plant Physiol Biochem.* **1990**, *28*, 37.
23. Ferrer, M. A.; Pedreño, M. A.; Muñoz, R.; Ros Barceló, A. *FEBS Lett.* **1990**, *276*, 127.
24. Badiani, M.; De Biasi, M. G.; Felici, M. *Plant Physiol.* **1990**, *92*, 489.
25. Chabanet, A.; Catesson, A. M.; Goldberg, R. *Phytochemistry* **1993**, *33*, 759.
26. Ricard, J.; Mazza, G.; Williams, R. J. P. *Eur. J. Biochem.* **1972**, *28*, 566.
27. Hapiot, P.; Pinson, J.; Neta, P.; Francesch, C.; Mhamdi, F.; Rolando, C.; Schneider, S. *Phytochemistry* **1994**, *36*, 1013.
28. Hayashi, Y.; Yamazaki, I. *J. Biol. Chem.* **1979**, *254*, 9101.
29. Bakovic, M.; Dunford, H. B. *Biochemistry* **1993**, *32*, 833.
30. Mäder, M.; Ungemach, J.; Schloss, P. *Planta* **1980**, *147*, 467.
31. Lagrimini, L. M. In *Biochemical, Molecular, and Physiological Aspects of Plant Peroxidases*; Lobarzewski, J., Greppin, H., Penel, C., Gaspar, Th., Eds; University of Geneva: Geneva, 1991; pp 59-67.
32. Ros Barceló, A. *Protoplasma* **1995**, *186*, 41.
33. Abeles, F. B.; Biles, C. L. *Plant Physiol.* **1991**, *95*, 269.
34. Morales, M.; Pedreño, M. A.; Muñoz, R.; Ros Barceló, A.; Calderón, A. A. *Food Chem.* **1993**, *48*, 391.
35. Polle, A.; Otter, T.; Seifert, F. *Plant Physiol.* **1994**, *106*, 53.
36. Sato, Y.; Sugiyama, M.; Komamine, A.; Fukuda, H. *Planta* **1995**, *196*, 141.
37. Pedreño, M. A.; Ros Barceló, A.; Sabater, F.; Muñoz, R. *Plant Cell Physiol.* **1989**, *30*, 237.
38. Marklund, S.; Ohlsson, P. I.; Opara, A.; Paul, K. G. *Biochim. Biophys. Acta* **1974**, *350*, 304.
39. Rasmussen, C. B.; Dunford, H. B.; Welinder, K. G. *Biochemistry* **1995**, *34*, 4022.

Chapter 8

Laccases Associated with Lignifying Vascular Tissues

Jeffrey F. D. Dean[1,2], Peter R. LaFayette[2], Clayton Rugh[1], Alexandria H. Tristram[2], J. Todd Hoopes[2], Karl-Erik L. Eriksson[2], and Scott A. Merkle[1]

[1]Daniel B. Warnell School of Forest Resources, University of Georgia, Athens, GA 30602–2152
[2]Department of Biochemistry and Molecular Biology, University of Georgia, Athens, GA 30602–7229

Lignin is formed *via* the oxidative polymerization of monolignols within the plant cell wall matrix. Peroxidases, which are abundant in virtually all cell walls, can oxidize the monolignols in the presence of H_2O_2, and have long been held to be the principal catalysts for this reaction. Recent evidence shows, however, that laccases (*p*-diphenol: O_2 oxidoreductases, EC 1.10.3.2) secreted into the secondary walls of vascular tissues are equally capable of polymerizing monolignols in the presence of O_2. Full-length and partial laccase gene clones recovered from several different species suggest that these enzymes are widely distributed in the plant kingdom, and may generally occur as multiple isozymes within acidic and basic isoform classes. Although expression of antisense laccase gene constructs in transgenic plants has yet to bring about demonstrable alterations in lignin content, ectopic expression of laccase activity in yellow-poplars regenerated from transformed embryogenic cell lines resulted in a severely dwarfed phenotype having fused organs and hardened tissues.

Lignin biogenesis may be considered to involve three compartmentalized stages: intracellular synthesis of the monolignol precursors (*p*-coumaryl, coniferyl, and sinapyl alcohols); transport and secretion of the monolignols; and extracellular polymerization of monolignols in a process coordinated with the deposition of other cell wall components (*1, 2*). Until recently, models of the extracellular component of this pathway usually depicted a process where cell wall-bound enzymes oxidized the secreted monolignols, and the resulting phenoxy radicals produced in these reactions diffused away to couple in a random fashion. Universally abundant in lignifying plant tissues (*3*), peroxidases have been the enzymes most commonly associated with monolignol polymerization since Higuchi and Ito demonstrated that plant peroxidases could form dehydropolymerisates (DHPs) from coniferyl alcohol and H_2O_2 *in vitro* (*4, 5*). Numerous studies supported this contention by correlating shifts in peroxidase activity with tissue lignification *in vivo* (*e.g. 3, 6-10*). However, to this day, unambiguous evidence for the involvement of specific peroxidases in lignin deposition has yet to be secured (*11*).

Evidence for Laccase Involvement in Lignification

Although peroxidases have been the focus of considerable attention, the first demonstrations of *in vitro* polymerization of coniferyl alcohol to form a lignin-like DHP actually used fungal laccases (*12, 13*). Freudenberg and coworkers subsequently demonstrated slow polymerization of monolignols in oxygenated extracts from *Araucaria excelsa* (*14*), and more rapid polymerization using a laccase partially purified from Norway spruce (*Picea abies*) cambium extracts (*15*). From this, Freudenberg contended that the combined activities of laccases and peroxidases were required for lignin biosynthesis. However, after finding that a laccase purified from Japanese lacquer tree (*Rhus vernicifera*) sap could not form a DHP from coniferyl alcohol, Nakamura (*16*) suggested that laccases were not involved in lignin deposition. Subsequent histochemical studies in which syringaldazine, a chromogenic substrate for fungal laccases, was oxidized when applied to lignifying tree stems in the presence, but not the absence, of H_2O_2 further suggested that laccases were either absent or inactive in these plant tissues, and that peroxidases were the principal oxidative catalysts for lignin deposition (*8*). These two observations led most researchers to conclude that plant laccases play no role in lignin biosynthesis. However, until recently, no functions had been advanced for plant laccases other than as a hardening agent for resins released from wounded tissues or as part of a plant defense mechanism (*17*). It should be noted, however, that we now know that *Rhus* laccases **can** oxidize monolignols, and that the technique of cryosectioning used in the histochemical studies of Harkin and Obst can inactivate xylem-localized laccases, many of which appear to be cold-labile, at least *in situ* (*18*).

An indirect line of support for the involvement of laccase in lignification comes from investigations showing that plants grown under copper-deficient conditions contain less lignin (*19, 20*). The laccase isolated from *Rhus vernicifera* (Japanese lacquer tree) sap was one of the first enzymes demonstrated to require a metal, copper, for activity (*21, 22*), and it remains one of the most intensively studied of all metalloenzymes (*23*). Suspension-cultured sycamore maple (*Acer pseudoplatanus*) cells were shown to secrete an inactive, partially metallated laccase when grown in a medium containing sub-optimal levels of copper (*24*), and the stems of *Zinnia elegans* plants grown under copper-deficient conditions were under-lignified and contained less extractable laccase activity (*25*). Figure 1 (*26*) illustrates the tendency of under-lignified *Zinnia* plants to grow in a prone position. However, a direct link between laccase and lignification could not be drawn from these results since lower specific activity has also seen for other copper-containing enzymes, such as ascorbate oxidase (*27*), produced under conditions of copper deficiency.

In a return to Freudenberg's pioneering work, a laccase purified from *A. pseudoplatanus* suspension-cultures was demonstrated to produce DHPs from each of the three monolignols (*28, 29*). Immunolocalization studies by Faye and coworkers (*30*) using polyclonal antibodies showed that the same enzyme was localized specifically to lignifying xylem and epidermal cell walls in sycamore maple trees. Furthermore, it was shown that spinach, cabbage, turnip, pear, banana, poplar and bean extracts contained polypeptides that cross-reacted with anti-laccase polyclonal antibodies, suggesting that laccases might be distributed across a wide range of plant species (*30, 31*). Recent observations from several laboratories have lent further support to the involvement of laccases and other oxygen-dependent phenoloxidases in the lignification of plant tissues by demonstrating that the activities of these enzymes are more tightly correlated with lignin deposition in vascular tissues than are the activities of peroxidases (*31-36*). The evidence supporting laccase involvement in lignin biosynthesis has been reviewed recently (*18, 37*).

Figure 1. Representative *Zinnia elegans* plants grown in vermiculite for ten weeks and watered with Hoagland's solution (*26*) having either normal (0.014 ppm) or reduced (0.0014 ppm) copper content. Although biomass levels were on the average equivalent for plants grown under the two conditions, plants grown on reduced copper (right) tended to grow in a prone position.

Plant Laccase Genes

As a result of these studies, there has been renewed interest in the study of plant laccases, particularly with respect to their potential as targets for the genetic manipulation of lignin deposition (*11, 38*). Using mRNA from suspension-cultured sycamore maple cells, a cDNA encoding a plant laccase was isolated and characterized (*39*). Analysis of the laccase gene sequence correlated well with the protein data previously accumulated for the enzyme in that the inferred amino acid sequence encoded a protein having a low pI (*ca.* 4.5), numerous (*16*) potential glycosylation sites, and complete conservation of the His and Cys residues that are known to act as ligands for the four copper atoms contained in the laccase catalytic center. Alignment of the *Acer* laccase gene sequence with those found for other laccases and blue copper oxidases, as shown in Figure 2, makes it clear that the relative positions of the ligand residues in each of the four copper-binding domains have been strictly conserved the primary sequence. Figure 3 is a schematic model, adapted from Messerschmidt and Huber (*40*) and based on the crystallographic structure of ascorbate oxidase, showing the relative orientations of the conserved amino acid ligands and the copper atoms in the laccase catalytic center. Comparison of the *Acer* gene sequence against the GenBank database indicated that fragments of laccase genes from *Arabidopsis* and rice have been identified as dESTs in the genome mapping programs for those species.

Laccase activity is relatively easy to detect in high-ionic strength buffer extracts of lignifying xylem from a variety of woody species, including yellow-poplar (*Liriodendron tulipifera*). *Liriodendron* made an attractive model system for efforts to determine the role of laccase in lignin deposition because a highly efficient transformation/regeneration system based on embryogenic cell cultures was readily available for this species (*41*). In addition, *Liriodendron* is a large tree, the abundance of which in the southeastern U.S. assured a ready supply of lignifying xylem as starting material. Initial efforts to detect the *Liriodendron* laccase gene using the *Acer* cDNA as a probe were unsuccessful as the probe did not cross-hybridize with either the mRNA or DNA at moderate stringency. Fortunately, it was possible to use degenerate oligonucleotide primers based on the copper-binding domains that are highly conserved in all laccases to amplify PCR products from cDNAs corresponding to the laccase mRNAs expressed in yellow-poplar xylem. To date, four different cDNAs have been isolated and characterized (LaFayette *et al.*, in preparation). All four cDNA sequences showed conservation of the amino acid residues that act as copper ligands, and all encoded multiple potential glycosylation sites (17, 18, 18, and 19, respectively). Eight of the potential glycosylation sites were completely conserved between all of the *Acer* and *Liriodendron* gene sequences.

Although the coding sequences for the *Liriodendron* cDNAs were highly conserved (>85% identity at the amino acid level), both the 5′- and 3′-untranslated regions varied greatly between the four cDNAs. Southern blot analyses using probes generated from unique sequences in the untranslated regions show that each laccase isozyme gene exists as a single copy in the yellow-poplar genome, but blots probed with the full-length coding sequence suggest that at least 6-8 closely related laccase genes exist in the yellow-poplar genome (*42*). None of the genomic sequences have been cloned yet, and there is no information concerning allelic variation. Sequence comparisons between the sycamore maple and yellow-poplar cDNAs showed an amino acid identity of *ca.* 45%, which indicated why the sequences did not cross-hybridize, but also showed that the genes were more closely related to one another than they were to any other genes currently in the sequence databases.

Laccase Isozymes and Isoform Classes

In contrast to the *Acer* enzyme, the amino acid sequences predicted from the yellow-poplar cDNAs suggest the *Liriodendron* enzymes have high pIs (*ca.* 9.0). This was particularly intriguing in light of the fact that, with the sole exception of the *Acer*

Ascorbate oxidase residue #		û	58	99	443	498	
Copper Site Ligands		û	2 3	3 3	1 2 3	313 1 1	
	Acer Laccase	TIHWHGVKMP	GTLWWHAHSD	QNHPMHLHGFSF	DNPGVWFLHCHFERHT.TWGMA		
PLANT	*Liriodendron* Laccase	TLHWHGVRQL	GTLFWHAHIS	ESHPLHLHGFNF	DNPGVWFMHCHLEVHT.SWGLK		
	Cucumis AAO	VIHWHGILQR	GTYFYHGHLG	EIHPWHLHGHDF	DNPGVWAFHCHIEPHL.HMGMG		
	Cryphonectria Laccase	TIHWHGIRQL	GTSWYHSHFS	LPHPIHLHGHDF	TNPGAWLMHCHIAWHV.SAGLG		
	Neurospora Laccase	SIHWHGMHQR	GTSWYHSHFS	LPHPIHLHGHDF	DNPGSWLMHCHIAWHV.SGGLS		
	Basidiomycete Laccase	SIHWHGFFQH	GTFWYHSHLS	FPHPFHLHGHTF	DNPGPWFLHCHIDFHL.EAGFA		
	Coriolus Laccase	SIHWHGFFQK	GTFWYHSHLS	APHPFHLHGHAF	DNPGPWFLHCHIDFHL.EAGFA		
FUNGI	*Phlebia* Laccase	TIHWHGFFQH	GTFWYHSHLS	GPHPFHLHG-Res.	SIHWHGIILP	GTYWYHSHSG	MTHPIHLHGMWS

Figure 2. Amino acid sequence alignments of the four copper-binding domains in the *Acer* and *Liriodendron* laccases, and other blue copper oxidases. The residues acting as ligands for copper atoms are shaded and numbered (on the second line) according to the electromagnetic signature of the copper atom to which they bind (Type-1, -2, or -3). The sequences are ordered from top to bottom according to their degree of similarity to the *Acer* laccase sequence as determined by the *PILEUP* software module in the Wisconsin Sequence Analysis Package (Genetics Computer Group, Madison, WI).

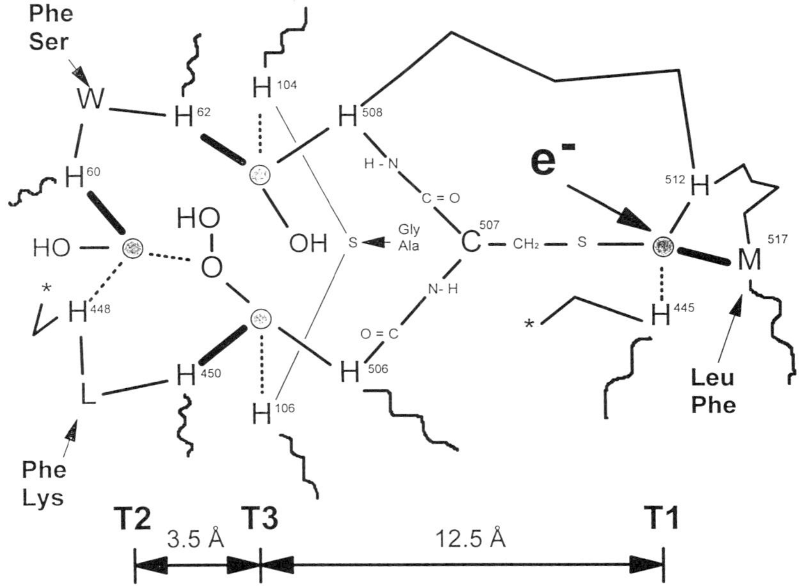

Figure 3. Schematic diagram showing the relative orientations of the four copper atoms and their amino acid ligands in the catalytic center of a typical blue copper oxidase. The Type-1, -2, and -3 copper atoms and their approximate distances of separation are denoted across the bottom. For purposes of clarity, a break, denoted by asterisks, was made in the polypeptide chain linking His^{445} (T1 copper site) with His^{448} (T2 copper site). Naturally occurring amino acid substitutions at select highly conserved sites are noted by arrows. The diagram is based on the crystallographic structure determined for zucchini (*Cucurbita pepo*) ascorbate oxidase (PDB:1AOZ), and adapted from (*39*).

enzyme, all of the plant laccases so far purified and characterized have high pIs. Even more intriguing was the observation that the high-pI and low-pI laccase classes differed in having either a Leu or Met residue, respectively, associated with one of their copper-binding sites (Figures 2 and 3). The dESTs encoding apparent laccase fragments from *Arabidopsis* and rice, as well as a cDNA fragment isolated from *Zinnia* (A. Tristram, unpublished data), also specify high-pI products and contain Leu, rather than Met, residues in the Type-1 copper site. This amino acid substitution has been predicted to modulate the redox potential of Type-1 copper sites in such a way that the Leu-containing, high-pI laccases produced in yellow-poplar xylem should be stronger oxidizers than the enzyme produced by cultured sycamore maple cells (*40, 43*). As shown in Figure 4, high-pI and low-pI laccases are both expressed in the lignifying stems of *Zinnia*; however, the activities show different temporal expression patterns (*25*). The high pI laccase is expressed in tracheary elements (xylem vessels), while the low-pI laccases only become detectable in *Zinnia* stem extracts concurrent with the onset of lignin deposition in libriform (xylem) and sclerenchymal (phloem) fibers (*25*). Efforts are currently underway to determine whether tissue-specificity underlies the temporal differences seen in the expression pattern for the different pI laccase isoform classes in *Zinnia* stems.

Potential Functions for Laccase Isozyme Variations

Ye and Varner (*44*) used tissue printing to show a tissue-specific expression pattern for two very different lignin-associated *O*-methyltransferases (OMTs) expressed in *Zinnia* stems. In that case, an OMT active on hydroxycinnamate:CoA thioesters was expressed in xylem vessels at early stages of lignification, while an OMT active on free hydroxycinnamates was expressed later when fibers were undergoing lignification. On this basis, the authors suggested that a second biosynthetic pathway for lignin precursors might be expressed in fiber cells. Observations that laccase isozymes may be expressed in a similar pattern suggest that the two pathways may also be maintained for the extracellular polymerization reactions. Some interesting implications become apparent when this two-pathway scheme for lignin biosynthesis is considered in light of the proposal by Logan and Thomas (*45*) describing an anatomical basis for variations in lignin composition. Homopolymeric guaiacyl lignin is synthesized in the tracheary elements, vessels, and vessel fibers that are common to all vascular plants, while heteropolymeric guaiacyl:syringyl lignin is found in the libriform fibers common to angiosperms, as well as sclerenchymal fibers found in a wide variety of species. As sinapyl alcohol, the precursor to syringyl lignin, can be oxidized more readily than coniferyl alcohol, the precursor to guaiacyl lignin, there may be a correlation between expression of particular laccases and the lignin precursor to be deposited. That is, low-pI, low redox potential laccases may be associated with syringyl lignin deposition, although such a strategy does not fit with any of the currently accepted hypotheses regarding the non-specificity of oxidation reactions associated with lignin deposition.

In addition to questions concerning the physiological rationale for the expression of high- and low-pI laccase isoform classes, a further question arises in regard to the apparently simultaneous expression of multiple laccase isozymes within a given tissue. That is, why were multiple high-pI laccase isozymes expressed in the lignifying xylem tissue from *Liriodendron*, and why are multiple low-pI laccase isozymes expressed simultaneously in suspension-cultures of *A. pseudoplatanus* cells (*29*; L. Faye, Université de Rouen, personal communication, 1994)? An interesting, albeit highly speculative, possibility is suggested by the recent identification of a so-called dirigent protein by Lewis and coworkers (*46* and chapter 22, this volume). These researchers showed that addition of the dirigent protein purified from *Forsythia* stem extracts to a laccase preparation provided for the stereospecific coupling of coniferyl alcohol to produce only (+)-pinoresinol, rather than a racemic mixture of multiple guaiacyl dimers. Might it be that individual laccase isozymes interact preferentially with different dirigent proteins to produce unique stereospecific

Figure 4. Isoelectric focusing gel (pH 3-10) showing separation of high-pI (*ca.* 9.0) and low-pI (*ca.* 4.0) laccase isoform classes extracted from *Zinnia* stems. The four lanes contain materials eluted from a DEAE-IEC column using a NaCl step gradient as follows: (1) flow-through; (2) 120 mM; (3) 160 mM; and (4) 220 mM. The gel was stained for laccase activity using 1,8-diaminonaphthalene, and subsequently counter-stained for protein with Serva Blue W. Multiple bands are likely to represent laccase isozymes.

compounds? In such a scenario, laccase isozymes might be considered as catalytic subunits, whereas dirigent proteins would function as regulatory subunits of a biosynthetic complex. In this way, phenolic precursors could be channeled into specific pathways for the production of any of the manifold secondary metabolites produced by plants. Such a system could also underlie the need plants have to produce so many different laccase isozymes. It is anticipated that heterologous expression of discrete laccase isozymes will provide material that will allow this possibility to be probed.

Transformation of Plants with Laccase Genes

To prove that the yellow-poplar cDNAs actually encoded laccases, rather than some other closely related enzyme, the genes were placed behind a CaMV 35S promoter in a cassette vector and used to transform suspension-cultured tobacco (BY-2) cells *via* microparticle bombardment. Calli that survived antibiotic (kanamycin) selection were tested for laccase activity by plating small pieces on solid medium containing the synthetic laccase substrate, 2,2′-azinobis-(3-ethylbenzthiazoline-6-sulfonate) (ABTS), as shown in Figure 5A, or, with greater sensitivity, by placing pieces in the wells of microtiter plates containing liquid culture medium supplemented with ABTS. Wild-type cells did not oxidize ABTS, even in the absence of H_2O_2 scavengers, such as catalase. However, transformed cell lines expressing the sycamore maple laccase and each of the four yellow-poplar enzymes have been recovered. Of potential interest was the observation that transformants expressing large amounts of laccase activity were obtained more frequently when the sycamore maple laccase construct was used in bombardment experiments. It may be that the codon usage or glycosylation pattern of the yellow-poplar enzymes is more difficult for the tobacco cells to reproduce, but it is also possible that the higher redox potential inferred for the yellow-poplar enzymes could put a negative selective pressure on cells producing high levels of these enzymes. For instance, decreased auxin levels due to increased oxidation might limit growth of transformed cells producing high levels of yellow-poplar laccase. However, this possibility has not yet been tested. The latter possibility may be interesting in light of the fact that *in vivo* this enzyme appears to be expressed only in terminally differentiated vascular cells that have stopped growing and are entering into a process similar to programmed cell death (*47*).

Table I. Pyrolysis-Mass Spectrometric Analysis of *Liriodendron* Cell Wall Phenolics

Sample	Syringyl[a]/Guaiacyl[b]	Syringyl+Guaiacyl (%)
Mature Wood	2.0 / 0.1	27.3
Regenerants:		
Untransformed	2.6 / 0.5	14.0
Vector Transformed	2.4 / 0.3	15.2
Antisense Transformed	2.2 / 0.7	14.2

[a] Syringyl units determined as the sum of 2,6-dimethoxyphenol, 2,6-dimethoxy-4-methylphenol, 4-ethyl-2,6-dimethoxyphenol, 2,6-dimethoxy-4-vinylphenol, syringaldehyde, and *trans*-2,6-dimethoxy-4-propenylphenol.

[b] Guaiacyl units determined as the sum of guaiacol, 4-methylguaiacol, 4-ethylguaiacol, vanillin, and *trans*-isoeugenol.

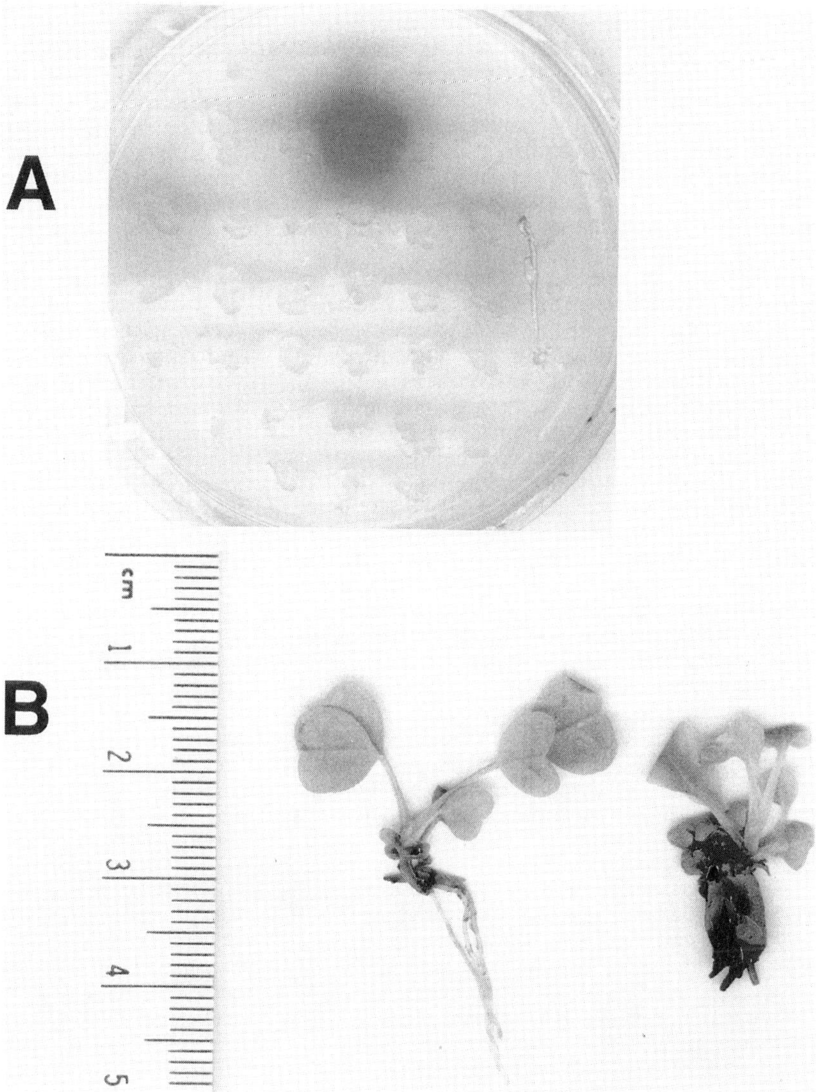

Figure 5. Expression of laccase activity in transgenic plant tissues. In panel A, laccase activity expressed by transformed tobacco calli was detected using ABTS. Kanamycin-resistant cells grown for three days on MS medium containing 0.1% ABTS formed zones of green product around calli that expressed laccase activity. All calli oxidized ABTS strongly upon addition of H_2O_2. Panel B shows yellow-poplar regenerated from embryogenic cells transformed with either the *Acer* laccase gene (right) or a control vector (left). Ectopic expression of the *Acer* laccase led to fused, rigidified tissues and a dwarf phenotype.

In initial efforts to test the commercial potential for producing trees with less lignin by down-regulating laccase expression, an antisense vector harboring the first 500 nucleotides of one *Liriodendron* laccase cDNA was used to transform embryogenic yellow-poplar cells. Approximately 60 transgenic cell lines were recovered and shown by RT-PCR to express varying levels of the antisense laccase construct. Attempts have been made to regenerate plants from several cell lines that displayed good growth characteristics in suspension culture and, although several of the lines produced embryos, only a single cell line has produced enough regenerants for lignin analysis. Despite encouraging preliminary results suggesting a slightly reduced lignin content in these regenerated trees, recent pyrolysis-mass spectrometric analyses performed by Dr. Hans Jung, shown in Table I, demonstrated that the lignin composition and content in the transgenic lines was unchanged from wild-type. Very recently, trees have been regenerated from three additional transformed cell lines, and preparations are currently being made to analyze the lignin content of their tissues.

In parallel with the attempts to down-regulate laccase in *Liriodendron* using antisense gene constructs, the sense expression constructs used to demonstrate laccase activity in transformed tobacco cell lines were used to transform embryogenic yellow-poplar cells. Transformed cell lines were selected on the basis of their antibiotic resistance, and several that also expressed the *Acer* laccase were detected by virtue of their ability to oxidize ABTS. These same cell lines were also distinguished by the tendency of their liquid culture medium to turn more or less brown in direct proportion to the level of laccase activity they released into the medium. Somewhat surprisingly, no *Liriodendron* cell lines expressing any of the *Liriodendron* laccases have been identified to date, perhaps lending further support to the possibility that the putative high redox potential of the high-pI laccase isoforms is particularly deleterious to the normal growth of suspension-cultured cells from at least some plant species. (Micronutrient levels in the culture medium were modified so that copper deficiency was not a factor in whether or not laccase activity was detected.)

On the other hand, cell lines expressing the *Acer* laccase retained the capacity to produce embryos, and these were subsequently germinated and transferred to soil. In addition to the previously noted change in culture medium color, embryos produced from transformed cell lines were significantly more rigid and harder than wild-type embryos, and the transgenic embryos tended to fuse and clump together. These phenotypic changes continued even after germination. Thus, as shown in Figure 5B, germinants continued to grow in a polar fashion, but remained more rigid (harder to section), were dwarfed, and displayed fused organs. Histochemical analyses of tissue sections from the transgenic germinants indicated that laccase activity was being expressed in all cell walls, but was particularly abundant in the nascent vascular tissues. Laccase activity was just barely detectable in the nascent vascular tissues of wild-type germinants, and not at all in any other tissues. Staining with phloroglucinol and permanganate demonstrated that lignin was still produced specifically in the vascular initials of transgenic plants, although there was also some stain-positive material deposited just under the epidermis in the vicinity of the root/stem transition zone. Although histochemical stains for lignin did not highlight materials deposited in the walls of other cell types, epi-fluorescence microscopy showed that there was a general increase in the fluorescence of all cell walls, suggesting the possible deposition of other phenolic materials, or at least a change in the phenolic components of the cell walls in transgenic plants. Further analyses of the composition of cell walls from transgenic *Liriodendron* expressing the *Acer* laccase are underway to determine the basis for tissue rigidification.

Concluding Remarks

Not so many years ago, DNA was considered to be an amorphous component of little consequence floating in the interior of cells. Yet, from today's perspective, it seems almost unbelievable that biology could have advanced so far without recognizing the structural elegance and fundamental nature of this macromolecule. Then, as now, it was new technology that provided critical data which stimulated revolutionary ideas that have in turn unlocked one new discipline after another, until no aspect of the biological sciences can be considered in disregard of DNA.

While few would suggest that the development of new concepts concerning the structure and biosynthesis of lignin and secondary plant phenolic components holds the same potential to revolutionize biology as the breakthroughs in our understanding of DNA, there are some similarities in the advance toward understanding that would suggest a revolution nonetheless to be underway. Lignin was originally referred to as the 'encrusting' substance in wood (*38*), and even today it is sometimes casually described as a 'random' polymer. However, sophisticated analytical techniques have provided an increasingly more detailed picture of the structure and complexity of lignin, as well as a greater appreciation of the diversity of secondary phenolic products produced by plants. From this work, it has become obvious that, while lignin structure may appear 'irregular,' it is certainly not 'random.'

From a perspective in which an ordered structure for lignin is likely to follow function, many new directions for inquiries into the biosynthesis of this polymer become apparent. For instance, so little new insight has been gained in the past 40 years concerning the role of phenoloxidases in lignin deposition that there must have been a fundamental problem with the conceptual models used as the basis for experimental design. For instance, most of the past *in vitro* studies performed with phenoloxidases employed mixtures of isozymes from a variety of sources, but recent evidence suggests that not only do many of these isozymes have unequal ability to oxidize substrates, but they also may interact with other proteins to produce monolignol radical coupling products of specific structure. This concept, alone, will drastically alter future approaches to work in this area. No doubt access to large amounts of individual phenoloxidase isozymes produced from cloned genes will prove helpful for demonstrating that phenoxy radical coupling can be controlled by ancillary proteins. Given sufficient understanding of the components controlling these phenoloxidase-catalyzed coupling reactions, there are seemingly unlimited possibilities for generating new applications and products *via* gene manipulation.

Acknowledgments

The authors thank the U.S. Department of Energy (grant DE-FG05-95ER20182) and the Georgia Consortium for Technological Competitiveness in Pulp and Paper (award PP96-FS3) for financial support, as well as Dr. Hans Jung for performing the pyrolysis-MS analyses of cell wall materials from transgenic plants.

Literature Cited

1. Boudet, A. M.; Lapierre, C.; Grima-Pettenati, J. *New Phytol.* **1995**, *129*, 203-236.
2. Whetten, R.; Sederoff, R. *Plant Cell* **1995**, 7:1001-1013.
3. *Molecular and Physiological Aspects of Plant Peroxidases.* Greppin, H., Penel, C., Gaspar, T., Eds.; University of Geneva: Geneva, Switzerland.
4. Higuchi, T. In: *Biochemistry of Wood*; Kratzl, K., Billek, G., Eds.; Pergammon Press: New York, NY, 1959; pp 161-188.
5. Higuchi, T.; Ito, Y. *J. Biochem.* **1958**, *45*, 575-579.
6. Bruce, R. J.; West, C. A. *Plant Physiol.* **1989**, *91*, 889-897.
7. Church, D. L.; Galston, A. W. *Plant Physiol.* **1988**, *88*, 679-684.
8. Harkin, J. M.; Obst, J. R. *Science* **1973**, *180*, 296-298.
9. Hepler, P. K.; Rice, R. M.; Terranova, W. A. *Can. J. Bot.* **1972**, *50*, 977-983.
10. Lagrimini, L. M. *Plant Physiol.* **1991**, *96*, 577-583.
11. Boudet, A. M.; Grima-Pettenati, J. *Mol. Breed.* **1996**, *2*, 25-39.
12. Freudenberg, K. *Holzforschung* **1952**, *6*, 37-42.
13. Higuchi, T. *J. Biochem.* **1958**, *45*, 515-528.

14. Freudenberg, K.; Harkin, J. M.; Rechert, M.; Fukuzumi, T. *Chem. Ber.* **1958**, *91*, 581-590.
15. Freudenberg, K. *Nature* **1959**, *183*, 1152-1155.
16. Nakamura, W. *J. Biochem.* **1967**, *62*, 54-61.
17. Mayer, A. M.; Harel, E. *Phytochemistry* **1979**, *18*, 193-215.
18. Dean, J. F. D.; Eriksson, K.-E. L. *Holzforschung* **1994**, *48*, 21-33.
19. Graham, D. *Aust. J. Plant Physiol.* **1976**, *3*, 229-236.
20. Downes, G.; Ward, J. V.; Turvet, N. D. *Wood Sci. Technol.* **1991**, *25*, 7-14.
21. Bertrand, M. G. *C.R.A.S. Paris* **1897**, *124*, 1355-1358.
22. Keilin, D.; Mann, T. *Nature* **1939**, *143*, 23-24.
23. Reinhammar, B. In *Copper Proteins and Copper Enzymes*; Lontie, R., Ed.; CRC Press: Boca Raton, FL, 1984; pp 1-35.
24. Bligny, R.; Gaillard, J.; Douce, R. *Biochem. J.* **1986**, *237*, 583-588.
25. Hoopes, J. T.; Tristram, A. H.; LaFayette, P. R.; Eriksson, K.-E. L.; Dean, J. F. D. *Plant Physiol*. **1995**, *108S*, 73.
26. Hewitt, E. J. *Sand and Water Culture Methods Used in the Study of Plant Nutrition*; Technical Communication No. 22 of the Commonwealth Bureau of Horticulture and Plantation Crops; Commonwealth Agricultural Bureaux: Farnham Royal, England, 1952; Table 30A.
27. Pitari, G.; Chichiricco, G.; Marcozzi, G.; Rossi, A.; Maccarrone, M.; Avigliano, L. *Plant Physiol. Biochem.* **1993**, *31*, 593-598.
28. Sterjiades, R.; Dean, J. F. D.; Eriksson, K.-E. L. *Plant Physiol.* **1992**, *99*, 1162-1168.
29. Sterjiades, R.; Dean, J. F. D.; Gamble, G.; Himmelsbach, D. S.; Eriksson, K.-E. L. *Planta* **1993**, *190*, 75-87.
30. Driouich, A.; Lainé, A.-C.; Vian, B.; Faye, L. *Plant J.* **1992**, *2*, 13-24.
31. Chabanet, A.; Goldberg, R.; Catesson, A. M.; Quinetszely, M.; Delaunay, A. M.; Faye, L. *Plant Physiol.* **1994**, *106*, 1095-1102.
32. Davin, L. B.; Bedgar, D. L.; Katayama, T.; Lewis, N. G. *Phytochemistry* **1992**, *31*, 3869-3874.
33. Bao, W.; O'Malley, D. M.; Whetten, R.; Sederoff, R. R. *Science* **1993**, *260*, 672-674.
34. Liu, L.; Dean, J. F. D.; Friedman, W. E.; Eriksson, K.-E. L. *Plant J.* **1994**, *6*, 218-224.
35. McDougall, G. J.; Stewart, D.; Morrison, I. M. *Phytochemistry* **1994**, *37*, 683-688.
36. Udagama-Randeniya, P. V.; Savidge, R. A. *Trees* **1995**, *10*, 102-107.
37. O'Malley, D. M.; Whetten, R.; Bao, W. L.; Chen, C. L.; Sederoff, R. R. *Plant J.* **1993**, *4*, 751-757.
38. Dean, J. F. D.; Eriksson, K.-E. L. *Holzforschung* **1992**, *46*, 135-147.
39. LaFayette, P. R.; Eriksson, K.-E. L.; Dean, J. F. D. *Plant Physiol.* **1995**, *107*, 667-668.
40. Messerschmidt, A.; Huber, R. *Eur. J. Biochem.* **1990**, *187*, 341-352.
41. Wilde, H. D.; Meagher, R. B.; Merkle, S. A. *Plant Physiol.* **1992**, *98*, 114-120.
42. Eriksson, K.-E. L.; LaFayette, P. R.; Merkle, S. A.; Dean, J. F. D. *Proceedings of the 6th International Conference on Biotechnology in the Pulp and Paper Industry*; Vienna, Austria, 1996; in press.
43. Karlsson, B. G.; Aasa, R.; Malmström, B.; Lundberg, L. G. *FEBS Lett.* **1989**, *253*, 99-102.
44. Ye, Z. H.; Varner, J. E. *Plant Physiol.* **1995**, *108*, 459-467.
45. Logan, K. J.; Thomas, B. A. *New Phytol.* **1985**, *99*, 571-585.
46. Davin, L. B.; Wang, H.-B.; Crowell, A. L.; Bedgar, D. L.; Martin, D. M.; Sarkanen, S.; Lewis, N. G. *Science* **1997**, *275*, 362-366.
47. Mittler, R.; Lam, E. *Plant Physiol*. **1995**, *108*, 489-493.

Chapter 9

Coniferyl Alcohol Oxidase: A New Enzyme Spatio-Temporally Associated with Lignifying Tissues

R. A. Savidge[1], P. V. Udagama-Randeniya[2], Yijun Xu[1], V. Leinhos[3], and H. Förster[4]

[1]Faculty of Forestry and Environmental Management, University of New Brunswick, Fredericton, New Brunswick E3B 6C2, Canada
[2]Department of Zoology, University of Colombo, Colombo 3, Sri Lanka
[3]Institut für Ernährung und Umwelt, FSU Jena, Naumburger Straβe 98, 07743 Jena, Germany
[4]Pharmazeutisch-Biologische Fakultät, Institut für Allgemeine Botanik, FSU Jena, Bereich Pflanzenbiochemie, Mittelhäuser Straβe, 99189 Kühnhausen, Germany

Coniferyl alcohol oxidase (CAO), a catechol oxidase (*o*-diphenol:oxygen oxidoreductase, E.C.1.10.3.1) specifically associated with lignification during wood formation in conifers, is described and its activity is compared with that of peroxidase. CAO activity is not affected by provision of H_2O_2, and although CAO is an O_2-requiring enzyme able to oxidize substrates in common with laccase, CAO is distinct from laccase in catalytic rate, amino acid composition, response to effectors, copper content and type, protein size, and immunodetectability. Peroxidase activity, readily detected in phloem, cambium and xylem, was inversely correlated with lignification. In contrast, CAO activity was restricted to the lignifying zone. Peroxide and superoxide anion were found in the phloem and cambial zone, respectively, but a pool of H_2O_2 in support of lignification could not be detected in lignifying xylem or elsewhere. Additional evidence has indicated that coniferin hydrolysis may be the rate limiting step in the supply of coniferyl alcohol for oxidation by CAO.

During investigations into peroxidase involvement with lignification, coniferyl alcohol oxidase (CAO) was serendipitously discovered in 1990 as an oxygen-requiring *p*-anisidine-oxidizing enzyme associated, both spatially and temporally, with active lignification during wood formation in conifers (*1-7*). CAO is a tenaciously bound cell-wall enzyme. Pulverized lignifying walls retain CAO activity even after being thoroughly extracted with molar salt solutions containing detergent (*3-5*). Thus, initially, CAO could be obtained from differentiating tracheids as a soluble preparation only by digesting the lignifying walls with a cellulase-pectinase mixture.

In addition to CAO, there have been reports of 'laccase' or 'laccase-like' enzymes associated with lignification (*8-12*). The evidence to date, however, indicates that those enzymes are distinct from CAO, CAO having been recently characterized as a catechol oxidase (*7*). It remains unclear whether CAO or laccase, or even peroxidase, actually participates in the catalysis of lignin polymerization in conifers. Here, the activity of peroxidase is compared and contrasted with that of CAO, and the matter of substrate supply in support of lignification is also examined.

Background to the CAO Discovery

The physiological research of one of us (RAS) has, since 1980, had the aim of identifying a mobile tracheid-differentiation factor (TDF) produced in conifer needles (*13-15*). The TDF has been identified tentatively as *myo*-inositol 1,4,5-*tris*phosphate (*15*); however, much research remains to be done before any conclusions on the TDF can be made. Nevertheless, it seems clear that, when exported from the needle into the stem, the TDF controls secondary-wall formation in cambial derivatives as they differentiate into tracheary elements (TEs).

Lignin deposition is one of several sub-cellular processes (including bordered-pit development, cellulose and hemicellulose production, and protoplasmic autolysis) that contributes to secondary-wall formation during TE differentiation in conifers. During differentiation of a cambial derivative, each process is initiated at a different point in time, indicating independent regulation of each. Lignification begins only after TE differentiation is well under way. Knowing this, and having evidence that TE differentiation was initiated by the TDF, the first attempt to isolate and characterize the TDF focused on detecting early differentiation features, such as bordered-pit development (*16, 17*) and cellulose deposition (*18, 19*), and lignin considerations were deliberately avoided. However, the TDF proved to be elusive in anatomically based bioassays (*15, 16*); consequently, the focus became biochemically oriented. As described below, the discovery that springtime appearance of coniferin in the cambium is a biochemical indicator of impending TE differentiation unexpectedly led us squarely into lignin biochemistry!

The Conifer Cambium as an Ideal Tissue for Biochemical Investigations. The stem of a conifer is an ideal object for biochemical investigation, not only because several discrete developmental zones become sequentially distinguishable by microscopy, but also because those different zones can be readily harvested for investigation using the straight-forward procedure of bark peeling. The zones can best be described in terms of how they arise (*20*). In late winter, the still-dormant cambial zone (CZ) appears in cross section as a radially narrow (5-20 μm) zone of small diameter cells sandwiched between phloem and mature latewood TEs. The bark will not peel from the wood at that time, and the CZ can only be obtained by paring away the bark with a sharp knife until a very thin translucent layer remains on the mature wood. The ability to peel the bark commences in late spring, immediately before the resumption of cell-division activity in CZ cells. At this time, the CZ adheres to and can be gently scraped from the wood surface as a pure meristematic tissue, free of contaminating phloem or xylem. Following resumption of cambial cell division activity (or 'cambial reactivation'), the innermost CZ cells—those adjoining the latewood—cease dividing and commence enlarging in the radial direction through the process of primary-wall growth. At that developmental stage, the enlarged cambial derivatives are thin-walled, >90% water, and in no way resemble woody elements. Hence, a zone of radial expansion (RE) between the previous year's latewood TEs and the radially narrow CZ cells can be readily distinguished. From the appearance of the RE throughout the remainder of the growing season, the bark can be easily stripped from the wood, the separation invariably occurring through actively enlarging cambial derivatives in such a way that the CZ adheres to the inner bark while the RE remains adhering to the mature or maturing wood (*21*). As the season progresses into late spring, cell-division activity continues and cells of the CZ and RE accumulate, in such a way that each of the two zones increases in radial width. Depending on the position in the tree stem, those enlarged RE cells adjoining the latewood sooner or later (beginning first in the foliated stem regions) commence bordered-pit development, a clear indication that TE differentiation has been initiated. This is followed by general secondary-wall (S1) polysaccharide deposition and the formation of yet another zone, the 'SL', where S1, S2 and S3 polysaccharide deposition and lignification occur. Following appearance of the SL, TE differentiation continues to be initiated in RE cells, progressing outward through the radial files cell by cell. The SL cells adjoining latewood TEs are the first to undergo protoplasmic autolysis and mature into earlywood TEs, and TE maturation also progresses centrifugally (*1, 2, 4, 5, 17, 21, 24*).

Discovery of Precocious Coniferin Accumulation in Relation to Lignification. In attempts to isolate and characterize the TDF, it was decided to determine whether any UV-absorbing endogenous metabolites of the cambial region could be identified as indicators of TE differentiation. Metabolic profiling of the different zones (CZ, RE, RE + SL) by reversed phase (C_{18}) high performance liquid chromatography (HPLC) revealed a conspicuously diagnostic peak, not present in the dormant CZ, arising at the time of CZ reactivation, before RE formation (*22*). The concentration of this metabolite increased in the CZ and was even greater in the RE zone at the time of RE formation. With appearance of the SL, the metabolite became the dominant peak in both the CZ and the RE+SL (*22*). The HPLC metabolite was collected, chemically characterized, and found to be *E*-coniferin (*23*). Subsequent work using a variety of methods (*24-27*), including ^{1}H NMR spectroscopy (Figure 1), repeatedly confirmed that the appearance of coniferin in the cambium was spatio-temporally associated with cambial growth in conifers. Coniferin is, by far, the dominant aromatic glucoside in the cambial region, making its quantitation by ^{1}H NMR spectroscopy straightforward (Figure 1).

As indicated in Figure 2, coniferin is derived from coniferyl alcohol by transglucosylation. Coniferyl alcohol generally is considered to be the monolignol precursor for guaiacyl lignin formation. The HPLC metabolic-profiling investigations made it apparent that coniferin, and hence coniferyl alcohol, biosynthesis occurred well before lignification was initiated (*22-24*). It was deduced, therefore, that a mechanism for diverting coniferyl alcohol into coniferin in the absence of concomitant lignin deposition must exist in the cambial region of conifers (*22-24, 27*). In other words, although there is abundant evidence that exogenous coniferin can support lignification (*23-29*), the presence of endogenous coniferin in a tissue is of itself not indicative of attending lignification.

In relation to the TDF research, the finding of precocious coniferin production, combined with the observation that coniferin was not present in dormant CZ, constituted evidence that a regulatory factor was initiating the gene expression underlying the monolignol pathway to guaiacyl lignin at the time of CZ reactivation, evidently in anticipation of impending lignification. On the other hand, the fact that coniferin accumulated in the RE (Figures 1b and 3)—rather than being utilized in concomitant lignification—indicated that an additional regulatory factor, one controlling conversion of coniferin into lignin, must also be involved in the control of TE differentiation.

Coniferin Metabolism in Relation to Coniferyl Alcohol Availability. The coniferyl alcohol supply to the lignifying wall from the coniferin pool requires β-glucosidase activity (Figure 2). Such activity was obtained in the cambial region as an acidic enzyme(s) specific to coniferin (*30, 31*). In a preliminary report (*27*), we also described how uridine diphosphoglucose:coniferyl alcohol glucosyltransferase activity varied in parallel with the rate of coniferin accumulation during seasonal cambial growth (Figures 3a and 3b).

Coniferin-specific β-glucosidase activity was detected in the RE, prior to SL appearance, as well as in the RE+SL (*30*). Thus, on the assumption that there is not a sub-cellular compartmental barrier between the site of coniferin generation and β-glucosidase activity, the potential exists for coniferin to be hydrolyzed to coniferyl alcohol at the RE developmental stage, before lignification has commenced. On the other hand, the activity of coniferin-specific β-glucosidase was never found to be strong in RE, being most readily obtained from the RE+SL (*30*). In contrast, UDPG:coniferyl alcohol glucosyltransferase activity was readily detected in the active CZ, before RE or SL appeared, as well as in the RE and SL later, as shown in Figure 3. We investigated the possibility that the glucosyltransferase might also

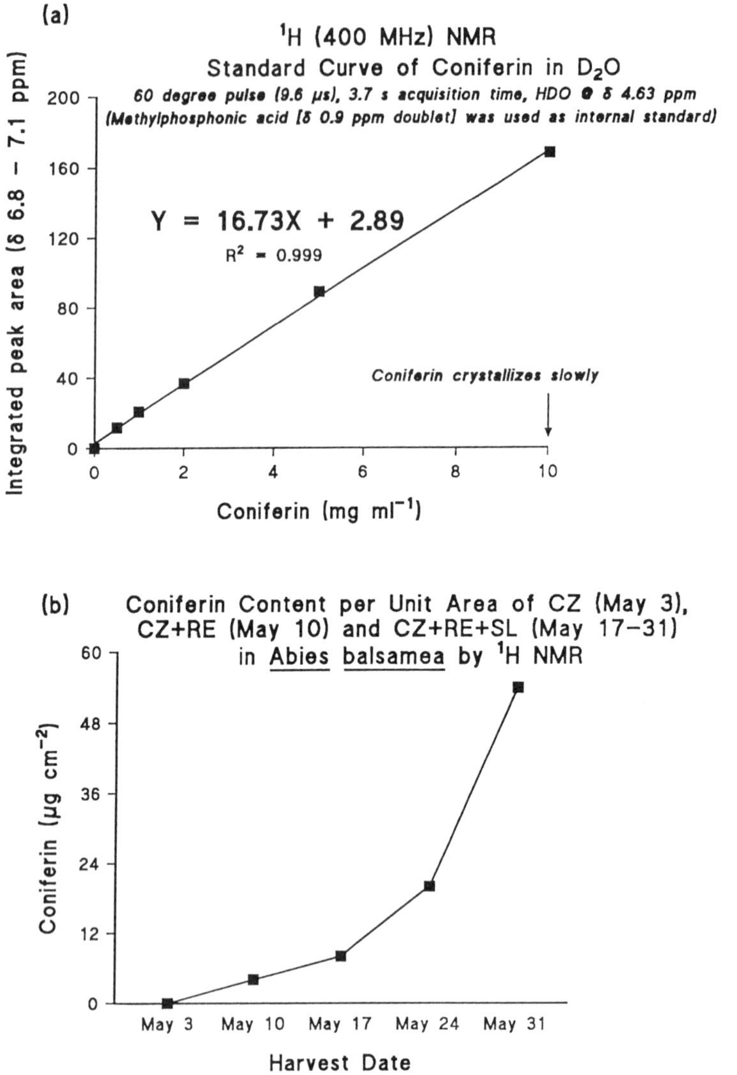

Figure 1. Rapid estimation of coniferin content in *Abies balsamea*, using ^1H NMR spectroscopy (400 MHz). Using destructive sampling of individual trees on each date, bark was peeled from the wood; the cambial surface area was measured; the exposed inner bark face and xylem surface were then scraped together into liquid N_2. After pulverisation (mortar and pestle), the powders were weighed and extracted (1 mL g^{-1} fresh weight, 4°C, 4 h) in buffer (per 100 mL D_2O:1.08 g Tris; 1.01 g boric acid; 1.90 g $EDTANa_4$; 0.10 g SDS; adjusted to pH 8.3 with acetic acid), using 1.00 mg of methylphosphonic acid as internal standard. The extracts were centrifuged (20 000 x g, 4°C, 30 min.) and the supernatants analyzed. The areas of aromatic peak (δ 6.8 - 7.1 ppm) were integrated and the coniferin content interpolated from a standard curve (a). The results (b) show a low level of coniferin in the RE (May 10), the level increasing as the SL appeared (May 17-31).

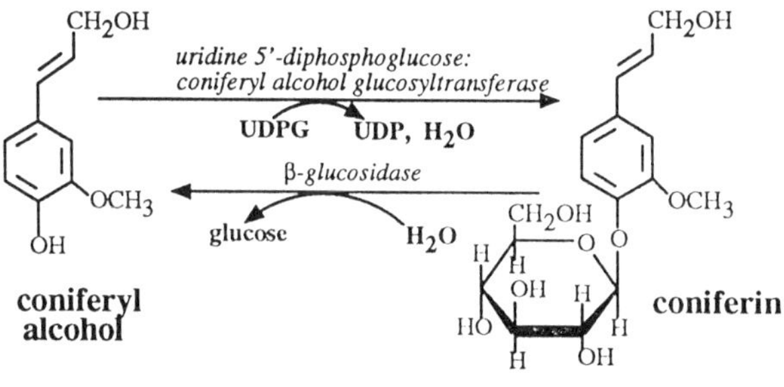

Figure 2. Coniferyl alcohol—coniferin interconversion as it occurs in conifers.

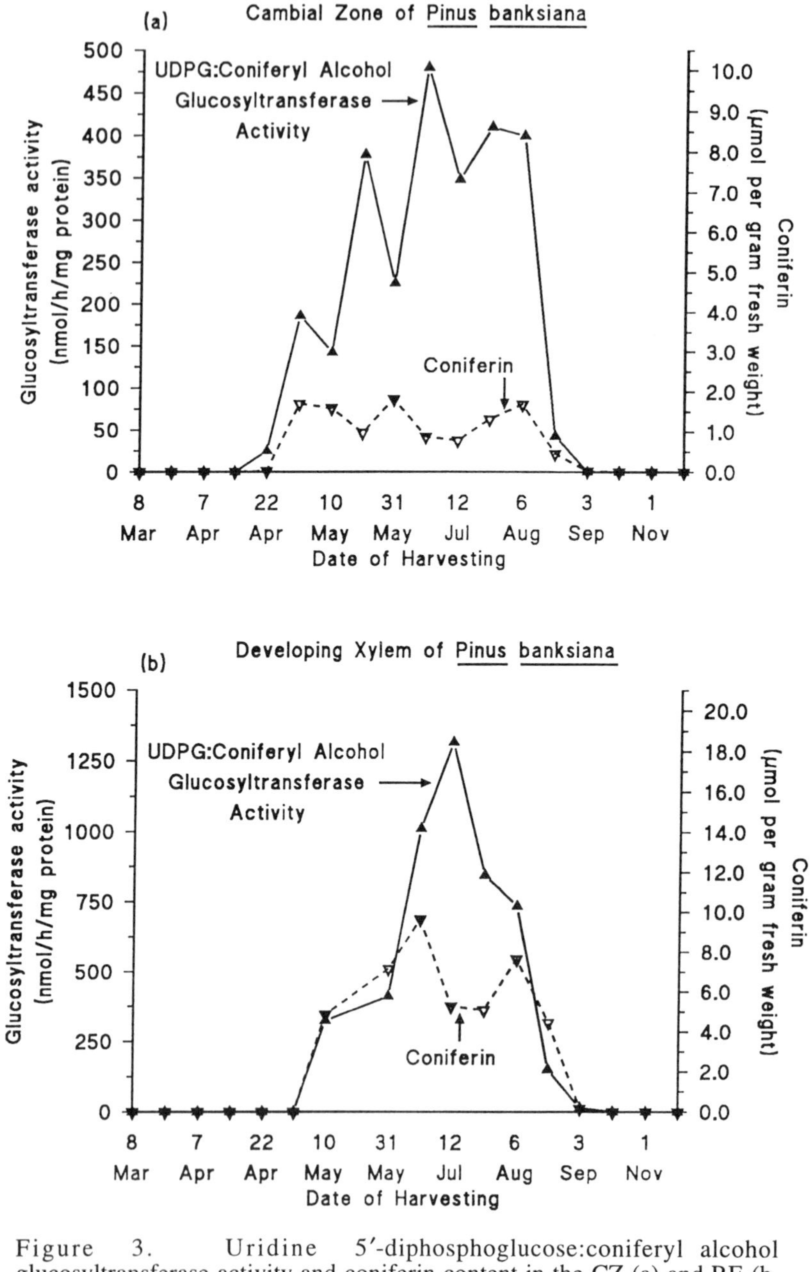

Figure 3. Uridine 5′-diphosphoglucose:coniferyl alcohol glucosyltransferase activity and coniferin content in the CZ (a) and RE (b, May 10) and RE+SL (b, May 31-Sept. 3) of *Pinus banksiana*. The presence of both coniferin and glucosyltransferase activity was qualitatively associated with the period of cambial growth.

function as a β-glucosidase; however, no evidence was found for this under the conditions tested. Thus, our evidence to date indicates that the catalytic environment in the cambial region of conifers favors coniferin accumulation, and that the factor(s) controlling coniferin-specific β-glucosidase activity (*e.g.* gene expression, RNA editing, post-translational processing, half-life of the enzyme) may be the key to regulating the rate, duration and extent of lignification. *In vitro* cambial culture evidence was obtained that an auxin-cytokinin phytohormone combination may have a role in activating coniferin specific β-glucosidase activity (*23*). Investigations into auxin-cytokinin regulation of coniferin-specific β-glucosidase activity remains to be performed.

While searching for coniferin-specific β-glucosidase, we also investigated the possibility that coniferin may itself be polymerized directly into lignin, in the absence of preceding hydrolysis. However, no evidence was found for this. For example, using crude horseradish peroxidase preparations, coniferin was found to be totally resilient to oxidation by peroxidase/H_2O_2. The same was found with CAO.

Peroxidase and H_2O_2 in Relation to Lignification

Once it was known that coniferin accumulated precociously, we considered that limiting peroxidase activity might be the explanation for coniferyl alcohol not being utilized directly in lignin deposition. This hypothesis was favored by the information that peroxidase was the only enzyme in vascular plants capable of catalyzing lignification (*32-34*). Various studies had reported correlations between peroxidase activity and lignification, and the questions remaining seemed to be concerned with precisely which peroxidase(s) was involved in lignin polymerization and how the requisite H_2O_2 was generated (*35-44*).

As an aside, it should be noted that the guaiacyl lignin content of softwoods (*i.e.* conifer wood) is about 30% by weight. Thus, in one gram of dry wood, approximately 1.67 mmol coniferyl alcohol must be polymerized into lignin. On the simplistic assumption that one H_2O_2 molecule is required for peroxidase oxidation of every two molecules of coniferyl alcohol into a dilignol, then 0.84 mmol equivalents (1.67 ÷ 2) of the H_2O_2-source molecule must be oxidized or, possibly, reduced in order to supply the requisite H_2O_2 for the first oxidative step in lignification. As explained above, cambial derivatives in the process of lignification can be rapidly harvested from sapling conifers in large quantity, and we therefore thought that H_2O_2 should be readily quantifiable. Thus, we undertook the investigations described below into endogenous peroxidase and H_2O_2.

Peroxidase Histochemistry. Peroxidase activity was observed histochemically in relation to progressive development of active CZ, followed by RE and SL. The species investigated were *Abies balsamea* (L.) Mill., *Larix laricina* (Du Roi) K. Koch, *Picea rubens* Sarg., *Pinus banksiana* Lamb., and *Pinus strobus* L. Sections containing live cambium were cut by hand from different positions in the tree, mounted in water, and examined in the microscope. In the presence of exogenous H_2O_2 (10 mM), with *p*-anisidine-HCl (20 mM) as histochemical substrate, the phloem and dormant cambium exhibited readily detectable peroxidase activity. Following springtime renewal of cambial cell division activity, both primary and secondary cell walls in cambium, RE, SL and also mature wood zones developed red to purple coloration rapidly, indicating active peroxidase to be abundantly present. All stem samples, regardless of species, stage of cambial growth, or position within the tree, displayed the ubiquitous presence of active peroxidase during all seasons.

Peroxidase Kinetics. Following the histochemical investigations, endogenous peroxidase activity was quantified (as horseradish peroxidase units, Figure 4) in relation to the seasonal formation of the active CZ, RE and SL zones. The method involved bringing stem segments from the forest to the laboratory, determining the developmental stage by microscopy, bark peeling, and scraping of the inner bark face

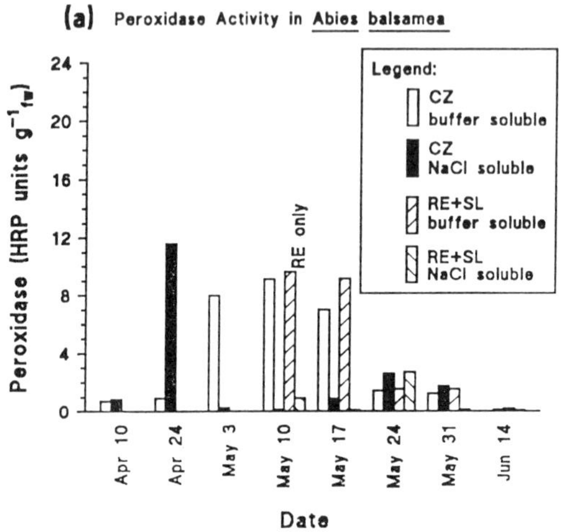

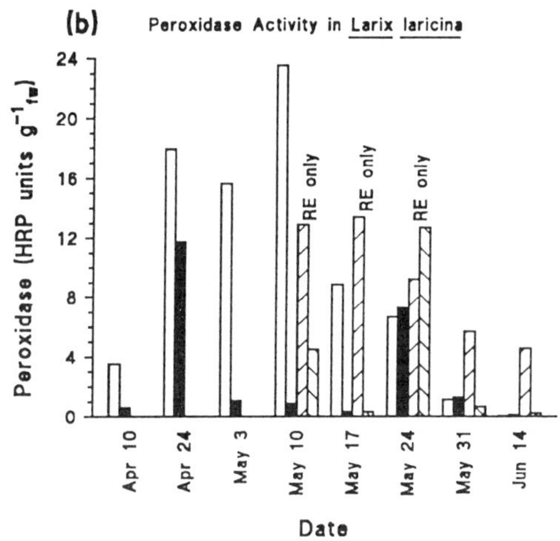

Figure 4. Peroxidase activity (as horseradish peroxidase units per gram fresh weight of tissue) in the cambial region of balsam fir (a), tamarack (b), red spruce (c) and jack pine (d) on successive dates from dormancy (April) to lignification of earlywood (May 24 - June 14). 'RE only': primary-walled radially expanded cambial derivatives present, but no SL could be detected. On dates preceding the 'RE only' stage, only CZ/phloem was analyzed; on dates following, RE+SL were analyzed together.

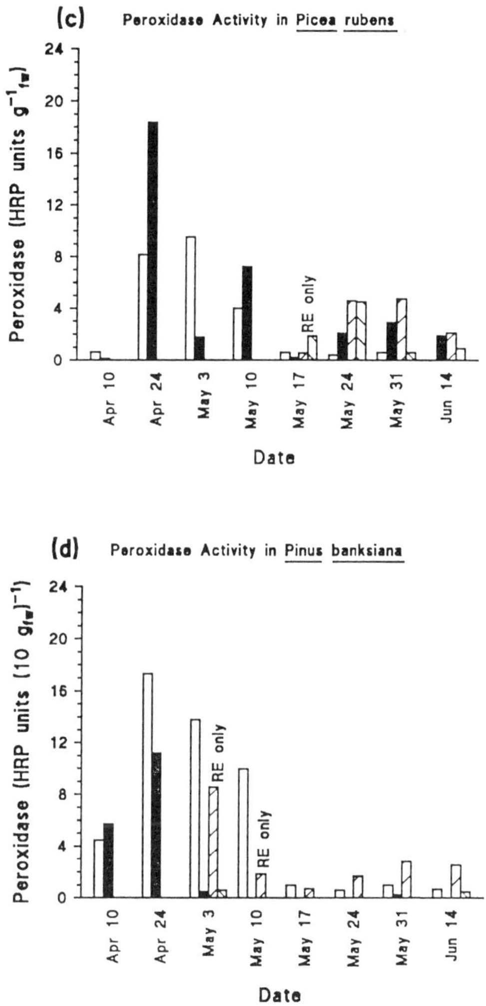

Figure 4. *Continued.*

(*i.e.* CZ, with some phloem) and exposed wood surface (*i.e.* RE, or RE+SL, depending on the harvest date) into liquid nitrogen. After pulverizing in liquid nitrogen, the frozen powder (100 mg) was extracted with phosphate-citrate (PC) buffer (pH 5.0, 50 mM, 1 mL) for 1 h at 4°C, then centrifuged (20 000 x g, 4°C, 15 min.). Aliquots (100 μL) of the supernatant ('buffer soluble') were analyzed in triplicate in PC buffer, with guaiacol (20 mM) and H_2O_2 (12.5 mM) as substrates, using zero-order kinetic approximations (416-488 nm, Hewlett-Packard HP8452 diode-array instrument, 2 nm/photodiode) at 25°C. The pellet was resuspended in NaCl (1 M) in PC buffer at 4°C, extracted for 1 h, centrifuged and the supernatant ('NaCl soluble') analyzed as described above.

Figure 4 shows the peroxidase activity found in four of the species (*P. strobus* is not shown, due to space limitations). On bars where 'RE only' is indicated, no SL was present on that harvest date. For each of the five species, peroxidase was abundantly available in the RE. Thus, the absence of active peroxidase was not obviously the explanation for coniferyl alcohol being converted into coniferin rather than lignin in RE. Unexpectedly, soluble peroxidase activity decreased, rather than increased, in association with the appearance of the SL (Figure 4). Peroxidase activity in the insoluble cell-wall residues was not quantified; however, by histochemical examination there clearly was some residual bound peroxidase present in NaCl-extracted RE and SL.

H_2O_2 Content of the Tissues. Using the same frozen (-196°C) powders which yielded the data of Figure 4, we investigated total peroxide (H_2O_2 and organic peroxides) content in the tissues. Cold trichloroacetic acid (TCA, 5.0%, 0.5 mL) was added to 100 mg frozen powder and the suspension centrifuged at 20 000 x g (4°C, 10 min.). Supernatant (250 μL) was pipetted into TCA (1.35 mL, 5%) followed by 0.4 mL of 50% TCA and 0.4 mL of ferrous ammonium sulfate (10 mM). Potassium thiocyanate (2.5 M, 0.2 mL) was added; the reaction mixture (pH 1.1) was vortexed, and color was permitted to develop for 10 min., at which time the absorbance at 480 nm was measured. Results were interpolated from a standard curve based on known quantities of H_2O_2.

Total peroxides in dormant CZ+phloem (early spring, no developing xylem present) of all five species were in the range of 0.02 - 0.09 μmol per 100 mg of tissue. This was in agreement with the histochemical observation that peroxidase substrates provided to freshly cut live sections in the absence of added H_2O_2 were oxidized, particularly in the region of the phloem, during cambial dormancy. With resumption of cell division activity in the CZ, the total peroxide values in the inner bark (CZ+phloem) declined to trace levels, and total peroxides in RE and RE+SL were below detectability in all species, even when large masses of tissues were analyzed. Our results contrast with what was reported for hardwoods, where substantial quantities of organic peroxide were reported in stems (*49*). The fact that peroxidase activity declined strongly in association with seasonal xylogenesis (Figure 4) is an indication that the parallel decline in instantaneous peroxide concentration was not due to heightened utilization of peroxide by peroxidase. Moreover, our instantaneous peroxide estimates do not seem to be in accord with our observation (see below) of IAA oxidase activity existing in RE and RE+SL tissues. Plausibly, the rate of peroxide decomposition by catalase became elevated in association with resumption of cambial growth; however, our limited observations on seasonal catalase activity did not support that hypothesis. Another possibility is that changes in the instantaneous peroxide content as measured in the tissue powders reflected changing rates at which superoxide anion (O_2^-) was being produced, since O_2^- transformation into H_2O_2 occurs extremely rapidly. Data relevant to this consideration are presented below.

Using KI, I_2 staining of starch can be demonstrated histochemically where lignification is occurring. This has been interpreted as evidence for H_2O_2 being associated with lignification (*45*). It also deserves consideration, however, that either O_2 (with or without an associated oxidase) or some other oxidizing agent may be at least part of the explanation. For example, we found evidence for I_2 production in thoroughly salt-washed, and even dead (boiled, ethanol-extracted), sections. The secondary walls of xylem elements harbor a diverse mixture of inorganic components (*46-48*), some of which may serve as catalysts to oxidize I^- to I_2. The possibility that deposits of inorganic materials in cell walls may participate in lignification also deserves consideration.

H_2O_2 Production by Peroxidase Functioning as IAA Oxidase. Indol-3-ylacetic acid (IAA) oxidase yields H_2O_2, and thus it is a potential peroxide source in support of peroxidase-catalyzed lignification (*50, 51*). It is well established that exogenous IAA promotes xylogenesis, and in the cambial region endogenous IAA was found to be concentrated in the RE and RE+SL (*21, 52*). However, within individual cells it remains uncertain precisely where IAA is localized; the plasma membrane and protoplasm are the most likely locations (*53, 54*). Cytosolic Ca^{++}-dependent IAA oxidase activity was readily detected in the cambial region of conifers, but activity was also noted in the cell-wall residue (*55*, Figure 5a).

As explained above, a minimum of 0.84 mmol (1.67 ÷ 2) IAA per gram of tissue would have to be oxidized if IAA in fact were the primary source of H_2O_2 in support of the first step in peroxidase-catalyzed lignification. The maximum instantaneous IAA content in RE+SL, as reported by combined gas chromatographic-mass spectrometric (GC-MS) estimation, was near 20 $\mu g\ g^{-1}$ dry weight, or 0.11 µmol (*1, 2, 21, 24, 52*). (It should be clear, however, that GC-MS estimates of IAA content in developing xylem remain crude in the sense that cells at all stages of xylem development—from RE through SL to fully lignified and autolyzing— are combined when the tissues are weighed and extracted. In the future, the IAA content in the region of lignification may be found to be very different from that in the RE or autolyzing regions.) Thus, with current information, it would be necessary that 0.11 µmol of IAA be continually oxidized and replaced, more than 7 600 times, in order to contribute 0.84 mmol H_2O_2. In addition, there would have to be a massive surplus of IAA remaining unoxidized and unavailable for basipetal transport to more basal regions of the tree stem (this would have to be the case unless IAA biosynthesis occurs in the cambial region; current evidence indicates that all cambial IAA is derived from buds and needles). Basipetal IAA transport rates are about 1 cm h^{-1}. Therefore, 7 600 replacements of 0.11 µmol IAA would require approximately 7 600 h, or 316 days. This does not agree with known rates of lignification within individual tracheids which, although generally slow, are nevertheless completed within a few weeks. It seems unlikely, therefore, that IAA is the principal agent for H_2O_2 production if peroxidase-catalyzed lignification actually occurs. Nevertheless, some contribution from IAA in the overall process cannot be excluded.

To investigate IAA oxidase activity, [1-^{14}C]IAA was incubated in solution with protein preparations from RE and RE+SL tissues. Soluble, or 'sol', protein (Figure 5) was obtained through phosphate buffer (pH 7.8, 0.05 M) extraction of tissue. Ionically bound, or 'ionic', protein (Figure 5) was obtained by re-extracting the tissue residue with 1 M NaCl/phosphate buffer. The low protein contents in these extracts were quantified by the Bradford method (*5*). The evolved $^{14}CO_2$ was trapped in alkali, and aliquots were examined with a scintillation counter. The results (Figure 5a) revealed that IAA oxidase activity in RE was no less than that in RE+SL, and that residual IAA oxidase activity in cell walls of RE ('insol', Figure 5a) was much greater than that in the walls of SL. Endogenous IAA varies seasonally, becoming elevated in association with development of the RE zone (*21*). Although the results in relation to a lignification mechanism were inconclusive, the findings were nevertheless interesting because they indicated the potential for H_2O_2 generation through IAA oxidase activity in the cambial region, despite the fact that we were unable to detect H_2O_2 directly during cambial growth.

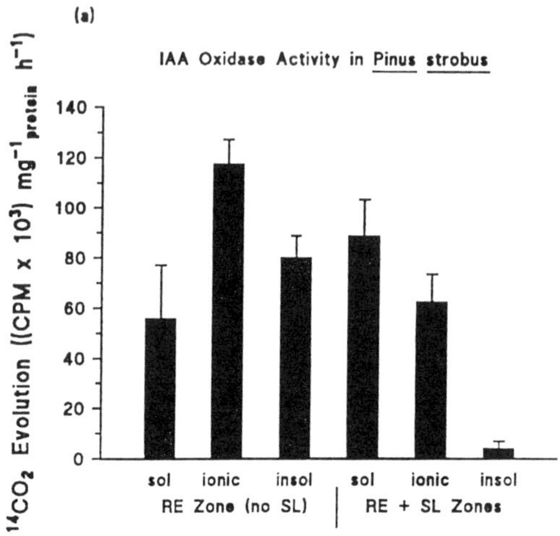

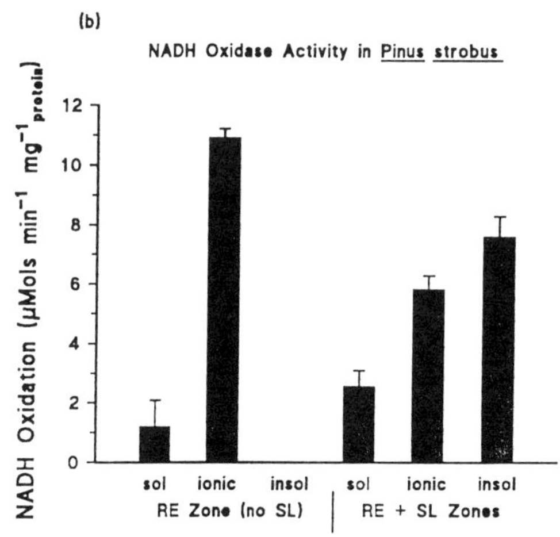

Figure 5. Evidence for IAA oxidase (a), NADH oxidase (b), dihydroxyfumaric acid oxidase (c) and peroxidase (d) activities in RE and RE+SL of *Pinus strobus* in early spring. The bars denote standard errors of the mean (n=3).

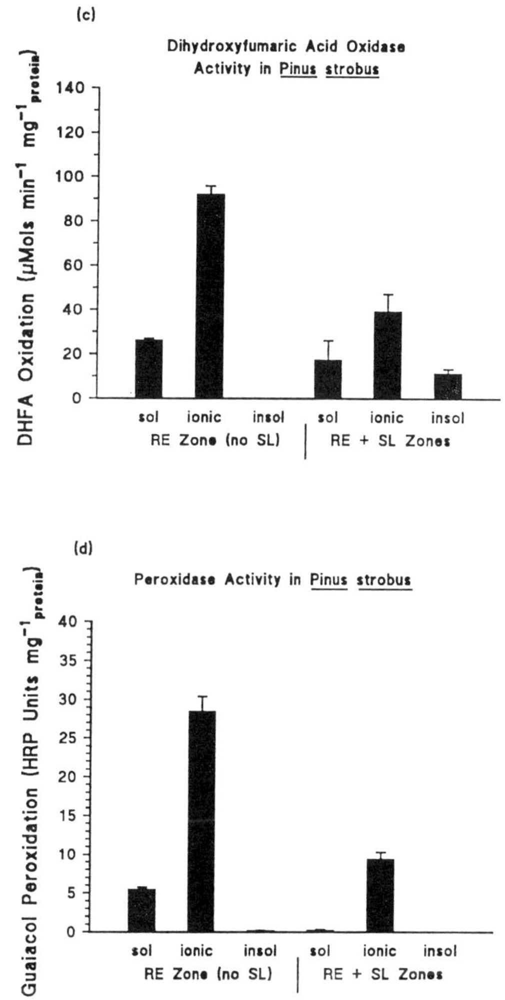

Figure 5. *Continued.*

Others Possible Sources of H_2O_2. No evidence for polyamine or glucose oxidation by proteins from conifer stems was found. Analyzing the preparations used for IAA oxidase (Figure 5a), NADH (Figure 5b) and dihydroxyfumaric acid (DHFA, Figure 5c) were oxidized by both soluble and salt-extracted RE and RE+SL. In the NaCl-insoluble cell-wall residue of RE+SL, NADH and DHFA were oxidized at a similar rate, approximately 8 µmol min^{-1} per mg of protein, but no oxidation of either NADH or DHFA by salt-extracted RE cell-wall residues was found. If NADH or DHFA can enter the SL cell wall, H_2O_2 could be generated within by these substrates. Figure 5d provides evidence for the level of peroxidase activity (20 mM guaiacol oxidation in the presence of 10 mM H_2O_2) in the same extracts. In the RE+SL, the enzyme(s) oxidizing guaiacol clearly was distinct from those oxidizing IAA and NADH (compare a, b and d of Figure 5).

H_2O_2 production from O_2^-, and Superoxide Dismutase Activity. It was considered that the histochemical and quantitative evidence for peroxides being present in the phloem might be due to the conversion of O_2^- to H_2O_2, followed by the rapid utilization of that H_2O_2 by peroxidase or catalase. Thus, the contents of O_2^- and superoxide dismutase (SOD) were investigated, again using the tissues already described. Superoxide anion (O_2^-) is converted into H_2O_2 and O_2 by the action of SOD thus: $2\,O_2^- + 2H^+ \rightarrow H_2O_2 + O_2$, and at physiological pH the reaction can also proceed spontaneously in the absence of SOD.

SOD activity was assayed by the method of McCord and Fridovich (*56*). Frozen powder (100 mg) was extracted in phosphate buffer (pH 7.8, 50 mM, 1.0 mL) containing 0.1 mM EDTA for 1 h at 4°C, then centrifuged at 20 000 x g (30 min., 4°C). Aliquots (100 µL) of the supernatants were assayed in the same buffer containing 10 µM cytochrome c and 50 µM xanthine. The reaction was started by addition of xanthine oxidase (90 µg protein, 0.44 U mg^{-1}). Rate constants were determined at 550 nm using a zero-order kinetic approximation, and the SOD activities in extracts were interpolated from a standard curve (0-10 U SOD, from horseradish). In all species, dormant phloem/CZ tissue exhibited very high SOD activity (200 - 500 U g^{-1}), and the activities remained high upon resumption of cell-division activity in the CZ. SOD levels in RE were similarly high; however, with the appearance of the SL, SOD activities in the combined RE+SL tissue declined sharply and often were negligible.

Superoxide anion has a half life in the millisecond range. To obtain an instantaneous estimate of the tissue's capacity to generate O_2^-, liquid-nitrogen-frozen powder (100 mg) was extracted with phosphate buffer (pH 7.8, 50 mM, 2.5 mL) containing 50 µM cytochrome c, 250 µg catalase, and 50 mM sodium benzoate for 10 min. at room temperature, then centrifuged (4 000 x g, 5 min., ambient temperature). Absorbances of supernatant aliquots were measured at 550 nm, using potassium superoxide as calibration standard. As indicated in Figures 6a and 6b, in *Abies balsamea* and *Larix laricina* essentially no O_2^- was produced by the RE and SL powders. In contrast, the inner bark phloem+CZ produced relatively high amounts of O_2^- in these same species. Thus, although a deficiency in O_2^--derived H_2O_2 might have been the explanation for coniferyl alcohol not being utilized by peroxidase at the RE stage, this reasoning was contradicted by the absence of O_2^- generation in the SL zone. In the other species (Figures 6c, 6d), O_2^- was found to be produced by RE or RE+SL powders, but the rate of O_2^- production was much higher in their phloem-CZ tissues. Results with *Pinus banksiana* (Figure 6d) and *Pinus strobus* (not shown) RE further indicated the capability for O_2^- production at the time when coniferin was accumulating.

It has been suggested that radial fluxes in water potential between xylem and phloem may be an important factor regulating cambial growth (*58*); however, it is not known whether O_2^- and/or H_2O_2 produced in phloem move radially into the RE+SL region of the stem. It is unlikely that O_2^- will cross an intact phospholipid membrane; moreover, O_2^- seemingly has too short a half life to be conveyed as such to sites of lignin deposition. On the other hand, H_2O_2 derived from O_2^- should move unimpeded through cambial cell membranes, and symplastic transport of H_2O_2 through ray-cell plasmodesmata may also occur (*58*). On the assumption that radial H_2O_2 movement is not constrained, and thus that H_2O_2 is supplied to developing xylem from the region of the phloem or cambial zone, our results indicate either that coniferyl alcohol is not at all available to wall-bound peroxidases or that those peroxidases are unable to utilize coniferyl alcohol. The former possibility seems the more probable. Presumably, coniferyl alcohol is rapidly converted to coniferin in a sub-cellular compartment devoid of oxidizing agents, and coniferyl alcohol derived from coniferin hydrolysis is carried into the wall by a specific transporter.

O_2, as well as H_2O_2, arises from O_2^- dismutation, and our results (Figure 6) therefore indicate that the CZ and adjoining phloem could be a major source of O_2 in support of cambial growth and vascular development. Long distance O_2 transport through the stem has previously been considered essential for maintenance of aerobic status, but it deserves consideration that O_2 generated by O_2^- dismutation could drive the activity of other enzymes, such as CAO.

Coniferyl Alcohol Oxidase

Current knowledge about coniferyl alcohol oxidase (CAO) in *Pinus strobus* is summarized in Table I. In addition to the data shown in Table I, work on substrates and inhibitors of CAO has been reported (*6, 7*). CAO activity was enhanced by sodium dodecyl sulfate (1 mM) and inhibited by tropolone and phenylhydrazine, providing supporting evidence for CAO being a catechol oxidase as opposed to a laccase (*7*).

Evidence for aromatic oxidase activity associated both in time and space with lignification, as detected by histochemical responses of 0.5 M NaCl-extracted sections to *p*-anisidine in the absence of added H_2O_2, was first reported in 1990 (*1, 2*). The spatio-temporal correlation between appearance of an oxidizing agent in cell walls and occurrence of lignification (as detected microscopically by UV illumination and histochemical stains) was evident over the stems of whole trees. Figure 7 presents the patterns of oxidizing activity found in the main stem of jack pine (*Pinus banksiana*) having RE only (May 9th) or also SL (May 25th), as determined by microscopy of live sections incubated on H_2O_2-free *p*-anisidine solution. Before initiation of SL, no *p*-anisidine oxidation occurred despite development of CZ and RE (Figure 7, May 9th). Oxidation on May 25th (Figure 7) occurred only in SL and those RE cells on the edge of the SL, and both the SL cell number and associated oxidizing activity decreased basally, disappearing in lower stem regions. It deserves emphasis that the cambium in the older (>6 years) stem regions had generated RE several weeks before either RE or SL development were initiated in younger foliated stem regions; nevertheless, the *p*-anisidine oxidizing activity first appeared in the upper crown of all species. By mid-June (not shown) the basal stem regions also contained the SL zone, at which time *p*-anisidine oxidizing activity was detected there. As the first early-wood tracheids underwent autolysis and maturated, the *p*-anisidine oxidizing activity in their cell walls disappeared. The developmental pattern of *p*-anisidine oxidation described here for white pine was confirmed with all five species during 1989 and again in spring 1990.

Subsequent work focused on isolating and characterizing the anisidine-oxidizing factor (*3-5*). To accomplish this, it was necessary to digest the cell-wall residue of the thoroughly extracted RE+SL using a pectinase-cellulase mixture, thus freeing wall-bound CAO (*5*). It was then found *in vitro* that CAO oxidized coniferyl alcohol to dehydrodiconiferyl alcohol and pinoresinol (*5*). Once recognized as a lignification-associated, oxygen-dependent coniferyl alcohol oxidase (*5*), a procedure was devised to purify the enzyme to electrophoretic homogeneity, and antibodies were produced against both native and deglycosylated CAO (*6*). Using purified CAO and monoclonal antibodies specific to CAO, evidence was obtained that CAO is similar in six conifer species (*Pinus taeda*, as well as the five addressed above) and is neither a laccase nor a peroxidase (*7*).

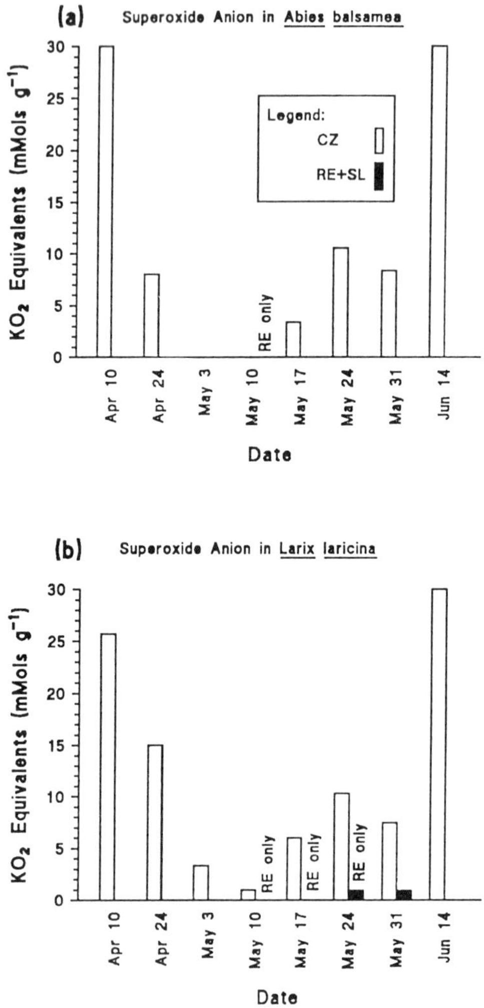

Figure 6. Superoxide anion content of balsam fir (a), tamarack (b), red spruce (c) and jackpine (d) on successive dates from dormancy (April) to lignification of earlywood (May 24 - June 14). 'RE only': primary-walled radially expanded cambial derivatives present, but no SL could be detected. On dates preceding the 'RE only' stage, only CZ/phloem was analyzed; on dates following, RE+SL were analyzed together.

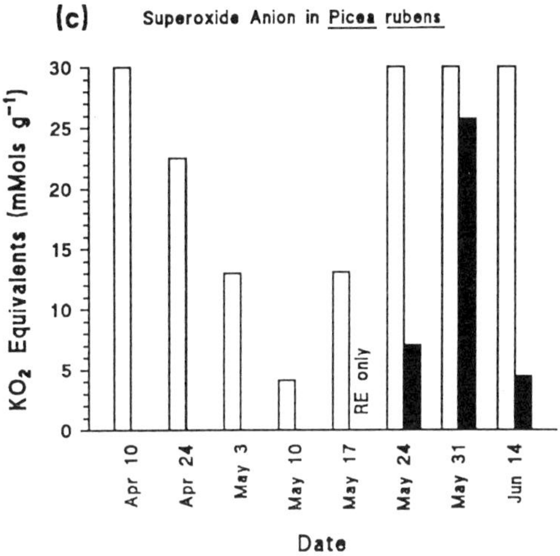

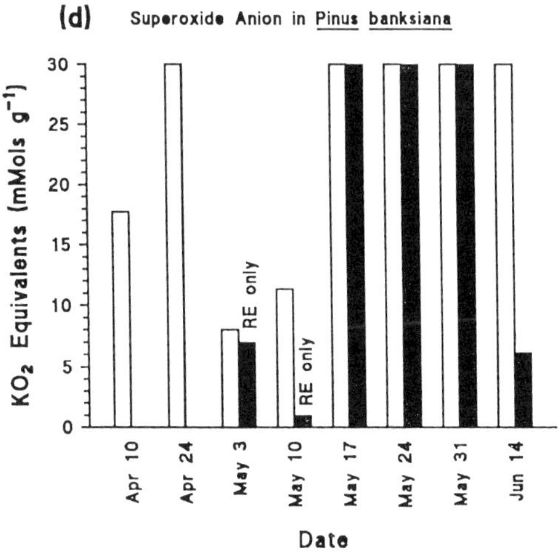

Figure 6. *Continued.*

Table I. Summary of Findings with CAO in Conifers

Characteristic	Result	Reference
1. Amino acid composition:	ASX (8.5), GLX (10.3), SER (6.3), GLY (9.8), HIS (2.2), ARG (4.7), THR (5.2), ALA (9.7), PRO (5.6), TYR (2.7), VAL (7.6), MET (2.8), ILE (4.7), LEU (9.4), PHE (4.9), LYS (5.5)	*7*
2. N-terminal sequence:	H_2N-X E L A Y S P P Y X P S	*7*
3. MOLECULAR WEIGHTS:		
Native CAO:	107 500, by SDS PAGE	*6*
Native CAO:	426 000, by native PAGE	*unpublished*
Deglycosylated CAO:	67 000, by SDS PAGE	*6*
4. Deduced glycan content:	38%	*6, 7*
5. Copper content:	1 atom per protein molecule	*6*
6. Copper type:	type 3 (EPR silent)	*7*
7. ACTIVITY STUDIES:		
Temperature optimum:	30°C	*6*
pH optimum:	6.3	*6*
Oxidation rate:	0.31 μmol h^{-1} mg^{-1} (70 μM coniferyl alcohol)	*7*
Oxidation products:	pinoresinol,	*3-5*
	dehydrodiconiferyl alcohol	*3-5*
Requirement for O_2:	N_2 inactivation	*3-5*
	CO inactivation	*unpublished*
	No H_2O_2 enhancement	*3-5*

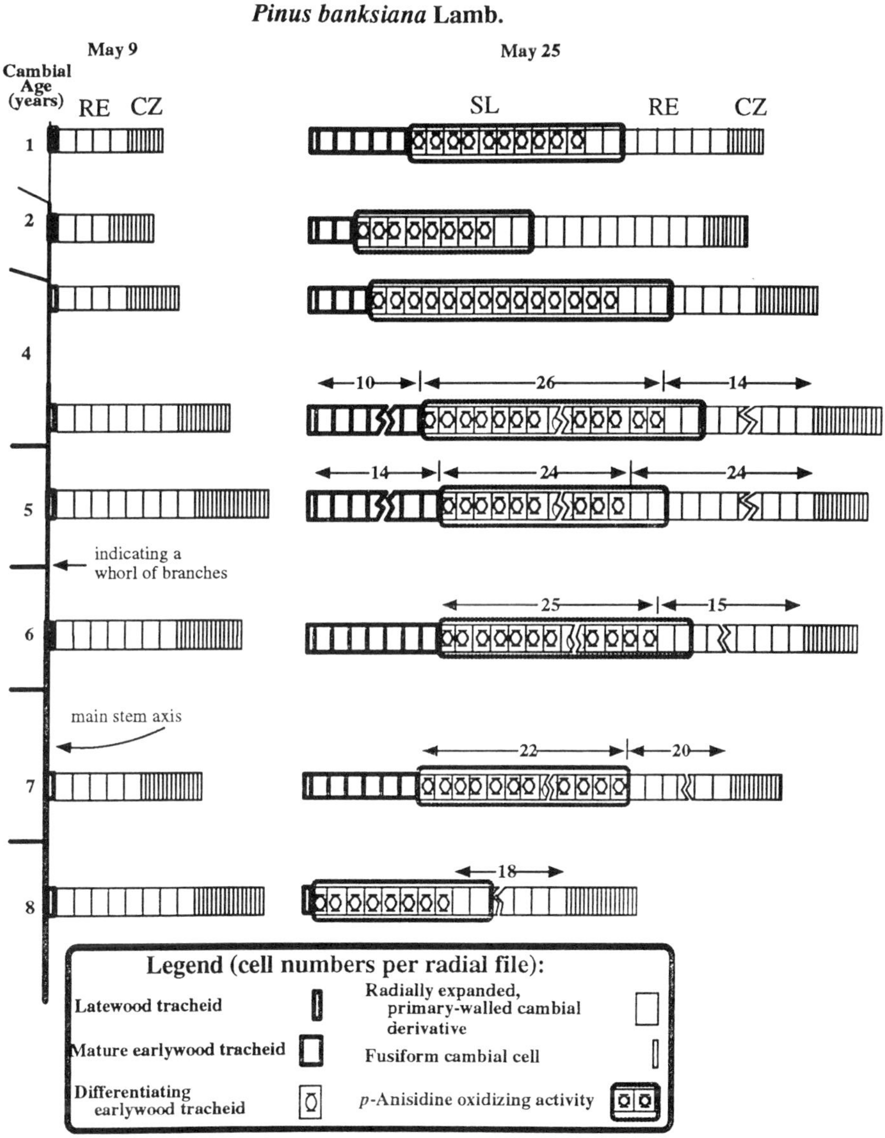

Figure 7. Pattern of emergence of *p*-anisidine oxidizing activity over the main axes of two 12-year-old jack pine trees on two successive dates. On May 9th, when the CZ was actively dividing and RE cells were accumulating, no *p*-anisidine oxidation was detected. On May 25th, *p*-anisidine oxidation was restricted to SL cells either actively lignifying or about to commence lignification.

Purification of CAO. We attempted to isolate the factor oxidizing *p*-anisidine by scraping RE+SL tissue from mature wood following bark peeling. For comparison, RE tissue was also collected from stems earlier in the season, before development of the SL zone. Both before and after phosphate-buffer/NaCl extraction, SL cell-wall residues contained a wall-bound factor oxidizing *p*-anisidine, whereas RE cell-wall residues were devoid of histochemically detectable *p*-anisidine oxidizing activity. No oxidizing activity was found, however, in $(NH_4)_2SO_4$ (90% saturation) precipitates from extracts (0.05 M phosphate-buffer (pH 7.0) / 0.5 M NaCl) of SL tissue, and the activity was not solublized from SL walls by either detergent (0.5% Triton X-100) or salt solutions (0.5 M NaCl, 2 M $CaCl_2$, 3 M LiCl). Incubation (16 h, room temperature) of the cell-wall residue of salt-extracted SL with a pectinase-cellulase mixture resulted in solubilization of >95% by weight of the material. After centrifuging this digest, the supernatant was still devoid of *p*-anisidine oxidizing activity. Repeated re-suspension (distilled H_2O) and centrifugation of the insoluble pellet yielded a particulate preparation enriched in bound protein that oxidized *p*-anisidine, *p*- and *o*-phenylenediamine, dopamine, gallic acid and *o*-dianisidine, but not guaiacol, in the absence of added H_2O_2.

The activity in the particulate preparation was solubilized in NaCl (0.5 M) and filtered through glass microfiber (Whatman GF/F). Upon diluting 1 mL of the filtrate (containing 5-6 mg of protein) ten-fold with 2-[*N*-morpholino]ethanesulfonic acid (MES) buffer (10 mM, pH 5.0, 4°C) and allowing the mixture to stand for 30 min. at 4°C, a precipitate was obtained that could be pelleted and washed with MES buffer. After dissolution in Tris-HCl buffer (pH 8.0, 50 mM), and dialysis against the same, CAO was purified to electrophoretic homogeneity (in both SDS-PAGE and IEF).

It was subsequently found that soluble, as well as cell-wall-bound, CAO activity could be obtained from SL by means of MES precipitation, as described above.

Electrophoretic Assessment of CAO pI. Early isoelectric focusing work led to CAO from *Pinus strobus* being described as having a pI of 9 (*4*); however, this was in error and was subsequently modified to pI = 6.5 (*5*). The other four conifer species investigated were found to have CAO activity at pI = 7.0 (*5*). These pI assessments, however, were based primarily on the use of *p*-phenylenediamine, chosen because it was relatively rapidly oxidized, thus enabling the focused bands to be quickly visualized. Using *p*-anisidine-HCl, it proved difficult to corroborate those pI's, there being only weak white pine CAO activity near pI 6.5 and most near pI 7.5 (*5*). *p*-Anisidine is much less readily oxidized by CAO than many other polyphenol oxidase substrates (*7*); however, because the initial histochemical evidence for CAO being associated with lignification resided in the use of *p*-anisidine-HCl, we imposed the requirement in subsequent work that only bands oxidizing *p*-anisidine be designated as CAO. This requirement was attended by the limitation that gels had to be incubated for a period of several hours to overnight. Although we eventually reassessed the pI of CAO at 7.6 for all six species (*6*), it could well be that isoenzymes do exist at pI 6.5 and 7.0, but were not seen using *p*-anisidine-HCl because they diffused out of the gel during the lengthy incubations.

Oxidation Products from Coniferyl Alcohol Following CAO Incubation. By combined gas chromatography-mass spectrometry (GC-MS), the two major products produced by CAO oxidation of coniferyl alcohol were dehydrodiconiferyl alcohol and pinoresinol (*5*). It remains uncertain whether further oxidation of these dilignols by CAO occurs. If not, although CAO may be an enzyme of lignification, there must be at least one more enzyme utilizing dilignols and involved in production of the lignin polymer.

CAO, Laccase or Catechol Oxidase? CAO oxidizes several laccase substrates in the presence of O_2; however, it is not a blue-copper protein. CAO was found by graphite furnace atomic absorption analysis to contain a single copper atom per molecule. By electron paramagnetic resonance spectroscopy, however, this copper could not be detected, indicating that it is a type 3 cupric ion (*6*), in agreement with the native PAGE evidence (Table I) that CAO may be a tetramer.

The kinetic activity of CAO, both *in vitro* and *in vivo*, appears to be very much lower than that of either laccases or peroxidases (when H_2O_2 is not limiting). On the other hand, the CAO rate may better approximate the very slow rate at which lignification occurs during wood formation *in vivo* (*57*). Slow catalysis is a characteristic of catechol oxidases. Using monoclonal antibodies directed against native and deglycosylated CAO, neither peroxidase nor laccase cross reacted (*7*). On the other hand, the polyclonal antibody preparation from which the monoclonals were selected did recognize laccases (*6*), indicating that there may be some epitopic features (*e.g.* the glycosidic chains) in common between the two proteins. Additional evidence, including native PAGE migration of CAO to Mr 426 000 (an indication that CAO could be a tetramer), amino-acid composition, phenylhydrazine and tropolone inhibition and SDS enhancement of CAO activity, led us to conclude that CAO may best be considered as a catechol oxidase, that is, as an *o*-diphenol oxidase (*7*).

Summary

In conifers, CAO is an oxygen-dependent oxidase capable of oxidizing coniferyl alcohol into dilignols, and the activity of CAO is specifically associated with lignification. Whether CAO actually participates in lignification remains uncertain; however, the same difficulty attends the presumed roles of peroxidase and laccase in production of the lignin polymer. Stable genetic transformants having inactivated gene expression for CAO, peroxidase, or laccase need to be produced in order to clarify the role of each enzyme in lignification.

Acknowledgments

The research was supported primarily by the Natural Sciences and Engineering Research Council of Canada with supplementary funding from the Canadian Forestry Service. H.F. was supported by the Ministry of Science, Research and Culture of Thuringia, Germany.

Literature Cited

1. Savidge, R. A. In *Fast Growing Trees and Nitrogen Fixing Trees;* Werner, D., and Müller, P., Eds.; Gustav Fischer Verlag; Stuttgart, 1990; pp 142-151.
2. Savidge, R. A. In *Proceedings, Division 5, IUFRO XIX World Congress;* Forestry Canada: Montreal, 1990; pp 65-76.
3. Savidge, R. A. *Polyphénols Actualités* **1992**, *8*, 12-13.
4. Savidge, R. A.; Randeniya, P. V. *Biochemical Society Trans.* **1992**, *20,* 229S.
5. Savidge, R. A.; Udagama-Randeniya, P. V. *Phytochemistry* **1992**, *31,* 2959-2966.
6. Udagama-Randeniya, P. V.; Savidge, R. A. *Electrophoresis* **1994**, *15*, 1072-1077.
7. Udagama-Randeniya, P. V.; Savidge, R. A. *Trees* **1995**, *10*, 102-107.
8. Bao, W.; O'Malley, D. M.; Whetten, R.; Sederoff, R. R. *Science* **1993**, *260,* 672-674.
9. Driouich, A.; Lainé, A.-C.; Vian, B.; Faye, L. *Plant J.* **1992**, *2*, 13-24.
10. Sterjiades, R.; Dean, J. F. D.; Eriksson, K-E. L. *Plant Physiol.* **1992**, *99*, 1162-1168.
11. Liu, L.; Dean, J. F. D.; Friedman, W. E.; Eriksson, K-E. L. *Plant J.* **1994**, *6*, 213-224.
12. McDougall, G. J.; Stewart, D.; Morrison, I. M. *Planta* **1994**, *194*, 9-14.
13. Savidge, R. A.; Wareing, P. F. *Planta* **1981**, *153*, 395-404.
14. Savidge, R. A.; Barnett, J. R. *J. Exp. Bot.* **1993**, *44*, 395-407.
15. Savidge, R. A. *Int. J. Plant Sci.* **1994**, *155*, 272-290.

16. Savidge, R. A. *Histochemical J.* **1983**, *15*, 447-466.
17. Leitch, M. A.; Savidge, R. A. *IAWA J.* **1995**, *16*, 289-297.
18. Savidge, R. A.; Colvin, J. R. *Can. J. Microbiol.* **1985**, *31*, 1019-1025.
19. Sternberg, L. S. L.; DeNiro, M. J.; Savidge, R. A. *Plant Physiol*. **1986**, *82*, 421-427.
20. Savidge, R. A. In *Oscillations and Morphogenesis*; Rensing, L., Ed.; Marcel Dekker: New York, NY, 1993; pp 343-363.
21. Savidge, R. A.; Heald, J. K.; Wareing, P. F. *Planta* **1982**, *155*, 89-92.
22. Savidge, R. A. *Can. J. Bot*. **1988**, *66*, 2009-2012.
23. Savidge, R. A. *Can. J. Bot*. **1989**, *67,* 2663-2668.
24. Savidge, R. A. *Forest Science* **1991**, *37*:953-958.
25. Leinhos, V.; Savidge, R. A. *Can. J. Forest Res*. **1993**, *23*, 343-348.
26. Leinhos, V.; Savidge, R. A. *Acta Horticulturae* **1994**, *381*, 97-103.
27. Förster, H.; Savidge, R. A. *Proc. Plant Growth Regulator Soc. Am*. **1995**, *22nd Ann. Mtg*., 390-395.
28. Terashima, N.; Fukushima, K. *Wood Sci. Technol.* **1988**, *22*, 259-270.
29. Xie, Y.; Terashima, N. *Mokuzai Gakkaishi* **1991**, *37*, 935-941.
30. Leinhos, V.; Udagama-Randeniya, P. V.; Savidge, R. A. *Phytochemistry* **1994**, *37*, 311-315.
31. Dharmawardhana, D. P.; Ellis, B. E.; Carlson, J. E. *Plant Physiol.* **1995**, *107*, 331-339.
32. Harkin, J. M.; Obst, J. R. *Science* **1973**, *180*, 296-297.
33. Higuchi, T. *Wood Sci. Technol.* **1990**, *24*, 23-63.
34. Lewis, N. G.; Yamamoto, E. *Annu. Rev. Plant Physiol. Plant Mol. Biol*. **1990**, *41*, 455-496.
35. Wolter, K. E.; Gordon, J. C. *Physiol. Plant*. **1975**, *33*, 219-223.
36. Goldberg, R.; Catesson, A.-M.; Czaninski, Y. Z. *Pflanzenphysiol*. **1983**, *110*, 267-279.
37. Imberty, A.; Goldberg, R.; Catesson, A.-M. *Planta* **1985**, *164*, 221-226.
38. Stich, K.; Ebermann, R. *Holzforschung* **1988**, *42*, 221-224.
39. Dalet, F.; Cornu, D. *Can. J. Bot*. **1989**, *67*, 2182-2186.
40. Pang, A.; Catesson, A.-M.,;Francesch, C.; Rolando, C.; Goldberg, R. *J. Plant Physiol.* **1989**, *135*, 325-329.
41. McDougall, G. J. *J. Plant Physiol*. **1991**, *139*, 182-186.
42. Dean, J. F. D.; Sterjiades, R.; Eriksson, K.-E. L. *Physiol. Plant.* **1994**, *92*, 233-240.
43. Polle, A.; Otter, T.; Seifert, F. *Plant Physiol.* **1994**, *106*, 53-60.
44. Alba, C. M.; Deforchetti, S. M.; Tigier, H. A. *Biocell* **1995**, *19*, 35-41.
45. Olson, P. D.; Varner, J. E. *Plant J.* **1993**, *4*, 887-892.
46. Bowers, L. J.; Melhuish, J. H. *Bull. Envir. Contam. Tox*. **1988**, *40*, 457-461.
47. Lövestam, G.; Johansson, E.-M.; Johansson, S.; Pallon, J. *Ambio* **1990**, *2*, 87-93.
48. Bailey, J. H. E.; Reeve, D. W. *J. Pulp Paper Sci*. **1994**, *20*:J83-J86.
49. Sagisaka, S. *Plant Physiol.* **1976**, *57*:308-309.
50. Ferrer, M. A., Pedreño, M. A.; Muñoz, R.; Barceló, A. R. *FEBS Lett.* **1990**, *276*, 127-130.
51. Ferrer, M. A.; Pedreño, M. A.; Muñoz, R.; Barceló, A. R. *Biochemistry Int.* **1992**, *28*, 949-955.
52. Little, C. H. A.; Pharis, R. P. In *Plant Stems: Physiology and Functional Morphology*; Gartner, B. L., Ed.; Academic Press: 1995; pp 281-319.
53. Sabnis, D. D.; Hirshberg, G.; Jacobs, W. P. *Plant Physiol.* **1969**, *44*, 27-36.
54. Ohmiya, A.; Hayashi, T.; Kakiuchi, N. *Plant Cell Physiol*. **1990**, *31*, 711-715.
55. Xu, Y.; Savidge, R. A. *Proc. Plant Growth Regulator Soc. Am.* **1995**, *22nd Ann. Mtg*, 396-401
56. McCord, J. M.; Fridovich, I. *J. Biol. Chem.* **1969**, *244*, 6049-6055.
57. Skene, D. S. *Ann. Bot*. **1969**, *33*, 253-262.
58. Savidge, R. A. *IAWA J.* **1996**, *17*, 269-310.

Chapter 10

Oxidative Coupling of Phenols and the Biosynthesis of Lignin

Gösta Brunow, Ilkka Kilpeläinen, Jussi Sipilä, Kaisa Syrjänen, Pirkko Karhunen, Harri Setälä, and Petteri Rummakko

Laboratory of Organic Chemistry, Department of Chemistry, University of Helsinki, P.O. Box 55, FIN-00014 Helsinki, Finland

Our understanding of the structure of lignin is based on our knowledge of its biosynthesis. Lignification in the plant cell wall is initiated by the enzymatic formation of phenoxy radicals from cinnamyl alcohol precursors. The radicals are thought to react without the intervention of enzymes. An essential element in this process is the cross coupling of radicals formed from the monomeric precursors with radicals formed on the polymer chain (end-wise polymerization). The rates of formation of C–C and C–O bonds are governed by the effective concentrations of the different radicals and by the regioselectivity of the coupling reaction. Finally, the stabilization reactions of intermediate quinone methides will, in many cases, determine the outcome of the coupling reaction.

Lignin has always baffled researchers in their attempts to determine its chemical structure. All methods of isolation destroy some parts of the biopolymer, making it difficult to find out what structures are artefacts and what structures are missing. All analytical methods yield fragmentary, sometimes contradictory, results. For this reason, the study of lignin biosynthesis, the charting of the mechanisms that govern the assembly of the macromolecule in the cell wall, has become such an important approach, more so than for other biopolymers. It is instructive to review briefly the main steps of the development of our current ideas on the structure of lignins and to see how new knowledge about the processes of phenol oxidation has contributed, and still contributes, to this development.

The Dehydrogenative Polymerization Theory

Although analytical evidence had shown that phenylpropane units form an important part of the structure, the first major breakthrough came when Erdtman in 1933 (*1*) proposed that the formation of a dimer in the oxidation of isoeugenol could be a model for the biosynthesis of lignin. This example highlights two important features of the dehydrogenative polymerization theory: the formation of carbon-carbon and carbon-oxygen bonds by oxidative coupling in the *ortho* position on the ring and in the β position of the side chain, and the reaction of nucleophilic oxygen with a quinone-methide intermediate. This concept opened up the chemistry of lignin and still forms the basis for our understanding of its chemical structure.

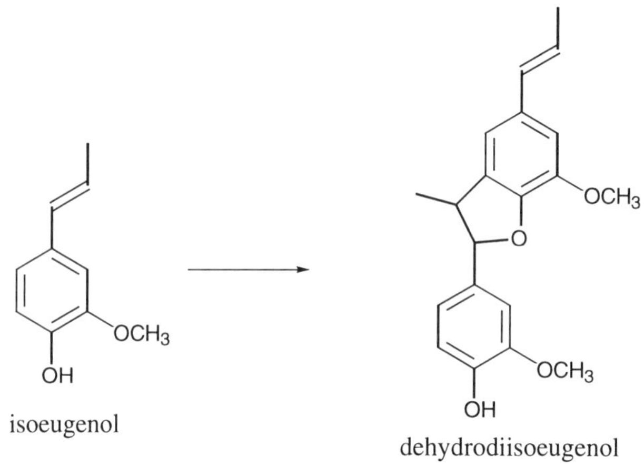

The idea was taken up by Freudenberg who, from the late 1930s onwards, developed it using coniferyl alcohol as the lignin precursor. He studied (*2*) the oxidation of coniferyl alcohol both with enzymes and with inorganic oxidants, and he obtained dehydrogenation polymers (DHPs) that in many respects resembled isolated lignins such as milled wood lignin from spruce, 'Björkman lignin'. The bonding patterns that are formed in the oxidation of coniferyl alcohol were revealed by stopping the reaction at a stage when it was possible to isolate dimeric and oligomeric products that lent themselves to exact structural determination. Freudenberg assumed that the bonding pattern in the dimers and oligomers is the same as in the polymer. This view has been modified in later work, as will be shown in later sections, but the basic idea, that lignin is formed by oxidative coupling, soon became established. It became apparent, through hydrolytic and other degradative studies, that most of the important linkages in softwood lignins could have been formed by oxidative coupling of coniferyl alcohol:

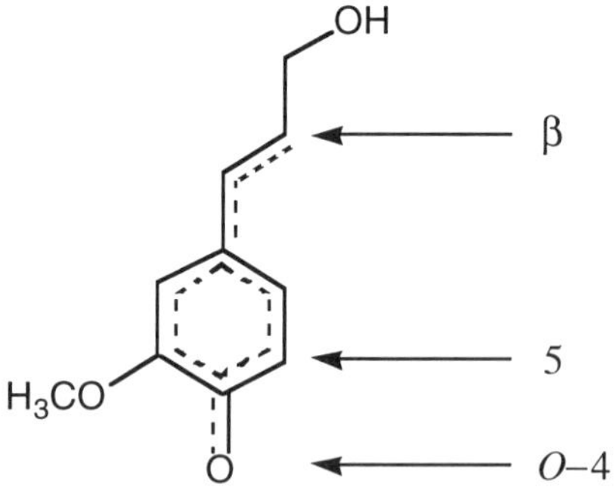

atomic centers to which linkages may be formed during oxidative coupling of coniferyl alcohol

It has to be pointed out, however, that the resemblance of a synthetic DHP to a lignin isolated from a plant cell wall rests on the same type of incomplete analytical data as does the structure of the lignin itself. In fact, with more advanced analytical techniques, the structural differences between DHPs and the corresponding lignins have become more and more apparent. The difficulties encountered by those who have attempted to prepare synthetic lignins show that a deeper understanding of the process of lignification is needed before it will be possible to duplicate it *in vitro*.

Regioselectivity and Stereoselectivity of Radical Coupling

The coniferyl alcohol radicals form C–C bonds and C–O bonds between one another, but the factors governing the distribution of these bond types are unknown. What we know is that the β-position is the most reactive, which means that in the most frequent coupling modes the β-carbon is involved in one of the moieties participating in the coupling reaction. The substitution pattern on the aromatic rings undoubtedly influence the selectivity of the coupling (*3*). As an example, the work of Sarkanen and Wallis (*4*) demonstrated that steric effects favor 'racemoid' coupling in the β–β dimerization of isoeugenol:

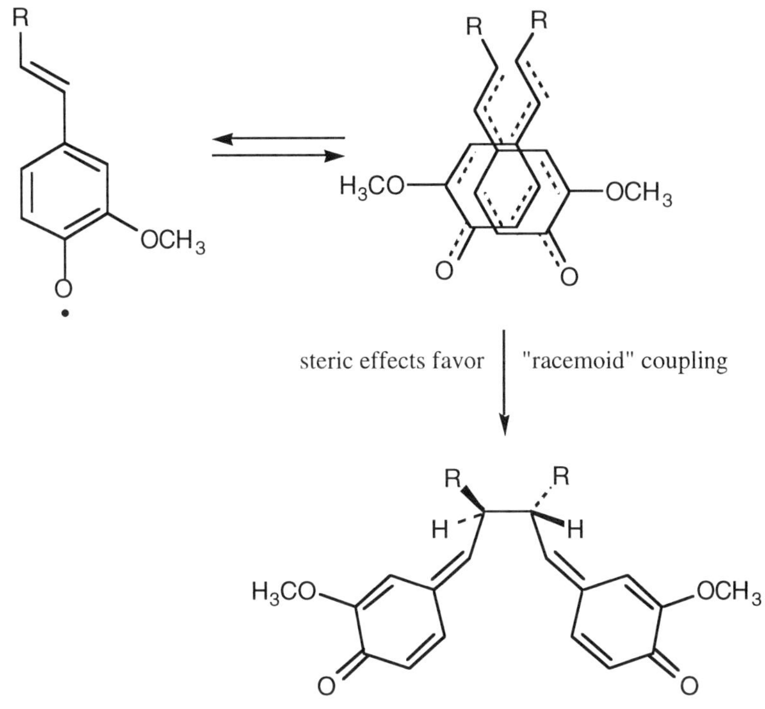

The steric requirements of a 'sandwich' transition state can be visualized, as shown above, as leading to a particular stereochemical outcome from coupling, but the factors determining the actual ratios of β–β′, β–5′ and β–*O*–4′ coupling are not known.

One factor that may influence the final distribution of bonding types is the occurrence of irreversible hydrolytic reactions. For instance, in β–*O*–4′ coupling, the formation of the final product involves the addition of water to an intermediate quinone methide, and this reaction requires an acid catalyst. The β–5′ and the β–β′ dimer formation, on the other hand, requires no water. Thus the availability of water and an acid catalyst may influence the formation of β–*O*–4′ dimers at the expense of other structures (*5*).

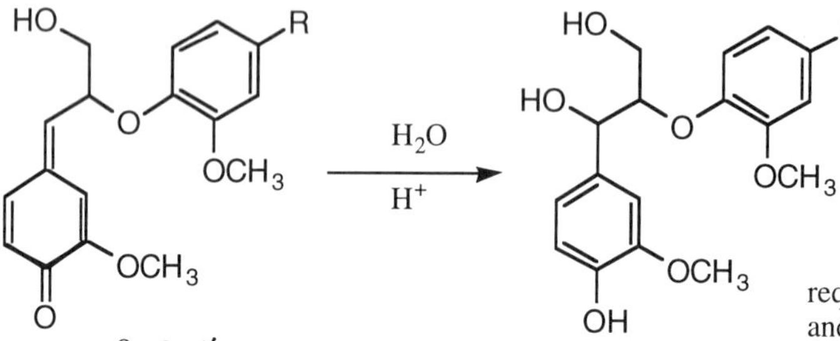

β–*O*–4′

requires water and acid catalyst

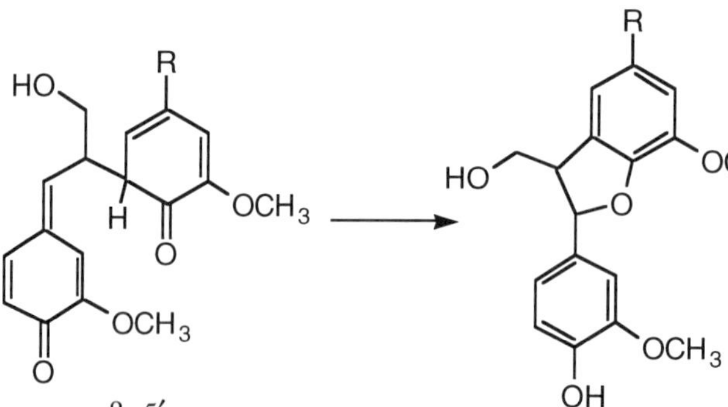

β–5′

requires no external nucleophile

Cross Coupling

In the mixture of dimers and polymers that are formed on the oxidation of coniferyl alcohol *in vitro*, the proportion of β–*O*–4′ structures is always lower than in natural lignins. When the coniferyl alcohol is added slowly to the oxidizing medium ('Zutropf'), the proportion of β–*O*–4′ structures rises, approaching that in natural lignins. This finding points to the importance of cross coupling between a coniferyl alcohol radical and a phenoxy radical generated on the growing polymer chain. This process, called 'end wise' polymerization by Sarkanen (*6*), is favored at low concentrations of coniferyl alcohol. At high concentrations of coniferyl alcohol, dimerization is favored, but when the concentration of coniferyl alcohol approaches zero, coupling between end-groups occurs, yielding 5–5′ biphenyl (and 5–*O*–4′) structures:

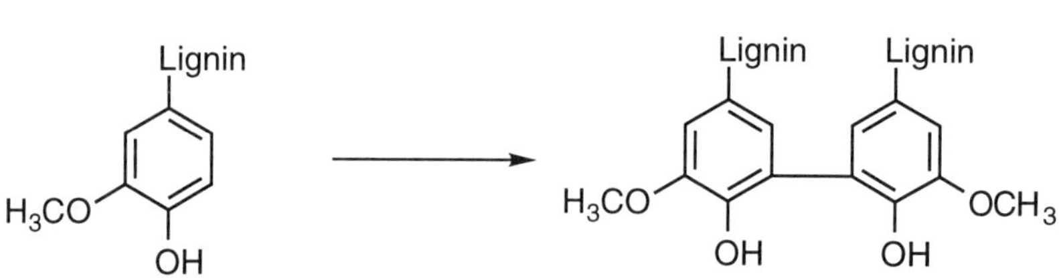

5-5′

We have thus identified another important factor controlling the process by which the structure of lignin is formed: the concentration of coniferyl alcohol in the zone where the growth of the lignin chain occurs. But why does cross coupling lead to higher amounts of β–*O*–4′ structures? In this case consideration of the transition state may provide a reasonable explanation. Perhaps the 'sandwich' transition state leading to the β–*O*–4′ structure may be more sterically favored than, for instance, the transition state leading to a β–5′ coupling:

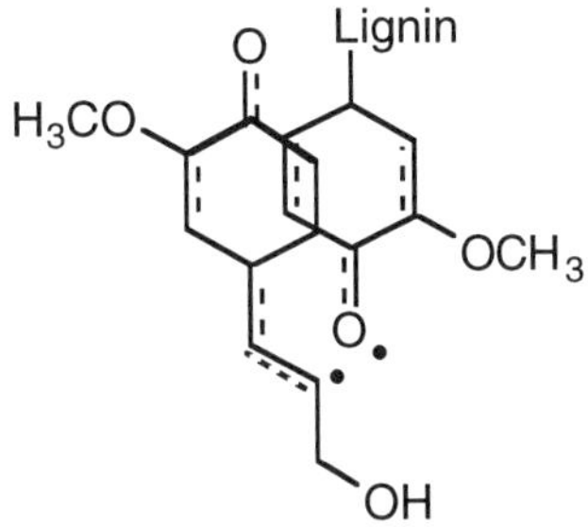

⟶ β–*O*–4′

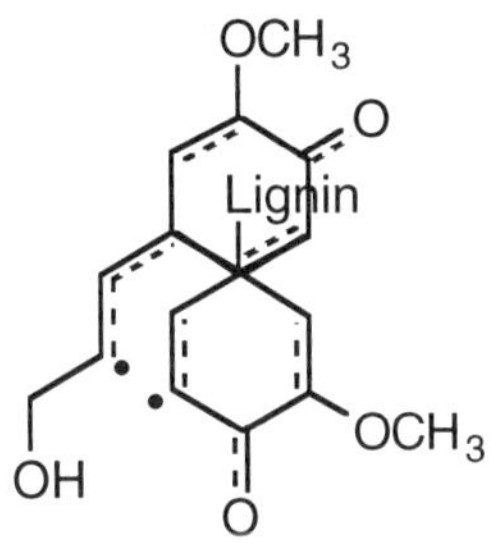

⟶ β–5′

Redox Potentials

Phenoxy radicals tend to dimerize and the dimerization is controlled by the reactivities and the concentrations of the radicals. For coupling between two different phenoxy radicals to occur (cross coupling), their respective reactivities and concentrations must, in combination, be such as to favor the process. One physical parameter which is important in this connection is the redox potential of the phenol. In the case of phenols important in lignin biosynthesis, very few measurements have been published. Figure 1 shows a rough estimate of the influence of substituents on the redox potentials of phenols. It has been constructed by extrapolation of data from two references (*7, 8*). The scale is not proportional; the data are intended to show the effects of *o*-methoxyl and *p*-(3-hydroxy)propenyl substituents on the oxidation potential of the phenol, and one 5–5′ biphenyl is included. [The numbers on the left are taken from ref. (*8*) and refer to the oxidation potential of the phenolate; the numbers on the right are approximations of substituent effects taken from ref (*7*).] When a mixture of phenols is oxidized, the phenols will react in the order shown, starting with the lowest on the scale. Phenols on the same level of the scale should form cross coupling products easily when equimolar mixtures are oxidized. Under other circumstances it is to be expected that the phenols react without significant cross coupling. Only if the phenol with the higher redox potential is present in large excess is cross coupling to be expected.

Assuming that cross coupling at equimolar concentrations occurs only between phenols of equal oxidation potential, it is possible to draw some important conclusions: in the case of softwood lignin, coniferyl alcohol has a lower oxidation potential than the phenols on the polymer that do not have a conjugated double bond. For cross coupling to occur, the concentration of the more reactive phenol must be kept low. The phenol with the higher oxidation potential has to be in large excess. The different redox potentials of the structural units in lignins can also lead to inhomogeneities. It has, for instance, been found that reaction wood and cell culture lignins are rich in *p*-hydroxyphenyl units, and that more than 90% of these units are

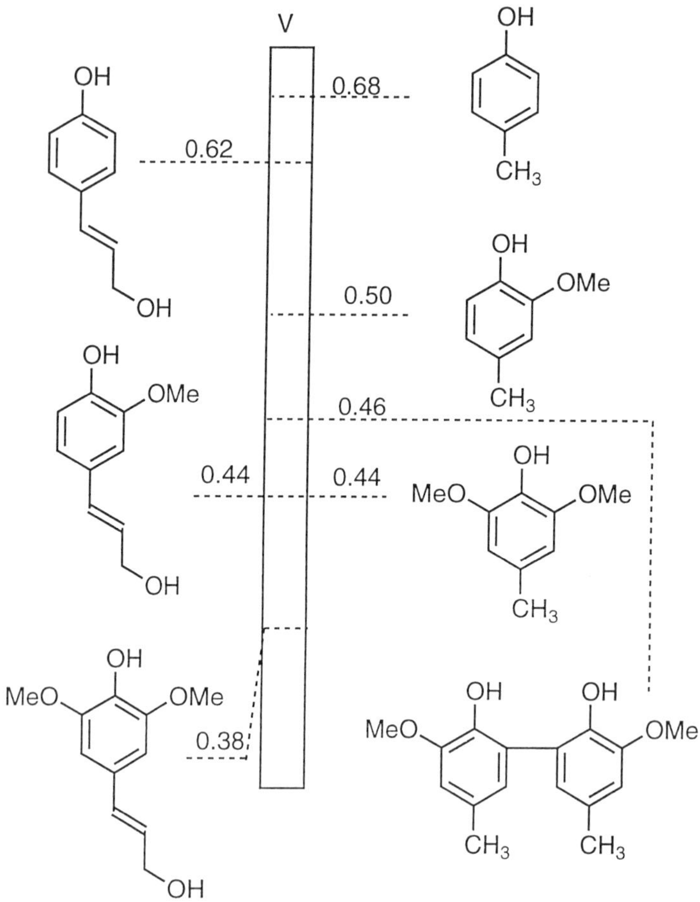

Figure 1. Estimates of substituent effects on the oxidation potentials of phenols relevant to the biosynthesis of lignin.

unetherified (*9*). This can be explained by the high oxidation potential of such units that makes them unreactive in phenolic coupling when they have become incorporated into the lignin.

Experiments on Cross Coupling

We have carried out a number of cross coupling experiments to demonstrate the constraints on cross coupling between phenols of unequal redox potential. Coniferyl alcohol and syringyl units without a conjugated double bond in the side chain can be assumed to have about equal oxidation potentials. Accordingly, oxidation of a mixture of coniferyl alcohol and the syringyl β–*O*–4′ model compound **1** yielded the cross coupling product **2** in good yield:

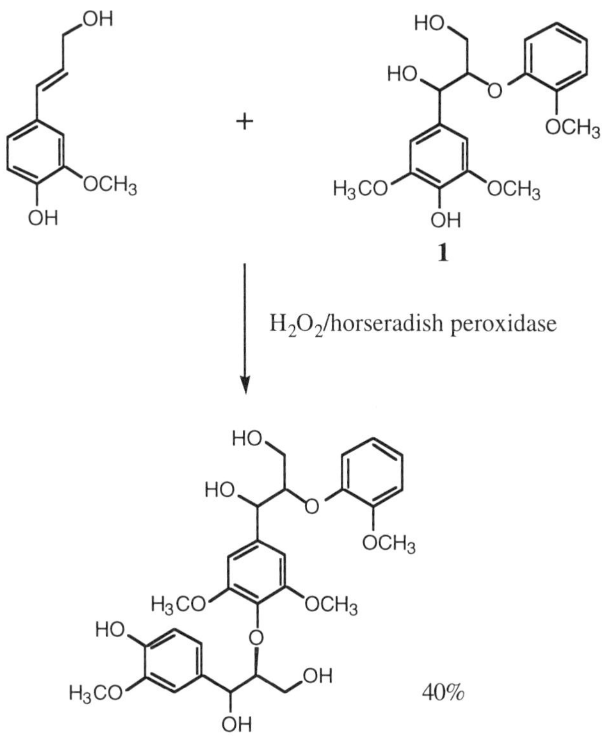

1

2

The oxidation of *p*-coumaryl alcohol together with a guaiacyl model compound yielded, among other cross coupling products, a similar β-ether but in lower yields. When this experiment is carried out with coniferyl alcohol and a guaiacyl model compound or sinapyl alcohol with a syringyl model compound, no cross coupling products were isolated. A rule of thumb can be formulated on basis of these results: when the monolignol and the dilignol have the same number of methoxyls, no cross coupling is obtained with equimolar mixtures; to obtain cross coupling, the monolignol must have fewer methoxyls than the dilignol.

To sum up, the concentration of coniferyl alcohol (or other cinnamyl alcohol) in the reaction zone is a major factor determining the structure of the growing polymer. It is possible that this concentration varies with time and in different regions of the cell wall, giving rise to inhomogeneity in the lignins formed in these regions. The following scheme shows how the competition between dimerization and cross coupling reactions can lead to different structures in the lignin.

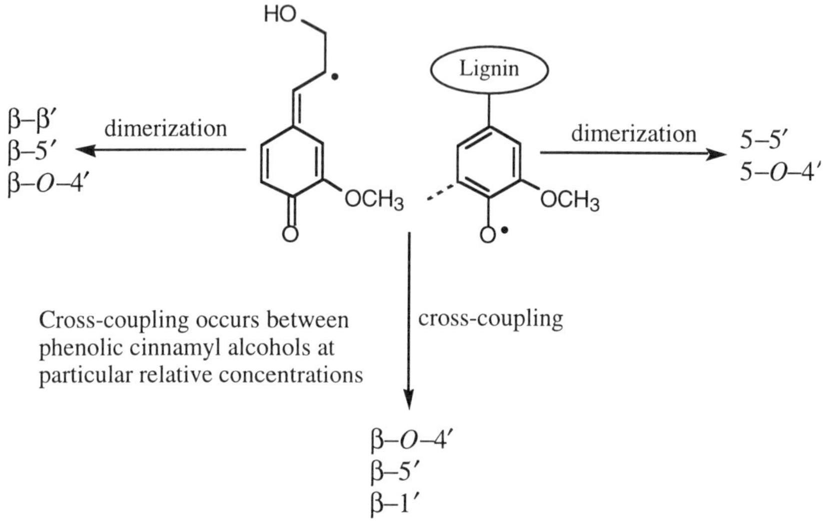

Knowledge of the mechanisms involved in the oxidative coupling reactions and the role of cross coupling *versus* dimerization allows us to predict under what conditions a given structure is formed. This gives us deeper insight into the structure of lignin and is an invaluable aid for interpreting analytical data. In the following sections we are going to demonstrate the value of this perspective by considering some structural units that have been studied in our laboratory.

The Non-cyclic Benzyl Aryl Ethers

The existence of non-cyclic benzyl aryl ethers or α–*O*–4′ bonds in lignins has been debated for some time (*10*). The evidence for their existence has been based on hydrolytic results from lignin preparations and on the fact that such structures are prominent among the oligomers isolated by Freudenberg (*11*). In our experiments we were able to repeat the results of Freudenberg: a quinone methide with a β-aryl ether in the side chain did indeed react with vanillyl alcohol forming a non-cyclic benzyl aryl ether:

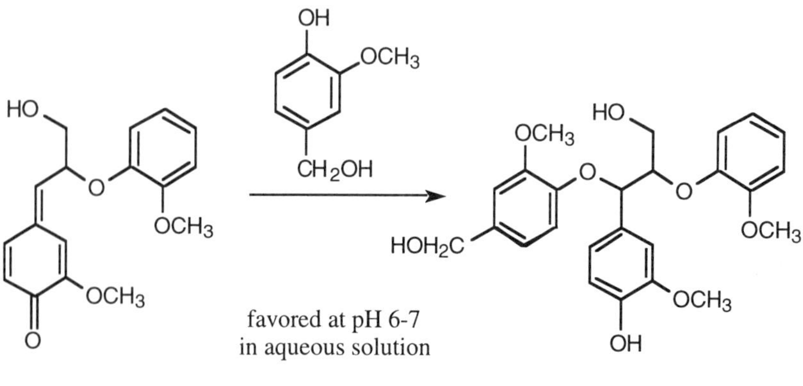

The problem is that attempts to find spectroscopic evidence for the existence of such structures in lignin have yielded mostly negative results (*12*). Minute amounts have been detected in a C-13 enriched sample of a poplar lignin (*13*) but no evidence for non-cyclic benzyl aryl ethers has been found in softwood lignins. A possible explanation for this apparent contradiction is that the biosynthesis of lignin occurs at a pH lower than that typically used in the dehydrogenation experiments. We have found that quinone methides in aqueous solution react with water in preference to a phenol when the pH is lower than 4 (Table I). The vanishingly small amounts that have been detected in NMR spectroscopic studies may thus be explained by a low pH during lignin biosynthesis. The reaction of the quinone-methide intermediate with water seems to be rate determining in the formation of β–*O*–4′ structures in lignins.

Table I. The Relative Rates of Addition of Nucleophiles to Quinone Methides

pH range	addition reaction	stereochemistry
5–7	carboxylic acids > phenols > water	*erythro* for acids and phenols, *threo* for water
3–5	water >> carboxylic acids	
<3	water	*erythro/threo* ratio approaches 1 with increasing acidity

It should be pointed out that, despite numerous attempts, no addition of carbohydrates to intermediate quinone methides has been observed in aqueous solutions. The question of how covalent bonds are formed between lignins and carbohydrates is still unresolved.

1,2-Diarylpropane or β–1′ Structures

In 1965 a novel structural unit was found in lignins (*14, 15*), the 1,2-diarylpropane or β–1′ unit. Compounds derived from the 1,2-diarylpropane structure have been found as prominent components upon acidolysis and thioacidolysis of many lignin preparations. Model compounds with this structure have been used extensively in lignin biodegradation studies. The biosynthetic mechanism for their formation involves cross coupling with displacement of one side chain. Determinations of the frequencies of such structures by NMR spectroscopy have revealed only small amounts in both softwood and hardwood lignins. The difficulties encountered in finding spectrometric evidence for this linkage has caused uncertainty in regard to the significance of the β–1′ structure in lignins. It is tempting to speculate that the β–1′ structures have accumulated in only some regions of the cell wall or that they may be formed on hydrolysis from some progenitorial feature in the lignin. An interesting observation was made by Gellerstedt and Zhang (*16*) when they isolated β–1′ type compounds from wood by mild acidolysis. After methylation the isolated compounds were mainly methylated at only one of the phenolic groups.

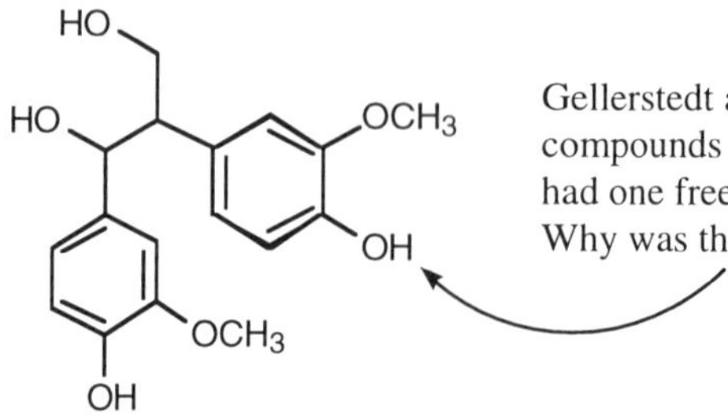

Gellerstedt and Zhang found that β–1′ compounds isolated after methylation had one free phenolic hydroxyl group. Why was this not methylated?

The existence in lignins of a true progenitor of the β–1′ structure has not been proved but, when cross coupling a syringyl model compound with sinapate ester, we found a structure that might explain the formation of β–1′ structures on acidolysis and the monomethylated compounds found in Gellerstedt and Zhang's experiments:

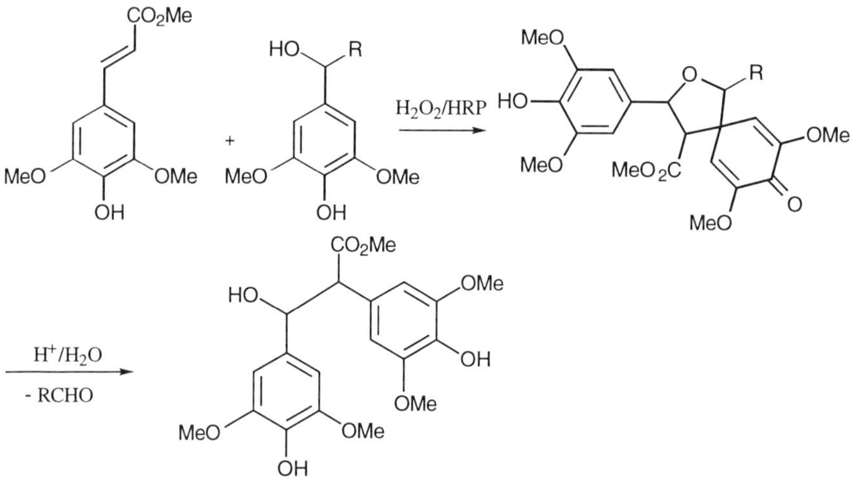

When methyl sinapate reacts with the syringyl type model compound, a tetrahydrofuran-3-spiro-4′-cyclohexadienone is formed. The resulting spirocyclohexadienone undergoes acid-catalysed hydrolytic cleavage of the α–*O* bond *via* a retro Prins reaction to form the corresponding 1,2-diarylpropane. A dienone-phenol rearrangement does not occur because of the presence of an α–*O* bond in the spirocyclohexadienone. If a similar cyclohexadienone is formed from coniferyl alcohol, it would not react on methylation and on mild acidolysis it would lose a side chain to form a β–1′ structure. The biosynthesis of the β–1′ structure most probably involves a cyclohexadienone intermediate of the type shown, but whether or not it is stable enough to survive in the lignin is not known at present.

The 5–5–*O*–4 Linkage, a New Structural Unit in Lignins

The 5–5′ biphenyl structure is prominent, especially in softwood lignins (*17, 18*). When a model compound with this structure was oxidized together with coniferyl alcohol, a novel cross coupled product with an 8-membered ring was obtained:

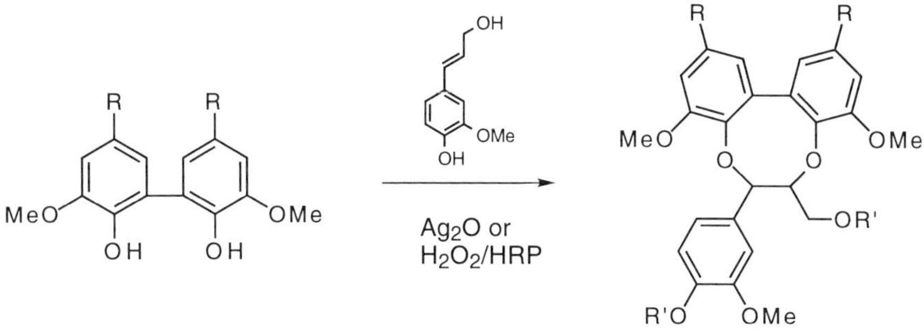

The NMR signals from an acetylated model compound (R = Pr, R' = Ac) were found to coincide perfectly with a group of signals observed in a two-dimensional NMR spectrum (HMQC) of an acetylated milled wood lignin from spruce (Figure 2) (*19*).

This dibenzodioxocin seems to represent a novel structure for a natural product. The amount of this structure present in softwood lignins can be estimated from the abundance of 5–5′ biphenyls and the fact that only part of the biphenyl structures are unetherified. Since the amount of 5–5′ structures is estimated at 24-26% (*18*) and free phenolic hydroxyls in biphenyls are 5-6%, the balance, 18 to 20%, may be 5–5–*O*–4 units. Their ease of formation as cross coupled products can be explained by the comparatively low oxidation potential of 5–5′ biphenyl structures compared to ordinary guaiacyl units. The ring formation is spontaneous: the intermediate quinone methide reacts instantaneously, even at low pH, with the adjacent phenolic hydroxyl (*20, 21*):

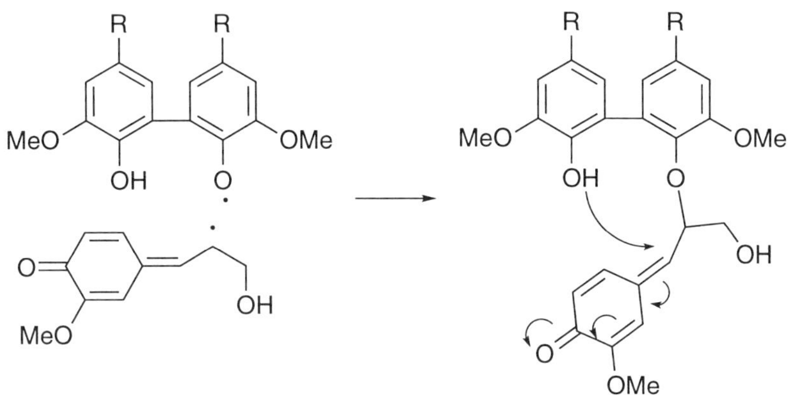

The ring is conformationally stable; the crystal structure (Figure 3) of an acetylated model compound (R = CH_2OAc) reveals that it is a ring with normal bond angles, the conformation being a so-called twist-boat-chair with a *trans* conformation of the α and β hydrogens :

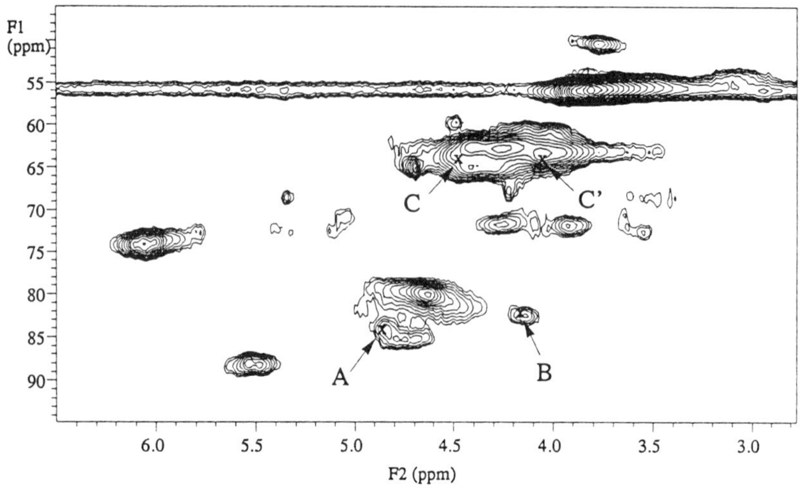

Figure 2. The side chain region of a HMQC spectrum of acetylated milled wood lignin from pine. The α, β and γ C/H correlations are marked A, B and C, respectively.

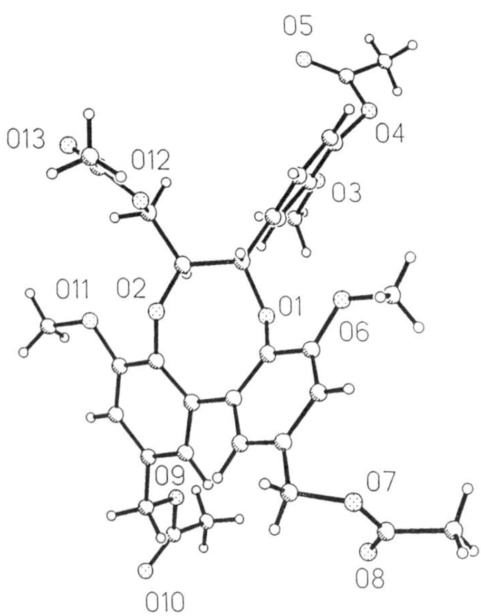

Figure 3. Crystal structure of a dibenzodioxocin model compound (see text).

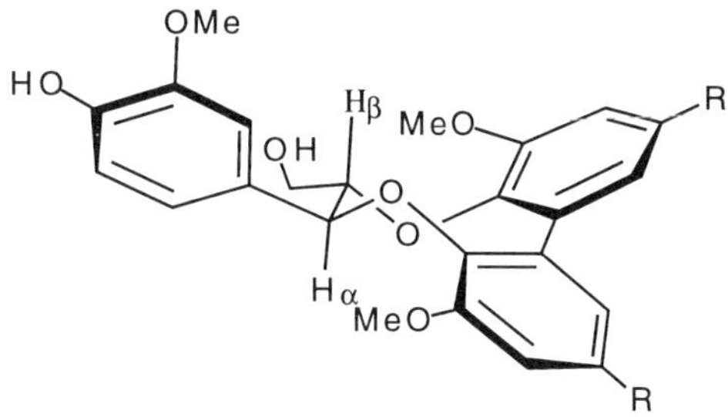

In model experiments, a diastereomer of this molecule, where the α and β hydrogens are *cis,* is also formed as a by-product:

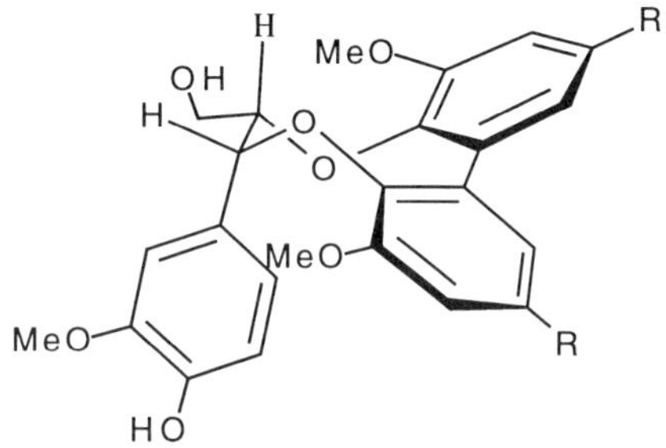

This isomer has not been found in softwood lignins, but some *cis* isomer can be found in cell culture lignin from spruce. The *trans* isomer seems to be thermodynamically the more stable one: when the *cis* form is dissolved in dilute alkali, it rearranges instantaneously into the *trans* form.

The dibenzodioxocin models all react with dilute acid. In the reaction, a new 7-membered ring is formed and a new phenolic group is released:

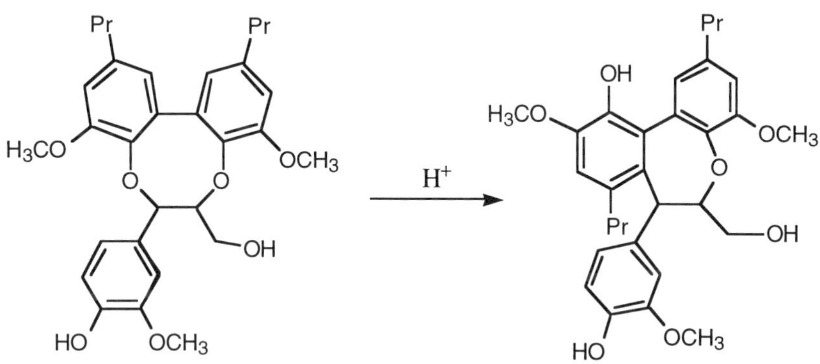

This reaction may account for the release of the phenolic hydroxyls that previously has been ascribed to the cleavage of non-cyclic benzyl aryl ethers. The dibenzodioxocins are, of course, cyclic benzyl aryl ethers and the study of their reactivities in pulping reactions and in biodegradation may yield valuable insight into the reasons for the behavior of lignin under various circumstances.

Implications of Oxidative Cross Coupling for the Biosynthesis of Lignin

To summarize, we can conclude that the concept of oxidative coupling of phenols still is a powerful instrument for understanding lignin biosynthesis, and further studies of these reactions will undoubtedly yield more explicit information. With our present state of knowledge we can already make some assertions.

We know that the differences in redox potentials between lignin precursors (monolignols), and between monolignols and the phenols in the polymer, make it possible for nature to fine-tune the structure of the lignin as it is deposited in the different parts of the cell wall. The formation of the ubiquitous β–*O*–4′ bonds found in natural lignins are a result of cross coupling between coniferyl alcohol and the phenolic groups in the growing polymer. The fact that the addition of water to the quinone-methide intermediate is rate determining and needs an acid catalyst raises the question of how this reaction is catalysed in nature. Acid catalysis is also implied by the virtual absence of non-cyclic benzyl aryl ethers in isolated lignins.

Oxidative coupling experiments have also helped to understand some conflicting results concerning non-cyclic benzyl aryl ethers and 1,2-diarylpropane structures.

As to the formation of dibenzodioxocins, this opens up questions concerning the interplay of 'bulk' and 'end wise' polymerization in the formation of lignin. 'Bulk' polymerization connotes a process where coniferyl alcohol dimerizes and the resulting dimers further undergo dehydrogenative coupling forming mainly 5–5′ and 5–*O*–4′ bonds. Coniferyl alcohol does not form 5–5′ bonds on dimerization, and therefore the formation of 5–5′ bonds is diagnostic for the occurrence of 'bulk' polymerization, which can be conceived to be favored when the concentration of coniferyl alcohol approaches zero in the reaction zone. In the polymer this means the connection of two chains in the polymer by a 5–5′ bond. This can be expressed using the formalism introduced by Sarkanen (*6*):

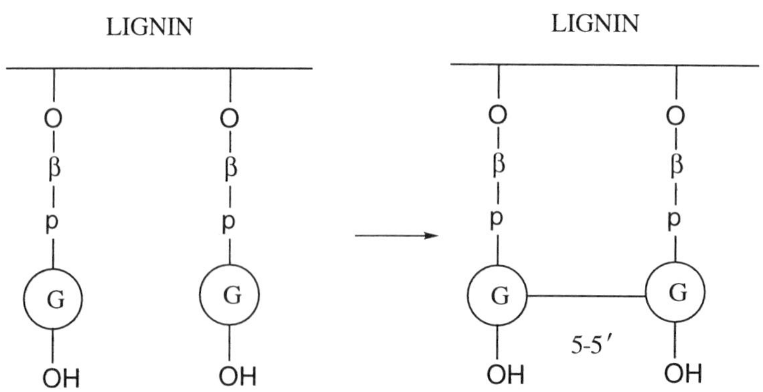

The new structural unit has a lower redox potential than its precursors and will be more reactive towards coniferyl alcohol radicals. On oxidative cross coupling a new polymer chain is started:

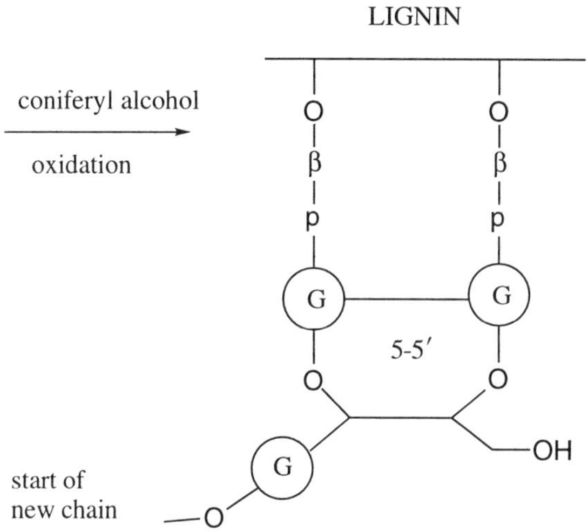

An interesting implication contained in this picture of events is that the concentration of coniferyl alcohol in the zone where the dehydrogenation occurs may fluctuate. In ordinary 'bulk' polymerization, coniferyl alcohol disappears after the formation of dimers. Further oxidation gives 5–5′ (and 5–*O*–4′) structures. After some time, coniferyl alcohol is again introduced, giving 'end wise' dehydrogenative polymerization structures. The first to be formed will be dibenzodioxocins and, when the 5–5′ biphenyls start to run low, ordinary β–*O*–4′ linkage production sets in. Nature thus may use the flow of coniferyl alcohol (in addition to pH) to control the structure of lignin in a manner that we still know very little about.

Conclusions: A Structural Scheme for Softwood Lignin

The combination of data from oxidative coupling experiments and 2-D NMR spectroscopic studies can be used to summarize our present knowledge on the salient points about the structure of lignin in a scheme showing two lignin fragments (Figure 4). This is necessarily a random collection of structural units similar to those published previously, with the addition of some new features such as the dibenzodioxocin structure. The scheme contains two fragments with a total of 25 aromatic units and is thus too small to depict the abundance of minor structural units accurately (the hypothetical precursor for the β–1′ structure is included as an illustration). The scheme does show that, according to our present knowledge, the most important branching points in the lignin polymer are the 5–5′ and the 5–*O*–4′ structures.

Literature Cited

1. Erdtman, H. *Annalen* **1933**, *503*, 283-294.
2. Freudenberg, K.; Neish, A. C. *Constitution and Biosynthesis of Lignin*; Springer-Verlag: Heidelberg, 1968; Vol. 2.
3. Armstrong, D. R.; Cameron, C.; Nonhebel, D. C.; Perkins, P. G. *J. Chem. Soc. Perkin Trans. II* **1983**, 581-585.
4. Sarkanen, K. V.; Wallis, A. F. A. *J. Chem. Soc. Perkin Trans. I* **1973**, 1869–1878.
5. Chioccara, F.; Poli, S.; Rindone, B.; Pilati, T. *Acta Chem. Scand.* **1993**, *47*, 610-616.

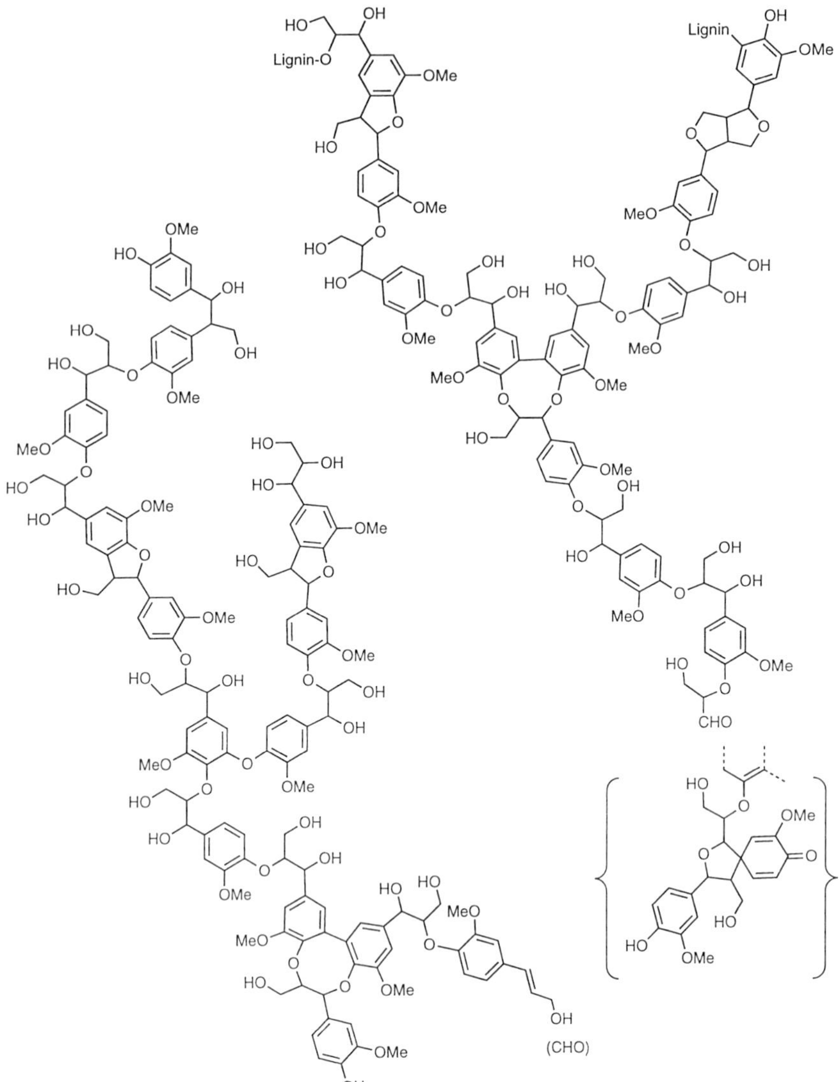

Figure 4. A structural scheme for softwood lignin fragments.

6. Sarkanen, K. V.; Ludwig, C. H. *Lignins: Occurrence, Formation, Structure and Reactions*; Wiley-Interscience: New York, NY, 1971.
7. Kratzl, K.; Claus, P.; Lonsky, W.; Gratzl, J. S. *Wood Sci. Technol.* **1974**, *8*, 35–49.
8. Jonsson, M.; Lind, J.; Reitberger, T.; Eriksen, T. E.; Merényi, G. *J. Phys. Chem.* **1993**, *97*, 8229–8233.
9. Lapierre, C.; Rolando, C. *Holzforschung* **1988**, *42*, 1–4.
10. Leary, G. J. *Wood Sci. Technol.* **1982**, *16*, 67–70.
11. Freudenberg, K.; Friedmann, M. *Chem. Ber.* **1960**, *93*, 2138–2148.
12. Ede, R. M.; Brunow, G. *J. Org. Chem.* **1992**, *57*, 1477-1480.
13. Kilpelainen, I.; Ammalahti, E.; Brunow, G.; Robert, D. *Tetrahedron Lett.* **1994**, *35*, 9267-9270.
14. Nimz, H. *Chem. Ber.* **1965**, *98*, 3153-3162.
15. Lundquist, K.; Miksche, G. E. *Tetrahedron Lett.* **1965**, *6*, 2131–2135.
16. Gellerstedt, G.; Zhang, L. *Nordic Pulp Pap. Res. J.* **1991**, *6*, 136–139.
17. Pew, J. C. *J. Org. Chem.* **1963**, *28*, 1048–1056.
18. Drumond, M.; Aoyama, M.; Chen, C.-L.; Robert, D. *J. Chem. Wood Technol.* **1989**, *9*, 421-426.
19. Karhunen, P.; Rummakko, P.; Sipilä, J.; Brunow, G. *Tetrahedron Lett.* **1995**, *36*, 169-70.
20. Karhunen, P.; Rummakko, P.; Sipilä, J.; Brunow, G. *Tetrahedron Lett.* **1995**, *36*, 4501-4.
21. Karhunen, P.; Rummakko, P.; Pajunen, A.; Brunow, G. *J. Chem. Soc. Perkin Trans. I* **1996**, 2303–2308.

Chapter 11

Biomimetic Initiation of Lignol Dehydropolymerization with Metal Salts

Lawrence L. Landucci and Sally Ralph

Forest Products Laboratory, U.S. Department of Agriculture Forest Service, One Gifford Pinchot Drive, Madison, WI 53705

Dehydropolymerization of *p*-hydroxycinnamyl alcohols with metal salt oxidants can lead to dehydropolymerisates (DHPs) that more closely resemble native lignin than those prepared by conventional techniques. Intermediate dimers, trimers, and tetramers obtained by this technique have provided the building blocks that are necessary for precise characterization of the DHPs by ^{13}C NMR spectroscopy. Individual dilignols have been used for specific preparation of higher oligolignols with readily definable (*i.e.* interpretable) features. Guaiacyl DHPs prepared with metal salts exhibit, in certain cases, distributions of interunit linkages more representative of native lignin than those prepared by conventional enzymatic techniques.

The use of reliable models for complex lignin macromolecules are essential in many areas of lignin, wood, and p!ant research. Dehydropolymerisates of *p*-hydroxy cinnamyl alcohols, commonly referred to as 'DHPs', are currently the best models available. DHPs are prepared by either oxidative coupling of coniferyl, sinapyl, or *p*-coumaryl alcohols and their mixtures through enzymatic catalysis involving peroxidase and hydrogen peroxide (*1, 2*).

There are many reasons for using DHPs rather than natural isolated lignins, such as a milled wood lignin (MWL). One is that the DHP structure is free of extraneous wood components such as extractives, tannins, proteins, etc. Also, it is very difficult to remove all of the carbohydrate material from isolated lignins, especially because a certain fraction of it is thought to be chemically bound to the lignin. Another significant advantage of DHPs is that they are relatively simple to label isotopically in specific positions by using appropriately labeled *p*-hydroxycinnamyl alcohols (*3-5*). The use of isotopically labeled DHPs is very important in the study of enzymatic delignification (pulping) and bleaching, and in microbial degradation studies (*6*). It is also possible to incorporate into DHPs minor structures (labeled or unlabeled) that are known to be present in natural lignins but are not formed from *p*-hydroxycinnamyl alcohols during dehydropolymerization. Experiments that utilize labeled DHPs supplement more difficult and time-consuming experiments involving the labeling of natural lignins by feeding labeled precursors to the plant (*7-11*).

DHP Types

DHPs can be formed from *p*-hydroxycinnamyl alcohols as illustrated in Figure 1. The most commonly produced DHP is that formed from coniferyl alcohol. The resulting guaiacyl- or G-DHP is generally used as a model for gymnosperm lignin. DHPs formed from a mixture of coniferyl and sinapyl alcohols or from all three alcohols are designated as guaiacyl-syringyl (GS) and guaiacyl-syringyl-*p*-hydroxyphenyl (GSH) DHPs, and are used as models for angiosperm and grass lignins, respectively. If pure *p*-coumaryl or sinapyl alcohols are used, DHPs which do not have any analog in Nature are obtained, but they are very useful in certain instances. One such example involves the sorting out of chemical shift assignments in NMR spectra associated with specific dimeric entities, which in natural lignin would be difficult to find and characterize because of their relatively low abundance.

Comparison of Conventional DHPs with MWLs.

The key step in both natural lignification processes and conventional laboratory dehydropolymerizations is the enzyme-initiated generation of a phenoxy radical that engenders a complex variety of non-enzymatic processes which include radical coupling reactions, nucleophilic additions to quinone methide structures, rearrangements, side-chain oxidations and eliminations to generate the complicated lignin macromolecule (*12*). Some of the more common dimeric entities that are formed by radical coupling reactions are illustrated in Figure 2. According to Adler (*13*), these six linkage types account for about 80% of the total in a gymnosperm lignin. The predominant linkage is the β–*O*–4′, which accounts for about half of the total. There are about 10% each of β–5′ and 5–5′, and smaller amounts of β–1′, β–β′ and 4–*O*–5′.

Although there are claims in the early literature (*1, 2*) that DHPs essentially duplicate natural lignins, they turn out to be rather poor models for actual lignin structure. This is especially true for guaiacyl DHPs. In typical G-DHPs the abundance of the β–5′ linkage is generally greater than or comparable to that of the β–*O*–4′ linkage, and the β–β′ linkages are much more abundant than in natural lignins. In addition, the presence of an appreciable quantity of α–*O*–4′ linkages in G-DHPs has been revealed by NMR spectroscopy.

The α–*O*–4′ linkage is not formed by radical coupling, but by the nucleophilic addition of a phenolic group to a quinone methide, as illustrated in Figure 3 (*1*). The characterization of this structural entity in DHPs depended heavily on the availability of NMR spectroscopic data from trilignols of this type (*14*).

Comparisons of the ^{13}C NMR spectra of MWLs with those of DHPs indicate that the latter are much simpler in structure and consist primarily of four linkage types: β–5′, β–β′, β–*O*–4′, and α–*O*–4′. The existence of other linkage types such as β–1′, 5–5′, and 4–*O*–5′ structures has been more difficult to substantiate with NMR spectroscopic techniques. This is primarily due to the lack of appropriate model compounds embodying these linkages and/or to their low abundance in both natural lignins and DHPs. Thioacidolysis studies have indicated their presence in small amounts in lignins and in even much smaller amounts in DHPs (*15*).

Comparison of the linkage distribution of a typical G–DHP with that of a pine MWL is shown in Table I. Clearly, the predominant linkage in the DHP is the β–5′ or phenyl coumaran linkage. Even in the best of the conventional preparations the abundance of the β–5′ linkage is comparable to that of the β–*O*–4′. The β–β′ linkage is also abnormally frequent in the DHP. However, perhaps the most serious discrepancy in the DHPs is a relatively large abundance of α–*O*–4′ linkages. This linkage appears to be the main branchpoint in G-DHPs in contrast with natural lignins where presumably the branchpoints occur primarily at 5–5′ linkages and also to a lesser extent at 4–*O*–5′ and β–1′ linkages.

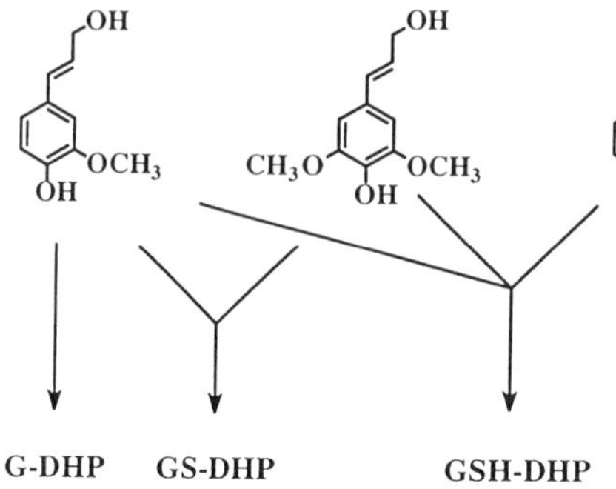

Figure 1. DHP types from coniferyl, sinapyl and coumaryl alcohols; G: guaiacyl; GS: guaiacyl-syringyl; GSH: guaiacyl-syringyl-*p*-hydroxyphenyl.

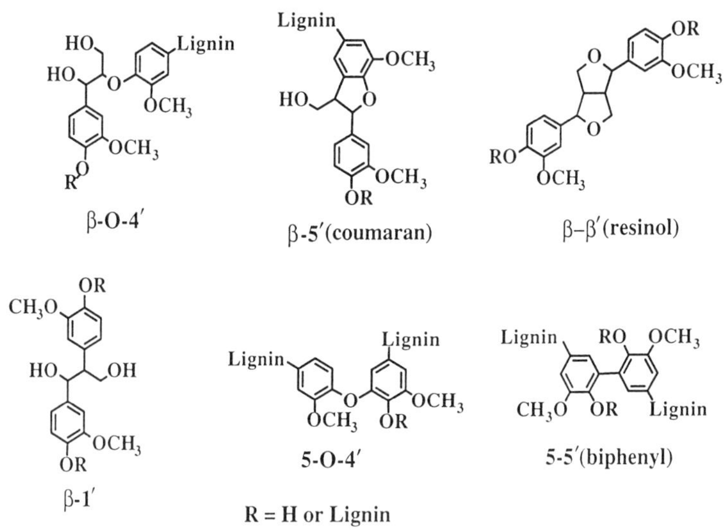

Figure 2. Common linkage types in lignin.

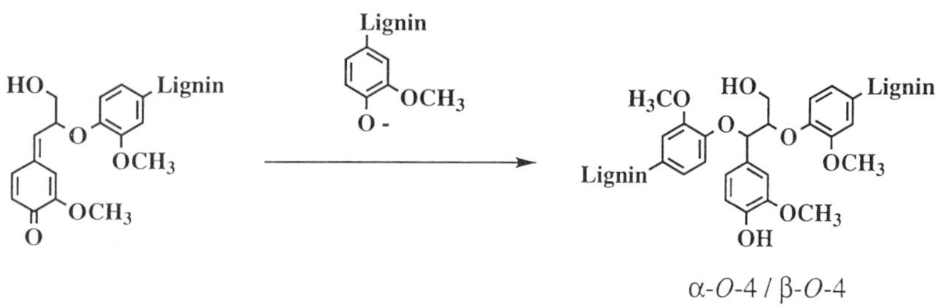

Figure 3. Formation of α–*O*–4′ linkages *via* quinone methides.

Table I. Comparison of Linkage Distributions in G-DHP and G-MWL

	Linkage (%)						
	β–*O*–4′	β–5′	β–β′	α–*O*–4′	5–5′	β–1′	4–*O*–5′
G-DHP	27	45	17	11	---	---	---
G-MWL	48	10	2	---	10	7	4

Another disadvantage of G-DHPs is that the content of coniferyl alcohol sidechains is enormous in comparison with a MWL where the unsaturated sidechain content is barely noticeable. This is illustrated in Figure 4 with both proton decoupled ^{13}C NMR spectra (all carbons) and DEPT spectra (CH's only) for a G-DHP and a pine MWL. The presence of the unsaturated sidechain in the G-DHP is indicated by the strong signal at about 135 ppm. In the corresponding spectrum of the MWL this signal is not apparent, not even in the DEPT spectrum which removes interfering quaternary carbon signals in this region.

Biomimetic Initiation of Dehydropolymerization.

The significant discrepancies between DHPs and natural lignins prompted an investigation of alternative methods of dehydropolymerization aimed toward improved fidelity in polymeric models for lignin. One promising method involves the use of metal salt oxidants (*16, 17*). It was found that an approach using transition metal salts as one-electron oxidants provided a convenient biomimetic route to phenoxy radicals analogous to the enzymatic generation of phenoxy radicals in conventional techniques. Subsequent radical coupling and other reactions, eluded to above, then ultimately lead to oligolignols and DHPs. These DHPs will be termed 'biomimetic' to distinguish them from DHPs prepared by conventional enzymatic techniques. The biomimetic approach is far more flexible than the corresponding enzymatic technique in that the reaction conditions are not constrained within the limits required by the enzyme. With the non-enzymatic system it is easy to vary the reaction conditions over a very large range. For example, temperatures from ambient to 100°C have been used; solutions from totally aqueous to totally nonaqueous; and in aqueous systems, the entire pH range was covered. In fact, with some metal salts the technique even works in the solid state. The advantage of this flexibility is that the reaction can be induced to generate a wide variety of product mixtures that differ in both molecular weight and linkage distributions.

The Building Blocks. In order to characterize the biomimetic DHPs, an intensive effort was made to isolate the intermediate building blocks of the polymer. Dilignol, trilignol, and tetralignol entities were of interest, as these relatively simple structures could be precisely characterized by NMR and mass spectroscopy, and signals in the ^{13}C spectra could be unambiguously assigned by an assortment of modern NMR experiments. The chemical shift data from these components is necessary for subsequent characterization of the DHPs. Oligolignols that have been isolated to date from a variety of reactions with coniferyl and sinapyl alcohols are shown in Figure 5. The abbreviated nomenclature associated with each of the structures in Figure 5 was developed in order to simplify discussion, and is presented in Table II.

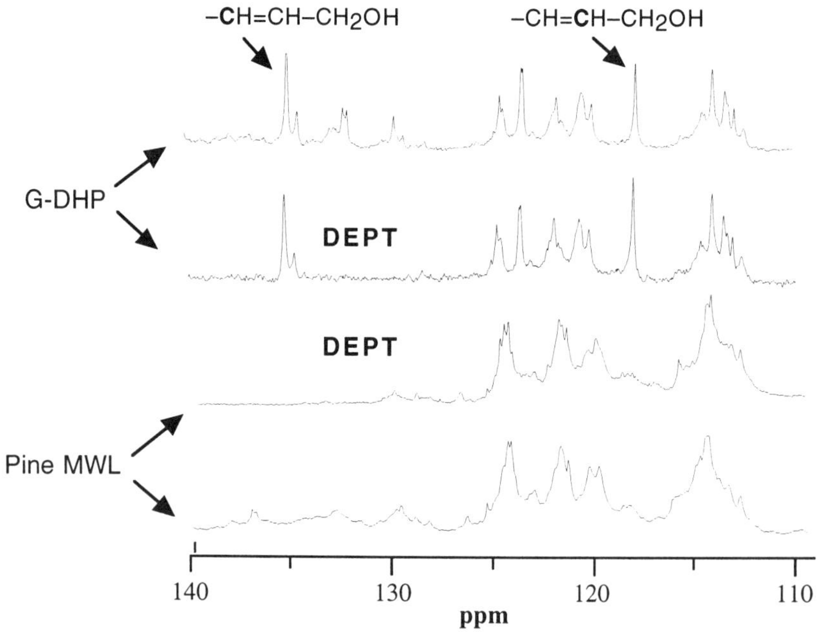

Figure 4. Proton decoupled and DEPT ^{13}C NMR spectra of a G-DHP and a pine MWL. [DEPT: distortionless enhancement polarization transfer].

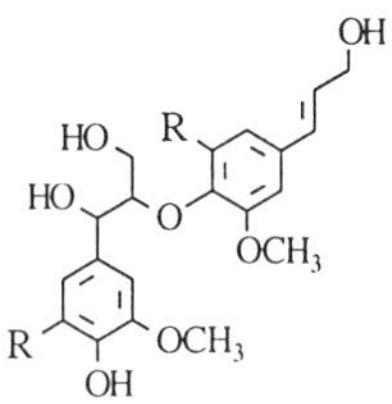

1 G-b-CA
2 G-b-SA
3 S-b-CA
4 S-b-SA

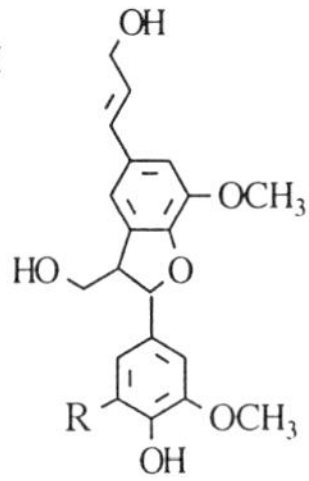

5 G-c-CA
6 S-c-CA

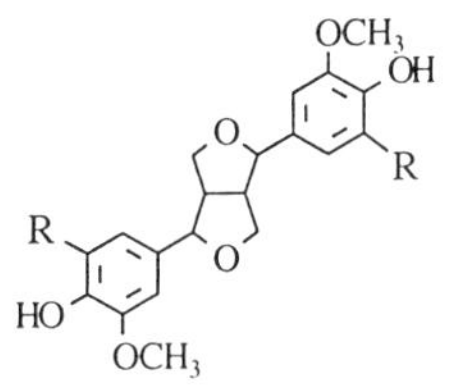

7 G-r-G
8 G-r-S
9 S-r-S

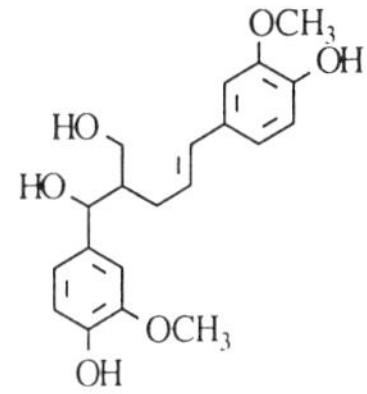

10 G-bg-G

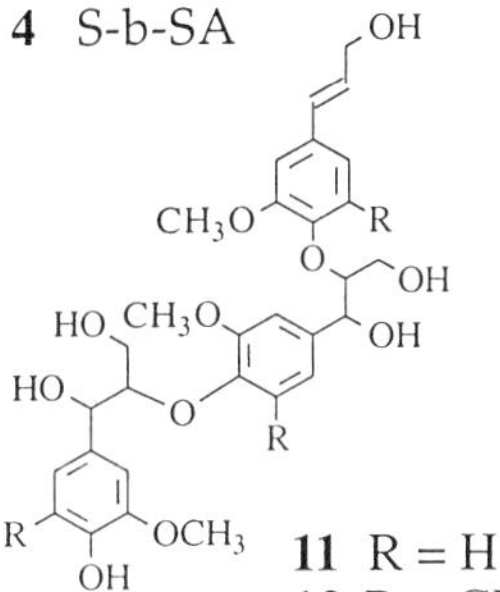

11 R = H G-b-G-b-CA
12 R = CH_3O S-b-S-b-SA

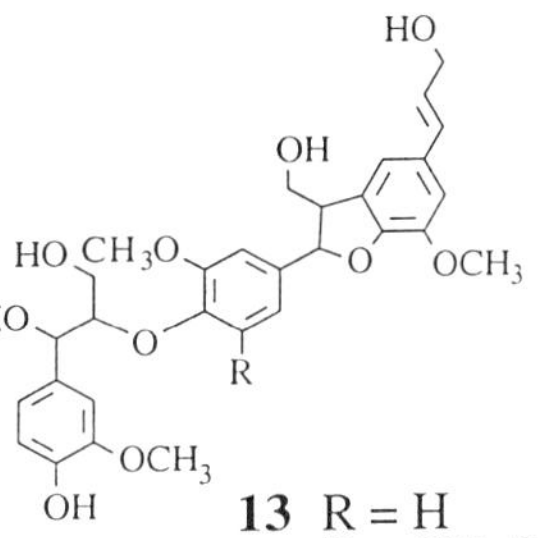

13 R = H G-b-G-c-CA
14 R = CH_3O G-b-S-c-CA

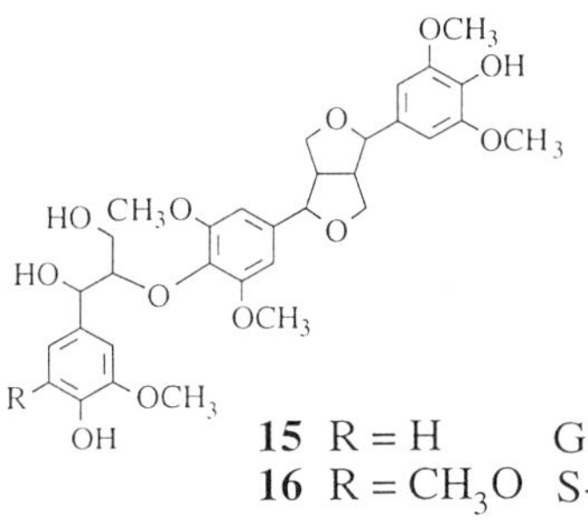

15 R = H G-b-S-r-S
16 R = CH_3O S-b-S-r-S

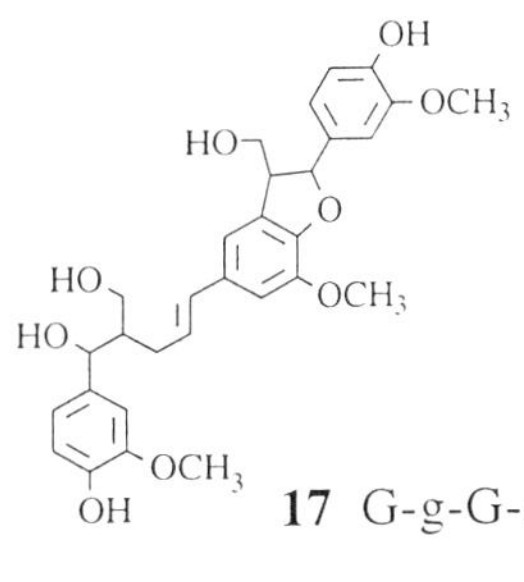

17 G-g-G-c-G

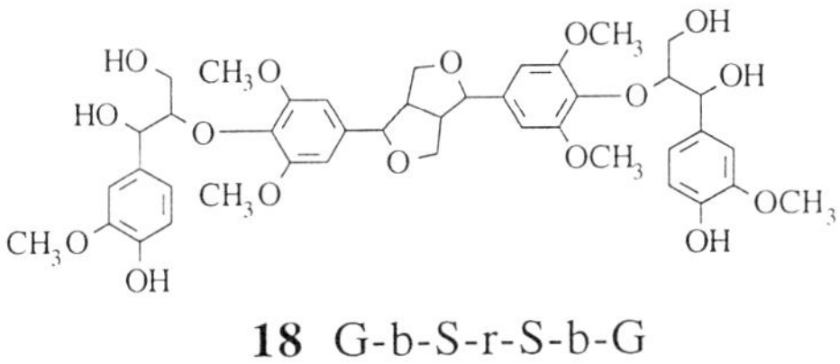

18 G-b-S-r-S-b-G

Figure 5. Oligolignols isolated from dehydropolymerization mixtures. (In **1-9**, R = H for G and CA groups, and R = OCH_3 for S and SA groups.)

Table II. Terminology of Abbreviated Structural Representations

Entity	Abbreviation
guaiacyl ring	G
syringyl ring	S
α–*O*–4′ linkage	a
β–*O*–4′ linkage	b
β–5′ (phenylcoumaran)	c
β–γ′ linkage	bg
β–β′ (resinol)	r
coniferyl alcohol end unit	CA
sinapyl alcohol end unit	SA
erythro	*e*
threo	*t*

All of the possible dilignols containing guaiacyl and/or syringyl rings and with either a β–*O*–4′ β–5′, or β –β′ linkage have been isolated from biomimetic dehydropolymerization mixtures (**1-9** in Figure 5). Although these dilignols are valuable for chemical shift assignments in higher oligolignols, they are not particularly good models for lignin or DHPs because it is unlikely that the combination of the unsaturated side chain and the free phenolic moiety would exist on the same C9 unit in either polymer. The smallest oligolignol that can be reliably used as a model for lignin or DHP is a trilignol. However, of the 24 possible trilignols (G, S, or GS) that contain at least one β–*O*–4′ linkage plus a β–5′ or β–β′ or another β–*O*–4′ linkage, only six (**11-16** in Figure 5) have been isolated in pure form from dehydropolymerization mixtures. The trilignols are generally much more difficult to isolate and purify because of the number of possible structures, their isomeric complexity, and because the trilignol content in the reaction mixtures usually ranges between 5 and 15%, with the exception of those from some of the reactions in which SA is present. With SA or CA/SA mixtures, the β–β′ linkage is sometimes predominant and di- and trilignols containing this linkage are isolated in high yield. The facile formation of some structures of this type is in keeping with previous reports of their presence in plant extracts (*18-20*).

Metal and Media Effects. The flexibility of using metal salts as oxidants in dehydropolymerization is illustrated by noting some of the results obtained under a variety of reaction conditions with various metal salts. In an effort to determine some of the controlling factors in these interesting reactions, several different metal/medium combinations were investigated (*16, 17*). One interesting series is illustrated in Figure 6 with the hexacyano complexes of Mn(III), Fe(III), and Co(III) in borate and NaOH solutions. With the iron compound in borate media, the three dilignols were obtained in comparable amounts. It should be noted that the β–*O*–4′ dilignol was primarily obtained in the *erythro* form (90%). In NaOH solution, quinone methide reactions dominated after initial radical coupling so the major products were α–*O*–4′ linked structures, the smallest being a trilignol. With the cobalt complex, a quantitative yield of a single compound was obtained. It was identified as a novel β–γ′ linked dilignol, **10** (Figure 5). Signals from this entity were not observed in any MWL or DHP that was examined, so the reaction was not further investigated. However, there was a recent report in the literature by Yoshida and coworkers where the S-S analog of this compound was isolated from a mixture produced when sinapyl alcohol was dehydrogenatively oxidized in the presence of pectin (*21*). In NaOH solution, the cobalt complex was unreactive and only starting material was recovered.

With the manganese complex in borate, the β–*O*–4′ dilignol was identified along with an unknown dilignol that is presently being identified. In NaOH solution, both the β–*O*–4′ and β–5′ dilignols were obtained. An interesting finding is that, contrary to the predominately *erythro* form in the iron reaction, a 50/50 mixture of *e* and *t* diastereomers were obtained.

When an equal molar mixture of coniferyl and sinapyl alcohols was subjected to the dehydropolymerization conditions optimized for lower oligolignols, very different results were obtained for different metal/medium combinations (*17*). For example, Table III summarizes the varying distribution of dilignols obtained in three systems.

Table III. Product Distribution in CA/SA Dilignol Fractions[a].

Dilignol	$Mn(OAc)_3$[b]	$K_3Mn(CN)_6$[c]	$Cu(OAc)_2$[d]
1 (G-b-CA)	6	6	8
2 (G-b-SA)	11	nd[e]	2
3 (S-b-CA)	11	nd[e]	1
4 (S-b-SA)	19	nd[e]	2
5 (G-c-CA)	5	7	5
6 (S-c-CA)	2	2	4
7 (G-r-G)	nd[e]	nd[e]	nd[e]
8 (G-r-S)	2	4	7
9 (S-r-S)	1	11	2

[a] Values are wt% of total acetylated products from ethyl acetate soluble material; [b] in pyridine; [c] in 0.1 M NaOH; [d] in 0.05 M borate buffer (pH = 9.2); [e] nd, not detected.

With the $Cu(OAc)_2$ system, the predominant dilignol was *erythro* G-b-CA, with only small amounts of the other possible combinations. With hexacyanomanganate in NaOH solution, the dilignol content consisted almost exclusively of 50/50 *e/t* G-b-CA. With manganese acetate in pyridine, all four β–*O*–4′ dilignols were obtained in comparable amounts. In this experiment the G-b-CA was almost exclusively *erythro*.

In addition to having some control over linkage types and distribution, the molecular weight distribution may be controlled somewhat by choosing appropriate reaction conditions. This is illustrated in Table IV with the polymerization of CA with manganese and copper systems. For example, with the manganese

Table IV. Effect of Metal/Medium on Molecular Weight Distribution.

Metal salt/Media	Degree of Polymerization					Major linkages
	1	2	3	4	≥5	
$Mn(OAc)_3/CH_2Cl_2/HOAc$ (fast addition)	2	78	6	6	5	β–*O*–4′ >>β–5′ >> β–β′
$Mn(OAc)_3/CH_2Cl_2/HOAc$ (slow addition)	5	19	2	4	70	β–*O*–4′ >>β–5′ >> β–β′
$Cu(OAc)_2$/borate	5	55	12	1	10	β–*O*–4′ ≈ β–5′ ≈ β–β′

acetate/methylene chloride/acetic acid combination, when the reactants were quickly mixed together, 78% of the total product was dilignols. On the other hand, if the CA was slowly added to the oxidant over a period of 8 h, 70% of the total product was high molecular weight, and only 20% of the total were dilignols. In between these two extremes was the copper borate system in which 55% of the total were dilignols. However, with the copper borate system, the β–*O*–4′, β–5′, and β–β′ linkages were roughly comparable, whereas with the manganese system the β–*O*–4′ linkages predominated over β–5′ and β–β′ which were present in only trace amounts.

It should be noted that an approximate indication of molecular weight can be obtained from an ^{1}H NMR spectrum of the sample. If relatively sharp signals like those seen in the upper spectrum of Figure 7 are present, it can be concluded that the material is perhaps composed mainly of di-, tri-, and tetralignols. This contrasts with the lower spectrum which is of a mixture containing predominantly polymeric material. Typically, with many of the metal systems, the crude product mixtures consist of material with molecular weights ranging from monomers to 20-mers, and they must be fractionated by preparative gel permeation chromatography.

Missing Building Blocks. In order to adequately elucidate the structures of the DHPs prepared by using metal salts, several additional tri- and tetralignols are needed. Some of these are undoubtedly present in the product mixtures already obtained, but their yields are too low or they could not be separated from numerous similar compounds. For example, trilignols that would be expected to be highly elusive in such complicated mixtures are those containing only β–*O*–4′ linkages. Considering all possible combinations of G and S rings, there are a total of 8 expected trilignols: G-b-G-b-CA, G-b-G-b-SA, G-b-S-b-CA, S-b-G-b-CA, G-b-S-b-SA, S-b-G-b-SA, S-b-S-b-CA, and S-b-S-b-SA. Of these components, only the pure guaiacyl (**11**) and the pure syringyl (**12**) trilignols (Figure 5) have been isolated.

Other missing building blocks are structures containing 5–5′, β–1′ and 4–*O*–5′ linkages. Some of these may not be overly important in elucidating the structures of many of the existing DHPs, since the latter are also thought to be lacking these linkages. However, as more improved DHPs are prepared that more closely resemble native lignin, these missing building blocks will become more important.

Tailored Syntheses of Oligolignols. Because of the difficulty of obtaining a sufficiently wide range of pure oligolignols from dehydropolymerization mixtures, it was necessary to consider modified approaches to obtain these vital components. As mentioned above, two of the main stumbling blocks are the difficulty of isolating β–*O*–4′ trilignols, because of the wide variety of possible trilignol structures, and the absence of certain linkages, such as 5–5′, β–1′, and 4–*O*–5′, in the mixtures. Both of these problems can conceivably be overcome by the dehydropolymerization of CA or SA in the presence of well characterized pure dilignols.

β–*O*–4 Trilignols. An initial attempt using this approach involved the reaction of CA with a 65/35 *e/t* mixture of G-b-CA in a $Mn(OAc)_3$ system (*22*). The trilignol fraction, which amounted to about 13% of the total, was composed mainly of G-b-G-b-CA. The complex pattern in the expanded β-carbon region of the ^{13}C spectrum in Figure 8 illustrates the stereochemical complexity of the trilignol. With a trilignol of this type there are 4 chiral centers giving rise to 16 optical isomers that include 8 distinct chemical isomers. Chemical shift assignments for the individual isomers in a case like this is generally not feasible because of the small differences involved. However, for the purposes of correlation with analogous lignin or DHP entities, these stereochemical considerations were somewhat redundant in that the small chemical shift differences between the various stereochemical configurations in an individual trilignol were generally not detected in the corresponding polymeric components because of much increased line widths in the spectra of the latter. It is only a problem when it comes to assigning precise chemical shifts for lower molecular weight components when the signals are still relatively narrow.

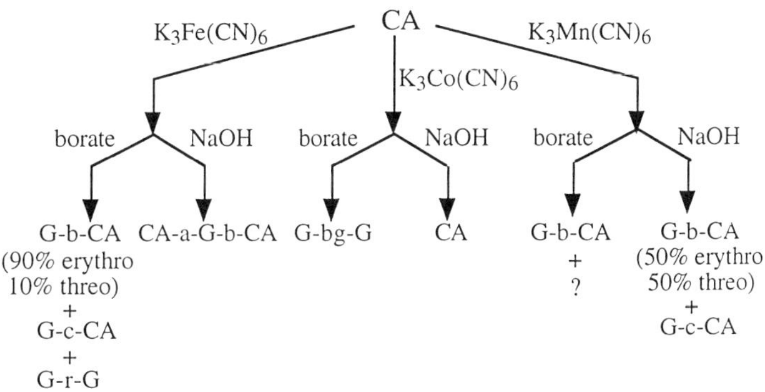

Figure 6. Dehydropolymerization of coniferyl alcohol with hexacyanometalates.

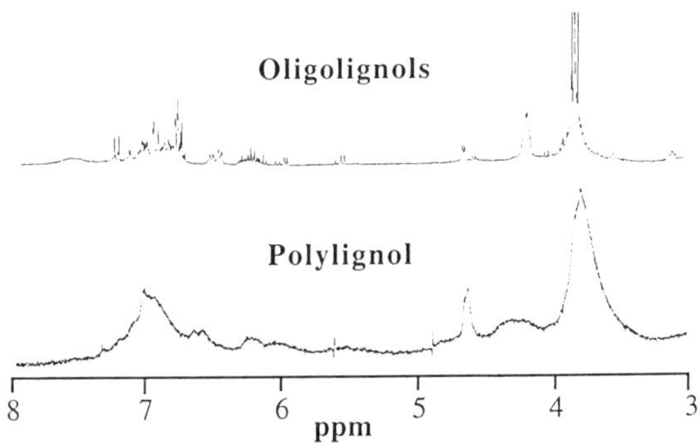

Figure 7. ^{1}H NMR spectra of oligolignols and polylignol (DHP).

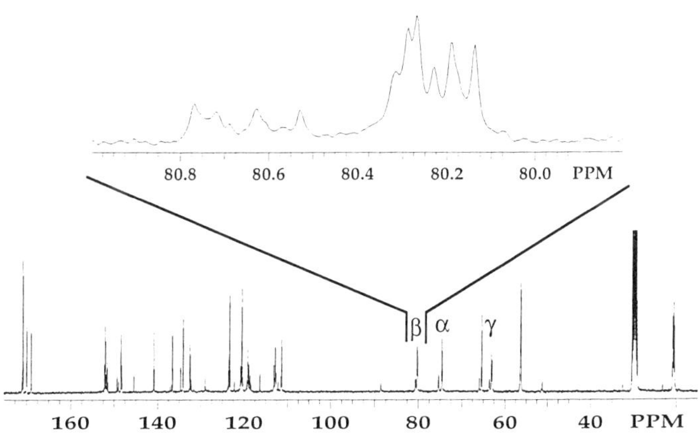

Figure 8. Isomers of trilignol G–b–G–b–CA.

If a stereochemically pure dilignol is used in the above experiment, the number of distinguishable isomers among the products is cut in half. Also, by imposing such a restriction on one of the linkages and two of the rings, a much simpler reaction mixture is obtained when we limit the self-coupling reactions of the monolignol. The use of dilignols also allows a greater flexibility in choice of structures for the terminal side chain of the resulting trilignol. A study of chemical shift substituent effects indicates that the terminal side chain typically has little effect on the chemical shifts of the internal C9 unit (*23*).

β–1/β–*O*–4 Trilignols. The coupling of a monolignol to a dilignol as a means of introducing a β–1′ linkage into a trilignol is illustrated in the contemplated scheme in Figure 9. In this example, the CA could conceivably add to either end of the dilignol to give two possible structures. A reaction such as this one would be expected to be quite complicated in that eight isomers could result from each coupling mode. However, the reaction could be simplified by using an isomerically pure dilignol. Also, the coupling reaction may possibly be induced to conform to only one of the pathways if the depicted precursor to G-b1-G (deoxyvanilloin) is used. Following the formation of the trilignol, formaldehyde addition and reduction (preferably by a stereoselective method) would afford the desired trilignol. This strategy is presently under investigation.

5–5/β–*O*–4 Trilignols. Another example of incorporating a linkage into a trilignol that has been difficult to obtain otherwise is illustrated in Figure 10 by the reaction of CA with the 5–5′ linked dilignol **21.** Interestingly, the expected trilignol **22** was not obtained; instead, the cyclic structure **23** was obtained which is analogous to one recently reported by Brunow and coworkers (*24*). The 5–5′ linkage has been particularly troublesome to characterize partly because of its rarity in conventionally prepared DHPs. Some results of work not yet published indicate an upper of 1 or 2% for the frequency of 5–5′ linkages in a G-DHP prepared by the so-called 'Zutropf' method. This is consistent with the thioacidolysis results obtained by Lapierre and coworkers who reported only a trace of 5–5′ linkages in this type of DHP. Other problems in the characterization of the 5–5′ entity is that there is extensive overlap of most of the pertinent signals with others, and that suitable etherified or partly etherified models such as **21** have not been available. It should be noted here that 3,4-dimethoxy (veratryl) derivatives are very poor models for β-ether linkages with respect to the observed ^{13}C NMR chemical shifts (*23*).

Polylignols (Biomimetic DHPs). Initial attempts to prepare DHPs with metal salts resulted in a variety of novel products (*16*). Similar to conventionally prepared G-DHPs, most of them were composed mainly of the four linkage types, β–*O*–4′, β–5′, β–β′, and α–*O*–4′. Relative contributions of these linkages in selected biomimetic G-DHPs are compared with a conventional G-DHP and a pine MWL in Table V. The wide range of results obtained with the biomimetic approach again attests to its inherent flexibility. With all four biomimetic DHPs, the predominant linkage is β–*O*–4′, in contrast to the predominance of β–5′ in the conventional DHP. G-DHP-1, obtained from a solid state reaction, contained a very high content of α–*O*–4′ linkages. In the biomimetic systems, in aqueous solution, the α–*O*–4′ content correlates somewhat with the pH. For example, G-DHP-2 is obtained with ferricyanide at pH 9 and the α–*O*–4′ content is 8%. In contrast, ferricyanide in NaOH solution (not shown in table) yields products with much higher α–*O*–4′ contents, and some do not contain any un-etherified benzylic carbons (*16*). It is thought that the formation of DHPs in the hexacyanomanganate/NaOH system also proceeds through α–*O*–4′ linked intermediates. However, as seen from the low α–*O*–4′ content in G-DHP-3, these linkages are cleaved during workup. This contention is supported by the 50/50 *e/t* character (expected ratio upon alkaline hydrolysis of α–*O*–4′ entities) of the β–*O*–4′ linkages in G-DHP-3, in contrast to a more typical 90/10 ratio in DHPs obtained at pH 9 or less.

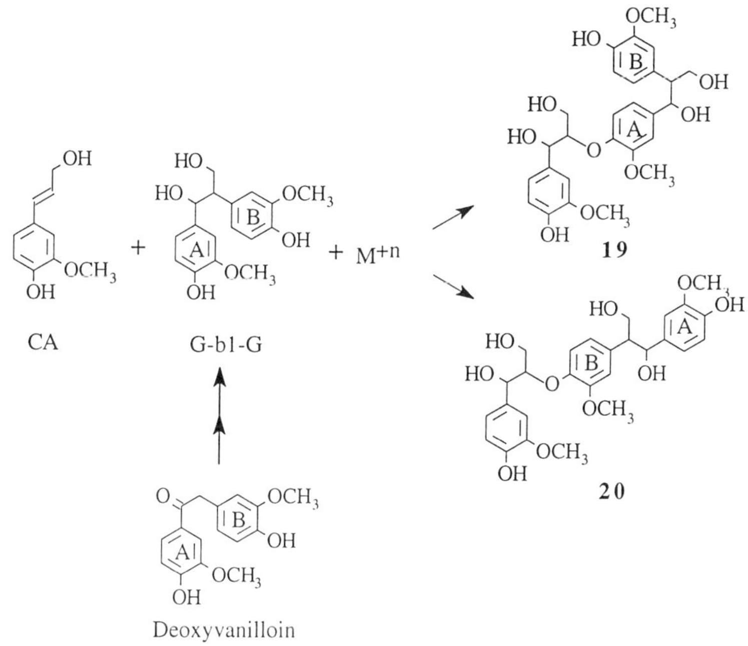

Figure 9. Formation of a β–1/β–*O*–4 trilignol.

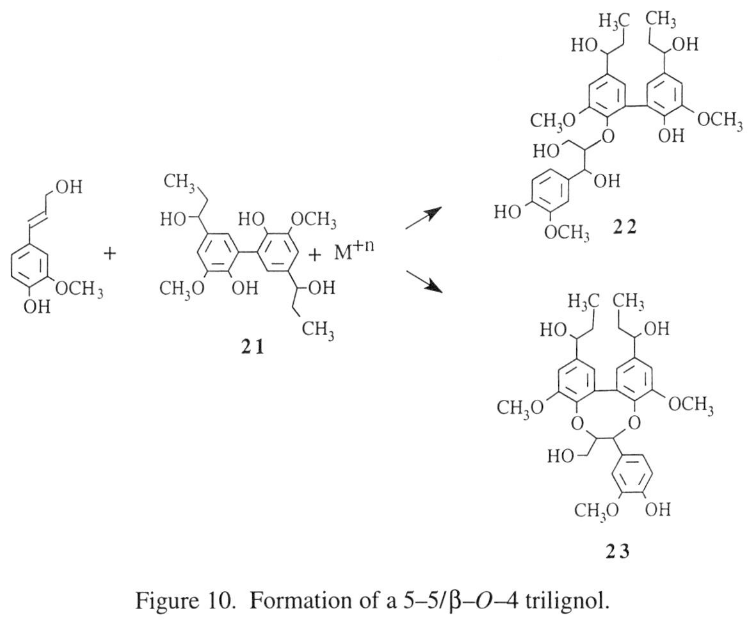

Figure 10. Formation of a 5–5/β–*O*–4 trilignol.

Table V. Comparison of Linkage Distributions in Selected DHPs and a MWL.

Substrate	Metal salt	Medium	Linkage			
			β–*O*–4′[a]	α–*O*–4′	β–5′	β–β′
Pine MWL	---	---	48	nd[b]	10	2
conventional G-DHP	---	H_2O/Peroxidase/H_2O_2	27	11	45	17
biomimetic G-DHP-1	$K_3Fe(CN)_6$	K_2CO_3, solid state	45	22	15	18
biomimetic G-DHP-2	$K_3Fe(CN)_6$	Borate buffer, pH 9	37	8	33	22
biomimetic G-DHP-3	$K_3Mn(CN)_6$	NaOH solution	42	3	29	25
biomimetic G-DHP-4	$Mn(OAc)_3$	Acetic acid	75	nd[b]	23	2

[a] With an un-etherified benzylic carbon; [b] nd, not detected.

Of the biomimetic DHPs, G-DHP-4 is the only one of the four in Table V that has realistic levels of β–5′ and β–β′ linkages (as compared to the MWL). The very high content of β–*O*–4′ in G-DHP-4 reflects the relative simplicity of DHPs in contrast to the MWL in which the four linkages in question only represent 60% of the total. By introducing other linkages such as 5–5′, β–1′, and 4–*O*–5′ into biomimetic DHPs of this type, it is expected that a closer resemblance to native lignins will be obtained.

Partial ^{13}C NMR spectra of the pine MWL, the conventional G-DHP, and the biomimetic DHP with the highest β–*O*–4′ content (G-DHP-4) are compared in Figure 11. Pertinent differences between the three materials not apparent from Table V are found in the content of unsaturated side chains and the benzylic hydroxyl frequency as reflected by the corresponding acetoxy carbonyl in the acetylated material. The α–carbon of CA sidechains is present as a very strong signal at about 134 ppm in the DHPs. The benzylic acetoxy carbonyl is the middle signal in the inset expansion at about 170 ppm. Clearly, the conventional DHP has a greater content of unsaturated side chains and a lower content of benzylic alcohol groups than does the biomimetic DHP. This difference can be largely attributed to the very high contents of β–5′ and β–β′ linkages in the conventional DHP, as reflected in the α–carbon signals of the corresponding C9 units between 86 and 88 ppm. More complete characterization and ^{13}C NMR signal assignments of selected biomimetic DHPs is presently in progress.

Conclusions

Dehydropolymerization of *p*-hydroxycinnamyl alcohols with metal salt oxidants results in the formation of a variety of oligolignols, which are valuable model compounds for the ^{13}C NMR characterization of DHPs and lignins. Novel guaiacyl DHPs can be obtained by this biomimetic approach that mimic natural lignin much better than those prepared by conventional techniques. A combination of the biomimetic with conventional syntheses allows introduction of rare linkages into oligolignols and will ultimately result in the capability of tailoring DHPs to certain specifications.

Acknowledgments

The authors gratefully acknowledge support from the USDA National Research Initiative Competitive Grants Program (Enhancing Value and Use of Agricultural and Forest Products) award #9403465.

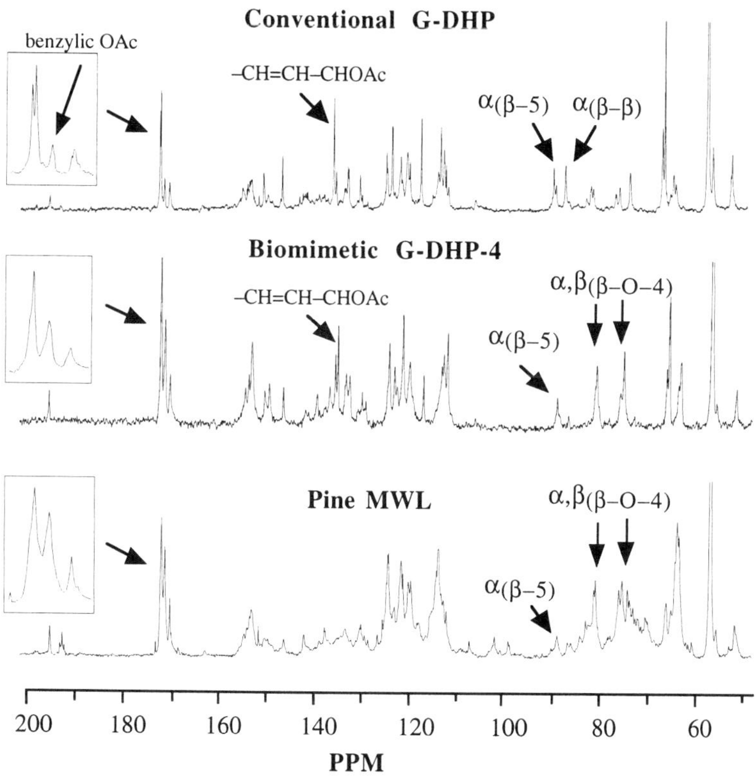

Figure 11. Partial ^{13}C NMR spectra of conventional G-DHP, biomimetic G-DHP, and pine MWL.

Literature Cited

1. Freudenberg, K. *Adv. Chem. Ser*. **1966**, *59,* 1-21.
2. Freudenberg, K. In *Constitution and Biosynthesis of Lignin;* Freudenberg, K., Neish, A. C., Eds.; Springer-Verlag: Berlin-Heidelberg, 1968; pp 47-129.
3. Newman, J.; Rej, R. N.; Just, G.; Lewis, N. G. *Holzforschung* **1986**, *40*, 369-373.
4. Lewis, N. G.; Newman, J.; Just, G.; Ripmeister, J. *Macromolecules* **1987**, *20*, 1752-1756 .
5. Terashima, N.; Seguchi, Y., *Proc. 4th Internat. Symp. Wood Pulp. Chem.*, **1987**, *2*, 1-4.
6. Hammel, K. E.; Jensen, K. A., Jr.; Mozuch, M. D.; Landucci, L. L.; Tien, M.; Pease, E. A. *J. Biol. Chem.* **1993**, *268*, 12274-12281.
7. Araki, H.; Terashima, N. *Mokuzai Gakkaishi* **1981**, *27*, 414-418.
8. Lewis, N. G.; Yamamoto, E.; Wooten, J. B.; Just, G.; Ohashi, H.; Towers, G. H. N. *Science* **1987**, *237*, 1344-1346.
9. Lewis, N. G.; Razal, R.; Yamamoto, E.; Wooten, J. B. *Proc. 5th Internat. Symp. Wood Pulp. Chem.* **1989**, *1,* 349-357.
10. Tomimura, Y.; Sasao, Y.; Yokoi, T.; Terashima, N. *Mokuzai Gakkaishi* **1980**, *26*, 558-563.
11. Xie, Y.; Terashima, N. *Mokuzai Gakkaishi* **1991**, *37*, 935-941.
12. Harkin, J. M. In *Oxidative Coupling of Phenols;* Taylor, W. I., Battersby, A. R., Eds.; Marcel Dekker: New York, NY, 1967; pp 243-321.
13. Adler, E. *Wood Sci. Technol.* **1977,** *11*, 169-218.
14. Quideau, S.; Ralph, J. *Holzforschun*g **1994**, *48*, 124-132.
15. Lapierre, C. In *Forage Cell Wall Structure and Digestibility;* Jung, H. G., Buxton, D. R., Hatfield, R. D., Ralph, J., Eds.; ASA-CSSA-SSSA: Madison, WI, 1993; pp 133-166.
16. Landucci, L. L. *J. Wood Chem. Tech*. **1995**, *15*, 349–368.
17. Landucci, L. L.; Luque, S.; Ralph, S. *J. Wood Chem. Tech*. **1995**, *15*, 493–513.
18. Omori, S.; Sakakibara, A. *Mokuzai Gakkaishi* **1974**, *20*, 48-49.
19. Omori, S.; Sakakibara, A. *Mokuzai Gakkaishi* **1979**, *25*, 145-148.
20. Houghton, P. J. *Phytochemistry* **1985**, *24*, 819-826.
21. Yoshida, S.; Tanahashi, M.; Shigematsu, M.; Shinoda, Y. *Mokuzai Gakkaishi* **1994**, *40*, 966-973.
22. Landucci, L. L.; Ralph, S. A. *Proc. 8th Internat. Symp. Wood Pulp. Chem.* **1995**, *II*, 51.
23. Landucci, L. L.; Ralph, S. A. *211th. ACS Natl. Meeting*: New Orleans, LA, 1996; poster presentation.
24. Brunow, G.; Sipilä, J.; Syrjänen, K.; Karhunen, P.; Setälä, H.; Rummakko, P. *Proc. 8th Internat. Symp. Wood Pulp. Chem.* **1995**, *I*, 33.

Chapter 12

Modeling Lignification in Grasses with Monolignol Dehydropolymerisate–Cell Wall Complexes

John H. Grabber[1], John Ralph[1,2], and Ronald D. Hatfield[1]

[1]**U.S. Dairy Forage Research Center, U.S. Department of Agriculture–Agricultural Research Service, Madison, WI 53706**
[2]**Department of Forestry, University of Wisconsin, Madison, WI 53706**

p-Hydroxycinnamyl alcohols are efficiently polymerized into primary walls from maize cell suspensions by H_2O_2 and wall-bound peroxidases to produce monolignol dehydropolymerisate–cell wall (DHP–CW) complexes. The structure and distribution of the synthetic lignins in complexes are similar to natural lignins formed in grass cell walls. This system was used to elucidate how ferulate-polysaccharide esters act as initiation sites for lignification, cross-link plant cell wall polymers, and restrict enzymatic degradation of structural polysaccharides. In addition, the effects of lignin composition and lignification conditions (pH and rate of precursor addition) on polysaccharide degradability were investigated. DHP–CW complexes are useful for modeling matrix interactions in lignified plant cell walls and for identifying means for improving the utilization of lignocellulosic materials for nutritional and industrial purposes.

Lignification plays an important role in plant growth and development by improving water conduction through xylem tracheary elements, enhancing the strength of fibrous tissues, and limiting the spread of pathogens in plant tissues (*1*). Lignin restricts the degradation of structural polysaccharides by hydrolytic enzymes, thereby limiting the bioconversion of fibrous crops into animal products or into liquid fuels (*2, 3*). Lignified dietary fiber also plays an important role in maintaining gastrointestinal function and health in humans (*4*). The effect of lignin on plant cell wall properties is thought to arise from the hydrophobicity of lignin, and its incrustation and attachment to other matrix components. The extent and importance of these interactions in affecting plant cell wall properties is poorly understood.

Lignification has been modeled *in vitro* through the dehydropolymerisates (DHPs) formed by the oxidative coupling of monolignols (*5-8*). The DHPs formed in the presence of carbohydrates, feruloylated oligosaccharides or proteins have been used to identify potential interactions between lignin and other cell wall components (*9-14*). Unfortunately, DHPs differ structurally from natural plant lignins (*15-17*) and they do not adequately reflect the three-dimensional structure of lignified cell walls. These limitations may be overcome in part by forming DHP–CW complexes using *in situ* peroxidases. Whitmore first used this approach in the late 1970's to study lignin-protein and lignin-carbohydrate interactions in cell walls isolated from the callus of *Pinus elliottii* (*18, 19*). Our objective was to further develop, evaluate, and utilize DHP–CW complexes as a means for studying lignin-matrix interactions in grass cell walls and for assessing the effect that such interactions have on the enzymatic degradability of cell walls.

Formation and Evaluation of DHP–CW Complexes

DHP–CW complexes are formed with cell walls isolated from cell suspensions of maize (*Zea mays*, cv. Black Mexican) (*20*). Cell walls from maize cell suspensions and similar tissues (*e.g.* maize coleoptiles) are composed of about 10% protein, 10% pectin, 50% hemicellulose (glucuronoarabinoxylans and heterologously linked glucans), and 25% cellulose (*21-23*). The cell walls also contain about 2% ferulic acid and 0.3% guaiacyl lignin (*20, 21*). Overall, the composition of the cell walls is typical of non-lignified primary walls in grasses.

As with other plant species, maize suspensions secrete peroxidases into the culture medium (Grabber, J.H., unpublished data). If lignin precursors are added to maize suspensions or isolated cells, secreted peroxidases would probably form DHPs in solution (*24*) or on the surfaces of cell walls (*25*) rather than within the wall matrix. Therefore, the cells are ruptured and extracted overnight with 100-200 mM $CaCl_2$ to remove extracellular and loosely bound peroxidases from the walls. However, even after extensive extraction with $CaCl_2$, a small quantity of peroxidase will gradually diffuse into the reaction medium during lignification. Cell walls isolated in this manner contain at least 9 peroxidase isozymes, primarily those of acidic pI (*20*). Multiple peroxidase isozymes are present in cell walls of plants and those of acidic pI are probably involved in lignification (*26*). Peroxidases in walls isolated from maize cell suspensions, like those isolated from maize internodes, will polymerize a variety of monolignols into DHPs, but their substrate affinity is greatest for ferulate and coniferyl alcohol, intermediate for *p*-coumarate, and low for sinapyl alcohol (Hatfield, R.D., unpublished data).

For complex formation, $CaCl_2$ extracted cell walls are thoroughly washed with water and suspended in 10-50 mM buffer solutions adjusted to pHs between 3.5 and 7.5 (*20*). Complexes may be formed in citrate, phosphate, HOMOPIPES or PIPES buffers. We generally use PIPES or HOMOPIPES buffers because they are not oxidized by hydrogen peroxide, they do not cause precipitation of calcium (needed for peroxidase activity and for maintaining the structural integrity of ionically bound pectins) and they do not form stable addition products with the quinone methide intermediates generated during the lignification process. In most cases, cell wall suspensions are lignified by adding a solution of monolignols and glucose over a 24 to 48 h period ('end-wise' polymerization). Monolignols are polymerized into cell walls by wall-bound peroxidases and hydrogen peroxide generated *in vitro* from glucose by glucose oxidase (added to the reaction medium prior to lignification). Since gluconic acid is a coproduct of this reaction, a relatively high buffer concentration (*e.g.* 50 mM) is required to maintain the pH. If desired, a dilute solution of hydrogen peroxide (*e.g.* 0.03%) may be added to the reaction medium in place of *in vitro* generated hydrogen peroxide.

Complexes with a Klason lignin content of up to 20% are readily formed with a variety of monolignols (*20*, Grabber, J.H., unpublished data). The maximum Klason lignin content observed in maize internodes is also about 20%. Incorporation of monolignols into cell wall-bound DHPs by 'end-wise' polymerization is usually over 90%. Yields are reduced either if the monolignol solutions include a high proportion of sinapyl alcohol (>60 mole %), or if hydrogen peroxide is added rapidly (< 1 h) to cell walls suspended in monolignol solutions ('bulk' polymerization), or if 'end-wise' polymerization is carried out under relatively acidic conditions (pH <4.5). Only in the latter case are the low yields attributable to peroxidase inactivation. The low yields of syringyl DHP–CW complexes are probably due to a low affinity on the part of the peroxidases for sinapyl alcohol. In the case of 'bulk' polymerization, DHPs are readily formed, but only about one-half is bound to the cell walls.

Thioacidolysis GC-MS, pyrolysis GC-MS and proton-decoupled ^{13}C-NMR analyses of the complexes and lignins isolated indicate that the synthetic lignins formed in DHP–CW complexes are structurally similar to natural lignins formed in grasses. In addition the DHPs, like natural grass lignins, are distributed throughout the wall matrix (*20*). Although many groups have attempted to produce DHPs which are representative of *in vivo* lignins, our findings suggest that DHP–CW complex formation may be the best system for modeling lignification of grass cell walls.

Applications of DHP–CW Complexes

Elucidation of Matrix Interactions in Cell Walls. Ferulic acid is attached as an ester **1** to the C5-hydroxyl of α-L-arabinosyl moieties of grass arabinoxylans. Recently our group (*27*) demonstrated that oxidative coupling mechanisms result in substantial amounts of 8–5′, 8–*O*–4′, and 8–8′ coupled diferulate esters (**2-4**) in addition to the previously reported 5–5′ coupled product **5**. We used nonlignified walls from maize cell suspensions and DHP–CW complexes to investigate cross-linked structures formed by ferulates in primary cell walls of grasses (*21*, Grabber, J.H., unpublished data). Cell suspensions were grown under normal conditions or with 40 μM 2-aminoindan-2-phosphonic acid (AIP, a selective inhibitor of phenylalanine ammonia lyase) to reduce ferulate deposition into cell walls. When nonlignified cell walls containing normal (17 mg g^{-1}) or low (5.3 mg g^{-1}) ferulate concentrations were incubated with an excess of H_2O_2, 40 to 55% of the ferulate monomers became oxidatively coupled by the wall-bound peroxidases into dehydrodimers. This remarkably high degree of dimerization, even in cell walls with extremely low ferulate concentrations, indicates that the spatial distribution of ferulates is somehow regulated in the maize walls to maximize dehydrodiferulate cross-linking of arabinoxylans. About one-half of the ferulate moieties were coupled by 8–5′ linkages; 8–*O*–4′, 5–5′, and 8–8′ coupled dehydrodimers each comprised 10 to 25% of the total. Ferulate monomers and dehydrodimers differed considerably in their propensity to become incorporated into lignin and to form hydrolyzable ether-linked structures with the lignin (Figure 1). These results indicate that the abundance and importance of ferulates as cross-linking agents has been greatly underestimated by solvolytic analysis of lignified plant tissues.

Oxidative coupling of ferulates to lignin was investigated further by synthetically lignifying cell walls containing γ-^{13}C labeled ferulate (Grabber, J.H.; Ralph, J., unpublished data). GC-MS analysis and long-range C–H correlation NMR spectroscopy studies of alkaline extracts from the complexes revealed that ferulate-coniferyl alcohol cross-products were coupled by 4–*O*–β′, 5–β′ and 8–β′ linkages (**6-8**). Very small amounts of 8–*O*–4′ and 8–5′ coupled products (**9, 10**) were detected by NMR spectroscopic analysis. In other work, only 4–*O*–β′, 5–β′ and 8–β′ coupled ferulate-lignin structures were detected by NMR spectroscopic analysis of lignins isolated from uniformly ^{13}C-labeled ryegrass (*28*). The virtual absence of 8–*O*–4′ and 8–5′ coupled ferulate-lignin cross-products in these studies indicates that ferulates couple almost exclusively to the β′-position of monolignols; coupling at the 5′ and *O*–4′ positions occurs only with dilignols and lignin oligomers in which the conjugated sidechain is no longer present. The selectivity of the reaction of ferulates with the β′-position in monolignols provides compelling evidence that ferulates act as initiation or nucleation sites for lignification. Similar approaches could be used to investigate the cross-linking of matrix polymers by other cell wall components such as amino acids and uronic acids.

Assessing the Impact of Cell Wall Modifications on Polysaccharide Degradability. Nonlignified cell walls and DHP–CW complexes were used to determine how reductions in ferulate cross-linking affect the degradability of structural polysaccharides by fungal enzymes (Grabber, J.H., unpublished data). In

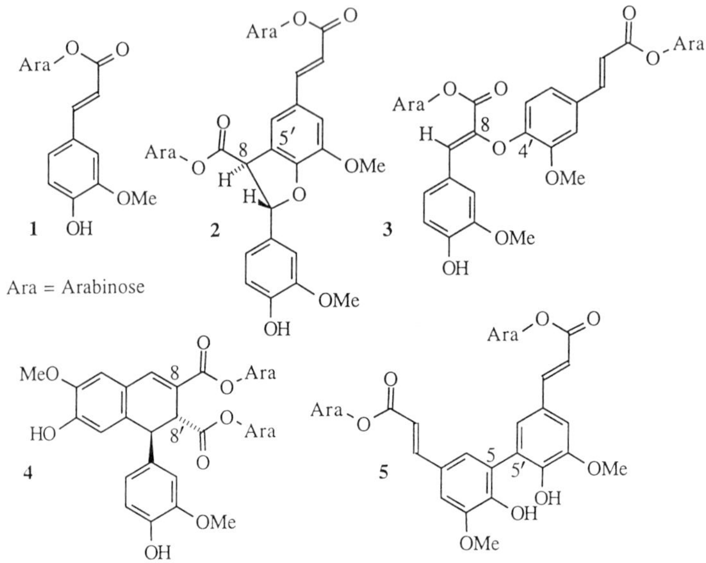

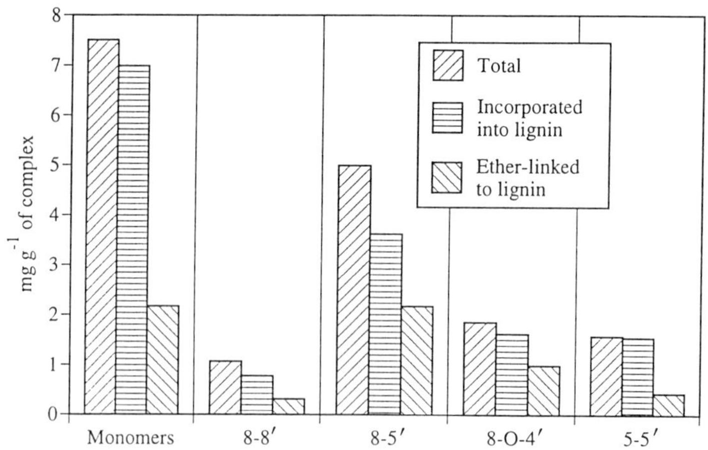

Figure 1. Incorporation of ferulate monomers and dehydrodimers into the lignins formed within DHP–CW complexes.

these studies, cell walls from normal or AIP treated cell suspensions were incubated with H_2O_2 or synthetically lignified with coniferyl alcohol/H_2O_2. Reductions in ferulate cross-linking primarily increased the rate of polysaccharide degradation in non-lignified cell walls and the extent of polysaccharide degradation in DHP–CW complexes (Figure 2). Similar results were observed when ferulate cross-linking was reduced by selective methylation of cell wall ferulates with diazomethane prior to complex formation. These results suggest that selection or genetic engineering of grasses for low levels of ferulate cross-linking will significantly improve the degradability of structural polysaccharides. Current studies in our laboratory are focused on determining the extent to which cell wall degradability is restricted by amino acid and uronic acid cross-linking of matrix components.

Lignins are formed by peroxidase/H_2O_2-mediated dehydropolymerization of *p*-coumaryl, coniferyl, and sinapyl alcohols (**11-13**). The incorporation of sinapyl alcohol into lignin is reduced in mutant or transgenic plants having low *o*-diphenol *O*-methyltransferase (OMT) activity (*29*). In mutant or transgenic plants where cinnamyl alcohol dehydrogenase (CAD) activity is reduced, *p*-hydroxy-cinnamyl aldehydes (**14, 15**) may become major components of the lignin (*30*). DHP–CW complexes were used as a biomimetic system for determining whether changes in lignin composition affect polysaccharide degradability by fungal enzymes (Grabber, J.H., unpublished data). Complexes formed with coniferyl aldehyde were about 40% less degradable than the complexes formed with coniferyl alcohol. Differences in degradability were only partially eliminated if complexes were incubated with high concentrations of fungal enzymes. When aldehyde groups were selectively reduced to alcohols (by ethanolic-sodium borohydride) prior to enzyme hydrolysis, however, degradability differences were eliminated. This suggests that high hydrophobicity of aldehyde-containing lignin might have been responsible for depressed cell wall degradation. Altering the type of *p*-hydroxy-cinnamyl alcohols used to lignify the walls did not affect subsequent cell wall degradability. Based on these results, we propose that improvements in cell wall degradability, previously attributed to changes in lignin composition (*29, 30*), were in fact due to other modifications in wall chemistry or ultrastructure.

The effect of lignin structure on cell wall degradability was investigated by forming DHP–CW complexes respectively by 'end-wise' and 'bulk' polymerization of coniferyl alcohol (Grabber, J.H., unpublished data). Analysis of the lignin structures by thioacidolysis indicated that DHPs in 'end-wise' complexes had a relatively linear configuration characteristic of natural grass lignins, whereas DHPs in 'bulk' complexes had a highly branched lignin configuration. The degradability of 'end-wise' and 'bulk' DHP–CWs by fungal enzymes was similar. These results indicate that lignin structure does not affect the enzymatic hydrolysis of cell walls.

The pH used to form DHP–CW complexes had considerable impact on polysaccharide degradability, particularly in complexes with high lignin contents (Grabber, J.H., unpublished data). Complexes formed at pH 5.5 were up to 20% more degradable than complexes formed at pH 4.0. The cause of these degradability differences is currently under investigation by our group.

Other Applications. Cell walls from cell suspensions and DHP–CW complexes would be valuable in other areas of research, including investigations on structural-functional relationships of dietary fiber as related to human health. Recently Eastwood and coworkers have used cell walls from cell suspensions of spinach to study the fermentation and metabolism of nonlignified walls by rats (*31, 32*). DHP–CW complexes provide a means of generating lignified fiber with various types of lignin-matrix interactions. Such materials would be valuable in small animal or *in vitro* studies aimed at elucidating the beneficial effects of lignified fiber on gastrointestinal function and health.

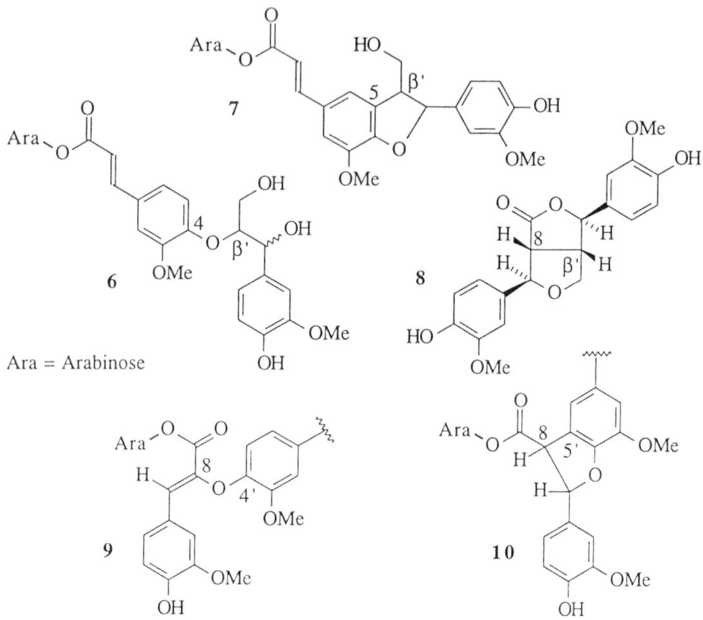

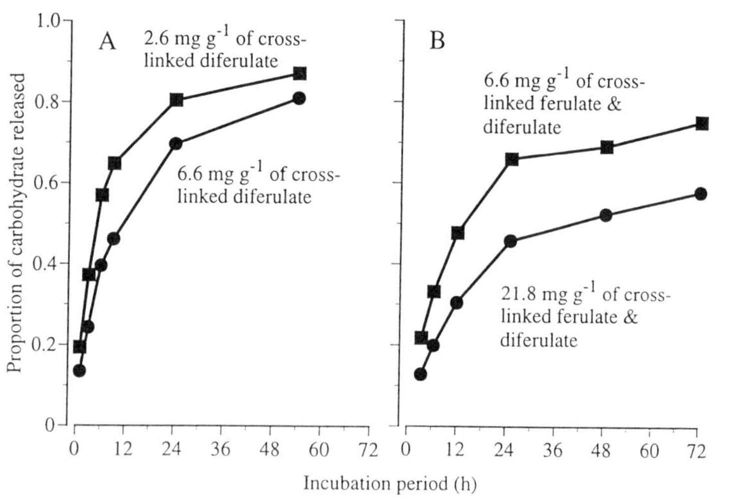

Figure 2. Effect of ferulate cross-linking on the release of carbohydrate from (A) nonlignified walls and (B) DHP–CW complexes during hydrolysis with fungal enzymes.

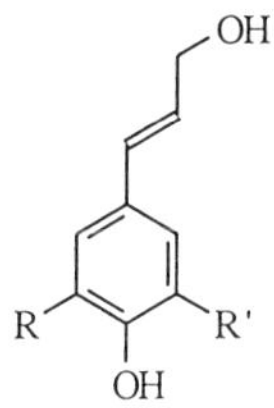

11; R = R' = H

12; R = H, R' = OMe

13; R = R' = OMe

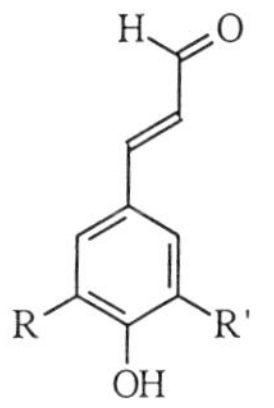

14; R = H, R' = OMe

15; R = R' = OMe

Limitations of DHP–CW Complexes

Because DHP–CW complexes are formed with primary walls, our system may accurately model lignification of cell types such as parenchyma which have only primary cell walls. Other cell types, like xylem and sclerenchyma, have thick secondary walls in addition to primary walls. Primary and secondary walls both contain noncellulosic polysaccharides, cellulose, structural proteins, hydroxycinnamic acids and lignin (*33-36*), but in differing proportions. Even in these tissues, primary walls are more heavily lignified and more resistant to degradation than secondary walls (*37*) and they limit penetration of degradative organisms and enzymes into the secondary walls (*38*). Since lignification occurs concurrently with secondary wall formation (*33*), it is not possible to adapt our system to cell walls with secondary thickenings because the cell walls must be isolated prior to complex formation.

Secondary cell wall formation and lignification can be stimulated in suspension or callus cultures by the judicious use of phytohormones or other elicitors (*39-42*). Although these systems are valuable for studying the biosynthesis of lignin, the prospect for manipulation of lignin composition or lignin-matrix interactions in these cultures is limited, since it would require reduced expression of a large number of specific enzymes. Developing enzyme inhibitors or genetically altered cultures to manipulate matrix interactions would be time consuming and problematic. As illustrated in the previous examples, DHP–CW complexes give us tremendous flexibility in manipulating and assessing how lignin-matrix interactions affect cell wall properties. Because of the perceived importance of primary cell walls in controlling wall properties, we believe that our system is useful for modeling matrix interactions in lignified walls and for identifying means for improving the utilization of lignocellulosic materials.

Acknowledgments

Partial support of this research by the USDA-ARS Administrator Funded Post Doctoral program, USDA-NRI competitive grants #92-37304-8057 (Plant Growth and Development) and #94-37500-0580 (Enhancing Value and Use of Agricultural and Forest Products), and the W.H. Miner Agricultural Research Institute, Chazy, N.Y., is gratefully acknowledged.

Literature Cited

1. Iiyama, K.; Lam, T. B. T.; Stone, B. *Plant Physiol.* **1994**, *104*, 315-320.
2. Brown, A. *J. Appl. Biochem.* **1985**, *7*, 371-387.
3. Jung, H. G.; Deetz, D. A. In *Forage Cell Wall Structure and Digestibility*; Jung, H. G., Buxton, D. R., Hatfield, R. D., Ralph, J., Eds.; ASA-CSSA-SSSA: Madison, WI, 1993; pp 315-346.
4. McDougall, G. J.; Morrison, I. M.; Stewart, D.; Hillman, J. R. *J. Sci. Food Agric.* **1996**, *70*, 133-150.
5. Freudenberg, K. In *Constitution and Biosynthesis of Lignin*; Freudenberg, K., Neish, A. C. Eds.; Springer-Verlag: Berlin, 1968; pp 47-116.
6. Gagnaire, D.; Robert, D. *Makromol. Chem.* **1977**, *178*, 1477-1495.
7. Lewis, N. G.; Newman, J.; Just, G.; Ripmeister, J. *Macromolecules* **1987**, *20*, 1752-1756.
8. Kirk, T. K.; Connors, W. J.; Bleam, W. D.; Hackett, W. F.; Zeikus, J. G. *Proc. Nat. Acad. Sci. USA* **1975**, *72*, 2513-2519.
9. Higuchi, T.; Ogino, K.; Tanahashi, M. *Wood Res.* **1971**, *51*, 1-11.
10. Terashima, N.; Seguchi, Y. *Cellulose Chem. Technol.* **1988**, *22*, 147-154.
11. Ohnishi, J.; Watanabe, T.; Koshijima, T. *Phytochemistry* **1992**, *31*, 1185-1190.
12. Evans, J. J.; Himmelsbach, D. S. *J. Agric. Food Chem.* **1991**, *39*, 830-832.

13. Ralph, J.; Helm, R. F.; Quideau, S.; Hatfield, R. D. *J. Chem. Soc., Perkin Trans. I* **1992**, 2961-2969.
14. McDougall, G. J.; Stewart, D.; Morrison, I. M. *Phytochemistry* **1996**, *41*, 43-47.
15. Erdtman, H. *Ind. Eng. Chem.* **1957**, *49*, 1385-1386.
16. Lewis, N. G.; Yamamoto, E.; Wooten, J. B.; Just, G.; Ohashi, H.; Towers, G. H. N. *Science* **1987**, *237*, 1344-1346.
17. Tollier, M.-T.; Lapierre, C.; Monties, B.; Francesch, C.; Rolando, C. *Proc. 6th Internatl. Symp. Wood Pulp. Chem.* **1991**, *2*, 35-41.
18. Whitmore, F. W. *Phytochemistry* **1978**, *17*, 421-425.
19. Whitmore, F. W. *Phytochemistry* **1982**, *21*, 315-318.
20. Grabber, J. H.; Ralph, J.; Hatfield, R. D.; Quideau, S.; Kuster, T.; Pell, A. N. *J. Agric. Food Chem.* **1996**, *44*, 1453-1459.
21. Grabber, J. H.; Hatfield, R. D.; Ralph, J.; Zon, J.; Amrhein, N. *Phytochemistry* **1995**, *40*, 1077-1082.
22. Kieliszewski, M.; Lamport, D. T. A. In *Self-Assembling Architecture*; Varner, J., Ed.; Proceedings of the 46th Symposium of the Society for Developmental Biology: St. Paul, MN, 1988; pp 61-77.
23. Carpita, N. C. *Plant Physiol.* **1984**, *76*, 205-212.
24. Brunow, G.; Kilpelainen, I.; Lapierre, C.; Lundquist, K.; Simola, L. K.; Lemmetyinen, J. *Phytochemistry* **1993**, *32*, 845-850.
25. Nakashima, J.; Takabe, K.; Saiki, H. *Mokuzai Gakkaishi* **1992**, *38*, 1136-1142.
26. McDougall, G. J. *Phytochemistry* **1992**, *31*, 3385-3389.
27. Ralph, J.; Quideau, S.; Grabber, J. H.; Hatfield, R. D. *J. Chem. Soc., Perkin Trans. I* **1994**, 3485-3498.
28. Ralph, J.; Grabber, J. H.; Hatfield, R. D. *Carbohydr. Res.* **1995**, *275*, 167-178.
29. Bernard-Vailhe, M. A.; Migne, C.; Cornu, A.; Maillot, M. P.; Grenet, E.; Besle, J. M. *J. Sci. Food Agric.* **1996**, *72*, 385-391.
30. Bernard-Vailhe, M.-A.; Cornu, A.; Robert, D.; Maillot, M.-P.; Besle, J.-M. *J. Agric. Food Chem.* **1996**, *44*, 1164-1169.
31. Gray, D. F.; Fry, S. C.; Eastwood, M. A. *Br. J. Nutr.* **1993**, *69*, 177-188.
32. Buchanan, C. J.; Fry, S. C.; Eastwood, M. A. *J. Sci. Food Agric.* **1995**, *67*, 367-374.
33. Terashima, N.; Fukushima, K.; He, L.-F.; Takabe, K. In *Forage Cell Wall Structure and Digestibility*; Jung, H. G., Buxton, D. R., Hatfield, R. D., Ralph, J., Eds.; ASA-CSSA-SSSA: Madison, WI, 1993; pp 247-270.
34. Iiyama, K.; Lam, T. B. T.; Meikle, P. J.; Ng, K.; Rhodes, D.; Stone, B. A. In *Forage Cell Wall Structure and Digestibility*; Jung, H. G., Buxton, D. R., Hatfield, R. D., Ralph, J., Eds.; ASA-CSSA-SSSA: Madison, WI, 1993; pp 621-683.
35. Ye, Z.-H.; Varner, J. E. *Plant Cell* **1991**, *3*, 23-37.
36. Hatfield, R. D. In *Forage Cell Wall Structure and Digestibility*; Jung, H. G., Buxton, D. R., Hatfield, R. D., Ralph, J., Eds.; ASA-CSSA-SSSA: Madison, WI, 1993; pp 285-313.
37. Chesson, A.; Stewart, C. S.; Dalgarno, K.; King, T. P. *J. Appl. Bacteriol.* **1986**, *60*, 327-336.
38. Wilson, J. R. In *Forage Cell Wall Structure and Digestibility*; Jung, H. G., Buxton, D. R., Hatfield, R. D., Ralph, J., Eds.; ASA-CSSA-SSSA: Madison, WI, 1993; pp 1-27.
39. Ingold, E.; Sugiyama, M.; Komamine, A. *Plant Cell Physiol.* **1988**, *29*, 295-303.
40. Hosel, W.; Fiedler-Preiss, A.; Borgmann, E. *Plant Cell Tissue Organ Cult.* **1982**, *1*, 137-148.
41. Campbell, M. M.; Ellis, B. E. *Phytochemistry* **1992**, *31*, 737-742.
42. Eberhardt, T. L.; Bernards, M. A.; He, L.; Davin, L. B.; Wooten, J. B.; Lewis, N. G. *J. Biol. Chem.* **1993**, *268,* 21088-21096.

Chapter 13

Cellulose and the Hemicelluloses: Patterns for the Assembly of Lignin

R. H. Atalla

Forest Products Laboratory, U.S. Department of Agriculture Forest Service
Madison, WI 53705–2698

The results of our most recent studies point to significant coupling between the tertiary structures of all three major constituents of the cell wall in native woody tissue. It is clear that the hemicelluloses are intimately involved in the aggregation of cellulose, and the hemicelluloses, together with cellulose, provide the pattern for assembly of the lignin. Our studies of the organization of lignin in native woody tissue, including Raman microprobe spectral analyses and measurements of photoconductivity and fluorescence, leave little question that the level of order in lignin is well beyond that expected from random free radical coupling in the absence of a stronger organizing influence. We believe that a dominant organizing influence is provided by strong associative interactions between lignin precursors and the polysaccharide matrix, and that the hemicelluloses in particular play a central role in the selective binding of specific lignin precursors. In this context the hemicelluloses are viewed as carriers of an important component of the information needed for the organization of lignin. The binding of the lignin precursors prior to their polymerization is expected to be one of the fundamental organizing influences that determines the primary structure of lignin. The precursors that are expected to participate in this include, in addition to the monolignols, a number of oligomers of lignin which are formed by membrane bound enzymes. Our proposal concerning the role of the hemicelluloses in the assembly of the cell wall suggests, for the first time, direct pathways for genotypic control of the assembly of lignin. Thus, variations in the structures of the hemicelluloses, the biosynthesis of which is controlled within the Golgi apparatus, are expected to result in corresponding systematic variations in the structure of the associated lignin. In consequence, it would appear that any effort to modify the assembly of lignin requires an orchestrated modification of the pathways for biosynthesis of the hemicelluloses, together with any programmed changes in the phenylpropanoid pathways for synthesis of mono- and oligolignols. Finally, our emerging view of the complexity and dynamic character of lignification in the developing cell wall matrix leads us to the view that the wall of the developing cell is a very active participant in both intracellular and extracellular processes.

The results of our studies of molecular organization in wood cell walls point to significant coupling between the tertiary structures of all three major constituents of the wall in native tissues. It is clear that the hemicelluloses are intimately involved in the aggregation of cellulose, and the hemicelluloses, together with cellulose, have a significant influence on the structure of lignin. Yet these patterns of correlation in organization at the nanoscale level are not addressed within the framework of the paradigms associated with the biogenesis of plant cell walls; the nanoscale level here is taken to encompass domains in the range of 1 to 10 nm. The purpose of this chapter is to examine those aspects of structure that cannot be rationalized within the framework of established paradigms, paying particular attention to key observations concerning the organization of lignin and the regulation of its assembly at the nanoscale level. Some proposals concerning relationships between the structure of lignin and the organization of the polysaccharide matrix are presented, which can account for the diversity of molecular organization observed in cell walls, and allow for the regulation of details of structure by the organism beyond any heretofore envisioned. Such regulation of these structural details is more consistent with the general patterns prevailing in biological systems and our hypotheses also provides a more appropriate framework for any current or future efforts to accomplish genetic modifications of woody species.

Historical Context

The prevailing paradigms with respect to assembly and molecular organization of the cell wall were established well over three decades ago, before many of the studies addressing both the structure and processes of assembly of cell wall constituents were even undertaken. The paradigms are based on a **two phase** model with the underlying premise that cellulose is the key structural component, and that lignin and the hemicelluloses are the matrix in which the cellulose is embedded (*1, 2*). In this context, lignin is usually regarded as the binding component, and the hemicelluloses are frequently seen as coupling agents between the lignin and the cellulose. This view has evolved from (i) early assimilation into the models of the cell wall, and the biosynthetic processes leading to its formation, as well as (ii) of many notions that have their origins within the branch of materials science dealing with artificial fiber/matrix composites. It also reflects the premise that each of the constituents is synthesized separately at a different location within the cell, with the polymerization of lignin during the final consolidation of the cell wall structure being viewed as the analog of the curing of the matrix polymer in artificial composites.

While for a number of decades the accepted model has been a useful one, it has serious shortcomings from both physical and biological perspectives. Perhaps the most serious flaw, from a biological perspective, is that the model has no provision for coordination of the deposition of the three different major constituents of the wall, nor does it afford plausible approaches for genetic control of cell wall organization to the degree suggested by phenotypic expression.

A number of recent developments have raised additional questions about the validity of the model of the cell wall from a physical perspective as well. These questions arise because the components of the cell wall are intimately integrated at a finer scale than is the case with synthetic composites. They are, in fact, integrated at a level approaching that of individual macromolecules, both in terms of the final structure and during its assembly. In addition, there is evidence of coupling in the assembly processes at intermediate stages preceding final consolidation.

Need for a New Model

From a physical perspective, the considerations which point to the need for re-examination of the accepted model are associated with scale. In most synthetic composites the amount of material at the surface is small compared to that within the boundary of an individual domain. The properties of the composites are then dominated by bulk properties of the components, with any modifications arising from the relatively weak (in terms of its contribution to the overall structure) coupling at the interfaces between the domains. In higher plant cell walls, in contrast, the cellulosic domains usually are such that more of the material is at the surface than in the bulk. For all other constituents, the domains are even smaller than they are for cellulose. Thus, in these nanoscale composites, the coupling interactions at the interfaces between the domains may well be the dominant determinants of the properties of the aggregates. Or, indeed, the integration of the constituents at the nanoscale level may not even admit an explicit definition of an interface. That this is not simply an abstract matter is well demonstrated by recent observations of the effects of hemicelluloses on the aggregation of cellulose, as well as the long established effect of polysaccharide matrices on the structures of dehydrogenation polymers (DHPs) of lignin precursors. In the studies of the effects of hemicelluloses, it was shown that, when nascent cellulose produced by *Acetobacter xylinum* aggregates in the presence of hemicelluloses, its structure is very similar to higher plant celluloses, suggesting that one of the functions of the hemicelluloses is the regulation of cellulose aggregation (*3, 4*). The effects of polysaccharide matrices on the patterns of assembly of monolignols in dehydrogenation polymers also point to a clear influence of microenvironment and association on the primary structures and interunit linkages of the resulting polylignols (*5-7*).

The considerations arising from a biological perspective are perhaps even more compelling because they are associated with regulation of molecular organization in the cell wall matrix. Current models of biosynthesis implicitly incorporate the premise that the primary structures of the constituents of cell wall polysaccharides are determined by intracellular or membrane bound biosynthetic activities (*8, 9*), while the primary structure of lignin is defined by a random free radical coupling reaction that occurs extracellularly, *i.e.* outside of the plasma membrane. Thus, the level of control exercised over the primary structure of the polysaccharides seems to far exceed the level exercised over the structure of lignin. Under these circumstances, lignin would seem to be exceptional among constituents of biological tissues. The final structure of lignin, it would follow, would be determined by the kinetics of the various possible reactions between the different free radicals, with the regulation of the assembly process occurring only through the influence of microenvironment on the kinetics. When one considers the degree to which the details of biological structures much more complex than lignin are under precise program control of the cell through a hierarchy of biochemical processes (*10*), free radical polymerization with its susceptibility to perturbation of kinetics and, hence, structure by the microenvironment seems quite primitive in contrast. An alternative model for the assembly of the lignin, one wherein the specific details of structure are under more direct control of the cell, would seem to be more consistent with the pattern of assembly of structural tissues in biological systems in general. It would also be in order to incorporate into any new model a possible degree of distribution in the processes of primary structure formation for the different constituents, rather than confining consideration to mechanisms whereby each of the macromolecular components is assembled in final form at one single location. Such a model could provide a more effective basis for organizing the information concerning the diversity of structures known to occur in native lignins.

A Comprehensive Model for the Assembly of Lignin

The more comprehensive model proposed here has grown from the realization that, for many cell wall constituents, the control and regulation of structure must be hierarchic so that it occurs at multiple levels, in different stages and at different locations. Thus, while the major component steps in the formation of the primary structures are carried out within the cell by membrane-bound enzymes in different biosynthetic apparatuses, at the cell surface or in its interior, further modification appears to be necessary prior to assembly into the cell wall matrix. At the first level, the primary structures are formed and optimized for solubility and freedom of movement within the cytoplasm to **prevent premature aggregation or irreversible associations**. Upon diffusion out through the cell membrane, or deposition by exo-cytosis, some further modification by extracellular or cell membrane-bound enzymes would then be necessary to impart to the constituents the characteristics prerequisite for participation in the processes of the assembly, integration and consolidation of the matrix. For some constituents, particularly the polysaccharides, the assembly and consolidation involve entry into tightly bound insoluble structural aggregates that are the key to the mechanical stability of the polysaccharide matrix. For other constituents such as the precursors of lignin, integration into the matrix requires participation in additional coupling reactions.

Among the polysaccharides the instances that are most clearly illustrative and well established are those of mannans and glucomannans in the secondary walls of gymnosperms and of xylans in the secondary walls of angiosperms. These are polysaccharides that are just as insoluble as cellulose even in mildly alkaline aqueous media. Their solubility within the cytoplasm must, therefore, be attributed to initial biosynthesis with limited branching to prevent aggregation. The complementary steps prior to assembly into the extracellular polysaccharide matrix would be enzymatic removal of the branching after release outside the cell membrane. The limited branching is seen as essential to prevent crystallization within the cell boundaries, while the de-branching is seen as prerequisite for tight aggregation with cellulose in the cell wall. The availability of a wide array of glycosidases capable of shortening or removing the branching in the extracellular region adjacent to the plasma membrane is well established.

Though this chapter is not perhaps the best context for presenting it in detail, a similar argument can be made for synthesis of cellulose in **higher plants** within an internally located biosynthetic apparatus, followed by limited debranching at the sites of deposition from the cell membrane with concomitant organization of the molecules into fibrillar bundles. For such a working hypothesis, the membrane bound complexes which have been visualized as sites of cellulose biosynthesis would instead be viewed as the loci of an enzyme system that coordinates aggregation of the fibrillar bundles simultaneously with the debranching. Two decades ago, the isolation of a "soluble cellulose" was reported. It was later found to have a glucose unit attached as a branch residue at the 2 carbon on every third anhydroglucose unit (*11*); most traditional tests for the presence of cellulose cannot discern the presence of these soluble gluco-glycans. A more comprehensive discussion of the issues of assembly of cellulose will be presented elsewhere.

The assembly of lignin represents an even more compelling example of distributed processes or stages, separated in time and space, that result in the biogenesis of a key cell wall constituent. Thus, many of the monolignol precursors of lignin are produced within the cytoplasm as the glucosides, which, upon transmission to the exterior of the membrane, are hydrolyzed by glucosidases to release the primary precursor (*12, 13*). This view is further reinforced by recent direct evidence for the occurrence, in lignifying tissues, of glucosidases that are specific to the monolignol glucosides (*14, 15*). The interior biosynthetic apparatus would be viewed as producing an intermediate structure optimized for the environment defined by the cytoplasm; the monolignol glucosides have a high degree of solubility in aqueous media and should thus be well solvated in the cytosol. Upon secretion into the extracellular environment, the hydrophilicity of the precursors is altered by removal of the glucose to enhance the affinity of the monolignol for the polysaccharide aggregates. In addition to facilitating diffusion into, and association with, the polysaccharide matrix, the monolignol is now more susceptible to the free radical coupling or propagation reactions leading to the formation of lignin.

Another clear instance of distributed assembly is the formation of different dilignols by extracellular or membrane bound enzymes, as recently established by Lewis and coworkers (*16-19*). Though it will be further suggested below that this may also be an essential intermediate step in the biosynthesis of lignin, it is presented here as another example of the distribution of the formation of primary structure between interior biosynthetic apparatus and enzyme systems bound to the cell membrane or released in its immediate vicinity. Thus, here again, the interior apparatus would be viewed as producing an intermediate structure optimized for the environment defined by the cytoplasm, while the further modification would be regarded as the second phase of control of the biosynthetic process in which structures are further elaborated in response to the activity of the membrane bound enzymes.

This proposal concerning stages in the assembly of lignin exemplifies the notion of distributed assembly, which is so often characteristic of the formation of structure in biological systems. Such assembly systems afford the possibility of sequential control of the stages of synthesis and the level of coupling of the synthetic processes for polysaccharides and lignin that is reflected in cell wall matrix organization. Coupling of the processes would admit much more precise orchestration of the assembly of the cell wall than allowed in current models.

Before discussion of the mechanisms of such a coupling, however, it is important to review evidence of association between the polysaccharides and lignin. The reports (*5-7*) noted above, which indicate an influence of polysaccharide matrices on the primary structures of DHPs formed within them, point to a strong associative interaction between the polysaccharides and the lignin precursors. At the least, the interaction appears to organize the monolignols in a manner which results in a DHP that is different from the DHP arising under otherwise identical conditions but in the absence of the polysaccharides. That subtle differences in the polysaccharides or in the conditions of polymerization can result in significant differences in the DHPs argues for a strong associative interaction (*20*). More recently the work of Grabber *et al.* provided additional support for this view (*21, 22*).

Two other studies undertaken in our laboratory also argue for strong associative interactions. Theoretical modeling analyses of the interactions of mono-, di- and trilignols with the surface of a cellulosic aggregate point to rapid adsorption of the lignol from an aqueous medium onto the surface of the cellulose (*23*). These results leave little question that, at the nanoscale level, the dominant interactions between oligolignols and polysaccharides are associative. Perhaps more compelling, because they suggest coherence on an even larger scale, are the observations of photoconductivity in woody tissues (*24*). The measurements of photoconductivity, together with complementary analyses of the fluorescence spectra of the same tissues, point to a high degree of spatial organization of lignin in the native state. The photoconductivity also points to the existence of pathways for electron transport that have not heretofore been recognized as an inherent characteristic of the structures of native lignins. The level of organization suggested by these observations is not at all consistent with the type of structure anticipated from random free radical coupling reactions. Additional evidence pointing to the organizing influence of the cell wall matrix on the assembly of lignin may be derived from the work of Sarkanen and coworkers who observed that the presence of lignin during the polymerization of monolignols modified the course of the polymerization reaction (*25*).

The final piece of information pointing to the probability of a high degree of coupling in the assembly of cell wall constituents, is the observation that branches in hemicelluloses are specific in type, and are rarely more than three sugar units in length, and more commonly one or two units in length. This needs to be considered in light of the observations mentioned earlier that the hemicelluloses are integrated into the structure of cellulose fibrils during aggregation (*3, 4*).

When taken together, the different groups of observations suggest the plausibility of a paradigm wherein one of the key roles of the polysaccharides is to organize the structure of the lignin. Cellulose would provide the primary framework, and the hemicelluloses, integrated into its structure or adsorbed onto its surface, would have the various branches available to associate with specific lignin precursors. In this scheme it would be anticipated that the polysaccharide backbones and specific monosaccharide branches are designed to organize the monolignols, while disaccharide and trisaccharide branches are designed to selectively bind specific dilignols or trilignols. The attribution of such selectivity to sites for association on the polysaccharides and their oligosaccharide branches is quite consistent with their well established role in many biological recognition mechanisms. It is suggested that these associations may well be as specific as the nucleotide base pairings that play such an essential role in the ribosomal assembly of proteins. Completion of the assembly process would then be expected to proceed by a polymerization process governed by free radical propagation in addition to free radical coupling.

What is proposed here is a role for the hemicelluloses well beyond that of regulating aggregation of cellulose, as was proposed earlier (*3, 4*). In the present context the hemicelluloses are viewed as performing an even more essential function with respect to lignin. This is an organizing function not unlike that of the messenger RNA in the assembly of enzymes or other proteins (*10*), though the same absolute precision in control of the assembly process may not be required. Such an assembly process for lignin would, however, provide for control of the primary structure of lignin by a process not as susceptible to perturbation by macroscale variations in thermal and environmental conditions as would be expected in a polymerization reaction wherein the coupling mode kinetics and the microenvironment are the dominant determinants of primary structure.

Implications and Conclusions

Though the model presented here represents a significant departure from accepted views concerning the biogenesis of lignin, it provides a plausible rationale for patterns of structure distribution in native lignins. It also suggests an avenue for intracellular regulation of the assembly processes for lignin that is more in keeping with the hierarchic and sequential control of biochemical steps that is characteristic of biological systems (*26*). That such a level of regulation is probable is evidenced by the uniformity of the structure of lignin in adjacent cells within a particular tissue. In woody plants, this uniformity manifests itself in the uniformity of structure within a particular annular ring. If the structure of lignin were determined by unregulated kinetic and microenvironmental factors, rather than being under the central control of the cells, one would expect most trees grown at latitudes outside the tropics to have lignins that differ in structure between northern and southern exposure within the same annual ring: due to the difference in exposure to solar radiation, the two regions would be expected to be at different temperatures during daylight hours. While it is not clear that this possibility has been explored experimentally, it would seem very likely that such differences would have resulted in anatomical differences that would have been recognized long ago on the basis of visual macroscopic examination, and certainly upon subjection to the different staining reactions that have long been used to characterize lignification in plants. That such differences have never been observed or reported can be taken as *prima facie* evidence for central intracellular regulation of the assembly of lignin.

The only plausible avenue for intracellular regulation of the structure of lignins is through the provision of templates for their spatial organization prior to the development of covalent linkages through radical coupling reactions. The most plausible templates are the domains of the polysaccharide matrix and their interior surfaces formed through the interactions between cellulose and the hemicelluloses. Since mechanisms for intracellular control of synthesis of the polysaccharides have been articulated, and are clearly a reflection of the expression of genetic information, the pathway to regulation of the structure of the lignins is complete.

This pathway to regulation of lignin structure suggests new approaches for genetic modification of the assembly of lignin. This would be seen as a process requiring two complementary components wherein modification of pathways for the synthesis of phenylpropanoid structures is coordinated with modification of the assembly of the polysaccharides that provide the templates for organization of the lignin. Thus, any program directed at development of transgenic plants with different lignin structures must include a component addressing modifications of the hemicelluloses as well. While this adds another dimension to the complexity of such a program, it does offer the possibility of another avenue for regulation of the structure.

The significant departure from the accepted views about the biogenesis of lignin presented here can provide both an organizing principle for empirical observations concerning the structure and diversity of native lignins, and a guide for future explorations of the biosynthetic pathways. One of its attractive features is that it envisions assembly processes for lignin that are consistent with the hierarchic and sequential control of elementary biochemical steps that is characteristic of biological systems (*26*).

When taken together with our observations of photoconductivity (*24*) and the coherence of order that they imply, the model suggests that cell wall systems in higher plants are hierarchically organized **at the nanoscale level**, and that the constituents are coupled at secondary and tertiary levels of molecular structure that are not adequately captured by the traditional models describing cell walls as two phase composite systems. In addition, they point to the possibility that the cell wall matrix possesses functions in which electronic charge transport may play an important role. The intimate relationship between lignin, which can be an electron conductor, and cellulose, which is piezoelectric, suggests the possibility that the integrated molecular composite represents an electromechanical system with properties that can translate mechanical deformation of the plant cell wall into electrical signals that may elicit cellular responses to mechanical deformation. Conversely, the piezoelectric character of the composite may play a key role in the response of plant tissues to photochemically generated electrical potentials that are likely to trigger many of the complex responses to light.

In summary, our understanding of the functions of lignin and the cell wall matrix needs to be re-examined. It may well lead to the conclusion that the wall of the developing cell is one of the most dynamic mediators between intracellular and extracellular processes.

Acknowledgments

This work was supported by the USDA Forest Service and the Division of Energy Biosciences, Office of Basic Energy Sciences, of the US Department of Energy; both are gratefully acknowledged.

Literature Cited

1. Preston, R. D. *The Physical Biology of Plant Cell Walls;* Chapman and Hall: London, 1974.
2. Frey-Wyssling, A. *The Plant Cell Wall;* Gebruder Borntrager: Berlin, 1976.
3. Atalla, R. H.; Hackney, J. M.; Uhlin, I.; Thompson, N. S. *Int. J. Biol. Macromol.* **1993,** *15*, 109.
4. Hackney, J. M.; VanderHart, D. L.; Atalla, R. H. *Int. J. Biol. Macromol.* **1994,** *16*, 215.
5. Higuchi, T.; Ogino, K.; Tanahashi, M. *Wood Res.* **1971,** *51*, 1.
6. Tanahashi, M.; Higuchi, T. *Wood Res.* **1981,** *67,* 29.
7. Terashima N.; Fukushima, K.; He, L.-F.; Takabe, K. In: *Forage Cell Wall Structure and Digestibility;* Jung, H. G., Buxton, D. R., Hatfield, R. D., Ralph, J., Eds.; *ASA-CSSA-SSSA*: Madison, 1993; p 247.
8. Northcote, D. H. In *Biosynthesis and Biodegradation of Wood Components;* Higuchi, T., Ed.; Academic Press: London, 1985, p 87.
9. Northcote, D. H. *ACS Symp. Ser.* **1989**, *399*, 1.

10. Singer, M.; Berg, P. *Genes and Genomes*, University Science Books: Mill Valley, CA, 1991.
11. Sowden, L. C.; Colvin, R. J. *Can. J. Microbiol.* **1978**, *24*, 772.
12. Gross, G. G. In *Biosynthesis and Biodegradation of Wood Components;* Higuchi, T., Ed.; Academic Press: London, 1985, p 229.
13. Terashima, N.; Nakashima, J.; Takabe, K. *ACS Symp. Ser.* **1997**, this volume.
14. Dharmwardhana, D. P.; Ellis, B. E.; Carlson, J. E. *Plant Physiol.* **1994,** *107*, 331.
15. Ellis, B. E.; Dharmwardhana, D. P. *ACS Symp. Ser.* **1997**, this volume.
16. Davin, L. B.; Lewis, N. G. In *Phenolic Metabolism in Plants;* Stafford, H. A., Ibrahim, R. K., Eds.; Plenum Press: New York, NY, 1992, p 325.
17. Lewis, N. G.; Davin, L. B.; Gang, D. R.; Fujita, M.; Xia, Z.-Q.; Dinkova-Kostova, A. T. *ACS Symp. Ser.* **1997**, this volume.
18. Davin, L. B.; Bedgar, D. L.; Katayama, T.; Lewis, N. G. *Phytochemistry* **1992,** *31*, 3869.
19. Davin, L. B.; Wang, H.-B.; Crowell, A. L.; Bedgar, D. L.; Martin, D. M.; Sarkanen, S.; Lewis, N. G., *Science,* **1997**, *275*, 362.
20. Terashima, N.; Atalla, R. H.; Ralph, S.; Landucci, L. L.; Lapierre, C.; Monties, B. *Holzforschung* **1995,** *49*, 521 & **1996,** *50*, 9.
21. Grabber, J. H.; Ralph, J.; Hatfield, R. D.; Quideau, S.; Kuster, T.; Pell, A. N. *J. Agric. Food Chem.* **1996,** *44*, 1453.
22. Grabber, J. H.; Ralph, J.; Hatfield, R. D. *ACS Symp. Ser.* **1997**, this volume.
23. Houtman, C.; Atalla, R. H. *Plant Physiol.* **1995,** *107*, 977.
24. Abosharkh, B.; Atalla, R. H. *Proc. 8th Internat. Symp. Wood Pulp. Chem.* **1995**, *II*, 23.
25. Guan, S.-Y.; Sarkanen, S. *ACS Symp. Ser.* **1997**, this volume.
26. Grobstein, C. In *Hierarchy Theory;* Pattee, H. H., Ed.; George Braziller: New York, NY, 1973; p 29.

Chapter 14

Proposed Structure for Protolignin in Plant Cell Walls

N. Terashima[1], J. Nakashima[2], and K. Takabe[2]

[1]2-610 Uedayama, Tenpaku, Nagoya 468, Japan
[2]Faculty of Agriculture, Kyoto University, Kyoto 606-01, Japan

Non-destructive methods for the study of lignification in plant cell walls have revealed that, depending on the cell type and morphological region, the structure of protolignin is heterogeneous with respect to its monomer composition, interunit linkages, and its association with polysaccharides. The monolignols are supplied to lignifying cell walls in the order of *p*-coumaryl, coniferyl, and sinapyl alcohols, and their polymerization occurs in polysaccharide gels which are formed before the onset of lignification. In the early stages of lignification, the bulk type oligomers are proposed to be formed as globular molecules, which grow in subsequent stages (by endwise polymerization) to form larger molecules. In the final stage, cross-links between endwise-polymerized polylignols are proposed to give rise to a 'condensed endwise polymer'. We suggest that the overall shape of protolignin in the secondary wall is that of a 'twisted honeycomb'.

Lignin is present in plant cell walls as a three-dimensional (3D) heterogeneous macromolecule associated with cellulose and hemicellulose *via* both covalent bonds and noncovalent interactions (*1, 2, 3*). Determining the 3D heterogeneous structure of lignin is essential in understanding the chemical, physical, and biological behavior of this plant cell wall biopolymer. The lignin in the cell walls is often called protolignin in order to distinguish it from various isolated lignin preparations, because removal of lignin from its natural location in the cell wall involves, at the least, rupturing lignin-polysaccharide bonds and a reduction in molecular weight (*2*), which will result in the loss of information on the original 3D-heterogeneous structure(s). For example, even the most frequently employed lignin preparation, milled wood lignin (MWL), is obtained in a yield of about a half of all the lignin in wood (*4*), and the structure of MWL prepared from *Picea jezoensis* wood is not identical with that in LCCs (lignin-carbohydrate complexes) and the lignin remaining in the wood after extraction of MWL and LCCs (*5, 6*). Thus, MWL does not adequately represent protolignin (*7-9*). In this chapter, various methods useful for the study of protolignin structure are first briefly discussed, and the results obtained by these methods are then combined to give a tentative description of protolignin structures in the plant cell.

Non-Destructive Methods for Elucidation of Protolignin Structure

Various schematic models for lignin structure have been proposed based on the results obtained from destructive analyses such as oxidation, reduction, and solvolysis of plant samples, together with the isolation and identification of putative lignin fragments (*1, 10-13*). These methods provide information, however, only on the average number of structural units and linkage types contained in a part of the protolignin, since such destructive chemical methods do not quantitatively convert protolignin into monomeric or oligomeric fragments. Nor do they provide information on the 3D-heterogeneous structure of cell wall protolignin.

While studies on the dehydrogenative polymerisates (DHPs) prepared from monolignols *in vitro* have given enough useful information on the structure of lignin for a structural model to have been proposed (*1, 2*), the structures of DHPs differ greatly according to the polymerization conditions employed (*1, 2*). Moreover, it is difficult to know the actual polymerization conditions in the cell wall, including factors such as the effect of the preformed polysaccharides on lignification.

Such difficulties in protolignin structural analysis can be circumvented to a considerable extent by its direct examination in the cell wall *via* application of various non-destructive methods. These include techniques such as UV (*14, 15*), IR (*16, 17*), and Raman microspectroscopy (*18, 19*), solid-state NMR spectroscopy (*20, 21*), scanning electron microscope-energy disperse X-ray analysis combined with chemical treatment including bromination (*22, 23*) or mercuration (*24, 25*), transmission electron microscopy together with rapid freeze-fracture (*26, 27, 28*), and histochemical analyses including immunochemical techniques (*29*).

Other useful approaches utilize tracer methods including ^{3}H and/or ^{14}C (*30, 31*), and stable radioisotopes, ^{2}H or ^{13}C (*32, 33*). For example, by administering to a growing plant a suitable radiolabeled precursor, it is possible to selectively label lignin (*1, 34-36*) and polysaccharides in the cell wall (*35-38*). Selective radiolabeling of pectin in *Pinus thunbergii* (*36, 37*), and xylan in *Magnolia kobus* (*38*), and mannan in *Pinus thunbergii* and *Ginkgo biloba* (*39*), respectively, is achieved by feeding a growing tree stem with UDP-glucuronic acid-[glucuronyl-^{14}C], inositol-[2-^{3}H], mannose-[2-^{3}H] + α-aminooxy-β-phenyl propionic acid + 2-deoxyglucose. Selective radiolabeling of different monomer units in protolignin has also been achieved by administering to lignifying wood xylem the corresponding monolignol glucosides-[lignol-^{3}H, or -^{14}C] (*35, 36*). Since a cross-section of newly formed wood xylem includes a file of cells undergoing differentiation from the earliest stages in the cambial region to the final stages in mature secondary xylem tissue, analysis of the resulting microautoradiogram of the labeled cross-section permits the incorporation of labeled precursors into differentiating cell walls to be visualized (*40-44*).

Methodology for ^{13}C-enrichment at a specific carbon of protolignin in the cell walls of various plants and subsequent analysis by solid/solution-state ^{13}C-NMR spectroscopy have also been developed, and these have provided useful information on the structure of lignin in the cell wall (*45-53*).

In terms of growth and development, the lignified cell wall is formed *via* the successive deposition of pectic substances, hemicelluloses-cellulose, and lignin during cell-wall differentiation, the cell-wall assembly process itself being irreversible (*54*). The polysaccharides deposited prior to lignification may play an important role as templates for formation of the lignin macromolecule (*54*), since the polysaccharides and lignin are bound by both covalent and noncovalent interactions to form a lignin-polysaccharide complex (*1-3*). Thus, the most useful information regarding the 3D macromolecular structure of protolignin has been obtained by non-destructive analyses of the major components in the wall from the earliest stages to the final stages of lignified cell wall assembly.

Biogenesis of Cell Walls in Differentiating Xylem of Trees

Radiotracer experiments have indicated that there is a definite common pattern in the assembly processes of the major cell wall components (*40, 44, 54*). A conceptual illustration of the assembly process is given in Figure 1.

Formation of Cell Wall Layers and Deposition of Polysaccharides. The first step of cell wall formation in unlignified cambial cells involves the fusion of Golgi vesicles to form a cell plate, which is considered to be a hard gel consisting mainly of pectin cross-linked with calcium ions (*55*) (Figure 1). On both sides of the cell plate, primary walls are formed, consisting of cellulose microfibrils and hemicelluloses containing much arabinose and galactose (*56-59*). The cross-linking by calcium ions between hemicelluloses, and the hydrogen bonding between cellulose microfibrils and hemicelluloses (* in Figure 2), is envisaged to result in the formation of a hard gel. The inner side of the primary wall and then the outer, middle, and inner layers of the secondary wall (S1, S2, and S3) are thereupon formed sequentially. The cellulose microfibrils and hemicellulosic gel are both deposited at the same time so that the former are kept separated by the spacer of hemicellulosic gel. A part of the hemicelluloses is oriented parallel to the cellulose microfibrils (*3, 60, 61*), these being associated by hydrogen bonding (Figure 2). The width of the cellulose microfibril in *Pinus densiflora*, *Populus euramericana* and other hardwoods is approximately 4-6 nm (*62, 63*), and its traverse cross-section is considered to be square. However, the strong association between the hemicelluloses and cellulose microfibrils (* in Figure 2) will perturb the conformation of β-1,4-glucan chains located in the outermost part of the latter (Figure 2) and thus lower its average crystallinity (*64, 65*). In the S2 layer of conifer tracheids, the cellulose microfibrils are oriented almost parallel to the cell axis, though their orientation changes 15-30 degrees layer by layer (*66, 67*), and they are surrounded by a hemicellulose gel to form a 'twisted honeycomb' (*54*).

Deposition of Lignin. Lignification proceeds in three distinct stages and is preceded by the deposition of polysaccharides (Figure 1). The first stage of lignification occurs at the cell corner and cell-corner/compound middle lamella regions, following the deposition of pectic substances and just before formation of the S1 layer begins (*36, 40, 44, 54, 68*). The second phase is a 'slow' lignification stage when cellulose microfibrils and hemicelluloses (xylan and/or mannan) are deposited in the developing cell wall. Then the main lignification process occurs in the third stage following deposition of the cellulose microfibrils in S3 (*39, 40, 44, 54, 57*). In the cell corner and compound middle lamella regions, polymerization of lignols takes place in a rather short period of time in the presence of pectic substances. On the other hand, the polymerization of lignols in the secondary wall proceeds slowly in a xylan or mannan matrix (*44, 54*). This basic pattern of deposition of polysaccharides and lignin in differentiating cell walls seems to be generally uniform among the gymnosperms and angiosperms, including the monocotyledons (*44, 54, 69*).

No lignin deposition has been observed in any location without prior deposition of polysaccharides (*1*). The structure of the carbohydrate-free lignin released from suspension cultured spruce cells in a suitable nutrient medium is different from that of MWL (*70*). Thus, protolignin in the cell wall is always formed in presence of polysaccharides, and its mechanism of formation and structure are inferred to be affected by the latter.

Heterogeneity in Formation and Structure of Lignin. The most characteristic feature of protolignin is its heterogeneity as far as its formation and structure are concerned. Lignin and various kinds of other cellular components such as isoprenoids, tannins, flavonoids, lignans and alkaloids are secondary metabolites produced by specific cells in particular tissues of certain plants when the plant or cell has reached a certain level of growth or differentiation. Thus, the formation of monolignols depends on the type and stage of differentiation of the cell. As a result,

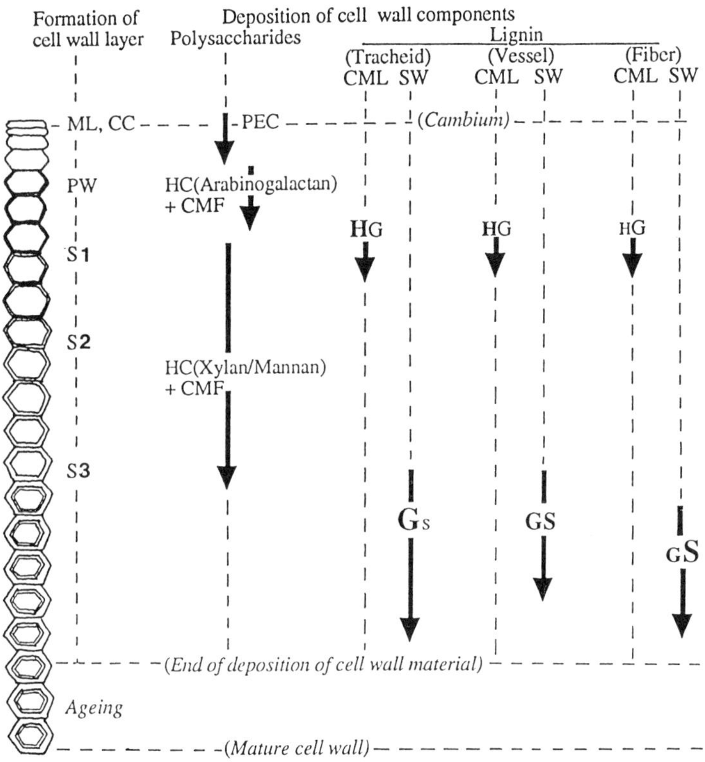

Figure 1. A conceptual illustration of the relationship between formation of cell wall layers and incorporation of polysaccharides and lignin. Cell wall layers are formed successively in the order of middle lamella (ML)/cell corner (CC), primary wall (PW), followed by the outer, middle, and inner layers of secondary walls (SW) (S1, S2 and S3 layers). Monolignol units are incorporated in the order of *p*-hydroxyphenyl (H), guaiacyl (G), and syringyl (S) units, and incorporated thus in three distinct stages. The first lignification takes place in the CC and compound middle lamella (CML) after the PW is formed and just before S1 formation starts. Thereafter slow (or no) lignification takes place during the deposition of cellulose microfibrils (CMFs) and hemicelluloses (HC) in S2. The most extensive deposition of lignin takes place in S2 and after S3 formation has started.

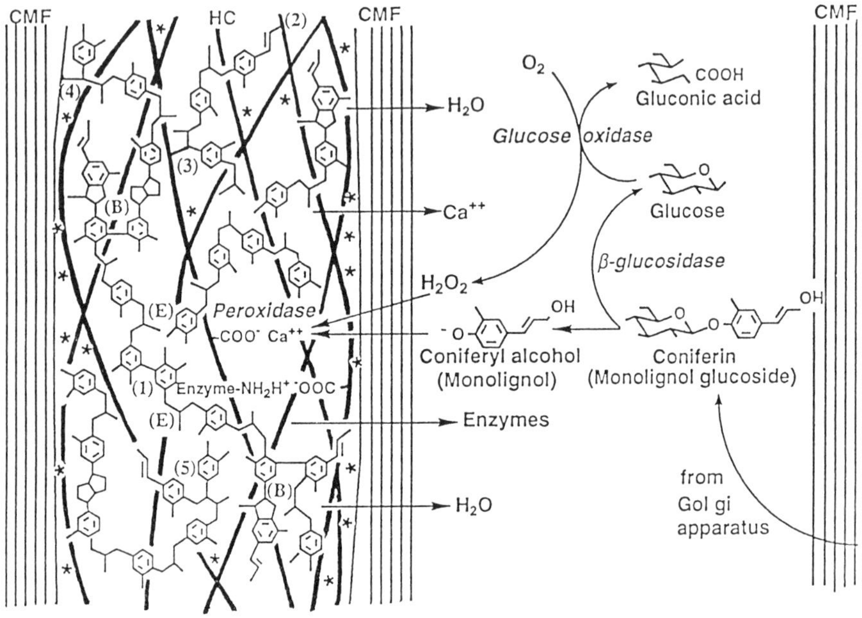

Figure 2. A conceptual illustration of the biogenesis of the lignified cell wall (see text).
*: hydrogen bond.

the polymeric structure of lignin is heterogeneous in many respects. This heterogeneous nature of protolignin is best understood from its contemplated mechanism of formation as follows.

Heterogeneity in Incorporation of Various Monomer Units into Protolignin. Microautoradiograms of differentiating xylem tissues of pine fed with ^{3}H-labeled monolignol glucosides indicated that three kinds of monomer residue are incorporated into protolignin in the order of *p*-hydroxyphenyl (H), guaiacyl (G), and syringyl (S) units (Figure 1) (*45, 54*). Because lignification takes place at the cell corners and in the compound middle lamella during the early stages, and in the secondary wall at later stages of cell development, the monolignols are deposited in distinct morphological regions to different extents. The distribution of silver grains on the microautoradiograms showed that H lignin is formed mainly in the compound middle lamella of differentiating pine tracheids (*44, 54, 71*). This is in accordance with the observation made on spruce wood that the methoxyl content of middle lamella lignin was lower than that of secondary wall lignin (*72*). The H units in compound middle lamella lignin may be of a highly condensed type, and its determination may be difficult to carry out by conventional destructive means. Indeed, this assumption is supported by the observation that artificial lignins prepared *in vitro* by dehydrogenative polymerization of monolignols containing *p*-coumaryl alcohol give either no or low amounts of *p*-hydroxybenzaldehyde upon nitrobenzene oxidation (*35, 73*). A microautoradiographic investigation of the distribution of H units in ginkgo (*Ginkgo biloba*) lignin has similarly shown the occurrence of condensed H units in the compound middle lamella region (*74*). The mature xylem cell wall of pine also gave lower amounts of *p*-hydroxybenzaldehyde on nitrobenzene oxidation than pine tissue which consisted mainly of immature cell walls (*75, 76*). G units are incorporated at all stages into lignins in the compound middle lamella and secondary wall of pine and ginkgo. On the other hand, S units were found mainly in the lumen side of the secondary wall, though the amount was very low in these gymnosperms (*71, 74*).

When the three kinds of ^{3}H-labeled monolignol glucosides were administered to the angiosperms, magnolia (*Magnolia kobus*), lilac (*Syringa vulgaris*), beech (*Fagus crenata*), and poplar (*Populus maximowiczii* X *Populus belorinensis*), the precursors were effectively incorporated into the newly-formed xylem in the order of H, G, and S units during different stages of cell wall formation (*77, 78*). That is, it followed the same trend as noted for formation of gymnosperm lignins, although the amount of S units was, of course, larger in angiosperms than in gymnosperms (*44, 54*). The H unit was again incorporated during the earliest stages of wall development mainly into cell corners and compound middle lamella lignin, and was not found in the secondary wall lignin. Since vessel walls lignify earlier than fiber walls, vessel wall lignin also contained larger amounts of H and G units than fiber wall lignin, with incorporation of G units continuing throughout all stages of vessel and fiber wall lignification. As expected, larger amounts of S units were incorporated into angiosperms than gymnosperms during middle and stages of lignification. That is, it was deposited mainly in the secondary walls of fiber, though a small part of it participated in the formation of middle lamella lignin. These observations are in accordance with those previously made by UV microscopic spectrophotometry (*14, 15, 79*), bromination-EDXA (*80*), and chemical characterization of cell wall fractions (*81-83*). Similar patterns of sequential deposition of H, G, and S lignins and morphological localizations have also been observed in rice plants (*84*). The general deposition pattern of the lignin-building units is schematically shown in Figure 1.

The observation that the deposition of the three kinds of monolignol occurs in the same order as their positions in the biosynthetic pathway may be understood by the assumption that the biosynthetic pathway becomes extended with the age of the cell. The dependence of its biosynthesis on the type and age of the cell may be one of the characteristic features of lignin as a secondary metabolite.

Heterogeneity in Distribution of Various Linkages During Growth of Lignin Macromolecule. Monolignol glucosides are probably biosynthesized in the Golgi apparatus and supplied to differentiating cell walls by exocytosis of Golgi vesicles (*43*), while the monolignols are presumably produced by the action of β-glucosidases on their glucosides (*1*) (Figure 2). The role of monolignol glucosides as direct intermediates in the pathway leading to lignification has been debated for many years, because monolignol glucosides seem to be absent from some angiosperms (*2*) except those belonging to Oleaceae and Magnoliaceae (*85*). However, when coniferyl alcohol was administered to differentiating cell walls, it was incorporated in an abnormal way as observed by either the radiotracer method (*86*) or electron microscopy (*87*), and indeed a coniferin-specific β-glucosidase has only been detected in the lignifying cell wall of pine (*88*). These observations support the hypothesis that monolignol glucosides participate in the pathway leading to lignification as a direct precursor of monolignols in lignifying cell walls (*1*).

The polymerization of monolignols takes place in a swollen gel of pectin or hemicelluloses to form lignin-pectin or lignin-hemicellulosic complexes (*1, 69, 89-92*) (Figure 2). One possible fate of the glucose liberated from the monolignol glucosides may be its oxidation to gluconic acid by oxygen in the presence of glucose oxidase, the oxygen being concomitantly reduced to hydrogen peroxide which, in turn, dehydrogenates monolignol to its phenoxy radical in the presence of peroxidase (PO).

The mode of polymerization is greatly affected by the relative numbers of radical species participating in the polymerization reaction (*1, 2*). The first products of the coupling of monolignol radicals are β–*O*–4′, β–5′, and β–β′ dimers; and, in the next stage of polymerization, bulk type oligomers are formed ((B) in Figure 2). As the number and size of the bulk dehydropolymerisate increase, the number of actively growing termini will increase, while the relative rate of supply of the monolignol radicals will decrease. This change will make endwise polymerization more prevalent in the subsequent stage of polylignol molecular assembly. As a result, the globular lignin macromolecule will be composed of a bulk polymer inside and an endwise polymer in the outer part ((E) in Figure 2). In the last stage, filling of the space between the cellulosic microfibrils and shrinking of the lignin-polysaccharide gel occurs (*69*), and the distance between globular polylignols will become short enough for new linkages to be formed between them. One of the major new linkages may be 5–5′ between β–*O*–4′ type endwise chains ((1) in Figure 2). In support of this idea, it has been reported that a DHP prepared under dehydrating conditions contains larger numbers of 5–5′ linkages than a conventionally prepared DHP (*89*). If two β–*O*–4′ endwise chains linked by 5–5′ bonds at their growing ends add one monolignol by α–*O*–4′/β–*O*–4′ linkages, an eight membered dibenzodioxocin ring structure will be formed. This type of linkage has been discovered in softwood lignin (*93, 94*).

The distribution of condensed units with carbon-carbon bonds at position 5 of the guaiacyl ring in protolignin can be observed by feeding a plant with coniferin-[aromatic ring-5-^{3}H] and inspecting the microautoradiograms of the lignifying tissue. If a new C–C or C–O bond is formed at position 5 of the guaiacyl ring, the ^{3}H at this position will be lost. On the other hand, when a woody plant is fed with coniferin-[aromatic ring-2-^{3}H], no removal of ^{3}H occurs; and the resulting autoradiogram shows the distribution of all guaiacyl units. The difference in distribution of silver grains between the autoradiograms of differentiating xylems of pine (*71*) and ginkgo (*78*) resulting from administration of the coniferin ^{3}H-labeled at the two positions indicates that the compound middle lamella lignin contains larger amounts of condensed structures than the secondary wall lignin; these results support the findings obtained by double radiolabeling (*35*). Thus, the content of condensed type structures is high in lignin formed at an early stage of cell wall formation in the middle lamella and cell-corner regions.

Lignin-Polysaccharide Bonds. It has been shown that carbohydrate esters of phenolic carboxylic acids are composed mainly of ferulic acid dimers formed by oxidative coupling, which also may contribute to the hardening of the cell wall in the early stage of its formation (*95, 96*). These phenolic acids, if esterified with carbohydrates, may also act as lignin anchors by their participation in polymerization of the lignols ((2) in Figure 2) (*97*), and may represent one type of linkage between lignin and polysaccharides in the cell wall.

Another type of lignin-polysaccharide linkage that could be formed is that resulting from addition of hydroxyl or carboxyl groups of polysaccharides to an intermediate quinone methide to form an α-benzyl ether or α-benzyl ester linkages (*1*) ((3) in Figure 2). Pectic substances and hemicelluloses are the chief polysaccharides participating in such reactions (*1, 2, 69, 89, 91, 92, 98*). However, the outermost cellulose chain of the cellulosic microfibrils may also participate in such bond formation ((4) in Figure 2), as proposed from results of chemical analysis (*99*). Phenolic hydroxyl groups of lignols can participate in this addition reaction to form α–*O*–4′ linkages ((5) in Figure 2); however, this addition reaction may hardly take place under conditions of a slow supply of monolignol to the polymerization site in the secondary wall. Indeed, the content of the α–*O*–4′ linkage in milled wood lignin is estimated to be low (*100*). The important factor affecting the addition reaction may be the relative amount of carboxyl or hydroxyl groups available to the quinone methides. In an early stage of lignification, the ratio of polysaccharides to quinone methides is high enough to permit frequent bond formation between lignin and polysaccharides.

Ultrastructural Assembly of Lignin and Polysaccharides in the Cell Wall.

Observation of lignifying pine cell walls by UV microscopy indicates that lignin deposition proceeds unevenly, in such a way that the lignifying regions move from the outer to the inner part of the cell wall as lignification progresses (*101*). Deposition of lignol on a water-swollen hydrophilic polysaccharide gel can be envisaged to make the gel hydrophobic, and water will then be displaced from the lignified outer part of the cell wall to the unlignified inner part (*54, 69*) (Figure 2). Calcium ions and enzymes including glucosidase and peroxidase will be replaced by lignols and in the process displaced to the inner side with the water (*54, 69*) (Figure 2). Thus, lignification proceeds from the outer to the inner parts of differentiating cell walls. This hypothesis was supported by observations of lignifying cell walls using electron microscopy combined with a quick-freezing technique which revealed that the innermost part near the plasma membrane contained much water (*27, 54*). The displacement of water in this manner will cause shrinkage of the swollen lignin-polysaccharide gel and, indeed, the formation mechanism of the cracked cell-wall structure of compression wood may be rationalized as a consequence of this shrinking (*102*). The shrinkage in the secondary wall will be greater in the direction perpendicular to the cell wall surface, since the hemicelluloses are hydrogen bonded (* in Figure 2) to the rigid cellulosic microfibrils which are oriented parallel to the cell wall surface (Figure 2). This anisotropic shrinkage of the volume available to the polylignols bound to the hemicellulosic gel may cause the orientation of aromatic rings in lignin molecules to become parallel to the cellulosic microfibrils (*54, 69*). Examination of the protolignin in the secondary wall by Raman spectrophotometry with a microprobe revealed that the aromatic rings are preferentially oriented parallel to the surface of the cell walls (*18, 19*). This may be conveniently explained by the anisotropic shrinkage of the swollen honeycomb-like gel of the lignin-hemicellulosic complex. However, it has been shown that oligolignols associated with cellulose microfibrils prefer a particular orientation (*103*), and formation of macromolecular lignin proceeds faster in the direction along cellulose microfibrils (*104*). So, even without anisotropic shrinkage, the intimate and regulated association of cellulose microfibrils, hemicelluloses, and lignin may cause the orientation of lignin aromatic rings to be parallel to the cellulose microfibrils which run parallel to the cell wall surface (Atalla, R. H., personal communication).

From the dimensional changes of the cell walls before and after delignification, and the results of electron microscopic observation, an interrupted lamella model has been proposed for the ultrastructure of S2 layers in which cellulosic microfibrils and the hemicellulose-lignin matrix form approximately 7-10 nm wide lamellae (*105-107*). Because the width of the microfibrils is 4-6 nm (*62-63*), the space available for the lignin-hemicellulosic complex in secondary walls is approximately 3-4 nm wide. In this context, Figure 3 shows the electron photomicrograph of secondary wall thickening of a tracheary element of *Zinnia elegans* near the final stages of lignification. The rapid-freeze fracture and freeze-etching technique used here enables us to visualize the architecture of the cell wall in almost its native state (*26, 28, 108*). As can be seen, the microfibrils are surrounded by the hemicellulose-lignin complex to form a honeycomb, and no lamella structure is observed. This supports the ultrastructural assembly of a honeycomb proposed for the cell wall of woody plants (*54*). The estimated width of the cellulose microfibrils is about 4-6 nm, and distance between them is about 3-4 nm. A study employing a structural model indicates that the distance between two aromatic rings in β–*O*–4′ linked oligolignols, or the distance between the aromatic ring and pyranose ring in β–*O*–4′ lignol/α–*O*–glucuronic acid esters, is about 0.3-0.4 nm when these compounds are placed in their most compact state (*54*). If the thin walls of honeycombs surrounding the cellulose microfibrils are composed of almost the same amounts of lignin and hemicellulose, it will contain about 4-7 monolignol units and 4-7 monosaccharide units between the microfibrils, where the aromatic rings of the lignin are oriented parallel to the surface of the microfibrils.

Proposed Structure for Protolignin in the Cell Wall.

The observations made on lignin formation in differentiating cell walls by non-destructive methods suggest the following mechanism of ultrastructural assembly of lignin and polysaccharides.

1. The lignin macromolecules in the cell corners and compound middle lamella regions may form a thick plate reinforced by pectic substances, hemicelluloses and randomly oriented cellulose microfibrils. On the other hand, the lignin in the secondary walls may be a thin-walled twisted honeycomb in which both lignin and hemicelluloses are intimately associated and held together by covalent and noncovalent interactions.
2. The structure of the protolignin macromolecule is heterogeneous with respect to monomer composition, inter-unit linkages, and association with polysaccharides. The structural heterogeneity within a single macromolecule depends largely upon the morphological region in the particular cell wall, type of cell, and plant species.
3. The structural heterogeneity, with respect to monomer composition, in protolignin results from the monolignols being supplied to the lignifying cell walls in the order of H, G, and S. That is, the polylignol formed at an early stage contains more H units, and that formed at a later stage contains more S units.
4. The structure of protolignin is heterogeneous with respect to the distribution of linkages. That is, in the early stage of polymerization, bulk type polymers are formed whereas, in the following stage of macromolecular assembly, endwise polymerization becomes more prevalent. As a result, the globular macromolecule is proposed to be composed of bulk polymer within and endwise polymer in its outer part. At the final stage of the process, cross-linkings between the endwise polymeric moieties give a condensed endwise polymer.
5. The structure of protolignin is also heterogeneous with respect to the morphological region in the cell wall. Protolignin in the cell corners and compound middle lamella regions contain larger amounts of condensed H and G units than that in the secondary wall, whereas the latter may contain larger amounts of uncondensed G and S units than that in the cell corners and compound middle lamella.

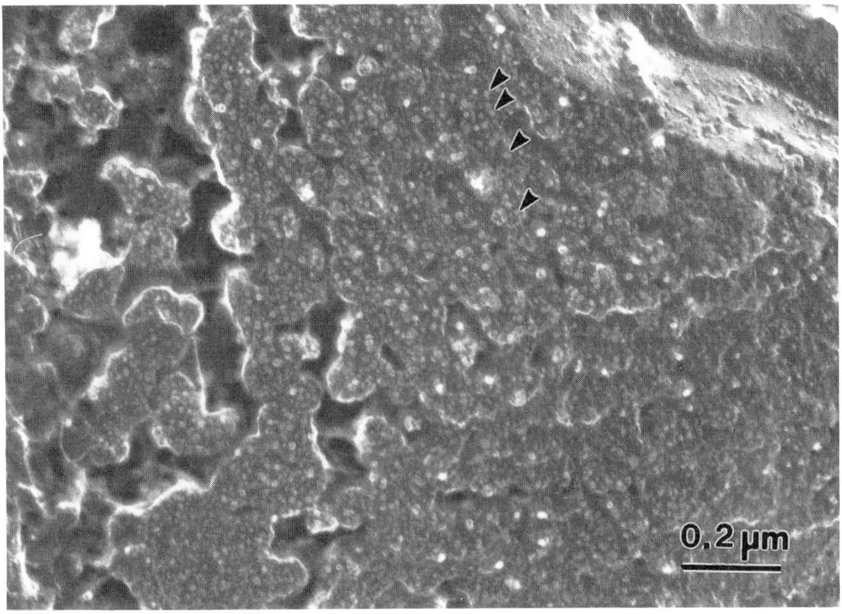

Figure 3. Electron micrograph of freeze-fractured secondary wall thickening of a tracheary element from *Zinnia elegans* near the final stages of lignification. Cellulose microfibrils (arrow heads) are almost perpendicular to fracture plane. They are surrounded by lignin-hemicellulose complexes to form twisted honeycomb structures. Their size (diameter of cross-section) is 4-6 nm. (Reproduced with permission from reference 108. Copyright 1997.)

6. During the polymerization of lignols, nucleophilic addition of carboxyl or hydroxyl groups on polysaccharides to quinone methide intermediates is envisaged to occur giving α-benzyl esters or ethers.

Concluding Remarks

The concepts enunciated about the structure of protolignin in this chapter constitute a working hypothesis which will be modified by further correction in the future as necessary. Nevertheless, the following summary can now usefully be made.

1. Protolignin is present in the cell wall as a lignin-polysaccharide complex formed in an irreversible manner. Therefore, non-destructive observations of the whole process of formation of this complex can provide useful information for understanding the structure of protolignin in the cell wall.
2. A most remarkable feature of lignin structure is its heterogeneous nature. The structure is not uniform within a single lignin macromolecule, and it is not the same with respect to morphological regions within the cell wall.
3. This heterogeneity is a natural and inevitable consequence of the unique process of lignin formation. The origin of this heterogeneity may be understood from the following considerations:

(i) Individual cells control the biosynthesis and assembly of polysaccharides and lignin to form lignified cell walls.

(ii) Lignin is a secondary metabolite, and the biosynthesis of polylignols largely depends on the type and age of the cell.

(iii) The polymerization of lignols in the differentiating cell walls takes place in the polysaccharide gel deposited before lignification. Different kinds of polysaccharides are deposited in each cell wall layer, being associated with the cellulose microfibrils which are oriented in different ways. These polysaccharides are proposed to act as templates for macromolecular lignin assembly and hence affect the formation of lignin structure.

(iv) The monolignols are supplied to the differentiating cell walls in the order of *p*-coumaryl, coniferyl, and sinapyl alcohols. The relative amounts of monomers change with the type and age of the cell and plant species. In general, the lignin produced at an early stage of cell wall formation in the compound middle lamella contains more condensed H and G units than that formed at a later stage in the secondary wall.

(v) During macromolecular lignin assembly, bulk type polymers are formed in an early stage on which endwise polymers subsequently grow. In the last stage, cross-linking between endwise polymers takes place to give a condensed endwise polymer.

4. The molecular weight and higher order structure of lignin are confined by cellulose microfibrils and the non-cellulosic polysaccharide gel in which the polymerization of lignols occurs. The compound middle lamella lignin is a high molecular weight thick mass, while the secondary wall lignin, associated intimately with hemicelluloses, is a thin film which surrounds the cellulose microfibrils like a twisted honeycomb. The thin film may contain 4-7 layers of monolignol units associated with hemicelluloses through covalent and noncovalent interactions.

Literature Cited

1. Freudenberg, K. In *Constitution and Biosynthesis of Lignin*; Freudenberg, K., Neish, A. C., Eds.; Springer-Verlag: Heidelberg, 1968; pp 47-116.
2. Sarkanen, K. V. In *Lignins—Occurrence, Formation, Structure and Reactions*; Sarkanen, K. V., Ludwig, C. H., Eds.; John Wiley & Sons: New York, NY, 1971; pp 10-155.
3. Fengel, D.; Wegener, G. In *Wood—Chemistry, Ultrastructure, Reactions*; Walter de Gruyter: Berlin, 1984; pp 26-238.

4. Björkman, A. *Svensk Papperstidn.* **1956**, *59*, 477-485.
5. Matsukura, M.; Sakakibara, A. *Mokuzai Gakkaishi* **1969**, *15*, 35-39.
6. Morohoshi, N.; Sakakibara, A. *Mokuzai Gakkaishi* **1971**, *17*, 354-361.
7. Whiting, P.; Goring, D. A. I. *Svensk Papperstidn.* **1981**, *84*, R120-R122.
8. Terashima, N.; Fukushima, K.; Imai, T. *Holzforschung* **1992**, *46*, 271-275.
9. Maurer, A.; Fengel, D. *Holzforschung* **1992**, *46*, 417-423.
10. Nimz, H. *Angew. Chem. Internat. Ed. Engl.* **1974**, *13*, 313-321.
11. Adler, E. *Wood Sci. Technol.* **1977**, *11*, 169-218.
12. Sakakibara, A. *Wood Sci. Technol.* **1980**, *14*, 89-100.
13. Glasser, W. G.; Glasser, H. R. *Pap. Puu* **1981**, *63*, 71-83.
14. Fergus, B. J.; Goring, D. A. I. *Holzforschung* **1970**, *24*, 118-124.
15. Musha, Y.; Goring, D. A. I. *Wood Sci. Technol.* **1975**, *9*, 45-58.
16. Abbott, T. P.; Palmer, D. M.; Gordon, S. H.; Bagby, M. O. *J. Wood Chem. Technol.* **1988**, *8*, 1-28.
17. Bolker, H. I.; Somerville, N. G. *Pulp Paper Mag. Can.* **1963**, *64*, T187-T193.
18. Atalla, R. H.; Agarwal, U. P. *Science* **1985**, *227*, 636-638.
19. Agarwal, U. P.; Atalla, R. H. *Planta* **1986**, *169*, 325-332.
20. Leary, G. J.; Morgan, K. R.; Newman, R. H. *Holzforschung* **1986**, *40*, 221-224.
21. Manders, W. M. *Holzforschung* **1987**, *41*, 13-18.
22. Saka, S.; Thomas, R. J.; Gratzl, J. S. *Tappi* **1978**, *61*, 73-76.
23. Saka, S.; Whiting, P.; Fukazawa, K.; Goring, D. A. I. *Wood Sci. Technol.* **1982**, *16*, 269-277.
24. Westermark, U.; Lidbrandt, O.; Eriksson, I. *Wood Sci. Technol.* **1988**, *22*, 243-250.
25. Eriksson, I.; Lidbrandt, O.; Westermark, U. *Wood Sci. Technol.* **1988**, *22*, 251-257.
26. Takabe, K. *Densi Kenbikyou* **1990**, *25*, 118-122.
27. Inomata, F.; Takabe, K.; Saiki, H. *J. Electron. Microsc.* **1992**, *41*, 369-374.
28. Nakashima, J.; Takabe, K.; Fujita, M.; Saiki, H. *Protoplasma* **1997**, *196*, 99-107.
29. Ruel, K.; Amber, K.; Burlat, V.; Joseleau, J. P.; Srebotnik, E.; Messner, K. In *Proc. Biotechnol. Pulp Paper Ind.*; Vienna, 1995; pp 575-581.
30. Saleh, T. M.; Leney, L.; Sarkanen, K. V. *Holzforschung* **1967**, *21*, 116-120.
31. Eberhardt, G.; Schubert, W. J. *J. Am. Chem. Soc.* **1956**, *78*, 2835-2837.
32. Freudenberg, K.; Jovanovic, V.; Topfmeier, F. *Chem. Ber.* **1962**, *95*, 2814-2828.
33. Gagnaire, D.; Robert, D. *Makromol. Chem.* **1977**, *178*, 1477-1495.
34. Kratzl, K. *Holz als Roh -u. Werkstoff* **1961**, *19*, 219-232.
35. Terashima, N. *ACS Symp. Ser.* **1989**, *399*, 148-159.
36. Terashima, N.; Fukushima, K.; Sano, Y.; Takabe, K. *Holzforschung* **1988**, *42*, 347-350.
37. Imai, T.; Terashima, N. *Mokuzai Gakkaishi* **1991**, *38*, 733-740.
38. Imai, T.; Terashima, N. *Mokuzai Gakkaishi* **1992**, *38*, 693-699.
39. Imai, T.; Goto, N.; Yasuda S.; Terashima, N. In *Proc. Ann. Meet. Jap. Wood Res. Soc.*; Kumamoto, Japan, 1996; pp 328-329.
40. Takabe, K.; Fujita, M.; Harada, H.; Saiki, H. *Mokuzai Gakkaishi* **1981**, *27*, 813-820.
41. Pickett-Heaps, J. D. *Protoplasma* **1968**, *65*, 181-205.
42. Fujita, M.; Harada, H. *Mokuzai Gakkaishi* **1979**, *25*, 89-94.
43. Takabe, K.; Fujita, M.; Harada, H.; Saiki, H. *Mokuzai Gakkaishi* **1985**, *31*, 613-619.
44. Terashima, N.; Fukushima, K. *ACS Symp. Ser.* **1989**, *399*, 160-168.
45. Lewis, N. G.; Yamamoto, E.; Wooten, J. B.; Just, G.; Ohashi, H.; Towers, G. H. N. *Science* **1987**, *237*, 1344-1346.
46. Lewis, N. G.; Razal, R. A.; Dhara, K. P.; Yamamoto, E.; Bokelman, G. H.; Wooten, J. B. *J. Chem. Soc. Chem. Commun.* **1988**, 1626-1628.
47. Lewis, N. G.; Razal, R. A.; Yamamoto, E.; Bokelman, G. H.; Wooten, J. B. *ACS Symp. Ser.* **1989**, *399*, 68-81.

48. Eberhardt, T. L.; Bernards, M. A.; He, L.; Davin, L. B.; Wooten, J. B.; Lewis, N. G. *J. Biol. Chem.* **1993**, *268*, 21088-21096.
49. Terashima, N.; Seguchi, Y.; Robert, D. *Holzforschung* **1991**, *45 suppl.*, 35-39.
50. Xie, Y.; Terashima, N. *Mokuzai Gakkaishi* **1991**, *37*, 935-941.
51. Xie, Y.; Robert, D.; Terashima, N. *Plant Physiol. Biochem.* **1994**, *32*, 243-249.
52. Xie, Y.; Yasuda, S.; Terashima, N. *Mokuzai Gakkaishi* **1994**, *40*, 191-198.
53. Terashima, N.; Atalla, R. H.; VanderHart, D. L. *Phytochemistry* **1997**, *46*, in press.
54. Terashima, N.; Fukushima, K.; He, L.-F.; Takabe, K. In *Forage Cell Wall Structure and Digestibility*; Jung, H. G., Buxton, D. R., Hatfield, R. D., Ralph, J., Eds.; ASA-CSSA-SSSA: Madison, WI, 1993; pp 247-270.
55. Whaley, W. G. In *The Golgi Apparatus*; Springer-Verlag: Wien, 1975, vol. 2.
56. Takabe, K.; Fujita, M.; Harada, H.; Saiki, H. *Mokuzai Gakkaishi* **1983**, *29*, 183-189.
57. Takabe, K.; Fukazawa, K.; Harada, H. *ACS Symp. Ser.* **1989**, *399*, 47-66.
58. Thompson, M. S.; Kremer, R. E.; Kaustein, O. A. *Tappi* **1968**, *51(3)*, 127-131.
59. Hardell, H.-L.; Westermark, U. *Proc. 1st Internat. Symp. Wood Pulp. Chem.* **1981**, *1*, 32-34.
60. Albersheim, P. *Sci. Am.* **1975**, *232(4)*, 81-95.
61. Page, D. H. *Wood Fiber* **1976**, *7*, 246-248.
62. Harada, H.; Kishi, A.; Sugiyama, J. *Proc. 2nd Internat. Symp. Wood Pulp. Chem.* **1983**, *4*, 6-10.
63. Sugiyama, J.; Otsuka, Y.; Murase, H.; Harada, H. *Holzforschung* **1986**, *40 Suppl.*, 31-36.
64. Atalla, R. H.; Hackney, J. M.; Uhlin, I.; Thompson, N. S. *Int. J. Biol. Macromol.* **1993**, *15*, 109.
65. Atalla, R. H. *Proc. 8th Internat. Symp. Wood Pulp. Chem.* **1995**, *1*, 77-84.
66. Kataoka, A.; Saiki, H.; Fujita, M. *Mokuzai Gakkaishi* **1992**, *38*, 327-335.
67. Abe, H.; Funada, R.; Otani, J.; Fukazawa, K. *Ann. Botany* **1995**, *75*, 305-310.
68. Imai, T.; Terashima, N. *Mokuzai Gakkaishi* **1992**, *38*, 475-481.
69. Terashima, N. *J. Pulp Paper Sci.* **1990**, *16*, J150-J155.
70. Brunow, G.; Ede, R. M.; Simola, L. K.; Lemmetyinen, J. *Phytochemistry* **1990**, *29*, 2535-2538.
71. Terashima, N.; Fukushima, K. *Wood Sci. Technol.* **1988**, *22*, 259-270.
72. Whiting, P.; Goring, D. A. I. *Wood Sci. Technol.* **1982**, *16*, 261-267.
73. Faix, O.; Schweers, W. *Holzforschung* **1975**, *29*, 48-55.
74. Fukushima, K.; Terashima, N. *Holzforschung* **1991**, *45*, 87-94.
75. Fukuda, T.; Terashima, N. *Mokuzai Gakkaishi* **1988**, *34*, 604-608.
76. Eom, T.; Meshitsuka, G.; Nakano, J. *Mokuzai Gakkaishi* **1987**, *33*, 576-581.
77. Terashima, N.; Fukushima, K.; Takabe, K. *Holzforschung* **1986**, *40 Suppl.*, 101-105.
78. Fukushima, K.; Terashima, N. *J. Wood Chem. Technol.* **1990**, *10*, 413-433.
79. Fergus, B. J.; Goring, D. A. I. *Holzforschung* **1970**, *24*, 113-117.
80. Saka, S.; Goring, D A. I. In *Biosynthesis and Biodegradation of Wood Components*; Higuchi T., Ed.; Academic Press: Orlando, FL, 1985; pp 51-62.
81. Hardell, H.-L.; Leary, G. J.; Stoll, M.; Westermark, U. *Svensk Papperstidn.* **1980**, *83*, 71-74.
82. Cho, N.-S.; Lee, J. Y.; Meshitsuka, G.; Nakano, J. *Mokuzai Gakkaishi* **1980**, *26*, 527-533.
83. Kim, Y.-S.; Meshitsuka, G.; Ishizu, A. In *Proc. 7th Internat. Symp. Wood Pulp. Chem.* **1993**, *3*, 1-5.
84. He, L.-F.; Terashima, N. *Mokuzai Gakkaishi* **1989**, *35*, 116-122.
85. Terazawa, M.; Okuyama, H.; Miyake, M. *Mokuzai Gakkaishi* **1984**, *30*, 322-328.
86. Tomimura, Y.; Sasao, Y.; Yokoi, T.; Terashima, N. *Mokuzai Gakkaishi* **1980**, *26*, 558-563.
87. Nakashima, J.; Takabe, K; Saiki, H. *Mokuzai Gakkaishi* **1992**, *38*, 1136-1142.

88. Dharmawardhana, D. P.; Ellis, B. E.; Carlson, J. E. *Plant Physiol.* **1995**, *107*, 331-339.
89. Terashima, N.; Atalla, R. H.; Ralph, S. A.; Landucci, L. L.; Lapierre, C.; Monties, B. *Holzforschung* **1996**, *50*, 9-14.
90. Meshitsuka, G.; Lee, Z.-Z.; Nakano, J. *J. Wood Chem. Technol.* **1982**, *2*, 251-267.
91. Eriksson, Ö.; Lindgren, B. O. *Svensk Papperstidn.* **1977**, *81*, 59-63.
92. Eriksson, Ö.; Goring, D. A. I.; Lindgren, B. O. *Wood Sci. Technol.* **1980**, *14*, 267-279.
93. Karhunen, P.; Rummakko, J.; Sipilä, J.; Brunow, G.; Kilpeläinen, I. *Tetrahedron Lett.* **1995**, *36*, 169-170.
94. Karhunen, P.; Rummakko, J.; Pajunen, A.; Brunow, G. *J. Chem. Soc. Perkin Trans. I* **1996**, 2303-2308.
95. Fry, S. C.; Miller, J. G. *ACS Symp. Ser.* **1989**, *399*, 33-46.
96. Ishii, T. *Carbohydr. Res.* **1991**, *219*, 15-22.
97. Yamamoto, E.; Bokelman, G. H.; Lewis, N. G. *ACS Symp. Ser.* **1989**, *399*, 68-88.
98. Iiyama, K.; Lam, T. B. T.; Meikle, P. J.; Ng, K.; Rhodes, D. I.; Stone, B. A. In *Forage Cell Wall Structure and Digestibility*; Jung, H. G., Buxton, D. R., Hatfield, R. D., Ralph, J., Eds.; ASA-CSSA-SSSA: Madison, WI, 1993; pp 621-684.
99. Lam, T. B. T.; Iiyama, K.; Nakano, J. *Mokuzai Gakkaishi* **1985**, *31*, 475-482.
100. Ede, R. M.; Kilpeläinen, I. *Proc. 8th Internat. Symp. Wood Pulp. Chem.* **1995**, *1*, 487-492.
101. Takabe, K.; Fujita, M.; Harada, H.; Saiki, H. *Mokuzai Gakkaishi* **1981**, *27*, 813-820.
102. Takabe, K.; Miyauchi, T.; Fukazawa, K. *IAWA Bulletin n. s.* **1992**, *13*, 283-296.
103. Houtman, C. J.; Atalla, R. H. *Plant Physiol.* **1995**, *107*, 977-984.
104. Donaldson, L. A. *Wood Sci. Technol.* **1994**, *28*, 111-118.
105. Kerr, A. J.; Goring, D. A. I. *Cellulose Chem. Technol.* **1975**, *9*, 563-573.
106. Ruel, K.; Barnoud, F.; Goring, D. A. I. *Wood Sci. Technol.* **1978**, *12*, 287-291.
107. Ruel, K.; Barnoud, F.; Goring, D. A. I. *Cellulose Chem. Technol.* **1979**, *13*, 429-432.
108. Nakashima, J.; Awano, T.; Takabe, K.; Fujita, M.; Saiki, H. *Plant Cell Physiol.* **1997**, *38*, 113-123.

Chapter 15

Template Polymerization in Lignin Biosynthesis

Simo Sarkanen

Department of Wood and Paper Science, University of Minnesota, St. Paul, MN 55108–6128

The primary structures of lignins are determined by the sequences *not* of the constituent monomer residues, which are very similar to one another, but *rather* of the interunit linkages, which are about ten in number. This configurational determinant has usually been thought to vary randomly among macromolecular lignin chains: the final step in the assembly of the biopolymer, namely the coupling of radical intermediates produced during the dehydrogenative polymerization of monolignols, is not under direct enzymatic control. However, it has now been found that, without covalent participation in radical coupling, lignin macromolecules *in vitro* cogently promote formation of high molecular weight species from the enzyme-catalyzed dehydropolymerization of monolignols. In lignifying tissues such an effect could provide the organization necessary for replicating sequences of interunit linkages along macromolecular lignins chains by a direct template polymerization mechanism. The configurations of the progenitorial lignin chains are presumably established by proteins which proffer arrays of dirigent sites comparable to those that prescribe regio- and stereoselectivity in the phenoxy-radical coupling processes responsible for lignan formation. Among the major families of structural proteins so far described in plant cell walls, the proline-rich proteins (PRPs) would most closely approach the plausible manifestation of such characteristics.

Despite their distinction in being the second most abundant group of biopolymers, lignins exhibit a number of bio/chemical oddities as far as commonly held views about their structures have been concerned. Prominent among these anomalies is the fact that the elucidation of primary structures for macromolecular lignin chains has never been effectively advocated in the literature. Indeed the concept is so alien to this context that it is necessary at the outset to clarify what meaning might most usefully be ascribed to the term "primary structure" in relation to lignins.

It is generally agreed that these aromatic biopolymers are formed in the cell walls of all vascular plants by the dehydrogenative polymerization of (usually) no more than three monolignols. Differing only in their aromatic methoxyl substitution patterns (*1*), the monomeric precursors appear to be sequentially incorporated into polymeric lignin structures in the same order—*viz.* *p*-hydroxycinnamyl (*p*-coumaryl) alcohol, 4-hydroxy-3-methoxycinnamyl (coniferyl) alcohol, and 4-hydroxy-3,5-dimethoxycinnamyl (sinapyl) alcohol—as they occur in their biosynthetic pathway (*2*).

The dehydrogenative coupling processes through which the monolignols are incorporated into macromolecular lignin chains result in the formation of as many as ten different interunit linkages. Although half of these are of the same 8−*O*−4′ alkyl aryl ether type (*3*), they span a much larger range of structures than the monolignol precursors themselves. Thus it is the sequence of interunit linkages rather than monomer residues that most appropriately delineates the primary structure of a macromolecular lignin chain.

It is quite extraordinary that no fewer than five enzymes have been variously implicated in the dehydrogenative coupling of monolignols to form lignins. Peroxidase (*4*), peroxidase and laccase (*5, 6*), laccase (*7, 8*), (poly)phenol oxidase (*9*), coniferyl alcohol oxidase (*10, 11*) and even cytochrome oxidase (*12*) have been thus cited, and indeed all seem to share an ability to oxidize monolignols. However, in *Zinnia elegans* tissues (*13*) and *Pinus taeda* suspension cultures (*14*), H_2O_2 generation (and thus, it may be presumed, peroxidase activity) exhibits a strong temporal correlation with lignin formation. Similarly incubation with H_2O_2 markedly increases, within only 1 h, the epitope concentration to antilignin antibodies in *Zea mays* coleoptile cell walls (*15*). Accordingly it seems reasonable to suppose that peroxidase is the enzyme that facilitates formation of dehydropolymerisates from monolignols *in vivo*. Nevertheless it should be borne in mind that the individual coupling processes through which lignin macromolecules are assembled from monolignol radicals are not generally considered to be under direct enzymatic control.

Dehydrogenative Coupling of Monolignols

There are some curious aspects to the regioselectivity of monolignol radical coupling that have remained poorly understood ever since they were first discovered. Over forty years ago it was reported that 8−5′ and 8−8′ dilignols far surpass 8−*O*−4′ linked dimers in product mixtures formed during laccase-catalyzed dehydrogenation of coniferyl alcohol by O_2 (*16*). Indeed this early finding has subsequently received ample confirmation (*17, 18*). Not long after the original result was published, however, it was found that 8−*O*−4′ linkages predominated during peroxidase catalyzed dehydrogenative coupling (by H_2O_2) of coniferyl alcohol with 8−5′, 8−8′ and 8−*O*−4′ linked dilignols (*19, 20*); yet the different dimers derived from the monomer alone under such conditions are formed in similar proportions to those produced in the presence of laccase and O_2 (*18*).

Clearly the unpaired electron densities on the atomic centers about to become linked do not exclusively determine the populations of coupling products formed from coniferyl alcohol alone or from the monolignol and dilignols. Otherwise a much larger proportion of 4−*O*−5″ linkages would be expected than is actually obtained from the dehydrogenative coupling of coniferyl alcohol with the 8−5′, 8−8′ and 8−*O*−4′ linked dimers. Thus the dehydropolymerization of monolignols is influenced in a very important way by noncovalent interactions between the species undergoing coupling.

Consequently it is not surprising that the frequencies of interunit linkages produced during the dehydrogenative polymerization of coniferyl alcohol *in vitro* when the monomeric precursor is added all at once ("Zulaufverfahren") to the enzyme-containing solution differ substantially from those formed when the monolignol is introduced gradually ("Zutropfverfahren") into the same dehydropolymerizing medium (*21*). Under the conditions engendered by Zutropfverfahren, the radical coupling of the monomers occurs primarily with the growing ends of macromolecular lignin chains (*22*), a situation that should reflect what transpires in plant cell walls.

The purest manifestation of Zutropfverfahren conditions might have been realized by allowing coniferyl alcohol and H_2O_2 to diffuse through a cellulose dialysis membrane into contact with horseradish peroxidase at pH 6.1—whereupon purportedly very high molecular weight dehydropolymerisates were formed (*23*); however continued enzyme-catalyzed oxidation after dehydrogenative coupling produced a relatively high frequency of carbonyl groups in the resulting polymeric material.

Seven years earlier the field of lignin biosynthesis had been momentarily stimulated by an interesting diversion. Not every lignin chemist (*24*) had been convinced that the underlying principles of macromolecular lignin assembly could be apprehended simply from the effects encountered under the limiting conditions of Zulauf- and Zutropfverfahren, respectively. Their reservations were briefly tempered by a report that, during the dehydrogenative coupling of coniferyl alcohol by Zutropfverfahren, intermediate "modules" were initially formed with degrees of polymerization 18-20, which then were further oxidized to macromolecular lignin-like species (*25*). Unfortunately this remarkable claim is difficult to evaluate because the size exclusion chromatographic conditions employed for characterizing the molecular weight distributions of the dehydropolymerisates promoted extensive intermolecular association (*26*).

More than one factor has been proposed to be responsible for the differences between the conditions existing in lignifying plant cell walls and those prevailing during Zutropfverfahren *in vitro*. First, the immediate precursor of the monolignol *in vivo* is claimed to be its 4-*O*-(β-D-glucopyranosyl) derivative, from which it is formed through the action of β-glucosidase; the glucose-oxidase catalyzed electron transfer to O_2 from the glucose thereby liberated produces the H_2O_2 which facilitates the subsequent dehydropolymerization of the free monomer (*2*). Second, this final step in macromolecular lignin assembly within plant cell walls occurs in the presence of cellulose, hemicelluloses and pectic substances, to which the incipient aromatic biopolymer can become, at least in principle, covalently bound. For example, in ryegrass ferulate moieties, which are bound through ester groups to α-L-arabinofuranose units of arabinoxylan, are also 8–8′ linked to some of the lignin monomer residues (*27*).

It seems that polysaccharides are also capable of exerting an influence upon the macromolecular configuration of lignins much more indirectly. When coniferyl alcohol was subjected *in vitro* to peroxidase-catalyzed dehydrogenative polymerization by H_2O_2 in the presence of pectin, mannan or dextran, the frequencies of linkages to the 5-positions of the guaiacyl residues in the dehydropolymerisates were increased (*28*). Nevertheless, when coniferin, the 4-*O*-(β-D-glucopyranosyl) derivative of coniferyl alcohol, was subjected consecutively to β-glucosidase, glucose oxidase–O_2 and peroxidase–H_2O_2, the resulting dehydropolymerisate appeared to resemble native softwood lignin more closely than that formed by direct monolignol dehydropolymerization under Zutropfverfahren conditions—even when carried out in the presence of pectin (*29*, *30*).

Secondary Structure of Lignins

It is difficult to understand how reasonable fidelity could be captured in ostensible representations for native lignin macromolecules synthesized *de novo* from monolignols *in vitro* with no input except judicious adjustments to homogeneous dehydropolymerizing solution conditions. No other biopolymer has been conceptually required to withstand the anarchy that such a dearth of information could augur. As far as the genesis of native lignin configurations is concerned, the question arises as to what the minimum necessary set of additional instructions would be, and what they would prescribe. Since no concerted efforts have been made to determine primary structures for lignins, it is not possible to speculate *a priori* how this very basic characteristic of macromolecular lignin chains could be perpetuated if it were anything but random.

Actually much more is evident about the secondary structures of lignins, which are probably established in a well-defined way (*31*) as a direct result of the macromolecular assembly process itself (*32*). Raman spectral analyses employing suitably polarized incident laser beams have shown that the aromatic rings of the lignin appearing in *Picea marina* tracheid cross-sections tend, in many places, to be parallel to the plane of the cell-wall lumen surface (*31*). Moreover, in lignifying *Pinus radiata* cell walls, potassium permanganate staining has revealed that lignin domains maintain a more or less uniform density as they expand (*32*). This finding attests to the existence of pronounced noncovalent interactions between the polymeric lignin chains: a crosslink density of only 0.052 (*33*) involving tetrafunctional branch points (*34*) in the macromolecular structure is too low to account for such an effect alone.

The mode of expansion of the lignin domains is closely related to the configuration of the carbohydrate matrix undergoing lignification. In the middle lamella region, discrete lignin "particles" expand uniformly until neighboring domains meet; only then are the intervening spaces filled in (Figure 1). The polysaccharides in the primary cell-wall region consist mainly of randomly oriented cellulose microfibrils that are not arranged into the kinds of lamellae which are typical of the secondary wall (*35*). In contrast to the situation in the middle lamella and primary wall, lignin deposition within the secondary wall takes place much more rapidly in directions that are aligned with, rather than perpendicular to, the axes of the cellulose microfibrils (Figure 2).

It has been proposed that the patterns of lignification are controlled by particular distributions of "initiation sites", which must have been incorporated into the cell wall during its formation if the cell protoplast is to have control over their placement (*32*). Since lignification begins in the middle lamella before spreading into the secondary wall, some feature of the overall process must preclude premature dehydrogenative polymerization of the monolignol, as it diffuses from the protoplast into the cell wall, until it is recognized by a congruent initiation site (*32*). Herein might lie the means whereby the protoplast could control the topochemical characteristics of the lignin being biosynthesized (*32*): the initiation sites could have been introduced into the various parts of the cell wall in specific nonuniform distributions that selectively recognize the different monolignols secreted during consecutive stages of lignification (*2*). If so, the precise nature of these "initiation sites" will be a subject of central importance for understanding macromolecular lignin assembly in mechanistic terms.

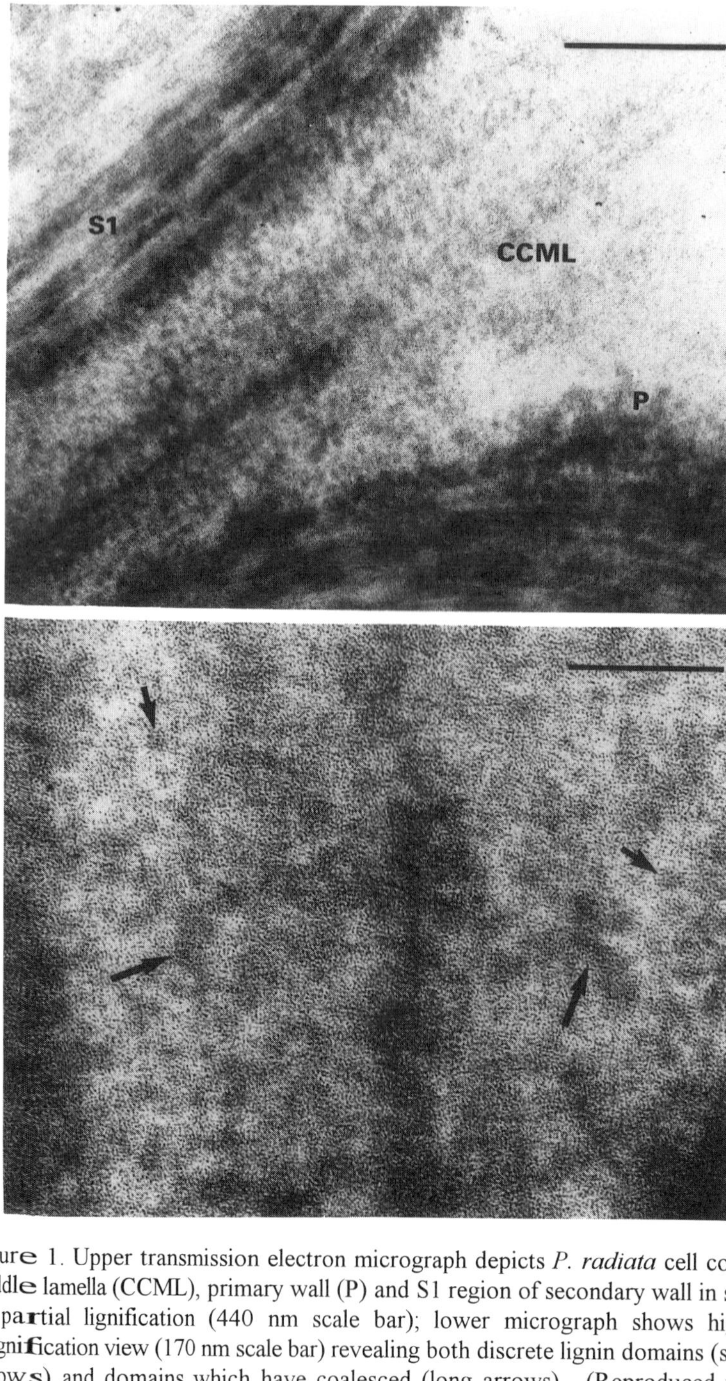

Figure 1. Upper transmission electron micrograph depicts *P. radiata* cell corner middle lamella (CCML), primary wall (P) and S1 region of secondary wall in state of partial lignification (440 nm scale bar); lower micrograph shows higher magnification view (170 nm scale bar) revealing both discrete lignin domains (short arrows) and domains which have coalesced (long arrows). (Reproduced with permission from ref. *32*. Copyright 1994 Springer-Verlag.)

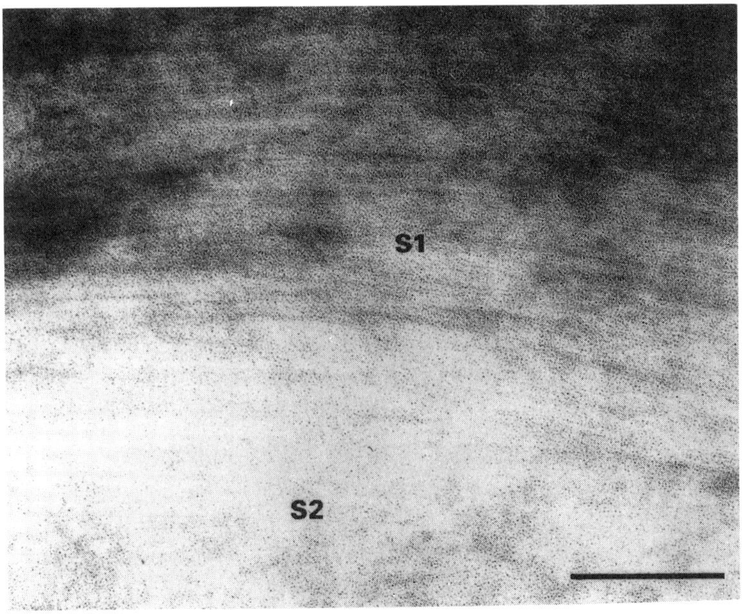

Figure 2. Transmission electron micrograph of relatively poorly lignified *P. radiata* S1/S2 boundary (250 nm scale bar). Lignin in inner S1 region exhibits preferential alignment with microfibrils. (Reproduced with permission from ref. *32*. Copyright 1994 Springer-Verlag.)

For the time being, the polymer chains of which lignins are comprised *in situ* are thought to be organized into compact domains where the aromatic rings tend to be parallel to one another. Furthermore, the observed molecular weight independence of monomolecular film thickness (*36*) and the molecular dimensions apparent in electron micrographs (*37*) both suggest that lignin derivatives consist of disk-like macromolecular components derived from lamellar parent structures which are themselves ~ 2 nm thick. The spatial arrangements of the monomer residues could either be established directly through particular orientational requirements for dehydrogenative coupling between monolignol and the growing polymer chain, or be an outcome of subsequent conformational changes in the lignin macromolecules themselves. Recent work has suggested that the former of the two alternatives, by taking advantage of a template polymerization mechanism, is responsible for the secondary structure of lignins *in vivo* (*38*).

Template Polymerization of Monolignols

Quite exacting conditions were used to determine whether macromolecular lignin species could in any way influence the dehydrogenative polymerization of coniferyl alcohol without direct covalent participation in the process. On account of their adequate solubilities in aqueous solutions around neutral pH, high molecular weight kraft (*39*) lignin components were adopted for these purposes. It is worth emphasizing that the configuration of the native biopolymer is well preserved in such species (*40*).

Horseradish peroxidase–H_2O_2 (*13, 14*) rather than laccase–O_2 (*7, 8*) was chosen to provide the necessary dehydropolymerizing conditions (*vide supra*). However, the identity of the enzyme itself was not expected to play a decisive role in controlling the *modes* of oxidative coupling of monolignol radicals in the presence and absence of high molecular weight lignin components.

Two macromolecular *Pinus banksiana* kraft lignin fractions were secured as plausible template species by ultrafiltration of a parent preparation (*38*). One embodied a (7:3 w/w) bimodal population of components with molecular weights of 51 500 and 568 000, respectively ($\overline{M}_w$ = 206 000). The other consisted of a chloroform-soluble set of 15 400 molecular weight components in which all aromatic hydroxyl groups had been methylated.

A protracted Zutropfverfahren was devised to examine whether these lignin components are capable of exerting an effect upon the molecular weight distributions of dehydropolymerisates formed from a monolignol *in vitro*. Thus coniferyl alcohol and H_2O_2 were very gradually introduced at a constant rate into aqueous 20% dioxane solution (pH 7.3) containing low levels of horseradish peroxidase activity in the presence and absence of the unmethylated lignin macromolecules (1.0×10^{-8} *M* initial concentration) for periods of time ranging between 20 and 80 h. During the process the (coupled and free) monolignol units in solution attained final molar concentrations which correspondingly varied from values 3100 to 8500 times greater than that of the pre-existent macromolecular lignin components.

The molecular weight distributions of the resulting dehydropolymerisates (Figure 3) demonstrated that, at low concentrations in homogeneous solution, the unmethylated softwood lignin macromolecules strongly promote the formation of high molecular weight entities from the peroxidase-catalyzed dehydrogenative coupling of coniferyl alcohol. Indeed the high molecular weight dehydropolymerisate components themselves seemed to enhance the effect: a marked acceleration in their rate of formation (akin to

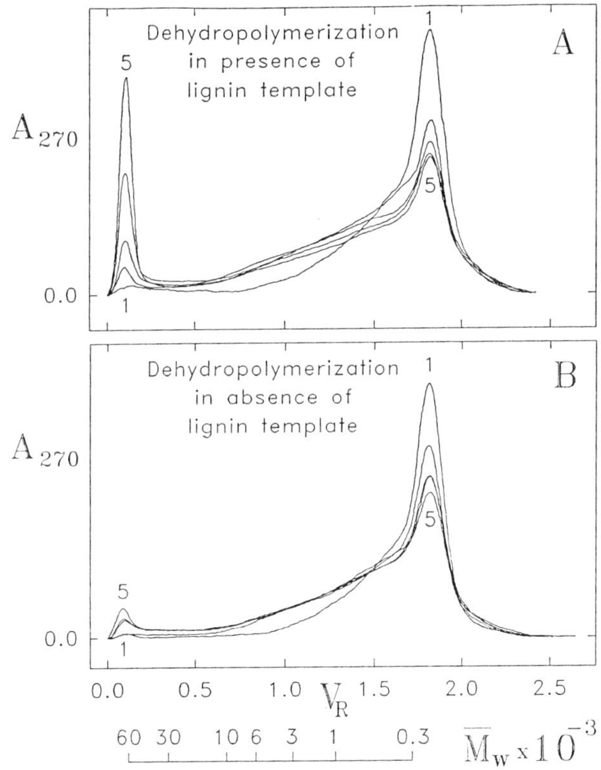

Figure 3. Molecular weight distributions of dehydropolymerisates successively formed under limiting Zutropfverfahren conditions from monolignol in (A) presence and (B) absence of unmethylated macromolecular lignin template after (1) 20 h, (2) 50 h, (3) 75 h, (4) 77.5 h and (5) 80 h. (Sephadex G100/aqueous 0.10 *M* NaOH elution profiles reproduced with permission from ref. *38*. Copyright 1997 Elsevier Science Ltd.)

autocatalysis) was observed during the later stages of the process between 75 and 80 h (Figure 3A) even though their molar concentrations remained far below those of the smaller species.

It is thus very improbable that the formation of the high molecular weight dehydropolymerisate species could have arisen from preferential radical coupling of the mono- (and oligo-) lignols with the growing ends of the unmethylated lignin macromolecules. Indeed an almost identical result was obtained when the coniferyl alcohol and H_2O_2 were gradually introduced at a constant rate into homogeneous aqueous 20% dioxane solution (pH 7.3) where the same horseradish peroxidase activity as before was being maintained in the presence of the *methylated* macromolecular lignin components (2.7×10^{-8} *M* initial concentration). In the 20 - 80 h interval during which the process was monitored, the coupled and free monolignol units together in solution reached molar concentrations which varied between values 1100 and 3100 times greater, respectively, than that of the methylated macromolecular lignin components.

It is clearly evident from the molecular weight distributions of the dehydropolymerisates (Figure 4) that, at comparably low levels, the methylated lignin macromolecules are about as effective as the *un*methylated macromolecular lignin components (Figure 3) in eliciting the formation of high molecular weight species during the peroxidase catalyzed dehydrogenative coupling of coniferyl alcohol. As before, an enhancement of the effect seemed to be engendered by the high molecular weight dehydropolymerisate components themselves: their rate of formation clearly underwent acceleration during the later stages of the process between 70 and 80 h (Figure 4A) even though their molar concentrations remained far below those of the smaller species.

The coupling between a coniferyl alcohol radical and another (mono- or oligolignol) radical occupying adjacent sites on a macromolecular lignin component competes, of course, with the corresponding process in open solution. In this respect the results depicted in Figures 3 and 4 were facilitated by the limiting conditions employed where there was little opportunity for coupling to occur away from the surfaces of pre-existing lignin macromolecules or macromolecular assemblies. As the radicals become more numerous, however, when the concentration of peroxidase is made larger, coupling in open solution would be expected to become more frequent. Such a trend was readily confirmed by increasing the enzyme concentration and concomitantly introducing the monolignol and H_2O_2 more rapidly into the dehydropolymerizing solution. Although the methylated lignin macromolecules still tended to promote formation of the highest molecular weight entities, the proportions of middle and lower molecular weight dehydropolymerisate species rose substantially under these circumstances (*38*).

The mechanism through which the macromolecular lignin components facilitate formation of high molecular weight dehydropolymerisate species is likely to be governed by relatively strong nonbonded orbital interactions with the (mono- and oligolignol) radicals: the unmethylated and methylated lignin macromolecules elicited remarkably similar effects (Figures 3 and 4) despite marked differences in their molecular weights ($\overline{M}_w$'s of 206 000 and 15 400, respectively) and solubility characteristics (*38*).

Implications for Lignin Biosynthesis

Clearly, even when radical coupling for them is absolutely precluded, lignin macromolecules can engender the template polymerization (*cf.* ref. *41*) of coniferyl alcohol

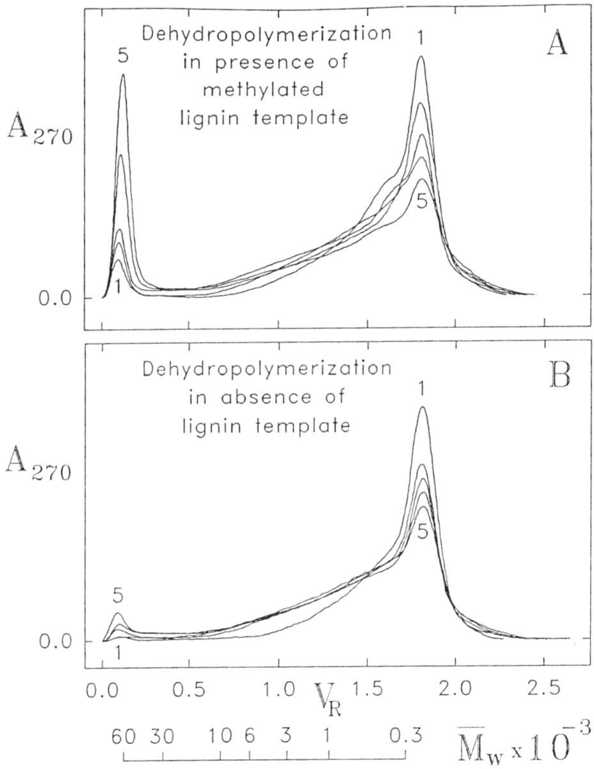

Figure 4. Molecular weight distributions of dehydropolymerisates successively formed under limiting Zutropfverfahren conditions from monolignol in (A) presence and (B) absence of methylated macromolecular lignin template after (1) 20 h, (2) 50 h, (3) 70 h, (4) 75 h and (5) 80 h. (Sephadex G100/aqueous 0.10 *M* NaOH elution profiles reproduced with permission from ref. *38*. Copyright 1997 Elsevier Science Ltd.)

in homogeneous solution under peroxidase-catalyzed dehydrogenative conditions (*38*). The low monolignol radical concentration prevailing in lignifying plant cell walls would tend to ensure the primacy of template polymerization *in vivo* because there would be little opportunity for coupling to occur away from the surfaces of macromolecular lignin chains. If the process is governed by nonbonded orbital interactions between the aromatic template moieties and the radicals about to undergo coupling, a generally parallel orientation of the benzene rings would be expected in the resulting biopolymer, much as observed in the lignin of *P. marina* tracheids (*31*).

To the extent that the relative positions of the aromatic rings in adjacent pairs of monomer residues are uniquely determined by the intervening interunit linkages, the biosynthesis of new polymer chains could entail direct replication of the primary structures of pre-existent lignin macromolecules (*38*). The consequences of post-coupling reductive processes, however, such as those encountered in divanillyl tetrahydrofuran, dihydrodehydrodiconiferyl alcohol, shonanin and isoshonanin (*14*), could not be passed on directly by a template polymerization mechanism, and it is indeed not clear that they actually become an integral part of the polymeric structure; certainly it would be surprising if they were distributed in an incidental manner along macromolecular lignin chains.

Origin of Lignin Primary Structure

Before describing how the primary structures of lignins could be propagated, some attention should be given to the likely extent, between different locations in plant cell walls, of any uniformity in the sequences of interunit linkages along macromolecular lignin chains. Traditional wisdom has supposed that there is no uniformity at all because the biosynthetic process itself ostensibly cannot exert any control over lignin primary structure. However this view has been called into question by the finding that macromolecular lignin assembly could involve the direct replication of polymer chain configuration through a template polymerization mechanism (*38*).

If so, the same lignin primary structure would certainly be preserved within any individual domain or "particle" (Figure 1) that has developed from a single "initiation site" (*32*). The question arises as to how far this uniformity extends among all lignin domains distributed between different locations in plant cell walls (Figures 1 and 2). The answer may be found in the nature of the initiation sites from which the respective lignin domains expand. Do these sites simply harbor an oxidative enzyme capable of dehydrogenatively polymerizing monolignols, or are they more complicated in character?

Insight into the configurations of the initiation sites for lignification in plant cell walls may be found in the mechanism through which regio- and stereoselectivity is prescribed for the critical phenoxy-radical coupling step during lignan biosynthesis (*17*, *42*). These processes are governed by dirigent proteins which, while lacking their own catalytically active centers, capture specific monolignol-derived free radicals in such a way that these unique substrates are juxtaposed into the correct relative orientations for regio- and stereoselective coupling (*42*). Thus the dirigent proteins possess active sites that are manifestly adapted for interacting in a highly distinctive way with the radicals formed by single-electron oxidation of specific monolignols. It would be very surprising if these types of active sites were relegated exclusively to the lignan biosynthetic pathway and not involved in macromolecular lignin assembly at all.

However, it is important to recognize that the same dirigent proteins, which govern regio- and stereospecific phenoxy-radical coupling during lignan biosynthesis, do not themselves participate in the dehydropolymerization of monolignols to form lignins. The dehydrogenative dimerization of monolignols cannot be the first step in any sequence of events that could produce dehydropolymerisates with the same frequencies of interunit linkages as found in native lignins. Rather dehydrogenative coupling of the monomers must be able to occur preferentially with the growing ends of macromolecular lignin chains (*22*) if the outcome from these assembly processes is to encompass what is presently known about the full range of dehydropolymerisate structures occurring *in vivo*.

Thus any protein(s) controlling macromolecular lignin configuration would be expected to proffer a contiguous array of monolignol-radical binding sites of the same types as those on the dirigent proteins that play a central role in lignan biosynthesis. This dirigent array would facilitate the assembly of a progenitorial lignin macromolecule with a specific primary structure formed through preferential dehydrogenative coupling between the individual monolignols, as they are released, and the growing end of the polymer chain. Once formed, the progenitorial lignin macromolecule will not begin to be displaced from the contiguous array of dirigent sites on the protein(s) as biosynthesis of the next dehydropolymerisate chain is initiated; its primary structure will instead be replicated through a direct template polymerization mechanism (*38*).

Many of the interunit linkages in lignins possess chiral carbon atoms, and dirigent proteins are capable of prescribing not only the regio- but also the stereospecificity inherent in the dehydrodimerization of monolignols to lignans (*42*). Thus it might be expected that a contiguous dirigent array of monolignol-radical coupling sites would produce some optical activity in macromolecular lignin chains, but it has long been claimed that there is no sign of this in the native biopolymer. Further comment about the matter should be deferred until the template polymerization process has been more fully characterized: there are a number of possibilities. For example, individual macromolecular lignin chains could be largely internally compensated as far as net optical activity is concerned, or their replication might partly involve enantiomeric configurations of interunit linkages that alternate from one strand to the next within any particular lignifying domain.

The important question is, of course, whether any contiguous dirigent arrays of monolignol-radical coupling sites have yet been discovered. Three kinds of structural proteins in plant cell walls have been distinguished according to their amino acid compositions and repeating sequences (*43*). These are the extensins or hydroxyproline-rich glycoproteins (HRGPs), proline-rich proteins (PRPs) and glycine-rich proteins (GRPs). Some remarkable findings have come to light through the use of polyclonal antibodies raised against a 33 000 molecular weight PRP from *Glycine max* (*44*) that is comprised almost completely of a decameric amino acid repeat, ProHypValTyrLysProHypValGluLys (*45*). By these means the localization of PRP epitopes has been correlated with lignin both spatially and temporally in developing cell walls of the *Zea mays* coleoptile (*15*) and in secondary walls of differentiating protoxylem elements in the *G. max* hypocotyl (*46*).

It has been suggested that PRPs may act as a scaffold for initiating lignification at their tyrosine residues (*46*). However, it is well known that polyphenols—including monomers such as propyl gallate—associate quite strongly with PRPs through noncovalent interactions with proline residues (*47*), the first in a ProPro sequence being a much preferred binding site. The distinct possibility arises, therefore, that there are favorable noncovalent interactions between PRPs and macromolecular lignin chains in plant cell walls.

At the time of writing nothing is known about how proline residues in polypeptide chains interact with free radicals derived from mono-, oligo- and polylignols; yet the question is potentially quite pertinent to whether the primary structures of macromolecular lignin chains could be transcribed from PRPs or other similar cell wall proteins. Here it may be of more than passing interest that the dirigent protein isolated from *Forsythia suspensa*, which is responsible for coupling two coniferyl-alcohol-derived free radicals to form (+)-pinoresinol (*42*), possesses a single ProProValGlyArg sequence near the middle of the polypeptide chain (N. G. Lewis, Washington State University, personal communication, 1997). This is reminiscent of the decameric repeat ProHypValTyrLysProHypValGluLys that almost completely comprises the PRP (*45*), antibodies raised against which have established the spatial and temporal co-localization of lignin and PRP epitopes in the *Zea mays* coleoptile (*15*) and *G. max* hypocotyl (*46*).

Ultimately, whether lignin primary structure is encoded in PRPs or elsewhere, an important requirement must be fulfilled during the biosynthesis and replication of the progenitorial lignin macromolecules in the respective lignifying domains of plant cell walls. Dehydrogenative polymerization will often require the coupling of newly formed monolignol radicals to occur preferentially with the growing ends of lignin chains (*22*) if long sequences of interunit linkages are to be assembled of the kind observed in the native biopolymer. How this is accomplished in the context of a template polymerization mechanism orchestrated by proteins and enzymes in the plant cell wall has become central to a fundamental understanding of lignin biosynthesis.

Acknowledgments

Acknowledgment for support of this work is made to the Vincent Johnson Lignin Research Fund at the University of Minnesota, and the Minnesota Agricultural Experiment Station.

Paper No. 974436804 of the Scientific Journal Series of the Minnesota Agricultural Experiment Station, funded through Minnesota Agricultural Experiment Station project No. 43-68, supported by Hatch Funds.

Literature Cited

1. Lewis, N. G.; Yamamoto, E. *Annu. Rev. Plant Physiol. Plant Mol. Biol.* **1990**, *41*, 455-496.
2. Terashima, N.; Fukushima, K.; He, L.-f.; Takabe, K. In *Forage Cell Wall Structure and Digestibility*; Jung, H. G., Buxton, D. R., Hatfield, R. D., Ralph, J., Eds.; American Society of Agronomy, Crop Science Society of America, Soil Science Society of America: Madison, 1993; pp 247-270.
3. Adler, E. *Wood Sci. Technol.* **1977**, *11*, 169-218.
4. Freudenberg, K.; Reznik, H.; Boesenberg, H.; Rasenack, D. *Chem. Ber.* **1952**, *85*, 641-647.
5. Freudenberg, K. *Nature* **1959**, *183*, 1152-1155.
6. Sterjiades, R.; Dean, J. F. D.; Gamble, G.; Himmelsbach, D. S.; Eriksson, K.-E. L. *Planta* **1993**, *190*, 75-87.
7. Higuchi, T. *J. Biochem.* **1958**, *45*, 515-528.
8. Bao, W.; O'Malley, D. M.; Whetten, R.; Sederoff, R. R. *Science* **1993**, *260*, 672-674.

9. Mason, H. S.; Cronyn, M. *J. Am. Chem. Soc.* **1955**, *77*, 491-492 and references therein.
10. Savidge, R.; Udagama-Randeniya, P. *Phytochemistry* **1992**, *31*, 2959-2966.
11. Udagama-Randeniya, P.; Savidge, R. *Electrophoresis* **1994**, *15*, 1072-1077.
12. Koblitz, H.; Koblitz, D. *Nature* **1964**, *204*, 199-200.
13. Olson, P. D.; Varner, J. E. *Plant J.* **1993**, *4*, 887-892.
14. Nose, M.; Bernards, M. A.; Furlan, M.; Zajicek, J.; Eberhardt, T. L.; Lewis, N. G. *Phytochemistry* **1995**, *39*, 71-79.
15. Müsel, G.; Schindler, T.; Bergfeld, R.; Ruel, K.; Jacquet, G.; Lapierre, C.; Speth, V.; Schopfer, P. *Planta* **1997**, *201*, 146-159.
16. Freudenberg, K.; Schlüter, H. *Chem. Ber.* **1955**, *88*, 617-625.
17. Davin, L. B.; Lewis, N. G. *An. Acad. bras. Ci.* **1995**, *67 (Supl. 3)*, 363-378.
18. Okusa, K.; Miyakoshi, T.; Chen, C.-L. *Holzforschung* **1996**, *50*, 15-23.
19. Freudenberg, K.; Nimz, H. *Chem. Ber.* **1962**, *95*, 2057-2062.
20. Freudenberg, K.; Tausend, H. *Chem. Ber.* **1963**, *96*, 2081-2085.
21. Freudenberg, K. *Angew. Chem.* **1956**, *68*, 508-512.
22. Sarkanen, K. V. In *Lignins—Occurrence, Formation, Structure and Reactions*; Sarkanen, K. V., Ludwig, C. H., Eds.; Wiley Interscience: New York, 1971; pp 95-163.
23. Tanahashi, M.; Higuchi, T. *Wood Res.* **1981**, *67*, 29-42.
24. Forss, K.; Fremer, K.-E. *Pap. Puu* **1965**, *47*, 443-454.
25. Wayman, M.; Obiaga, T. I. *Can. J. Chem.* **1974**, *52*, 2102-2110.
26. Himmel, M. E.; Mlynár, J.; Sarkanen, S. In *Handbook of Size Exclusion Chromatography*; Wu, C.-s., Ed.; Marcel Dekker: New York, 1995; *Chromatographic Science Series 69*, 353-379.
27. Ralph, J.; Grabber, J. H.; Hatfield, R. D. *Carbohydr. Res.* **1995**, *275*, 167-178.
28. Terashima, N.; Seguchi, Y. *Cellulose Chem. Technol.* **1988**, *22*, 147-154.
29. Terashima, N.; Atalla, R. H.; Ralph, S. A.; Landucci, L. L.; Lapierre, C.; Monties, B. *Holzforschung* **1995**, *49*, 521-527.
30. Terashima, N.; Atalla, R. H.; Ralph, S. A.; Landucci, L. L.; Lapierre, C.; Monties, B. *Holzforschung* **1996**, *50*, 9-14.
31. Atalla, R. H.; Agarwal, U. P. *Science* **1985**, *227*, 636-638.
32. Donaldson, L. A. *Wood Sci. Technol.* **1994**, *28*, 111-118.
33. Yan, J. F.; Pla, F.; Kondo, R.; Dolk, M.; McCarthy, J. L. *Macromolecules* **1984**, *17*, 2137-2142.
34. Pla, F.; Yan, J. F. *J. Wood Chem. Technol.* **1984**, *4*, 285-299.
35. Wardrop, A. B. In *The Formation of Wood in Forest Trees*; Zimmermann, M. H., Ed.; Academic Press: New York, 1964; pp 87-134.
36. Goring, D. A. I. *ACS Symp. Ser.* **1977**, *48*, 273-277.
37. Goring, D. A. I.; Vuong, R.; Gancet, C.; Chanzy, H. *J. Appl. Polym. Sci.* **1979**, *24*, 931-936.
38. Guan, S.-y.; Mlynár, J.; Sarkanen, S. *Phytochemistry* **1997**, *45*, 911-918.
39. Gierer, J. *Wood Sci. Technol.* **1980**, *14*, 241-266.
40. Garver, T. M., Jr.; Sarkanen, S. *Holzforschung* **1986**, *40 Suppl.*, 93-100.
41. Volpe, R. A.; Frisch, H. L. *Macromolecules* **1987**, *20*, 1747-1752 and references therein.

42. Davin, L. B.; Wang, H.-B.; Crowell, A. L.; Bedgar, D. L.; Martin, D. M.; Sarkanen, S.; Lewis, N. G. *Science* **1997**, *275*, 362-366.
43. Ye, Z.-H.; Song, Y.-R.; Marcus, A.; Varner, J. E. *Plant J.* **1991**, *1*, 175-183.
44. Marcus, A.; Greenberg, J.; Averyhart-Fullard, V. *Physiol. Plant.* **1991**, *81*, 273-279.
45. Datta, K.; Schmidt, A.; Marcus, A. *Plant Cell* **1989**, *1*, 945-952.
46. Ryser, U.; Schorderet, M.; Zhao, G.-F.; Studer, D.; Ruel, K.; Hauf, G.; Keller, B. *Plant J.* **1997**, *12*, 97-111.
47. Baxter, N. J.; Lilley, T. H.; Haslam, E.; Williamson, M. P. *Biochemistry* **1997**, *36*, 5566-5577.

Chapter 16

Cell Wall Cross-Linking in Grasses by Ferulates and Diferulates

John Ralph[1,2], Ronald D. Hatfield[1], John H. Grabber[1], Hans-Joachim G. Jung[1,3], Stéphane Quideau[1,4], and Richard F. Helm[1,5]

[1]U.S. Dairy Forage Research Center, 1925 Linden Drive West, Madison, WI 53706
[2]Department of Forestry, University of Wisconsin, Madison, WI 53706
[3]Department of Agronomy and Plant Genetics, University of Minnesota, St. Paul, MN 55108

Ferulate polysaccharide esters in grasses enter into free-radical coupling reactions in the cell wall. By radical dimerization of ferulates, polysaccharide-polysaccharide cross-linking is effected. A range of diferulate isomers are produced, only one of which had been quantitated in the past. Diferulates have been underestimated by factors of up to 20, belittling their contribution to functions in the cell wall. Both ferulates and diferulates participate in lignification reactions and become intimately bound with lignin. Under-quantitation is significant since it is not possible to release ferulate or diferulates from some of the structures generated. Overall ferulates play a significant role in cell wall development and impact upon polysaccharide utilization in grasses.

Cross-linking of plant cell wall components is expected to have a marked influence on various properties of the cell wall (*1, 2*). For example, in grasses, several authors have postulated the effects of isodityrosine and diferulate cross-links on cell-wall accessibility, extensibility, plasticity, digestibility and adherence (*3, 4*). The topic has been reviewed extensively (*1, 5-10*). In woody species, the cross-linking mechanism to receive most attention involves the direct attachment of polysaccharides to lignin *via* nucleophilic addition to intermediate lignin quinone methides during lignification. An American Chemical Society Symposium recently addressed some of these issues (*11*). The extent of such cross-linking and how the plant is able to exert control on this seemingly 'opportunistic' approach has yet to be fully determined.

It is well known that grasses have substantial amounts of hydroxycinnamic acids intimately associated with the cell wall, as has been detailed in a number of reviews (*1, 5-10, 12*). This chapter will concern itself only with ferulate, which has the major cross-linking role. It was recognized early on that ferulic acid, an ideally structured difunctional molecule, could have a role in cross-linking polysaccharides with lignin (Figures 1 and 2).

[4]Current address: Department of Chemistry, Pennsylvania State University, University Park, PA 16802
[5]Current address: Department of Wood Science and Forest Products, Virginia Polytechnic Institute and State University, Blacksburg, VA 24061–0346

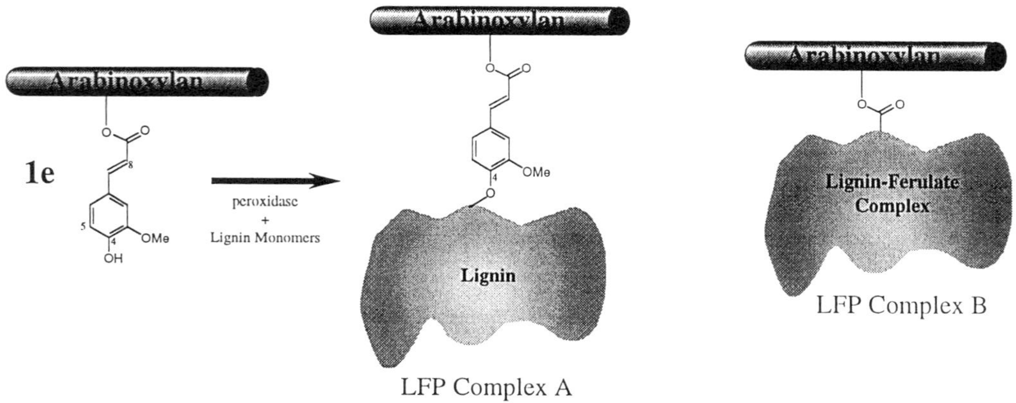

Figure 1. Ferulates, exported into the cell wall region as esters of polysaccharides, (primarily arabinoxylans in grasses), will cross-link with lignin via two mechanisms to yield a cross-linked lignin-ferulate-polysaccharide (LFP) complex from which the ferulate may or may not be releasable. The 'passive' mechanism (ferulate addition to lignin quinone methides) gives complexes (LFP complex A) from which the ferulate is fully releasable. The 'active' mechanism, where ferulate radicals cross-couple with lignin monomer or oligomer radicals, produces an LFP (Complex B) from which ferulate is only partially releasable.

Legend for all following structural Figures. Definitions of R, L, Ara.

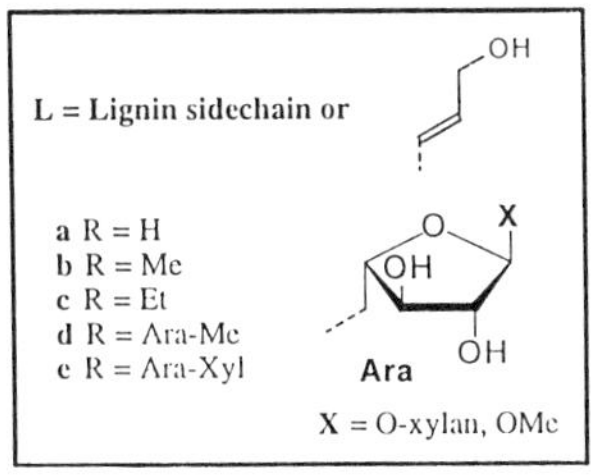

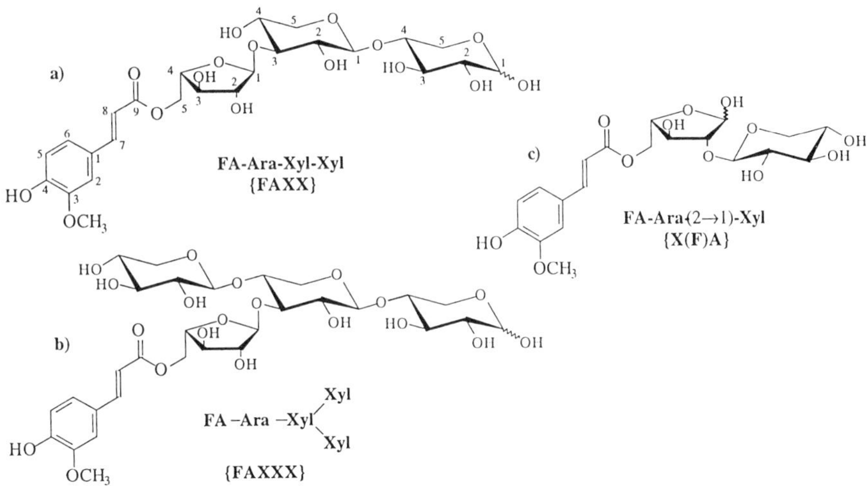

Figure 2. Ferulate-oligosaccharide esters isolated from grass cell wall following dissolution by mild acid or enzyme hydrolysates: (a) *O*-[5-*O*-(*trans*-feruloyl)-α-L-arabinofuranosyl]-(1→3)-*O*-β-D-xylopyranosyl-1→4)-D-xylopyranose (FAXX), (b) *O*-β-D-xylopyranosyl-(1→4)-*O*-[5-*O*-*trans*-feruloyl-α-L-arabinofuranosyl-(1→3)]-*O*-β-D-xylopyranosyl-1→4)-D-xylopyranose (FAXXX), and (c) 2-*O*-β-D-xylanopyranosyl-(5-*O*-feruloyl)-L-arabinose (FA-Ara-(2→1)-Xyl). Further xylose homologues of (c), X(F)AXX, and XX(F)A have also been reported.

Ferulate-esterified Polysaccharides

Ferulate **1** (Figure 3) was found linked to polysaccharides, particularly arabinoxylans, in a regiospecific manner (*13-22*) as has been reviewed in the literature (*6-9, 23, 24*). In grass arabinoxylans it is attached exclusively at C-5 of α-L-arabinofuranoside residues, which are themselves glycosidically linked to the 3-positions of a xylan backbone (*17, 19, 20, 24-27*). The originally reported (*16*) 2-attachment was corrected (*19*). The arabinosyl residue may be further linked at C-2 to another xylose residue (*28-30*) or even to another xylan chain (*27*). Presumably the attachment of ferulate to the arabinoxylan occurs intracellularly in conjunction with polysaccharide synthesis and the product is exported out into the cell wall (*10*). Isolation and characterization of FAXX, FAXXX, X(F)A (Figure 2), X(F)AXX, and XX(F)A from mild acid hydrolyzed or enzyme-degraded bamboo shoots (*17, 18, 25, 26*), bermudagrass (*19, 30*), bagasse (*13, 15*), barley straw (*20*), maize (*14, 29*), sugar beet (*31*), wheat bran (*16*), and fescue (*28, 32*) was invaluable in defining the structural characteristics. Ferulates attached to other (oligo)saccharides have been reported in other plants, particularly spinach (*31, 33*).

Polysaccharide Cross-linking by Ferulates: Ferulate Dimerization

The presence of ferulates attached to polysaccharide components provides a convenient mechanism for cross-linking polysaccharides. Ferulate dimerization is all that is required to effect polysaccharide-polysaccharide cross-linking. Two mechanisms have been identified. Photochemical [2+2] cycloaddition produces cyclodimers, the truxillic acids (*19, 34-38*). Such acids can be released from grass walls and, although most of the dimers involve *p*-coumaric acid, early speculations were that this cycloaddition was a major cross-linking mechanism (*19, 34-36*). As will be shown below, this was in large part due to the major portion of the other mechanism's products having been missed. It is hard to know the plant's role in effecting such cross-linking; it presumably has little control over the photo-induced dimerization other than to arrange for hydroxycinnamates to be sufficiently proximate.

Radical-mediated dimerization is the other mechanism, although it was not recognized as such earlier. A dehydrodimer of ferulic acid, the 5–5′-coupled dimer **10** (Figure 3), commonly referred to as 'diferulic acid', has been isolated and 'quantitated' (see later) from grasses by saponification for some 20 years (*39-47*). In 1989 Yamamoto, Bokelman and Lewis (*6*) remarked that radical mediation had not been proven in a manner to confer only this coupling. We too were concerned since radical dimerization by single electron metal oxidants or peroxidases/H_2O_2 produced little to none of the 5–5′-coupled dehydrodimer **10** (Figure 3) (*48*). The main product was always the 8–5′-coupled dimer **8**. This had been discovered and reported several times (*49-51*) and has been recently reconfirmed (*52*). The 8–8′ products **12** were also formed, as well as some 8–*O*–4′ product **9**. Our main concern was how the 5–5′-coupled dimer **10** could be formed in plants to the apparently complete exclusion of the other dimers. Such a phenomenon would appear to indicate very tight enzymatic control of coupling. As will be shown below, this is simply not the case.

Dehydrodiferulates: Dimerization of Ferulate Radicals

Figure 3 shows how ferulates would be expected to dimerize *in planta* or in *in vitro* systems. Again, a consideration of this figure confirms it to be remarkable that only the 5–5′-coupled product **10** would be formed, particularly in light of what we know about the related dimerization of coniferyl alcohol (*2, 53-58*). In fact, when a variety of grass wall materials were saponified and analyzed by GC/FID or GC-MS (Figure 4), it was found that the entire spectrum of expected coupling products (with the exception of the 4–*O*–5′-coupled dimer **17** was present (*48*). Furthermore, the 5–5′-coupled product **16** was not the major dimer. Other researchers were also beginning

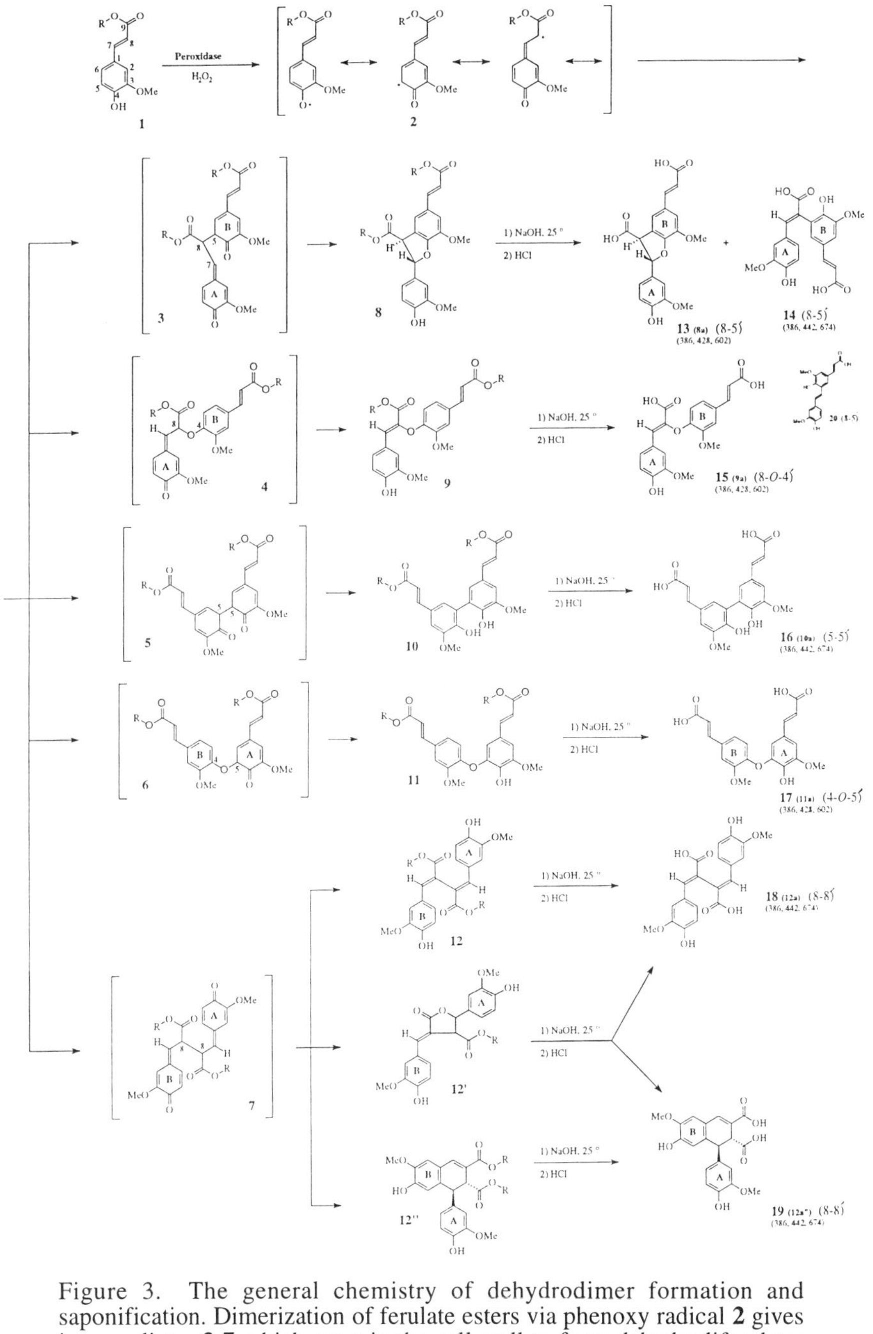

Figure 3. The general chemistry of dehydrodimer formation and saponification. Dimerization of ferulate esters via phenoxy radical **2** gives intermediates **3-7** which react in the cell wall to form dehydrodiferulates **8-12**. During chemical analysis **8-12** can be saponified to dehydrodiferulic acids **13-19**, also producing a small amount of compound **20**. The values in parentheses below the structures represent the nominal molecular masses for the parent compound, the methylated, and the trimethylsilylated derivatives.

to suspect that there were more dehydrodimers but the two reports about the subject (*59, 60*) made the assumption that such dimers had all to be 5–5′-coupled. Thus the orientational isomers of Stewart *et al.* (*60*) certainly do not exist as discrete entities, and it is easy to show (by photochemical treatment of the 5–5′-dimer) that the claims of *cis/cis* or *cis/trans* isomers (*59, 60*) are not the explanation; only traces of non-*trans/trans* isomers were isolated from plant materials that we have screened (*48*).

Each of the radical-coupling products in Figure 3 was independently synthesized and characterized (*48*). GC-FID response factors were obtained, and a method was developed for quantitating all of the dimers. Figure 5 shows the results of such quantitation for four samples. More recently, an HPLC method has also been developed (*61*); the use of diode array detection aids in structural identification. Under-quantitation of even the 5–5′-coupled dimer **16** by possible neglect of response factors (*36, 62*), along with the ignorance of the other dehydrodimers **13-15**, **17-20** may have led to up to a 20-fold underestimation of ferulate dehydrodimers (*48*). Dimers can account for as much as 70% of the total ferulate released from the wall. Clearly they play a far more significant role in wall development than previously considered. And it is not possible to quantitate all of the ferulate dehydrodimers; as we shall see below, they become inexorably covalently linked during lignification.

Diferulates are also being found in a variety of other plant materials. Waldron's group has shown very high levels in water chestnuts (*63, 64*), and suggests that they may be largely responsible for their crunchiness even after cooking. The light response of pine hypocotyls was tied to diferulate-mediated cross-linking of polysaccharides (*65*). Interestingly this group could find no 5–5′-coupled products, an observation more in line with the results of *in vitro* dimerizations. Previously, hypocotyl light responses in *Oryza* (*66*) and *Avena* (*67*) species had been noted and attributed to solely the 5–5′-dimer. Again, this may be an over-simplification, since we have recently shown that all dimers are present in their hypocotyls (*68*). Sugar beet has also been shown to be rich in diferulates (*69, 70*). Furthermore, elicitation of further cross-linking by treatment of beet pulps with peroxidase/H_2O_2 improves gelling properties (*70*). There is also a suggestion that, in a suspension culture system, cells compensated for reduced ferulate deposition (when grown in the presence of a phenylalanine-ammonia lyase inhibitor, 2-aminoindan-2-phosphonic acid) by increasing the extent of dehydrodimer formation (*71*).

Although clearly providing advantageous properties for plant growth and development, cross-linking has been thought to limit polysaccharide degradability (*5-8, 23, 72-74*), and this effect was essentially observed in a useful cell wall suspension cultured system (*75*). Higher levels of diferulates (but lower total ferulates) depress both the rate and extent of polysaccharide degradation (Figure 6) (*76, 77*).

Lignin-Polysaccharide Cross-linking: Ferulate Incorporation into Lignin

Verification that ferulates could be 'attached' to lignin was readily proven. Following room-temperature saponification, further ferulates were released by high temperature alkaline saponification under essentially wood pulping conditions (2-4 M NaOH, 170°C, 2 h) (*73, 78-84*). Ferulate release by acidolysis was also demonstrated (*85-87*). Lam, Iiyama and Stone provided a method of measuring ferulates that were both esterified and etherified, thus proving conclusively that ferulates were lignin-polysaccharide cross-linking agents (*78, 80, 81, 88, 89*). They found that all of the ferulate that was etherified was also esterified. This is expected if it is polysaccharide-ferulate esters that are exported into the cell wall. For us, it was the mechanism that was intriguing. Although various researchers had demonstrated that ferulates were 'attached' to lignin, there was little effort focused on the regiochemistry of attachment and the consequent mechanistic implications. Nearly all of the literature assumed a mechanism that we have described as 'opportunistic' or 'passive' (*9, 49, 90, 91*).

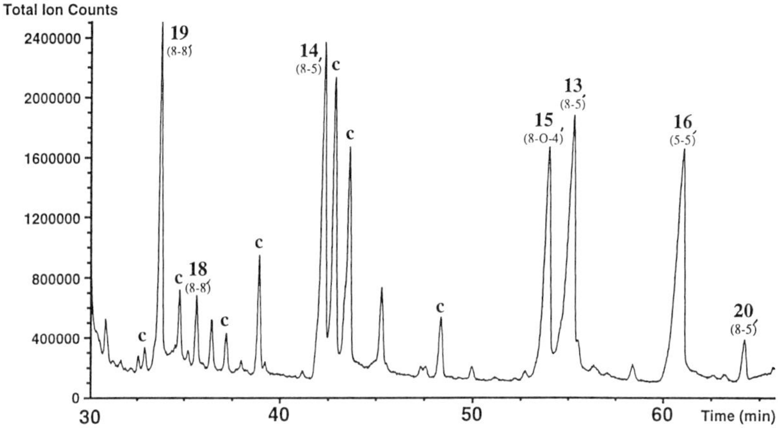

Figure 4. Total ion chromatogram from GC-MS of the dimer region of a saponified extract of primary cell walls from suspension cultured maize showing dehydrodiferulic acids **13**-**19** (with the exception of **17**). Peaks labeled **c** are assigned, without further authentication, to [2+2]-cyclodimers by observation of an *m/z* 338 peak in their mass spectra.

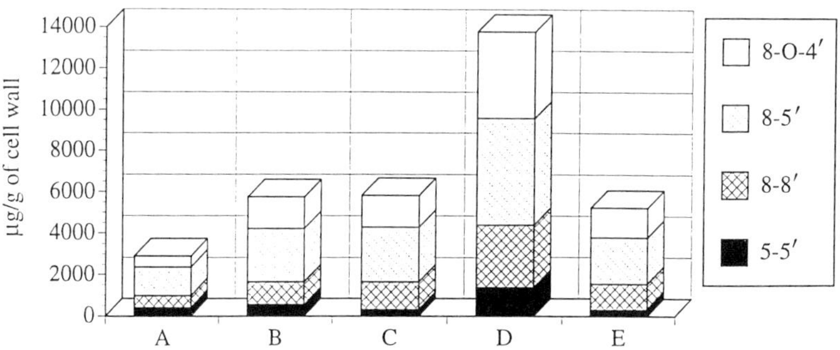

Figure 5. Composition (µg/g of cell wall material) of dehydrodimers deriving from the individual coupling modes (5–5′, 8–8′, 8–5′, and 8–*O*–4′) for various plant cell wall samples: (A) suspension-cultured maize, (B) orchardgrass parenchyma, (C) orchardgrass sclerenchyma, (D) switchgrass parenchyma, (E) switchgrass sclerenchyma.

'Passive' Attachment of Ferulates to Lignin

It has been generally assumed that the attachment of lignin to ferulates was through α-ethers **26** (Figure 7) (*78*), a reasonable mechanism that can readily be demonstrated by model studies (*86*). However, consideration of only this mechanism had the unfortunate consequence that a great deal of the diversity in cross-linking was overlooked and the quantity and importance of this cross-linking severely underestimated. This is because α-linked ferulates are fully releasable (although not necessarily quantitatively) as ferulic acid by alkaline or acidic solvolysis. As we shall see later, products of the other mechanism of lignin-ferulate attachment cannot be similarly released under these conditions.

There are additional chemical questions relating to plant control of cross-linking. The α-ferulate ethers are formed by nucleophilic attack of the ferulate phenolic hydroxyl on the α-carbon of quinone methides, intermediates formed during lignification. Recall, however, that quinone methides are produced as the product of radical coupling between a coniferyl alcohol radical and another such radical or a radical derived from a lignin dimer or higher oligomer (Figure 7). The 'passive' cross-linking mechanism denies ferulate participation in the radical-coupling process, and yet we have just seen that ferulates certainly homo-couple *via* analogous radical reactions. The ferulate must also compete for the quinone methide with other phenols that are present (including lignin monomers and oligomers), and water, a reaction which *in vitro* is capricious and low-yielding (*57, 91*), and extensive work by Sipilä indicates that it is a poor reaction under aqueous conditions (*92-94*). There are conditions that favor such ether formation in DHP's. However, it is now recognized that non-cyclic α-aryl ethers may be of very low abundance in secondary wall lignins. Prior estimates of 6-9% (*95*) are in fact attributing other cyclic structures to non-cyclic α-aryl ethers. Ede's elegant work (*96, 97*) demonstrated clearly that α-ethers are scarcely detectable in NMR spectra of milled wood lignins and that the presumed α-ethers can be attributed to dibenzodioxocins, a lignin structure recently identified by Brunow's group (*98, 99*). So, although ferulate may add more competitively than other phenols (this has not been tested), conditions seem to be against its competing effectively for the quinone methide. Why then would the plant choose such an uncontrolled and poor reaction for effecting what is presumably a very important process, lignin-polysaccharide cross-linking? Despite these problems, or perhaps because researchers have not sufficiently considered the alternatives, the predominant mechanism described and illustrated in the literature is this 'passive' mechanism. [We coined the term 'passive', as opposed to the 'active' mechanism described below, because ferulate must just sit around waiting for a quinone methide to which it then may chance to add. The alternative mechanism has ferulates actively participating in the lignification process.]

Active Incorporation of Ferulates in Lignins

The logical alternative (or concurrent) mechanism involves ferulate radicals **25** (Figure 8). Since feruloylated polysaccharides are present in the wall prior to lignification, peroxidases and hydrogen peroxide could couple those ferulates that are sufficiently proximate to produce a whole range of ferulate dimers, as described above. Peroxidases (or alternative oxidases) and hydrogen peroxide are required for lignification. The scant mention (*6, 12, 24*) that this mechanism has enjoyed recognized that β-ethers would be produced by such a mechanism, but obviously this is an over-simplification (*9*).

Biomimetic Lignification of a Ferulate-Polysaccharide Model

The first logical issue to address was whether ferulates would cross-couple with lignin monomers/oligomers and, if so, what types of products formed were. One way to obtain this type of information is *via* synthetic lignification, since this process will

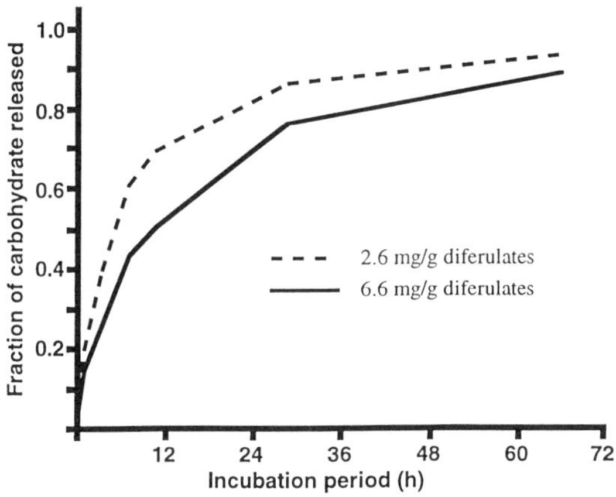

Figure 6. Total carbohydrate release from samples with differing diferulate levels.

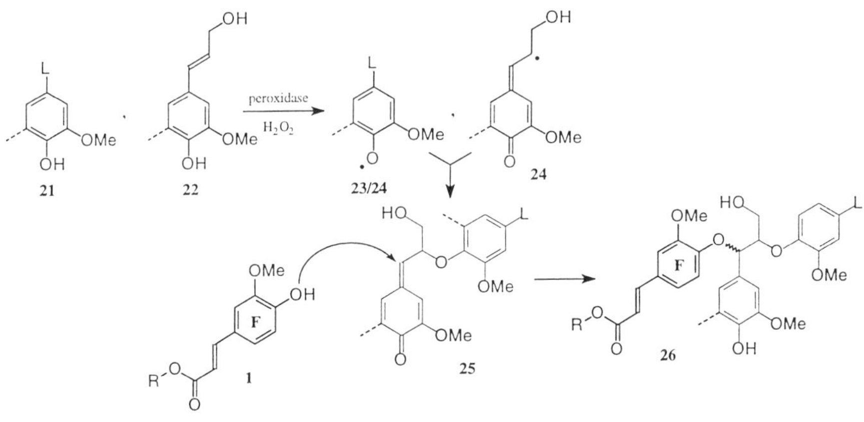

Figure 7. The popular 'passive' mechanism for incorporation of ferulates into lignins. Ferulate **1** does not enter into the one-electron oxidative coupling processes; these processes do, however, produce quinone methide intermediates **25** from coupling of a hydroxycinnamyl alcohol lignin monomer **22** with a lignin monomer or oligomer **21**. Ferulates nucleophilically add to the quinone methides, in competition with other nucleophiles in the cell wall including acids and water, to produce lignin-ferulate α-ethers (benzyl aryl ethers) **26**. Ferulate is fully releasable from these structures by high temperature base solvolysis. L = hydroxycinnamyl alcohol sidechain or generic lignin sidechain plus the remainder of the lignin oligomer/polymer.

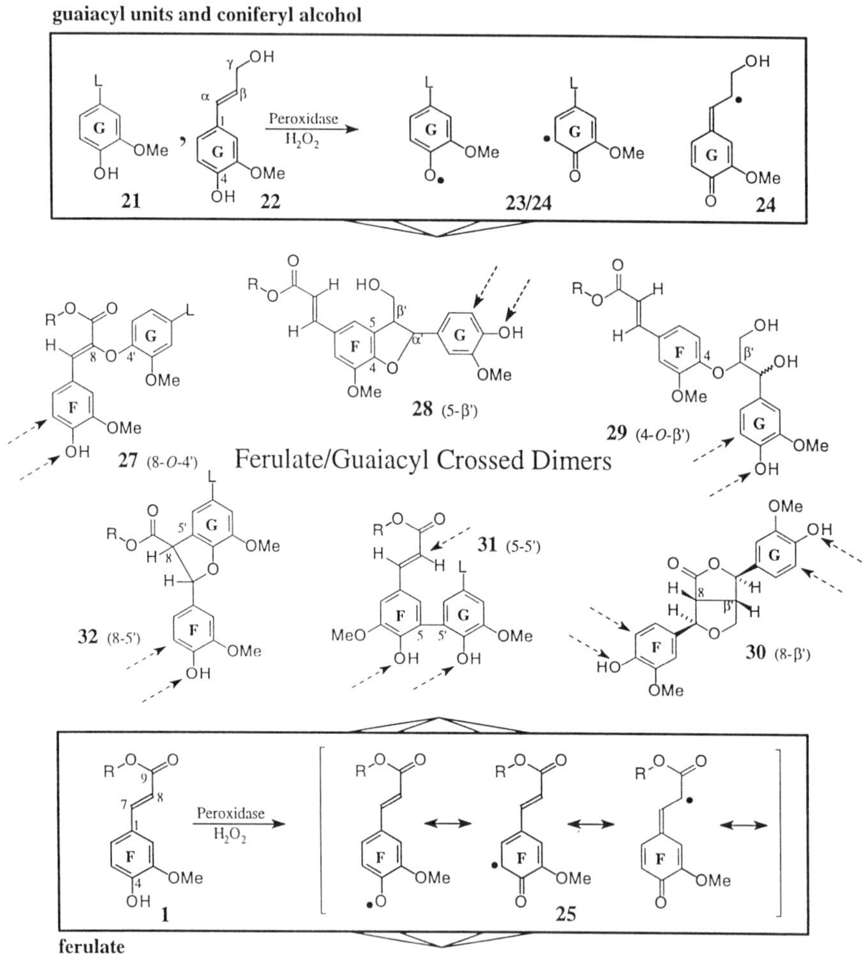

Figure 8. The 'active' mechanism for incorporation of ferulates into lignins. Radical cross-coupling products **27-32** formed by oxidative coupling of ferulate **1** with coniferyl alcohol **22** and guaiacyl (from coniferyl alcohol oligomers) units **23-24** (upper half), and **27'-30'** from coupling of ferulate **1** with sinapyl alcohol **22'** and syringyl (from sinapyl alcohol oligomers) units **23'-24'** (lower half). **F**, **G**, and **S** are to indicate aromatic rings originating from ferulate, guaiacyl, and syringyl units. R is defined in Figure 2, L = coniferyl alcohol sidechain or generic lignin sidechain plus the remainder of the lignin oligomer/polymer. Only compounds **29** and **29'** will release ferulate on high temperature base solvolysis. Dashed arrows indicate sites available for further polymerization.

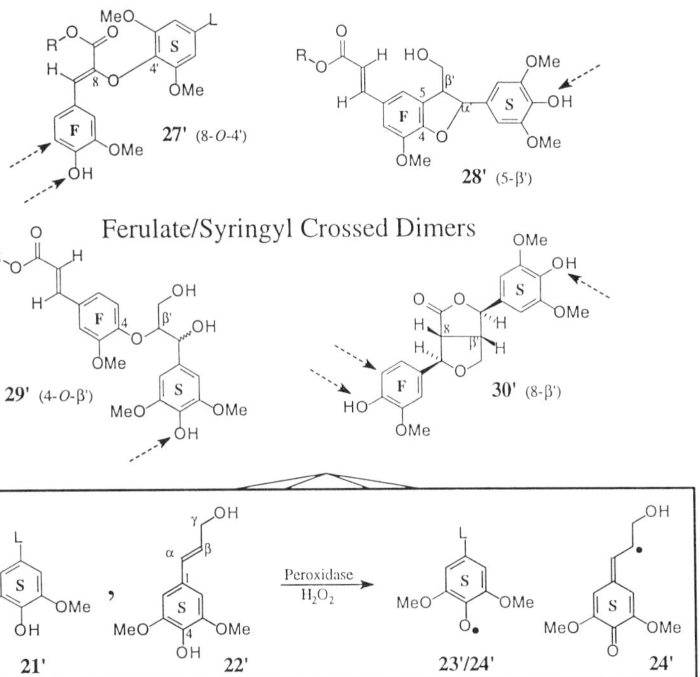

syringyl units and sinapyl alcohol

Figure 8. *Continued.*

produce a range of structures that are possible from such cross-coupling reactions, but not all of those will necessarily be produced *in planta*. The most diagnostic and unambiguous method to identify such structures is NMR spectroscopy. ^{13}C-Labeling of the C-9 (ester carbonyl carbon) of the ferulate model **1d** (Figure 9) allows ready identification of any ferulate product that is linked at the 8-position (*c.f.* β-position in lignin), *i.e.* in 8–β′ **30**, 8–5′ **32**, and 8–*O*–4′ **27** structures (see Figure 8). The ring-linked structures, 4–*O*–α′ **26**, 4–*O*–β′ **29**, and 5–β′/4–*O*–α′ **28**, are distinguishable from the three 8-linked structures but are not resolvable with this label—a 4-label would be valuable in that case but is significantly harder to introduce.

The model ferulate chosen (*49, 100*) (Figure 9) was one that mimics the ferulate attached to arabinoxylans in grasses. Thus the ferulate is attached to C-5 of an α-L-arabinofuranoside; a simple methyl group replaces the xylan chain. A DHP was prepared using *ca.* 10% of this [9-^{13}C]-labeled model **1d** and 90% coniferyl alcohol **22** (*49*). An inverse-detected long-range C–H correlation method, the HMBC experiment (*101*), was then used to correlate the labeled carbonyl carbons with protons within 3 bonds of the carbonyl. As seen in Figure 10, the (partial) HMBC spectrum is totally diagnostic, revealing that the single ferulate carbonyl resonance has produced some 4 groups of structures representing all of the expected bonding modes. Model data (*49, 102*) correspond exactly in both dimensions confirming the assignments. The 8–β′ structure **30**, in particular, conclusively illustrates that cross-coupling between ferulate (8-position) and coniferyl alcohol (β-position) has occurred—homocoupling products are symmetrical and have only one α-proton for example. This product will be highlighted below to show that similar processes indeed occur in plants. The model and DHP data provide the necessary database to search for evidence of active ferulate participation in lignification *in vivo*.

NMR Evidence for 'Active' Mechanisms in Ryegrass

Armed with the required spectroscopic data for the array of cross-linking products that are possible (as indicated by the DHP study), it is conceivable that plant materials could be probed for such structures. Unfortunately, ferulate is present at a very low level in lignins (*ca.* 2% of the lignin) and NMR spectroscopy is a notoriously insensitive technique. Still, no other method comes close to being as structurally revealing. Solution-state NMR spectroscopy is the only consideration since it alone has the necessary resolution for these types of structural studies. We determined that uniform labeling of plant material should allow detection of these compounds. The ferulate products were sufficiently resolved from other lignin/polysaccharide peaks and the two-dimensional methods, which provide up to five pieces of concurrent data for each structure, are highly diagnostic. The alternative, labeling of say the 9-carbon of ferulate and somehow incorporating this into the plant, is fraught with problems of uptake and non-natural metabolism.

Ryegrass was grown in an environment in which the CO_2 was *ca.* 15% enriched with ^{13}C (*90*). Lignins were isolated and NMR spectra, particularly the HMBC approach, proved revealing. The first obvious revelation is that the 8–β′-crossed product **30** is clearly visible in the ryegrass spectrum (Figure 11). The beautiful and characteristic correlations are unmistakable and the chances of some other unrelated compound having the same five pieces of matching NMR data is highly improbable. Thus, the mechanism of active incorporation of ferulates into plant lignins was proven. A closer look at the spectrum revealed something even more intriguing. Some of the expected (from the DHP) correlation peaks, and therefore their corresponding structures, were missing. As it turns out, these are the very structures that come *only* from coupling of ferulate with preformed lignin dimers or oligomers. Only those products that arise from coupling with a coniferyl alcohol monomer radical were present. This striking observation suggests, therefore, that ferulates are acting as nucleation sites for lignification. That is, they are the sites at which building of the lignin macromolecule starts in the wall (although we cannot say that

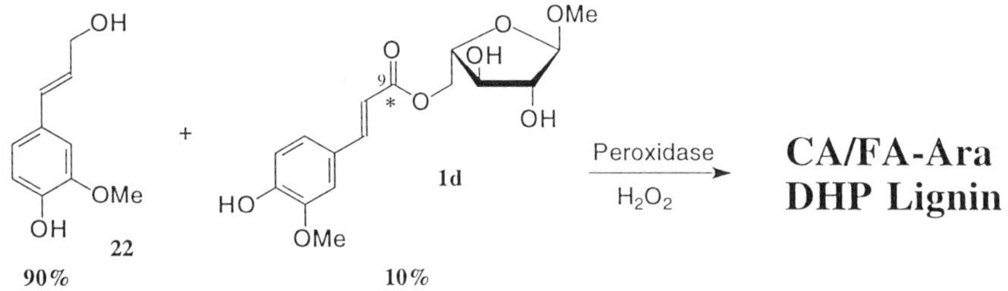

Figure 9. Preparation of FA-Ara/coniferyl alcohol copolymer DHP. The FA-Ara is strategically ^{13}C-labeled at C-9.

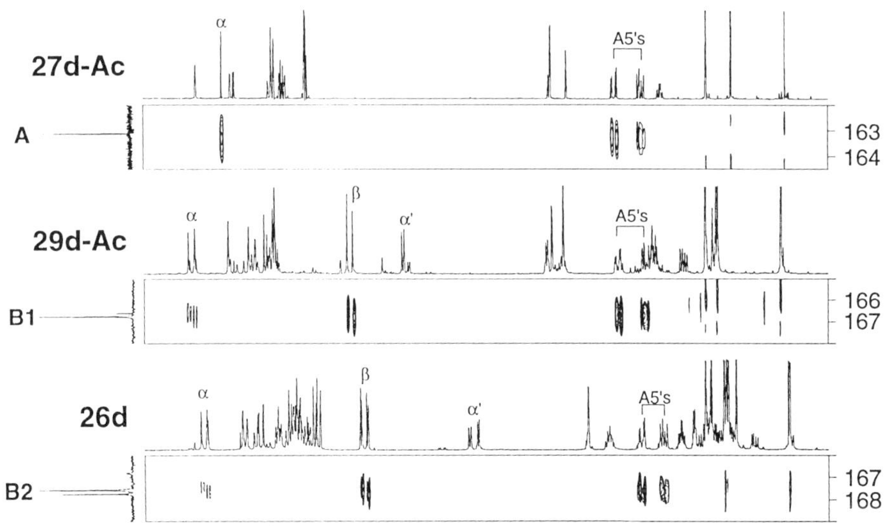

Figure 10. Portion of inverse-detected long-range 2D ^{13}C–^{1}H correlation spectrum of FA-Ara/coniferyl alcohol DHP showing just the carbonyl carbon region. The 1D carbon and proton spectra along the axes are from quantitative 1D experiments. Peak groupings are assigned to structures in Figure 8. Authentication of assignments is shown *via* similar regions of the HMBC spectra of model compounds. Regrettably, compounds **27d-Ac** and **29d-Ac** are fully acetylated and consequently do not model the parent compounds precisely, particularly with regard to the arabinosyl proton shifts. Compound **30** was not obtained in pure form; the spectrum slice shown is from a mixture of the required compound **30** along with pinoresinol and the dilactone.

Continued on next page.

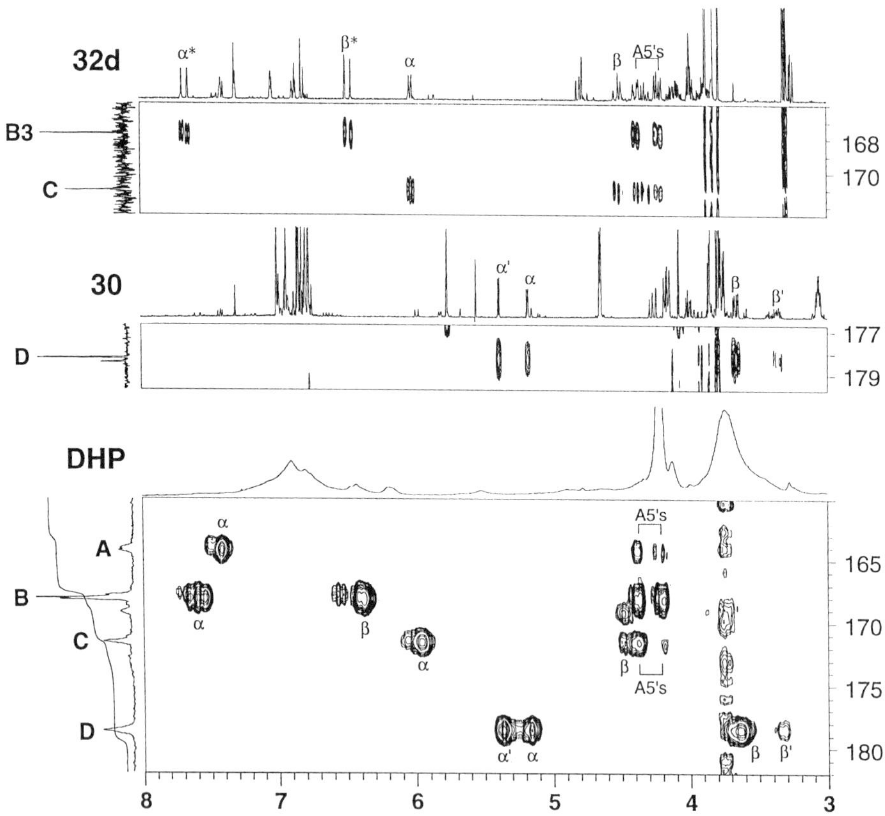

Figure 10. *Continued.*

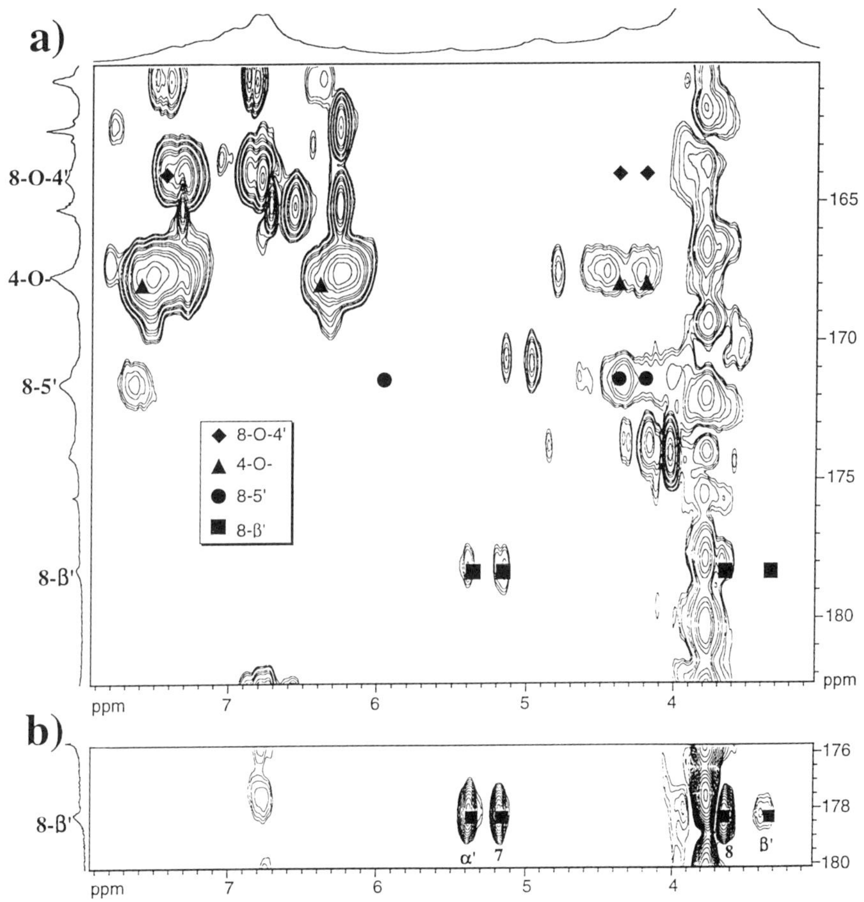

Figure 11. (a) Small section of a long-range ^{13}C–^{1}H correlation (HMBC) spectrum of uniformly ^{13}C-enriched ryegrass LPC with an 80 ms long-range correlation delay showing just the carbonyl carbon region. Processing was with gaussian apodization in t_2 as a compromise for all correlations (LB = -40, GB = 0.35). Regions are labeled corresponding to regions in a prior synthetic lignin-FA-Ara polymer (Figure 10). Overlaid are the data for similar correlations in the FA-Ara DHP. (b) The 8-β′-selection of a similar experiment run using a 110 ms long-range coupling delay which reveals the β- and 8-proton correlations more completely. Gaussian apodization was employed in t_2 (LB = -40, GB = 0.4). Numbering convention relates to that in the original ferulate (7-9 in the sidechain) and the original hydroxycinnamyl alcohol (α′, β′, γ′ in the sidechain). When describing a dimer by its linkage, *e.g.* 4–*O*–β′, the first character refers to the ferulate moiety and the primed character to the lignin moiety.

lignification must start at these sites). Similar mechanisms for lignin initiation had previously been proposed (*6, 23, 103, 104*).

All this seems to make more sense for plant growth and development. Now we have a mechanism whereby all accessible ferulate can become involved in cross-linking to lignin—there is no competition for quinone methides as in the 'passive' mechanism. Thus the plant has more direct control of this important process. Furthermore, it appears that the plant may utilize these ferulates to direct lignification to specific sites in the wall.

Further Evidence of 'Active' Mechanisms

A ferulate-coniferyl alcohol cross-coupling product **29a** released from grass straw was recently identified by Jacquet *et al.* (*105*). Such an observation already proves that radical mechanisms are in operation. Our group was humbled by not having thought to look for such a product. We had synthesized exactly these compounds earlier to provide NMR data (*91*), but had never considered that 'lignification' would finalize at the dimer stage. Since Jacquet's finding, we have confirmed that cross-coupling dimeric products are releasable from grasses. These include both isomers of **29a**, **30a** and a ring-opened analog of **30a** and an analog of **28** without the hydroxymethyl group (*106*).

Additional evidence comes from observations pertaining to the lignification of feruloylated primary cell walls. A maize suspension culture system has been described (*71, 76, 107*) that serves as an excellent model for wall lignification. It is a primary wall system that has laid down cellulose and hemicelluloses; essentially no lignin is present but the arabinoxylans are feruloylated. The extent of feruloylation can be manipulated by using the phenylalanine ammonia lyase inhibitor 2-aminoindan-2-phosphonic acid (*71*). The walls also contain a complement of peroxidases. Thus a biomimetic lignification, utilizing the walls' own peroxidases, can be effected by supplying lignin monomers and a source of hydrogen peroxide. The lignin produced in this manner is almost identical to native maize lignins as demonstrated by analytical thioacidolysis and NMR spectroscopy (*75*). It is assumed that slow diffusion of monomers into the wall containing the peroxidases is largely responsible for mimicking the natural process far better than traditional DHPs or even variants utilizing polysaccharide 'templates' and slow generation of lignin monomers (*108, 109*). It would be interesting to have similar analyses of Tanahashi and Higuchi's high molecular weight lignins produced in dialysis tubes with and without added polysaccharides (*110-113*). When the feruloylated cell walls are lignified, the amount of ferulic acid releasable by room-temperature saponification drops to as low as 1/20th of its original level (*71*). That is, up to 95% of the ferulate has become incorporated into the lignin. But only *ca.* 40% of this incorporated ferulate is releasable by high temperature base treatment. We have shown that the only structures that release ferulate are the 4–*O*–α′- and the 4–*O*–β′-ethers **26** and **29** (*71*). Thus, even if the entire released component were from the passive α-ether products **26**, the upper limit on the passive incorporation mechanism would be 40%. However, the corollary is that at least 60% is incorporated *via* the active mechanism. This mechanism would of course be expected to produce 4–*O*–β′-ethers **29** and so the passive component must be significantly lower. We are currently working on methods utilizing acetyl bromide to unambiguously distinguish 'active' from 'passive' incorporation mechanisms and quantitate the partitioning between them (*114*).

Lignin-Polysaccharide Cross-linking by Diferulates

Diferulates, no longer just the 5–5′-coupled dimer, have already cross-linked polysaccharide chains. Each of them has at least one phenolic group free. Presumably, diferulates can be incorporated into lignins *via* 'active' (and perhaps 'passive') mechanisms too (Figure 12). Grabber's work has provided significant

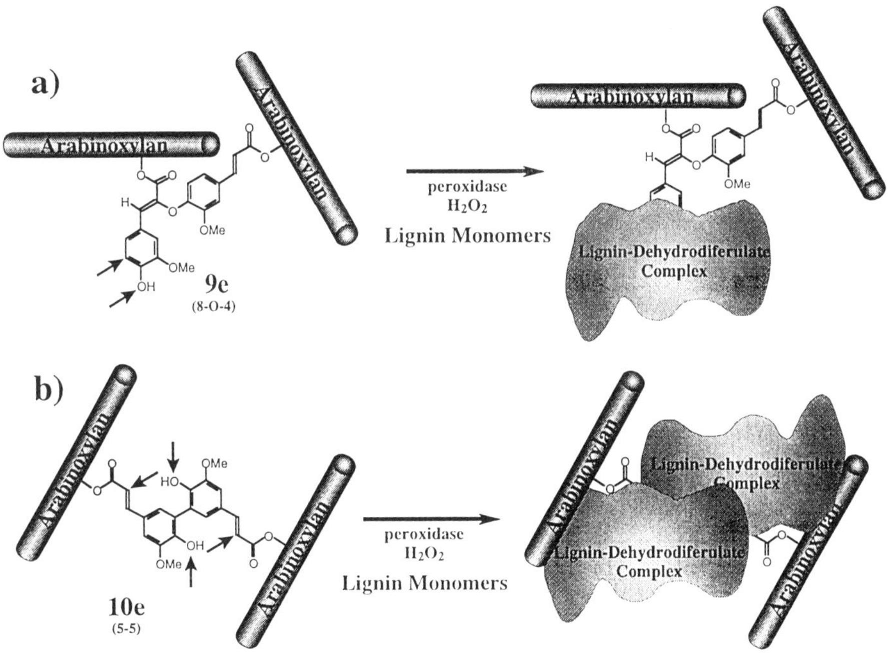

Figure 12. Ferulate dimers, already cross-linking polysaccharide chains, can also be incorporated into lignins via active mechanisms.

evidence that this is indeed the case. Just as ferulate release by room-temperature saponification was diminished by lignification, so was the release of all of the diferulates (*75*). A portion of the lignin-incorporated diferulates could be released by high temperature base treatment, the amounts depending on the particular diferulate concerned (*76, 106*).

Although not the most prevalent diferulate, the 5–5′-coupled structure has a special fascination. It is unique among the dimers of having the potential to couple with lignin (monomers) at both sidechain and phenolic positions (Figure 13). Since coupling at the sidechain 8-position still leaves the phenol remaining on that unit, there is the potential for 5–5′-coupled diferulate to couple to lignin molecule(s) at all four available positions, producing a multiple branching point and a very tightly coupled lignin-polysaccharide network (although 8–β′-coupled structures lose their connection to the polysaccharide—see structure **38**, Figure 13 and structure **30**, Figure 8). To understand how 5–5′-coupled diferulate **10** can be incorporated into lignins and, in particular, if it is possible to couple at the 8-positions, the strategically ^{13}C-labeled 5–5′-coupled diferulate **10d** was incorporated into a synthetic lignin DHP (*115*) (Figure 14). The DHP, even with the relatively modest loading of diferulate, had significantly different properties than normal DHPs. In particular, it was not soluble in 9:1 acetone-water for NMR spectroscopy. For NMR studies, the material was dissolved in DMSO-d_6 (2 parts) and diluted with acetone-d_6 (1 part) to lower the viscosity. ^{13}C-NMR showed that the single 9-resonance in the diferulate incorporated into some 4 different groupings of resonances (vertical projection on Figure 15a) as was the case for ferulate monomers (Figure 15b). Again, long-range correlations from the labeled 9-carbon(s) were totally diagnostic (Figure 15a) particularly when compared to the corresponding ferulate monomer experiment (Figure 15b). Although the difference in solvent caused some significant proton shifts, it is clear that 8–*O*–4′, 4–*O*–x′/5–x′, 5–β′, and 8–β′ coupling has occurred. In fact, the incorporation profile is remarkably similar to that of ferulate. As with ferulate, the diferulate obviously reacts with lignin monomers to give 8–β′ and 4–*O*–β′ structures (lignin monomers invariably react at the β position in cross-coupling reactions) and dimers/oligomers to give 8–*O*–4′ and 8–5′ structures. Clearly, then, 5–5′-coupled diferulate is capable of radical coupling reactions with lignin monomers and dimers to give the range of expected structures. How extensively each molecule is cross-coupled (*i.e.* how many of the four possible coupling sites on the dimer are occupied) is not possible to determine.

One unique feature of incorporation of the 5–5′-coupled dimer into lignins is the ability to form dibenzodioxocins **35** (*115*). Such 8-membered ring structures have recently been identified in lignin itself and are strikingly predominant (*98, 99*). Normally, incorporation of each of the diferulate's ferulate moiety into lignin would be considered to be two independent events. This unique structure allows a sequential incorporation of both moieties (Figure 14). If one or both moieties couple first at the 8-position, we assume that NMR spectroscopic data from existing models would be appropriate. However, if the ferulate dimer itself reacts directly at the β position (and cyclizes at α) of a coniferyl alcohol radical, the product is likely to exhibit significantly different NMR spectroscopic data. In an effort to see if such diferulate structures **35** are likely to make important contributions, models were prepared to obtain the NMR data required to find them in the DHP (*115*). The most direct way to synthesize these models was *via* radical-coupling methods. Thus, the 5–5′-diferulate **10b** or **10c** was reacted with an excess of coniferyl alcohol using Ag_2O as a single-electron oxidant. The purified yields of the required products were low (*ca.* 15-20% based on diferulate **10**) but the single-step preparation was significantly easier than the multi-step approach that did not use a radical coupling step. The data for these compounds agreed closely with those reported from similar compounds by Karhunen *et al.* (*99*). The spectra were fully consistent with the dibenzodioxocin structures; the HMQC, HMBC, and HMQC-TOCSY experiments were the most diagnostic. In particular, in the long-range ^{13}C-^{1}H correlation experiment (HMBC, not shown), Hα correlated nicely with an aromatic 4–*O*-carbon. This three-bond coupling correlation establishes that the α–*O*–4′ bond has formed.

At first the data from models dissolved in acetone and the DHP, which was in 2:1 DMSO-acetone, seemed to have little in common. However, the solvent shift afforded by the DMSO was remarkable. When models were dissolved in 2:1 DMSO-acetone, the shifts coincided directly with the corresponding shifts in the polymeric DHP (*115*). As seen in Figure 16, the dibenzodioxocin structure is clearly identified in both the HMQC and, more diagnostically, in the HMQC-TOCSY. The HMBC spectrum (not shown) is also consistent; Hα correlates with Hβ, A1, A2, A6, and C4.

Advances in NMR Spectroscopy for Detailing Plant Chemistry

As an aside, the NMR studies described in this chapter were all performed on what is now considered older generation instrumentation that is not fully digital, and without the use of pulsed field gradients. It was also conducted at 360 MHz, a quite adequate field strength but one far below the 600, 750 and now even 800 MHz instruments that are available. The differences in spectra obtained on these two generations of machines is quite spectacular (Figure 17). Even ignoring the obvious dispersion gain from the 750 MHz instrument, the freedom from T_1-noise artifacts in the digital/gradient system (bottom spectrum) is striking. Peaks close to the methoxyl region are no longer ambiguous. In addition the flat baseplanes and improved sensitivity allow us to look more closely down to the noise level to reveal many more potentially valuable correlations. The only drawback is that we now have a great many more assignments to make in this spectrum! The expectation is that a great many more questions will succumb to unambiguous answers *via* spectra such as these.

Conclusions

In past primary publications on ferulates and their cross-linking with other ferulates and with lignin, two misconceptions have been perpetuated. First, the only dehydrodiferulate reported was the 5–5′-coupled dimer. It is certainly not the major dehydrodimer in any plant sample recently analyzed but, fortuitously, may be one of the more important in effecting tight lignin-polysaccharide cross-linking due to the number of coupling sites available to it. Second, it has frequently been reported that the only cross-linking structure to lignin is an α-ether or benzyl aryl ether. Such ethers can be produced by the attack of the phenolic hydroxyl (in ferulate/diferulate) on lignin quinone methides, which result from the coupling of a lignin monomer with another monomer or a lignin oligomer. Adherence solely to such a mechanism has a number of significant drawbacks. It does not permit ferulate/diferulate access to radical mechanisms that are going on all around it and producing the very quinone methides with which they must react. In the case of the 5–5′-coupled diferulate, nucleophilic attack by the sterically hindered 4-phenol is likely to be a particularly poor reaction. In practical terms, it has the unfortunate consequence that quantitation of ferulates and diferulates (by their hydrolytic release) is assumed to reflect their involvement in cross-linking. As is hopefully abundantly clear from this chapter, ferulates and diferulates readily undergo radical coupling reactions and cross-react with lignin monomers by radical coupling mechanisms; whether they cross-couple with lignin oligomers *in vivo* has yet to be investigated—certainly in ryegrass there is evidence only for coupling with lignin monomers suggesting a nucleation-site role for ferulates and diferulates in lignification (*90*). Since some of the structures that ferulates/diferulates are involved in cannot be hydrolytically cleaved, current 'quantitation' methods significantly under-estimate the importance of these species. What is clear is that ferulates and diferulates are more abundant components in grass cell walls than previously recognized and have powerful roles in effecting polysaccharide-polysaccharide and lignin-polysaccharide cross-linking. The 5–5′-

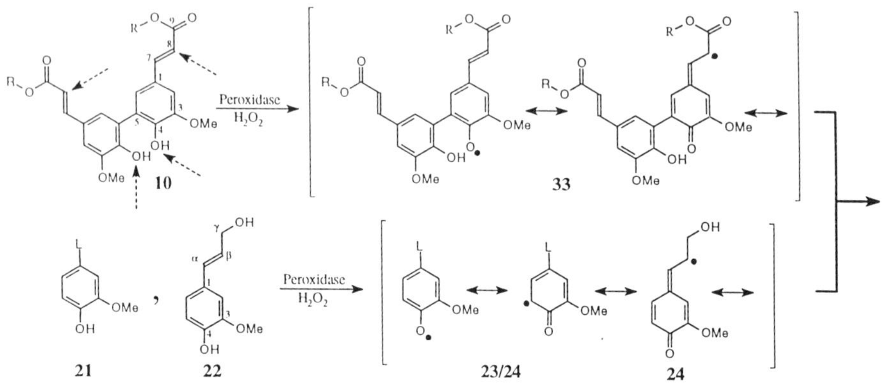

Figure 13. Peroxidase/H_2O_2 produces radical **33** from diferulate **10**, and radicals **23/24** from coniferyl alcohol **22** and pre-formed lignin dimers or oligomers **21**. The diferulate radical **33** cross-couples with a coniferyl alcohol radical, the latter invariably reacting at the β-position; initial 8–*O*–4′-coupling leads to dibenzodioxocin **35**, 8–β-coupling to the furofuranoid 'monoepoxylignanolide' (*102, 116*) **38**. Note that 8–β-coupling results in a loss of connectivity with the polysaccharide moiety, (structure **38**). Cross-coupling with a lignin oligomer can afford products from 8-coupling of the ferulate moiety; 8–5′-coupling leads to phenylcoumaran **40**, and 8–*O*–4′-coupling to styryl aryl ether **42**. Dashed arrows on the diferulate **10** and final products **35**, **38**, **40** and **42**, identify sites at which radical coupling reactions are possible.

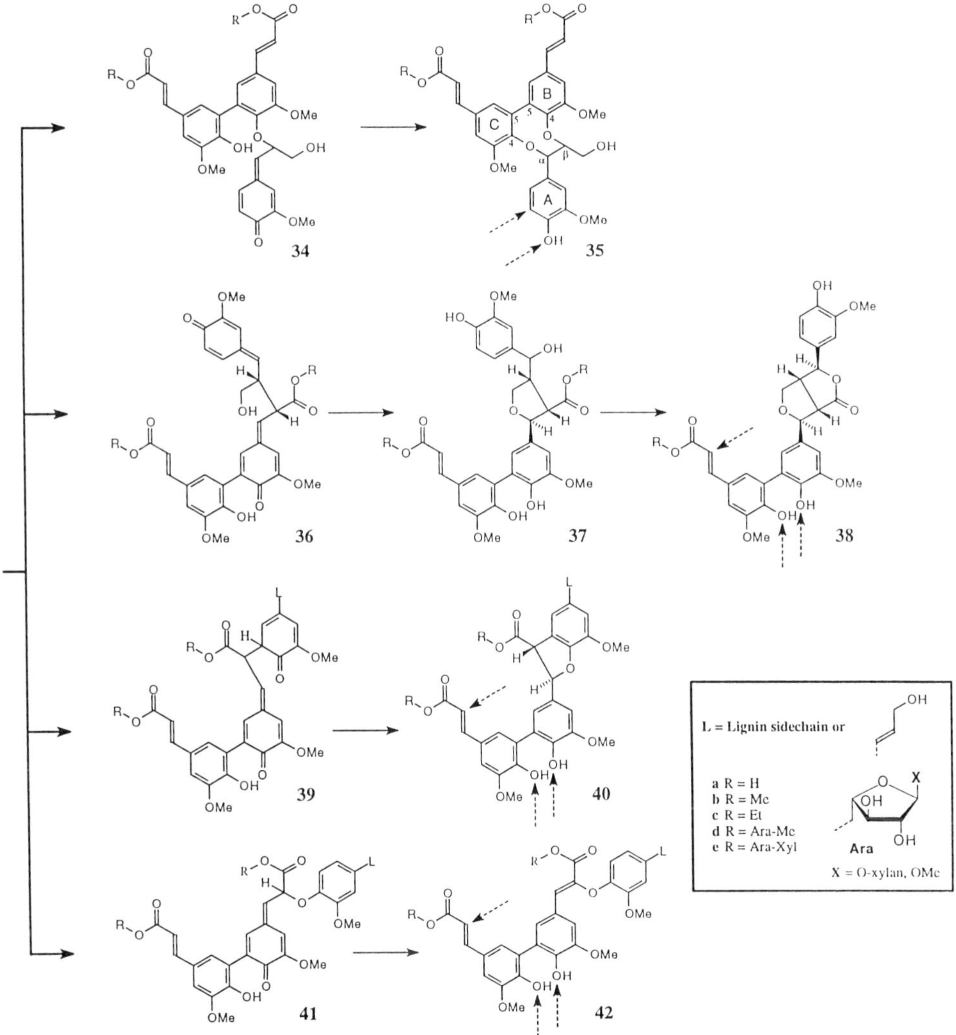

Figure 13. *Continued.*

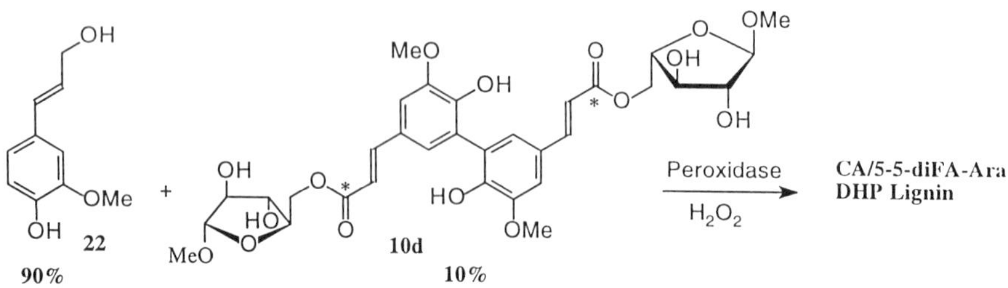

Figure 14. Preparation of 5,5′-diFA-Ara/coniferyl alcohol copolymer DHP. The diFA-Ara is strategically ^{13}C-labeled.

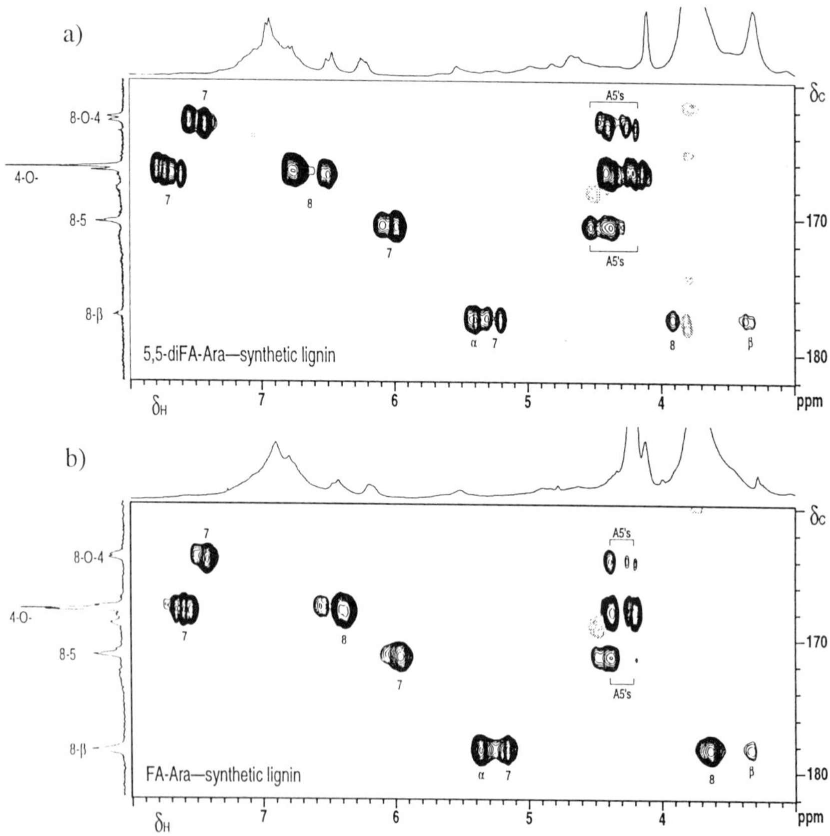

Figure 15. Partial HMBC spectra (carbonyl carbon region only) showing correlations of C-9 with protons within 3-bonds. The concurrent matching of several (up to 5) carbon and proton chemical shift data makes the assignments essentially unambiguous. The 5,5′-diFA-Ara DHP has some solvent-induced shifts caused by the DMSO. Nevertheless, as in the ferulate, clear evidence is seen for 8–*O*–4′, 4–*O*–x′, 8–5′ and 8–β′ cross-coupling structures. In fact, the incorporation profiles are remarkably similar.

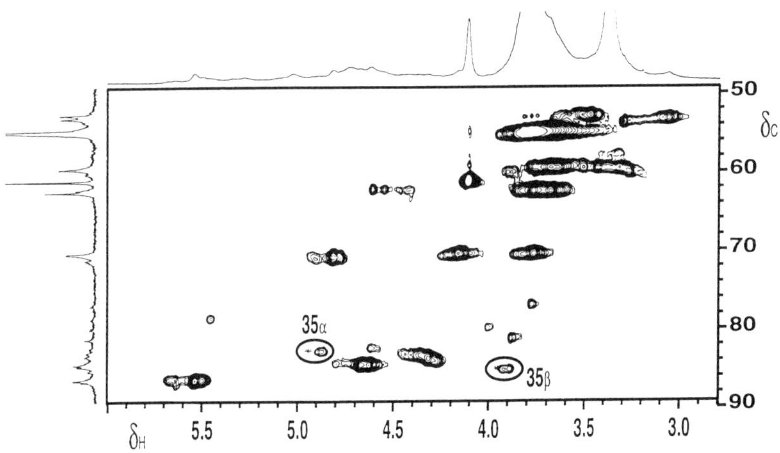

Figure 16. Partial (sidechain region) HMQC spectrum showing clear evidence for the dibenzodioxocin structures **35**. Correlations from models **35b/c** are shown as x's. HMBC and HMQC-TOCSY spectra (not shown) provided further confirmatory evidence by correctly correlating other resonances.

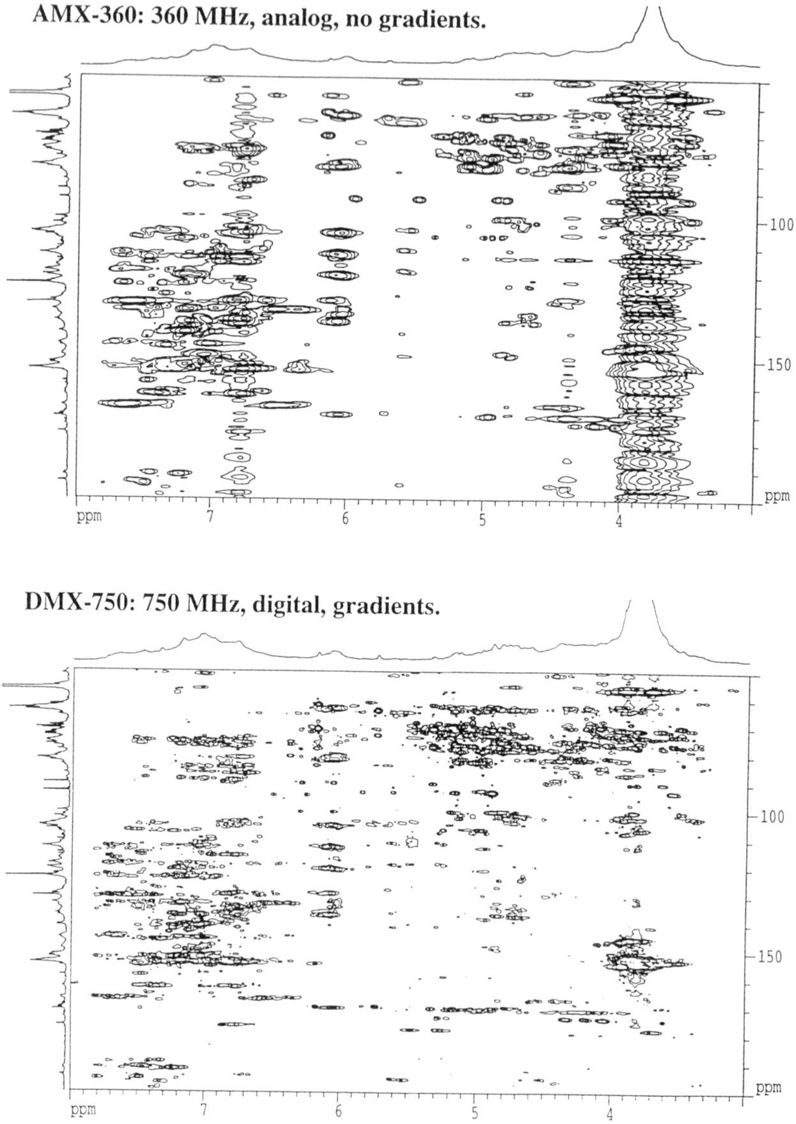

Figure 17. Comparison of HMBC spectra of labeled ryegrass lignins at different field strengths with and without gradients and digital acquisition. Top spectrum: AMX-360, non-digital, without gradients, 300 mg, 60 h. Bottom spectrum: DMX-750, fully digital, with gradients, 122 mg, 24 h. Note particularly the absence of so-called T_1-noise artifacts obfuscating the methoxyl region in the top spectrum (*ca.* 4 ppm on the proton scale), and the richness of detectable peaks due to the ability of approaching the flat base-plane more closely (bottom spectrum), and the enhanced dispersion resulting from the MHz difference.

dimer is particularly suited to this role, and the demonstration of the ability of such structures to form dibenzodioxocins during lignification is significant

Ferulate-mediated cross-linking of cell wall components is likely to be found to be increasingly important as more aspects of plant growth and development are related to this process. Although the chemistry is now becoming more clearly defined, a great deal remains unknown about the biochemistry, genetics, and regulation/control of these processes (*1*). With respect to genetic manipulation in grass plants for improved cell wall degradability, we suggest that current approaches based on lignin manipulation are likely to be less efficacious than an attack on the very mechanism that strongly limits degradability, *i.e.* lignin-polysaccharide cross-linking effected by ferulates. That is not to say that such approaches will be successful. This cross-linking process appears to be rather essential to the plant and it is conceivable that ferulate levels cannot be fully down-regulated without a severe impact on plant viability. However, woody plants make their cell walls without cross-linking mechanisms mediated by ferulate. The exploitation of genetic options for modifying ferulate-mediated cross-linking could prove to be extremely useful.

Acknowledgments

The authors are grateful to Fachuang Lu, Sally A. Ralph, and Lawrence L. Landucci for valuable discussions, and to Gundolf Wende for help with the ferulate-polysaccharide section. NMR studies at 750 MHz (Figures 16, 17) were carried out at the National Magnetic Resonance Facility at Madison (operation subsidized by the NIH Biomedical Research Technology Program under grant RR02301; equipment funded by the University of Wisconsin, NSF Academic Infrastructure Program under grant BIR-9214394, the NIH Shared Instrumentation Program under grants RR02781 and RR08438, the NIH Biomedical Research Technology Program under NIH grant RR02301, the NSF Biological Instrumentation Program under grant DMB-8415048, and the U.S. Department of Agriculture). We also gratefully acknowledge support through the USDA-National Research Initiative competitive grants, #90-27261-5617, #92-37304-8057, #94-02764 and #96-02587 in the Plant Growth and Development program.

Literature Cited

1. Bolwell, G. P. *Int. Rev. Cytol.* **1993**, *146*, 261-324.
2. Monties, B. L. In *Methods in Plant Biochemistry*; Harborne, J., Ed.; Academic Press: London, 1989; Vol. 1, pp 113-157.
3. Taiz, L. *Ann. Rev. Plant Physiol.* **1984**, *35*, 585.
4. Fry, S. C. *Phytochemistry* **1984**, *23*, 59-64.
5. Fry, S. C.; Miller, J. C. *ACS Symp. Ser.* **1989**, *399*, 33-46.
6. Yamamoto, E.; Bokelman, G. H.; Lewis, N. G. *ACS Symp. Ser.* **1989**, *399*, 68-88.
7. Hartley, R. D.; Ford, C. W. *ACS Symp. Ser.* **1989**, *399*, 137-145.
8. Jung, H. G.; Ralph, J. In *Microbial and Plant Opportunities to Improve Lignocellulose Utilization by Ruminants*; Akin, D. E., Ljungdahl, L. G., Wilson, J. R., Harris, P. J., Eds.; Elsevier: New York, NY, 1990; pp 173-182.
9. Ralph, J.; Helm, R. F. In *Forage Cell Wall Structure and Digestibility*; Jung, H. G., Buxton, D. R., Hatfield, R. D., Ralph, J., Eds.; ASA-CSSA-SSSA: Madison, WI, 1993; pp 201-246.
10. Bolwell, G. P. *Phytochemistry* **1988**, *27*, 1235-1253.
11. *211th ACS Natl. Mtg. Book of Abstracts* **1996**, *1*, abstr. Cell 51-57, 73-79
12. Monties, B. In *Proc. 6th Internatl. Symp. Wood Pulp. Chem.* **1991**, *1*, 113-121.
13. Kato, A.; Azuma, J.; Koshijima, T. *Chem. Lett.* **1983**, 137-140.
14. Kato, Y.; Nevins, D. J. *Carbohydr. Res.* **1985**, *137*, 139-150.
15. Kato, A.; Azuma, J.; Koshijima, T. *Agric. Biol. Chem.* **1987**, *51*, 1691-1693.

16. Smith, M. M.; Hartley, R. D. *Carbohydr. Res.* **1983**, *118*, 65-80.
17. Ishii, T.; Hiroi, T. *Carbohydr. Res.* **1990**, *196*, 175-183.
18. Ishii, T. *Phytochemistry* **1991**, *30*, 2317-2320.
19. Hartley, R. D.; Morrison, W. H., III; Himmelsbach, D. S.; Borneman, W. S. *Phytochemistry* **1990**, *29*, 3705-3709.
20. Mueller-Harvey, I.; Hartley, R. D.; Harris, P. J.; Curzon, E. H. *Carbohydr. Res.* **1986**, *148*, 71-85.
21. Chesson, A.; Gordon, A. H.; Lomax, J. A. *J. Sci. Food Agric.* **1983**, *34*, 1330-1340.
22. Nishitani, K.; Nevins, D. J. *Plant Physiol.* **1989**, *91*, 242-248.
23. Jung, H. G.; Deetz, D. A. In *Forage Cell Wall Structure and Digestibility*; Jung, H. G., Buxton, D. R., Hatfield, R. D., Ralph, J., Eds.; ASA-CSSA-SSSA: Madison, WI, 1993, pp 315-346.
24. Bacic, A.; Harris, P. J.; Stone, B. A. In *The Biochemistry of Plants*; Academic Press Inc.: New York, NY, 1988; Vol. 14, pp 297-371.
25. Ishii, T.; Hiroi, T.; Thomas, J. R. *Phytochemistry* **1990**, *29*, 1999-2003.
26. Ishii, T.; Hiroi, T. *Carbohydr. Res.* **1990**, *206*, 297-310.
27. Wende, G.; Fry, S. C. *Phytochemistry* **1997**, *44*, 1011-1018.
28. Wende, G.; Fry, S. C. *Phytochemistry* **1997**, *44*, 1019-1030.
29. Saulnier, L.; Vigouroux, J.; Thibault, J.-F. *Carbohydr. Res.* **1995**, *272*, 241-253.
30. Himmelsbach, D. S.; Hartley, R. D.; Bornemann, W. S.; Poppe, L.; van Halbeek, H. *Magn. Reson. Chem.* **1994**, *32*, 158-165.
31. Ishii, T. *Plant Cell Physiol.* **1994**, *35*, 701-704.
32. Wende, G.; Fry, S. C. *Phytochemistry*, **1997**, *45*, 1123-1129.
33. Fry, S. C. *Biochem. J.* **1982**, *203*, 493-504.
34. Ford, C. W.; Hartley, R. D. *J. Sci. Food Agric.* **1990**, *50*, 29-43.
35. Hartley, R. D.; Morrison, W. H., III; Balza, F.; Towers, G. H. N. *Phytochemistry* **1990**, *29*, 3699-703.
36. Hartley, R. D.; Morrison, W. H., III. *J. Sci. Food Agric.* **1991**, *55*, 365-375.
37. Hartley, R. D.; Whatley, F. R.; Harris, P. J. *Phytochemistry* **1988**, *27*, 349-351.
38. Ford, C. W.; Hartley, R. D. *J. Sci. Food Agric.* **1989**, *46*, 301-310.
39. Markwalder, H. U.; Neukom, H. *Phytochemistry* **1976**, *15*, 836-7.
40. Hartley, R. D.; Jones, E. C. *Phytochemistry* **1976**, *15*, 1157-1160.
41. Hartley, R. D.; Jones, E. C. *Phytochemistry* **1977**, *16*, 1531-1534.
42. Harris, P. J.; Hartley, R. D.; Lowry, K. H. *J. Sci. Food Agric.* **1980**, *31*, 959-62.
43. Kamisaka, S.; Takeda, S.; Takahashi, K.; Shibata, K. *Physiol. Plant.* **1990**, *78*, 1-7.
44. Tan, K.-S.; Hoson, T.; Masuda, Y.; Kamisaka, S. *Physiol. Plant.* **1991**, *83*, 397-403.
45. Tan, K.-S.; Hoson, T.; Masuda, Y.; Kamisaka, S. *Plant Cell Physiol.* **1992**, *33*, 103-108.
46. Eraso, F.; Hartley, R. D. *J. Sci. Food Agric.* **1990**, *51*, 163-70.
47. Ishii, T. *Carbohydr. Res.* **1991**, *219*, 15-22.
48. Ralph, J.; Quideau, S.; Grabber, J. H.; Hatfield, R. D. *J. Chem. Soc., Perkin Trans. 1* **1994**, 3485-3498.
49. Ralph, J.; Helm, R. F.; Quideau, S.; Hatfield, R. D. *J. Chem. Soc., Perkin Trans. 1* **1992**, 2961-2969.
50. Teutonico, R. A.; Dudley, M. W.; Orr, J. D.; Lynn, D. G.; Binns, A. N. *Plant Physiol.* **1991**, *97*, 288-297.
51. Chioccara, F.; Poli, S.; Rindone, B.; Pilati, T.; Brunow, G.; Pietikäinen, P.; Setä lä , H. *Acta Chem. Scand.* **1993**, *47*, 610-616.
52. Wallace, G.; Fry, S. C. *Phytochemistry* **1995**, *39*, 1293-9.
53. Harkin, J. M. In *Oxidative Coupling of Phenols*; Taylor, W. I., Battersby, A. R., Eds.; Marcel Dekker: New York, NY, 1967, pp 243-321.
54. Harkin, J. M. In *Chemistry and Biochemistry of Herbage*; Butler, G. W., Ed.; Academic Press: London, 1973; Vol. 1, pp 323-373.
55. Freudenberg, K. *Nature* **1959**, *183*, 1152-1155.

56. Davin, L. B.; Lewis, N. G. In *Phenolic Metabolism in Plants*; Stafford, H. A., Ibrahim, R. K., Eds.; Plenum Press: New York, NY, 1992; pp 325-375.
57. Quideau, S.; Ralph, J. *Holzforschung* **1994**, *48*, 12-22.
58. Landucci, L. L. *J. Wood Chem. Technol.* **1995**, *15*, 349-368.
59. van Huystee, R. B.; Zheng, X. *Phytochemistry* **1993**, *34*, 933-939.
60. Stewart, D.; Robertson, G. W.; Morrison, I. M. *Biological Mass Spectrometry* **1994**, *23*, 71-74.
61. Waldron, K. W.; Parr, A. J.; Ng, A.; Ralph, J. *Phytochemical Analysis* **1996**, *7*, 305-312.
62. Cohen, M. D.; Schmidt, G. M. J.; Sonntag, F. I. *J. Chem. Soc.* **1964**, 2000-2013.
63. Parr, A. J.; Waldron, K. W.; Ng, A.; Parker, M. L. *J. Sci. Food Agric.* **1996**, *71*, 501-507.
64. Waldron, K. W. *Biotechnology and Biological Sciences Research Council Business* **1996**, *April*, 10-11.
65. Sánchez, M.; Peña, M. J.; Revilla, G.; Zarra, I. *Plant Physiol.* **1996**, *111*, 941-946.
66. Shibuya, N. *Phytochemistry* **1984**, *23*, 2233-7.
67. Kamisaka, S.; Takeda, S.; Takahashi, K.; Shibata, K. *Physiol. Plant.* **1990**, *78*, 1-7.
68. Wende, G. **1997**, unpublished.
69. Micard, V.; Grabber, J. H.; Ralph, J.; Renard, C. M. G. C.; Thibault, J.-F. *Phytochemistry* **1997**, *44*, 1365-1368.
70. Oosterveld, A.; Grabber, J. H.; Beldman, G.; Ralph, J.; Voragen, A. G. J. *Carbohydr. Res.* **1997**, *300,* 179-182.
71. Grabber, J. H.; Hatfield, R. D.; Ralph, J.; Zon, J.; Amrhein, N. *Phytochemistry* **1995**, *40*, 1077-1082.
72. Jung, H. G.; Valdez, F. R.; Abad, A. R.; Blanchette, R. A.; Hatfield, R. D. *J. Anim. Sci.* **1992**, *70*, 1928-1935.
73. Jung, H. G.; Valdez, F. R.; Hatfield, R. D.; Blanchette, R. A. *J. Sci. Food Agric.* **1992**, *58*, 347-355.
74. Jung, H. G.; Ralph, J.; Hatfield, R. D. *J. Sci. Food Agric.* **1991**, *56*, 469-478.
75. Grabber, J. H.; Ralph, J.; Hatfield, R. D.; Quideau, S.; Kuster, T.; Pell, A. N. *J. Ag. Food Chem.* **1996**, *44*, 1453-1459.
76. Grabber, J. H.; Ralph, J.; Hatfield, R. D. *ACS Symp. Ser.* **1997**, see *Chapter 12*, this volume.
77. Grabber, J. H.; Ralph, J.; Hatfield, R. D., in preparation for *J. Sci. Food Agric*.
78. Iiyama, K.; Lam, T. B. T.; Stone, B. A. *Phytochemistry* **1990**, *29*, 733-737.
79. Iiyama, K.; Kasuya, N.; Lam, T. B. T.; Stone, B. A. *J. Sci. Food Agric.* **1991**, *56*, 551-560.
80. Lam, T. B. T.; Iiyama, K.; Stone, B. A. *Phytochemistry* **1992**, *31*, 1179-1183.
81. Lam, T. B. T.; Iiyama, K.; Stone, B. A. *Phytochemistry* **1992**, *31*, 2655-2658.
82. Jung, H. G.; Vogel, K. P. *J. Sci. Food Agric.* **1992**, *59*, 169-176.
83. Jung, H.-J. G.; Buxton, D. R. *J. Sci. Food Agric.* **1994**, *66*, 313-322.
84. Deetz, D. A.; Jung, H.-J. G.; Buxton, D. R. *Crop Sci.* **1996**, *36*, 383-388.
85. Scalbert, A.; Monties, B.; Lallemand, J. Y.; Guittet, E.; Rolando, C. *Phytochemistry* **1985**, *24*, 1359-1362.
86. Scalbert, A.; Monties, B.; Rolando, C.; Sierra-Escudero, A. *Holzforschung* **1986**, *40*, 191-195.
87. Sharma, A.; Brillouet, J.-M.; Scalbert, A.; Monties, B. *Agronomie* **1986**, *6*, 265-271.
88. Lam, T. B. T.; Iiyama, K.; Stone, B. In *1991 International Symposium on Forage Cell Wall Structure and Digestibility;* Madison, WI, 1991; p A4.
89. Lam, T. B. T.; Iiyama, K.; Stone, B. *Proc. 6th Internatl. Symp. Wood Pulp. Chem.* **1991**, *2*, 29-33.
90. Ralph, J.; Grabber, J. H.; Hatfield, R. D. *Carbohydr. Res.* **1995**, *275*, 167-178.
91. Helm, R. F.; Ralph, J. *J. Agric. Food Chem.* **1992**, *40*, 2167-2175.
92. Sipilä , J.; Brunow, G. *Holzforschung* **1991**, *45*, 9-14.

93. Sipilä , J.; Brunow, G. *Holzforschung* **1991**, *45*, 275-278.
94. Sipilä , J.; Brunow, G. *Holzforschung* **1991**, *45*, 3-7.
95. Adler, E. *Wood Sci. Technol.* **1977**, *11*, 169-218.
96. Ede, R. M.; Kilpeläinen, I. *Res. Chem. Intermediates* **1995**, *21*, 313-328.
97. Ede, R. M. In *Proc. 8th Internatl. Symp. Wood Pulp. Chem.* **1995**, *1*, 487-494.
98. Karhunen, P.; Rummakko, P.; Sipilä, J.; Brunow, G.; Kilpeläinen, I. *Tetrahedron Lett.* **1995**, *36*, 169-170.
99. Karhunen, P.; Rummakko, P.; Sipilä, J.; Brunow, G.; Kilpeläinen, I. *Tetrahedron. Lett.* **1995**, *36*, 4501-4504.
100. Hatfield, R. D.; Helm, R. F.; Ralph, J. *Anal. Biochem.* **1991**, *194*, 25-33.
101. Bax, A.; Summers, M. F. *J. Am. Chem. Soc.* **1986**, *108*, 2093-2094.
102. Ralph, J.; Helm, R. F.; Quideau, S. *J. Chem. Soc., Perkin Trans. 1,* **1992**, 2971-2980.
103. Delmer, D. P.; Stone, B. A. In *The Biochemistry of Plants*; Academic Press Inc.: New York, NY, 1988; Vol. 14, pp 373-420.
104. Fry, S. C. In *Oxford Surveys of Plant Molecular and Cell Biology*; Milflin, B. J., Ed.; Oxford Univ. Press: New York, NY, 1985; Vol. 2, pp 1-42.
105. Jacquet, G.; Pollet, B.; Lapierre, C.; Mhamdi, F.; Rolando, C. *J. Agric. Food Chem.* **1995**, *43*, 2746-51.
106. Grabber, J. H.; Ralph, J., in preparation for *Phytochemistry*.
107. Grabber, J. H.; Ralph, J.; Hatfield, R. D.; Quideau, S.; Kuster, T. A. *211th ACS Natl. Mtg. Book of Abstracts* **1996**, *1*, abstr. Cell 20.
108. Terashima, N.; Atalla, R. H.; Ralph, S. A.; Landucci, L. L.; Lapierre, C.; Monties, B. *Holzforschung* **1996**, *50*, 9-14.
109. Terashima, N.; Atalla, R. H.; Ralph, S. A.; Landucci, L. L.; Lapierre, C.; Monties, B. *Holzforschung* **1995**, *49*, 521-527.
110. Tanahashi, M.; Higuchi, T. *Mokuzai Gakkaishi* **1990**, *36*, 424-428.
111. Tanahashi, M.; Aoki, T.; Higuchi, T. *Mokuzai gakkaishi* **1981**, *27*, 116-124.
112. Tanahashi, M.; Higuchi, T. *Wood Res.* **1981**, *67*, 29-42.
113. Tanahashi, M.; Aoki, T.; Higuchi, T. *Holzforschung* **1982**, *36*, 117-122.
114. Lu, F.; Ralph, J. *ACS Symp. Ser.* **1997**, see *Chapter 20,* this volume.
115. Quideau, S.; Ralph, J., submitted to *J. Chem. Soc., Perkin Trans. 1*.
116. Weinges, K.; Nader, F.; Künstler, K. In *Chemistry of Lignans*; Rao, C. B. S., Ed.; Andhra University Press: Visakhapatnam, India, 1978; pp 1-37.

Chapter 17

Improvement in NMR Structural Studies of Lignin Through Two- and Three-Dimensional NMR Detection and Isotopic Enrichment

D. Robert[1], E. Ämmälahti[2], M. Bardet[1], G. Brunow[3], I. Kilpeläinen[2], K. Lundquist[4], V. Neirinck[1], and N. Terashima[5]

[1]DRF/SCIB CEA, 17 rue des Martrs 38054 Grenoble Cedex, France
[2]Institute of Biotechnology, University of Helsinki, P.O. Box 45, FIN-00014, Helsinki, Finland
[3]Department of Chemistry, University of Helsinki, P.O. Box 6, FIN-00014, Helsinki, Finland
[4]Chalmers University of Technology, S-41296 Göteborg, Sweden
[5]Faculty of Agriculture, Nagoya University, Nagoya 464, Japan

2D and 3D NMR spectra of uniformly ^{13}C enriched lignin and specifically labeled (^{13}C and ^{199}Hg) lignin improve substantially the prospects for lignin structural analysis. The 2D INADEQUATE ^{13}C NMR spectrum of ^{13}C enriched aspen milled wood lignin allows observation, within the polymeric lignin structure, of chemically distinct β–*O*–4′ diastereoisomers. In this same sample, using 2D-HMQC, 2D-HOHAHA and 3D-HMQC/HOHAHA pulse sequences, one can observe minor but important structures such as β–1′, α–*O*–4′ and eight membered ring dibenzodioxocin units, through ^{1}H-^{1}H, ^{1}H-^{13}C, and ^{1}H-^{13}C-^{13}C-^{1}H correlations. It was shown that selective ^{13}C labeling of the propyl side chain of hardwood and softwood lignin, through exogenous administration of labeled coniferin, does not appear to perturb lignin biosynthesis and therefore provides a convenient means for putting an NMR probe into effect. Mercuration of lignin provides selective labeling of the aromatic ring which can be correlated with the lignin condensation pattern. ^{1}H-^{199}Hg HMQC NMR experiments allow assignments to be made of the mercurated sites in model compounds.

Classical one-dimensional ^{13}C NMR spectroscopy is a very efficient tool for lignin structural analysis. Nevertheless it suffers from disadvantages such as important signal overlaps in many spectra, as well as indirect signal assignments, which rely on comparison with model compounds. In attempts to circumvent these problems, we employed various 2D and 3D NMR pulse sequences on ^{13}C enriched or selectively ^{13}C and ^{199}Hg labeled lignin samples. The multidimensional NMR techniques increase by polarization transfer the sensitivity of nuclei with low sensitivity, and the large range of the carbon chemical shifts in lignin makes ^{1}H-^{13}C and ^{13}C-^{13}C correlation methods efficient for structural analysis. ^{13}C enrichment provides signal enhancement which allow detection of very minor signals and make possible the use of NMR pulse sequences which would otherwise be too time consuming as far as data acquisition is concerned. In the case of selective labeling, the isotope serves as a structural probe and allows direct structural assignments to be made.

In the present work we selected an uniformly ^{13}C enriched aspen milled wood lignin (MWL) sample, as well as softwood and hardwood MWL samples selectively labeled either on the aliphatic chain or the aromatic ring with ^{13}C and ^{199}Hg, respectively. We report the structural information gained from using the 1D, 2D and 3D NMR pulse sequences and indicate how they are useful for stereochemical and biosynthetic studies of lignin and for observing condensed units.

NMR Spectroscopic Detection of Stereoisomers in Lignin Structures

Despite the presence of asymmetric carbons in the aliphatic sidechains, no optical activity has ever been observed in lignin structures and so far lignin has been thought to be completely racemic. This lack of optical activity makes it nearly unique among biopolymers and supports the argument that there is little direct enzymatic control over its biosynthesis, which very likely occurs through the enzyme initiated radical polymerization of cinnamyl alcohols. One of the most important and well identified structures in lignin is the arylglycerol-β-aryl structure, the so called β–*O*–4′ linkage. Due to the presence of two asymmetric carbons on the propyl sidechain, diastereoisomers are to be found in lignin. Indeed they have been identified among lignin degradation products (*1*).

In lignin itself β–*O*–4′ diastereoisomers and *erythro*/*threo* ratios have been observed by ^{1}H (*2-4*), ^{29}Si (*5*), ^{13}C (*6-8*), ^{19}F (*9*) and ^{31}P (*10*) NMR spectroscopy. Dimeric and trimeric model compounds, indispensable for facilitating NMR spectral assignments in lignin samples, have also been studied (*11-13*). A set of acetylated β-ether/β-ether trimers, giving a more representative reflection of lignin, have been synthesized and all the sidechain protons of each of the 8 chemically distinct isomers were distinguished through ^{1}H 2D NMR spectroscopic experiments (*14*).

Due to the different possible combinations of linkages between syringyl and guaiacyl units and the existence of phenolic and nonphenolic moieties, β–*O*–4′ diastereoisomers are present in a variety of distinct chemical forms. In this article we present an attempt (*15*) to detect by ^{13}C NMR spectroscopy as many as possible of these diastereoisomers in their different chemical forms within lignin structure; we try as well to estimate the relative ratios between *erythro* and *threo* forms. For this purpose a 2D INADEQUATE pulse sequence was selected for studying a sample of uniformly ^{13}C enriched aspen MWL.

^{13}C 2D-INADEQUATE Experiment. This homonuclear pulse sequence uses double quantum transitions in *J*-coupled ^{13}C-^{13}C spin systems. The correlation peaks symmetrically placed about the diagonal correspond to directly bonded carbon atoms (*8*, *16*). The 2D INADEQUATE experiments were conducted at 100.13 MHz with a Bruker AM 400 spectrometer. The double quantum coherence was obtained using a 1/4J delay (5.2 ms) corresponding to an average C-C coupling of 48 Hz. The spectra were obtained from an 18% solution of 11% ^{13}C enriched lignin in DMSO-d_6. The effective ^{13}C sensitivity is a critical problem in the detection of the ^{13}C-^{13}C coupling; the enrichment enhances this sensitivity and thereby makes it possible to carry out the INADEQUATE experiment in a reasonable amount of time.

***Erythro*/*threo* Ratio of β–*O*–4′ Diastereoisomers in Lignin Structures.** Depending upon the guaiacyl or syringyl nature of the two linked aromatic rings, there are four possible combinations of monomer residues, GG, GS, SG and SS, and, depending upon whether or not these units are phenolic, there are 8 different types of dimeric structures that may be encountered. Each of them has an *erythro* and a *threo* form, providing 8 pairs of diastereoisomers that together constitute the set of 16 β–*O*–4′ structures represented in Figure 1. A selection of these 16 β–*O*–4′ dimers has been synthesized and their stereochemistry confirmed by x-ray crystallography (*17*).

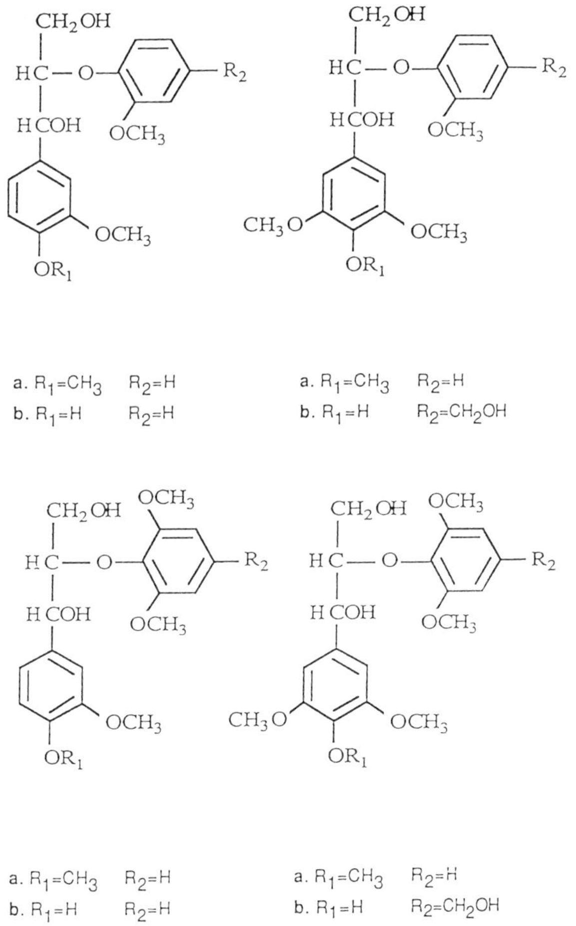

Figure 1. Four groups of β–*O*–4′ dimers with phenolic and nonphenolic groups. Each of the 8 dimeric structures exists in diastereoisomeric *erythro* and *threo* forms.

The 2D INADEQUATE ^{13}C spectrum of the ^{13}C enriched MWL sample revealed most of the ^{13}C-^{13}C bonds found in the main dimeric lignin structures and thus provides a direct reflection of the lignin carbon skeleton (*8*, *16*). In addition the Cβ–Cα carbon correlations assigned to the β–*O*–4′ dimers represent a set of four cross peaks (*15*), clearly discerned from four maxima in the density contours (Figure 2), which are centered at 83.5 -72.2 ppm (I), 84.5-71.4 ppm (II), 86.0-72.5 ppm (III) and 87.2-71.9 ppm (IV). One observes a fairly consistent and precise distribution of the Cβ-Cα chemical shifts for the 16 β–*O*–4′ model compounds among these four cross peaks.

Stereochemical studies of lignin are severely restricted by its lack of optical activity. The 2D INADEQUATE NMR pulse sequence with a ^{13}C enriched lignin, facilitating the observation of diastereoisomers, partly overcomes this limitation. With the experiment, not only can one trace directly the ^{13}C-^{13}C skeleton of the lignin macromolecule but one can also detect, without isolation, within the lignin polymeric structure 8 pairs of diastereoisomeric forms of GG, GS, SG and SS β–*O*–4′ substructures, whether etherified or not.

Biosynthesis of Lignin Monitored by ^{13}C NMR Spectroscopy.

Together with other laboratories, we have also investigated the possibility of using ^{13}C-enrichment as an NMR probe for monitoring lignin biosynthesis. To incorporate this isotope into the lignin, three different samples of coniferin selectively labeled with ^{13}C in each of the propyl sidechain positions (Figure 3) were synthesized and administered exogenously to growing stems of pine and ginkgo trees. The corresponding MWL isolated from the newly formed xylem was then analyzed by ^{13}C NMR spectroscopy (*18-20*). The isotopic enrichment in the first place served to monitor the fate of the precursor through enhanced signals in the lignin spectra, and also provided a check as to whether coniferin could be incorporated into lignin without incurring any structural artefacts.

^{13}C NMR Analysis. ^{13}C spectra of the selectively enriched and natural abundance ginkgo (Figure 4) and pine MWL lignins were recorded using the same routine pulse sequence with continuous ^{1}H decoupling conditions (*19*, *20*). These conditions are acceptable for the comparison of signal intensities if the corresponding carbons are chemically similar and are bonded to the same number of protons. To introduce satisfactory quantitation one needs to use an appropriate inverse gated proton decoupling pulse sequence (*21*), which usually requires a larger amount of sample and is more time consuming. The combined intensities of aromatic carbon signals 22, 25 and 27 serve as a common reference for comparing signal intensity. The enhanced carbon signals observed in the spectra of Figure 4 belong to the major, **A**, **B**, **C**, and minor, **D**, **E**, **F**, and **G**, substructures depicted in Figure 5. Corresponding to the major substructures, signals 29, 30 and 36 have enhanced intensities in the spectrum of Figure 4(b) (Cβ labeled ginkgo lignin) compared to that of Figure 4(c) (unlabeled ginkgo lignin). They are respectively assigned to C-β in β–*O*–4′ units **A** with R_1 = H (signal 29), in β–*O*–4′/α–*O*–4″ units **A** with R_1 = aryl (signal 30), and in β-5′ units **B** (signal 36), and in β-β′ units **C** (signal 36).

Structural Information Gained from ^{13}C Labeling. As previously demonstrated, the prominent enhancement of signal 29 in Figure 4(b) further affirms that the content of β–*O*–4′ substructures in the newly formed lignin is the highest, a characteristic feature of protolignin in the cell wall. Thus it was shown by Terashima through micro-autoradiography (22) that in such a case the formation of lignin is most active in the newly formed xylem tissue.

In addition to this expected signal enhancement, other structures in very low amounts, but of structural importance, were observed. Enhanced signal 38 (Figure 4(a) assigned to the Cβ methylenic carbon in dihydroconiferyl alcohol **D** (R_2 = OCH_3), provides evidence for the presence of this minor structure in lignin. This

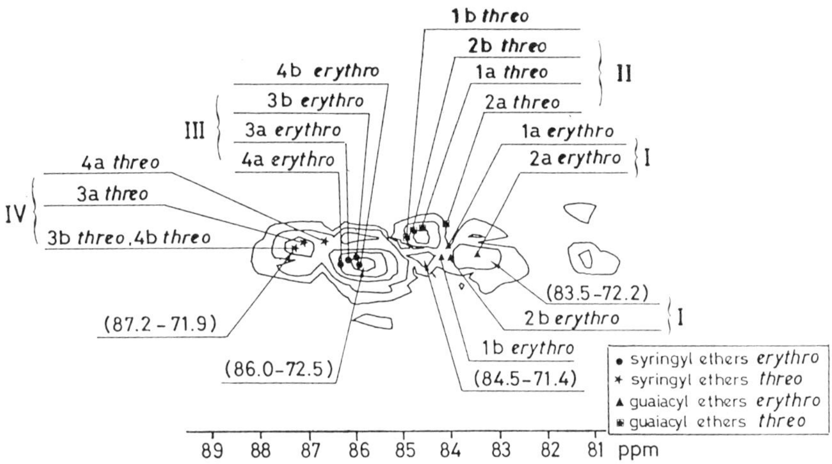

Figure 2. 2D INADEQUATE ^{13}C spectrum of uniformly ^{13}C enriched aspen lignin (MWL): assignments of 16 Cβ–Cα correlation cross peaks to the 16 β–*O*–4′ structures presented in Figure 1.

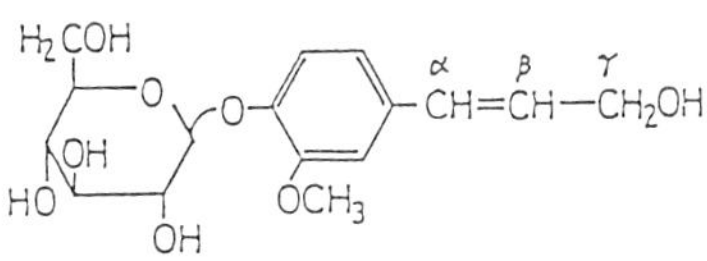

Figure 3. (α-^{13}C), (β-^{13}C) and (γ-^{13}C)coniferins.

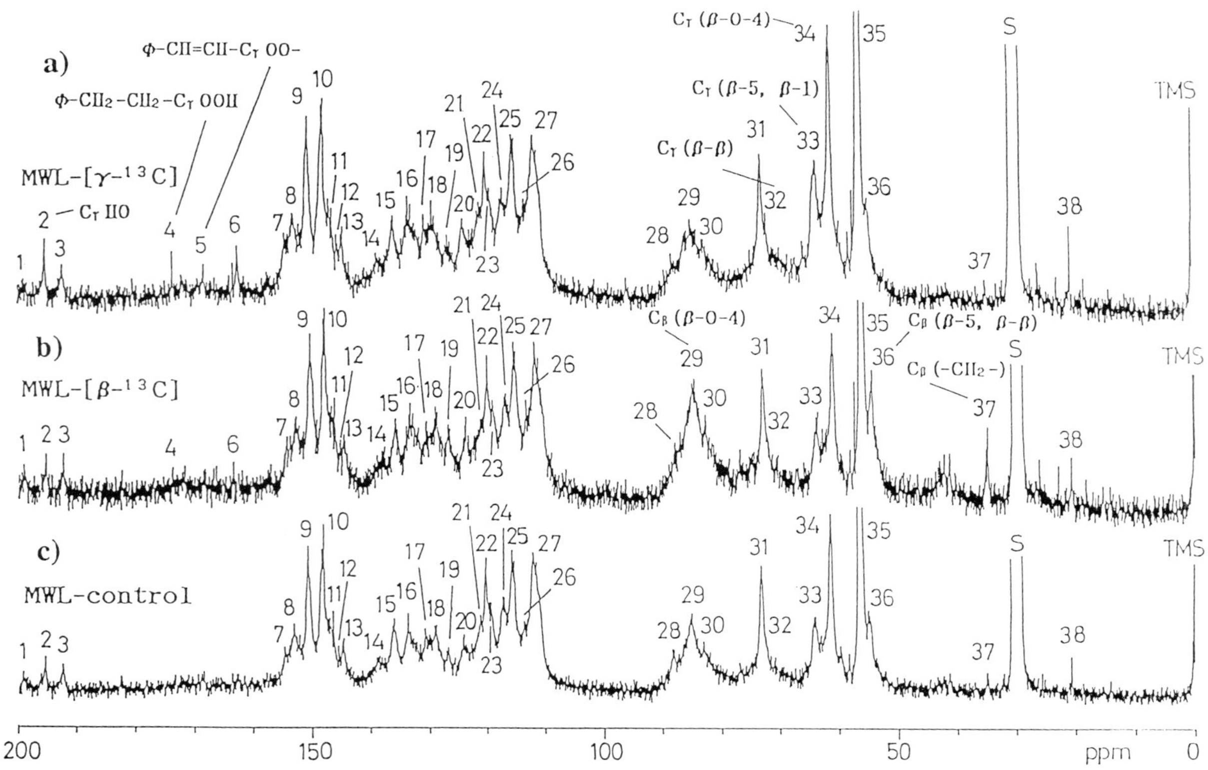

Figure 4. ^{13}C NMR spectra of milled wood lignin from ginkgo administered with (A) [β-^{13}C]coniferin, (B) [γ-^{13}C]coniferin and (C) natural abundance carbon-13 precursor (control).

reduced moiety is rarely observed, but recently it was detected in the extracellular precipitate of *Pinus taeda* cell cultures as well as in another reduced structure, dihydrodehydroconiferyl alcohol (*23*). It confirms that oxidoreductive processes are occurring in the final stage of lignin biosynthesis. In the spectrum of Figure 4(a), enhanced signals 2, 4 and 5 are assigned respectively to Cγ in coniferaldehyde **F** ($R_2 = OCH_3$), the phenylpropionic acid moiety **E**, and the cinnamic acid structure **G**.

Selective Labeling of Natural Precursor as a Structural NMR Probe. Within the NMR sensitivity limits no differences were observed in the spectra of MWL from plants to which coniferin had been administered. It suggests that metabolism of coniferin, which has been administered exogenously to the growing plant, does not perturb lignin formation in the cell wall. In this respect the present study suggests the suitability of using coniferin for labeling protolignin in the cell wall. The labeling method has the proposed advantage of providing a natural selective ^{13}C enrichment. The different possible locations of labeled carbons in the monolignol moiety of coniferin provides a set of probes for investigating lignin structure by ^{13}C NMR spectroscopy.

Mercuration as a Means for Probing 'Condensed' Structures in Lignins by NMR Spectroscopy

The so-called 'condensed' structures in lignins include mainly 5–5′ and β–5′ units, and to a lesser extent 4–*O*–5′ diaryl ethers, diarylmethane bonds and other C–C linkages between the aromatic ring and the aliphatic sidechain. These very stable covalent bonds are difficult to cleave without complete degradation of the polymeric network. Condensed units can be isolated from lignin degradation products and characterized by chemical methods such as permanganate oxidation (*24*) and thioacidolysis (*25*). In our study we propose an attempt to observe these structures within the polymeric structure itself, rather than among its degradation products (*26*). For this purpose ^{199}Hg has been chosen as a means for using NMR spectroscopy to probe the possible condensation patterns in lignin. Mercuration of the aromatic ring (*27*) with mercuric acetate, in methanol/acetic acid at 95°C, proceeds through an electrophilic substitution mechanism (*28*) at the free aromatic positions, with the mercurio-acetate species acting as strong electrophiles. The intermediate aromatic carbocation formed, the so-called Wheland intermediate, is mesomerically stabilized, and the substituents already on the aromatic ring determine the position of substitution as that having arisen from the more stable intermediates (*29*). The substituted positions can be observed and assigned by ^{199}Hg and ^{13}C NMR spectroscopy, using direct and indirect detection methods. In a preliminary step five monomeric lignin model compounds were mercurated (Figure 6), *viz* vanillin **2a**, syringaldehyde **3a**, vanillyl alcohol **4a**, syringyl alcohol **5a** and veratryl alcohol **6a**, to facilitate NMR spectroscopic analysis of the mercurated lignin samples.

NMR Studies. Among the many mercury isotopes only Mercury-199 has a nuclear spin of I = 1/2; it has a 16.84% natural abundance, a gyromagnetic ratio (γ) of 4.7912 10^7 rad s^{-1} T^{-1}, a sensitivity/^{13}C of 5.42 and a Larmor frequency (ν) of 44.801 MHz at 5.85 T (250 MHz for the proton).

Direct ^{199}Hg NMR Spectroscopic Detection. Direct observation and assignments are rendered difficult by a very large frequency range of 3000 to 4000 ppm and a chemical shift that is very sensitive to chemical environment, solvent and temperature. As a consequence there is still a lack of data about ^{199}Hg parameters (*30*). Nevertheless direct detection is possible in mercurated monomeric model compounds. In the direct NMR detection of ^{199}Hg, δ-values were measured at 71.565 MHz in DMSO-d_6 as solvent with neat dimethylmercury as an external reference in a Bruker AM400 spectrometer equipped with a 10 mm probe using a spectral width of 83000 Hz, 32k size memory, 90° pulse and an average of 14500 transients.

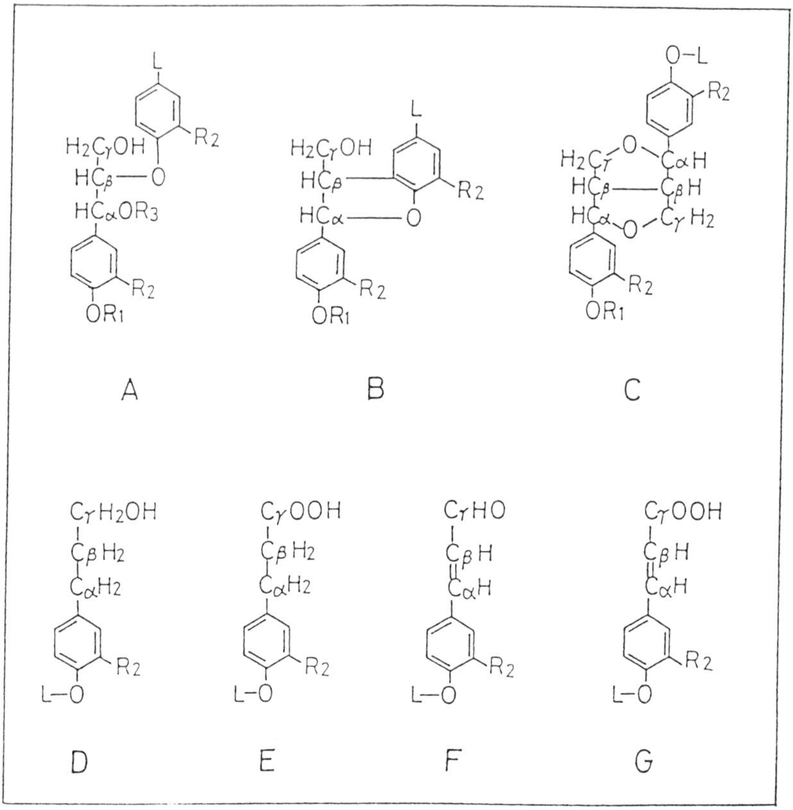

Figure 5. Main dilignol substructures and minor monolignol structures found in lignins administered with β- and γ- labeled ^{13}C coniferins; L = lignin.

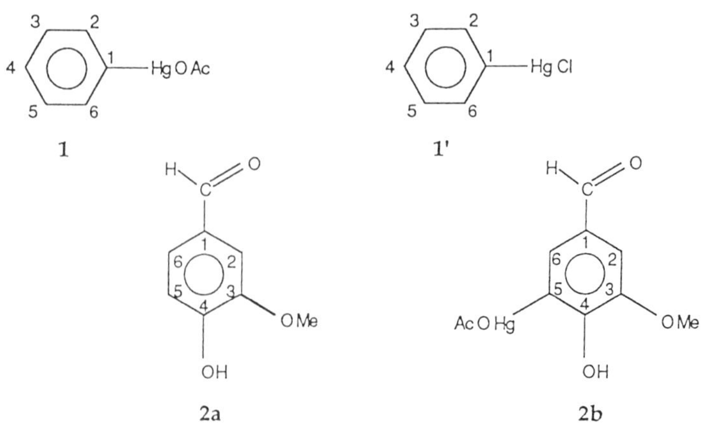

Figure 6. Mercurated lignin model compounds: assignments of the mercuration sites.

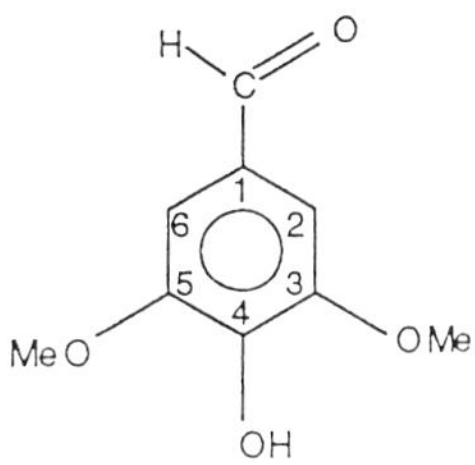

3a

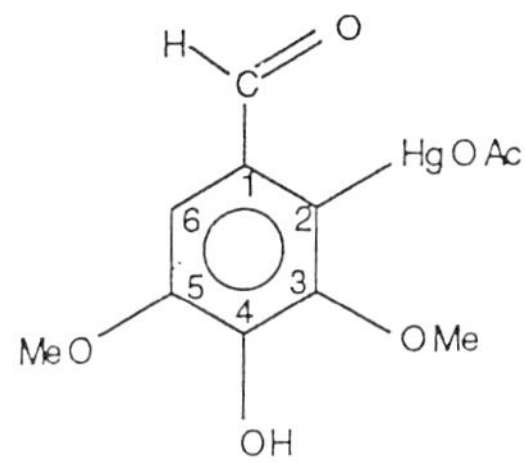

3b

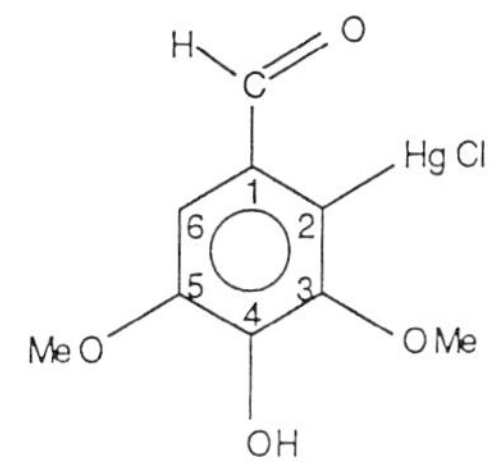

3b'

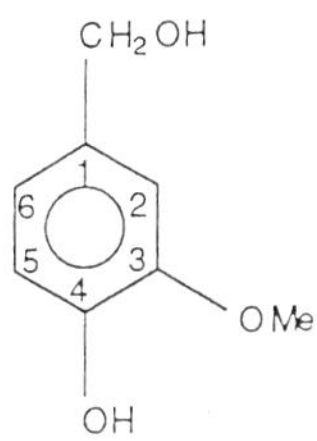

4a

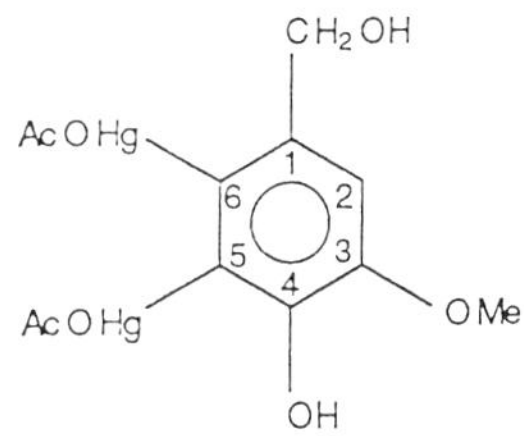

4b

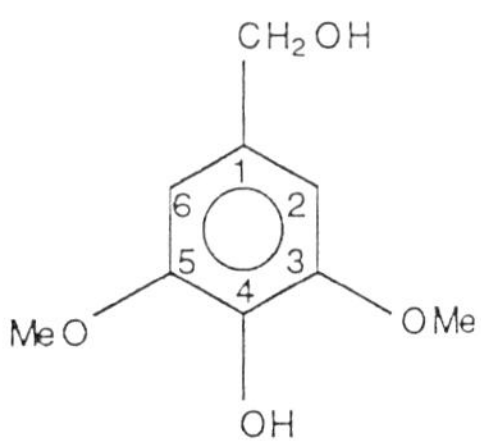

5a

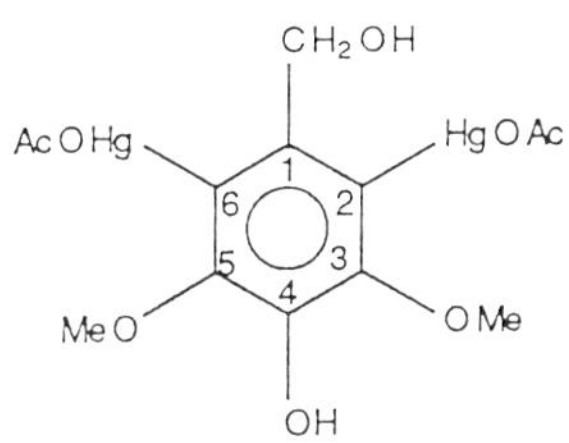

5b

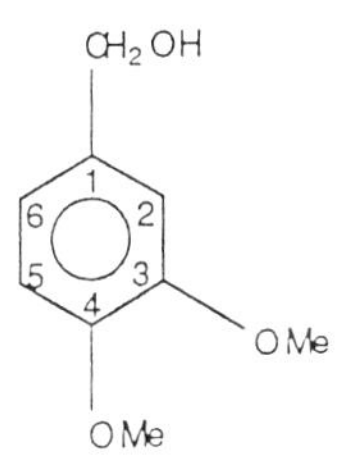

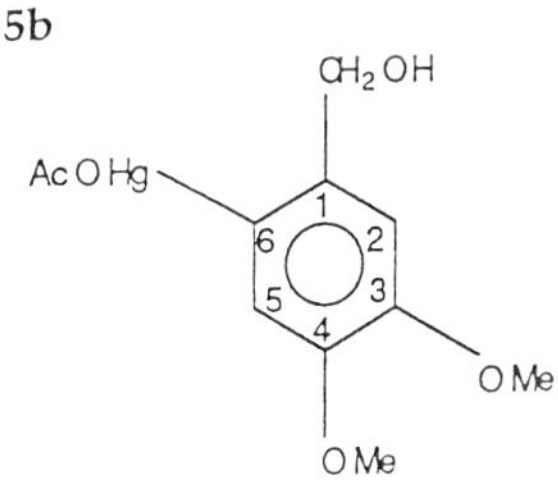

Figure 6. *Continued.*

1-D and 2-D Indirect ^{199}Hg Detection. When direct detection is not possible or too time consuming, one can use indirect NMR detection of nuclei that are appropriately coupled to protons. This method (*31*) relies on the detection of insensitive nuclei or low-abundant nuclei, like ^{199}Hg, through the observation of a sensitive nucleus, like the proton, to which their spins are coupled. A polarization transfer occurs during a 1/2J refocusing delay, J being the value of the $^{n}J^{1}H-^{199}Hg$ coupling constant with n being the number of bonds between the proton and mercury nuclei.

The 1-D $^{1}H/^{199}Hg$ HMQC NMR spectra were recorded with a 250 MHz spectrometer using a HMQC (*32*) pulse sequence. We chose a very seldom used 1-D inverse detection sequence (REVOBS.) from the Bruker program which does not require any special excitation unit (*33*) when using a 10 mm broadband probe to acquire 256 transients with a 3 s relaxation delay. The spectra edited for each possible $^{n}J^{1}H-^{199}Hg$ value unambiguously reveal the existence of protons coupled to mercury, and the corresponding J and n values give the assignment of the aromatic mercurated site(s).

The 2-D $^{1}H/^{199}Hg$ HMQC spectra give, in addition to the proton spectrum, the ^{199}Hg chemical shifts on the second axis which can be compared to the values obtained from direct observation. Spectra were recorded with a Bruker AM400 spectrometer using sweep widths of 1100 Hz and 1400 Hz. The resulting data matrix of 2k x 128k was multiplied by a pseudo-echo function in both dimensions and zero filled to 2k x 1k.

Mercurated Lignin Model Compounds (Figure 6). ^{199}Hg signals observed by direct detection (*31*) are found at -1435 ppm in compound **2b**, -1341 ppm in **3b**, -1446.5 ppm for $\delta_{Hg(6)}$ and -1485.4 ppm for $\delta_{Hg(5)}$ in **4b**, -1330.5 ppm in **5b** and -1355 ppm in **6b**. ^{199}Hg brings a downfield shift for nearly all the aromatic carbons, ranging from 36 to 14 ppm for the *ipso* carbons and from 2 to 8 ppm for the *ortho* carbons. Monosubstitution at the C5 for vanillin and C2 or C6 positions for syringaldehyde are easily observed from the ^{13}C spectra.

In the case of vanillyl and veratryl alcohols, it is the 1-D NMR indirect spectra which show unambiguously and easily a dimercuration of vanillyl alcohol at the C5 and C6 carbons and a monomercuration of veratryl alcohol at the C6 carbon (*34*). Figure 7 shows the HMQC spectra edited for each possible $^{n}J^{1}H-^{199}Hg$ value, namely 200 Hz for a ^{3}J, 77 Hz for a ^{4}J and 20 Hz for a ^{5}J corresponding to successive 1/2J refocusing delays of 0.0025, 0.01 and 0.025 s. Only ^{3}J and ^{4}J couplings appeared, revealing that a mercury monosubstitution occurred at the C6 atom. This difference between the two alcohols indicates the importance of alkylation at the C4 oxygen position which obviously plays a role in the substitution mechanism and therefore in lignin structure and reactivity.

As a whole, then, the mercurated model compounds provided an important set of NMR data concerning the $^{n}J^{1}H-^{199}Hg$, $^{n}J^{199}Hg-^{13}C$ and $^{3}J^{199}Hg-^{199}Hg$ coupling constant values.

NMR Analysis of Mercurated Lignins. There are not many lignin mercuration studies (*27*) and the chemical mechanisms involved in the formation of these derivatives are not well known. In our first excursion into this field we considered two lignin samples from hardwood and softwood, which were mercurated under the same experimental conditions, and characterized them by ^{13}C NMR spectroscopic analysis. The set of ^{13}C chemical shifts and other ^{199}Hg parameters determined from the model compounds (*33*, *34*) made possible the carbon assignments in the mercurated lignin spectra. In the case of mercurated Birch milled wood lignin (MWL) (Figure 8b) we observed a collapse of signals corresponding to the G6, G5 and G2 tertiary aromatic C-H in G units and the S2/S6 in S units, the decrease in intensity being more prominent for the S than for the G units. Concomitantly new signals appeared which can be assigned, according to the model compound studies, to S3/S5 and G5 carbons having a mercury substituent at the *ortho* position. Quantitative analysis indicated that 0.55 mercury atoms are substituted *per* aromatic

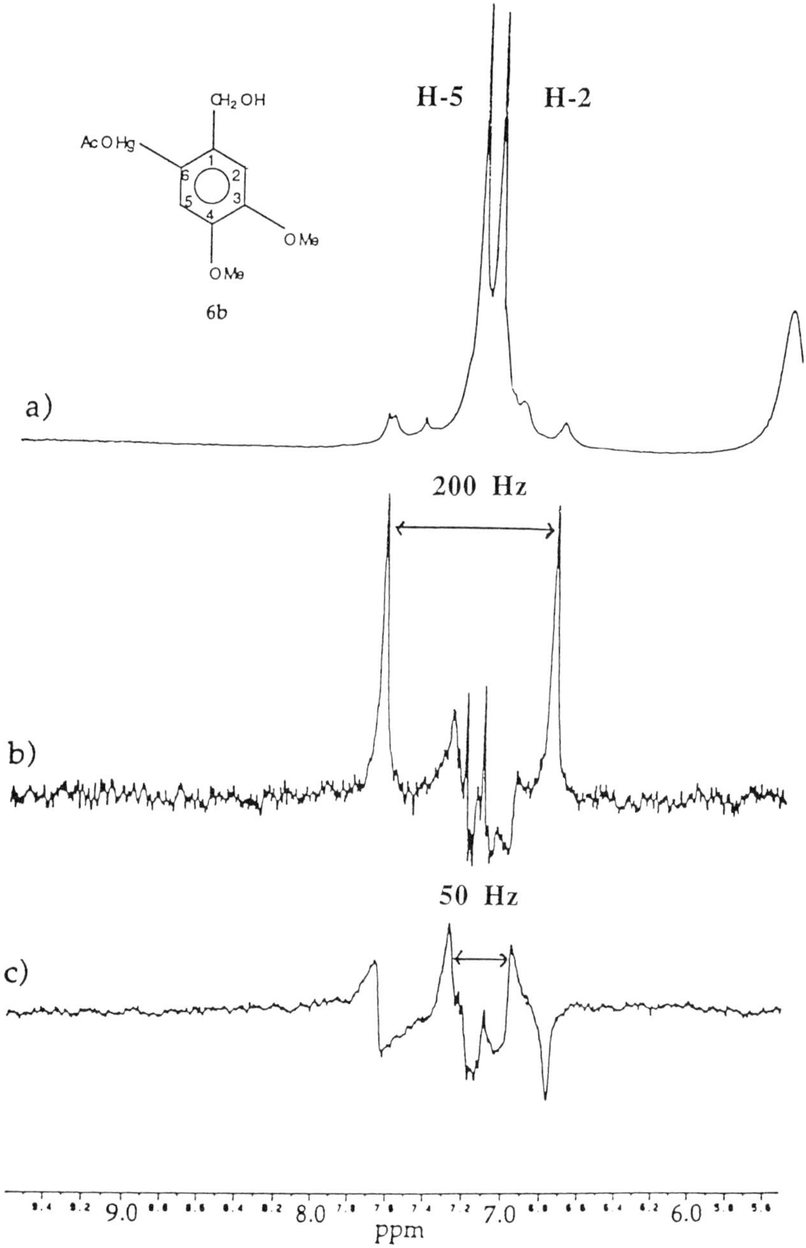

Figure 7. Mercuration site assigned by indirect ^{199}Hg NMR detection in 6-acetoxymercurio veratryl alcohol. (a) Total aromatic ^{1}H spectrum and two detected ^{n}J ^{1}H-^{199}Hg couplings in 1-D HMQC spectra corresponding to (b) ^{3}J = 200Hz and (c) ^{4}J = 50 Hz.

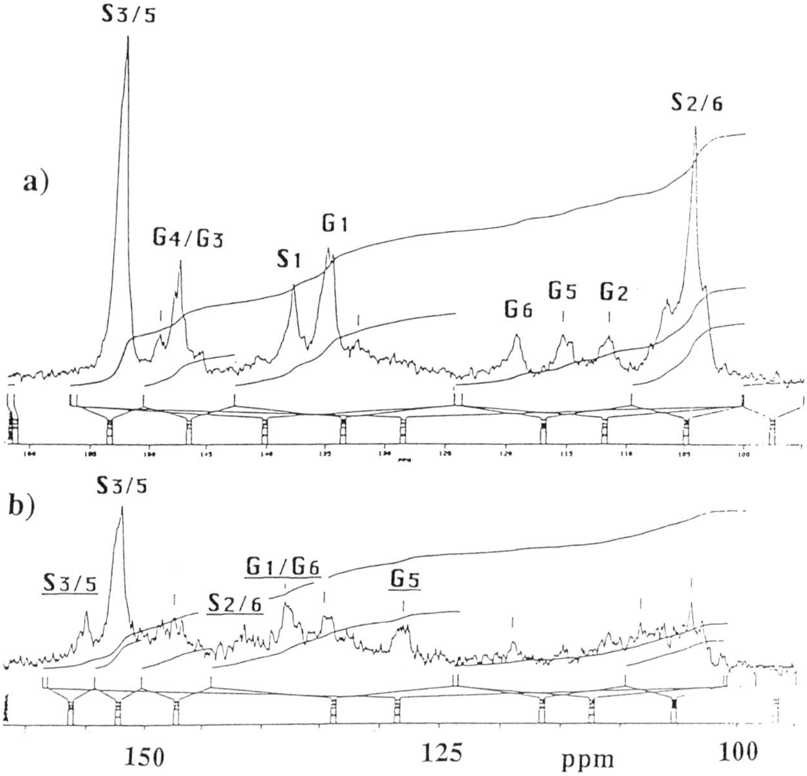

Figure 8. ^{13}C NMR spectra (aromatic part) of a) birch MWL and of mercurated Birch MWL after b) 2 hours mercuration. New signals appearing in b) compared to a) are underlined.

ring. Under the same mercuration conditions only 0.26 aromatic carbons possessed mercury substituents in a pine MWL sample; softwood lignin being more condensed than hardwood lignin, this finding demonstrates the effect of condensed structures which limit the number of free sites available for mercuration and/or the reactivity of the aromatic ring toward electrophilic substitution.

Mercuration of lignins in conjunction with ^{199}Hg NMR indirect detection are worthwhile tools to consider for investigating the condensation patterns of aromatic rings in lignin structures. The present study was an attempt to observe directly in the polymeric lignin structure the correlation between the number of condensed units and limitations to substitution reactions on the aromatic ring. Indeed it has been shown that syringyl lignins, which are less condensed than guaiacyl lignins, exhibited higher mercuration yields, even though they have more methoxyl substituents. To establish more reliable and precise conclusions, the yield of the mercuration reaction has to be improved and the electrophilic substitution mechanisms better controlled and understood. This technique thus comes as a complement to chemical methods.

Application of 2D and 3D HMQC-HOHAHA NMR Spectroscopy in Structural Studies of Lignins

The successful application of 2D HMQC and 3D HMQC/HOHAHA NMR experiments to lignin samples have provided reliable assignments of propyl sidechain protons and carbon nuclei. Two-dimensional techniques like 2D HMQC giving ^{1}H-^{13}C correlations by inverse detection (*12-13*, *35*), or the combination of 2D HOHAHA giving ^{1}H-^{1}H correlations with the 2D HMQC (*36*) pulse sequence, offer the possibility of tracing out most of the sidechain spin systems present in lignins (Figure 9) and assigning them unambiguously. These experiments have the great advantage of being applicable to non ^{13}C-enriched lignin samples. However, even in two dimensional NMR spectra, some of the correlations overlap and it is not possible to trace out the whole spin system. So, to overcome these limitations, a 3D NMR HMQC/HOHAHA experiment has been conducted (*37*, *39*) with an acetylated ^{13}C-enriched MWL, isolated in a previous study (*8*) from a hardwood species, *Populus tremula*, that had been grown in a $^{13}CO_2$ enriched atmosphere .

NMR Pulse Sequences. All NMR spectra were obtained using a Varian Unity 500 spectrometer for acetylated samples in $CDCl_3$ solutions held in 5 mm diameter tubes. The 2D HMQC ^{1}H-^{13}C correlation spectra were measured as described earlier (*13*). The 2D ^{1}H-^{1}H correlation HOHAHA spectra were acquired as described by Wijmenga (*38*). The spectral widths in F2 and F3 were set to 6 kHz; F1 was set to 25 kHz. The delay for polarization transfer between ^{1}H and ^{13}C corresponded to an assumed 140 Hz J coupling constant, while the mixing times used were 30, 60 and 100 ms.

2D HMQC and 2D HOHAHA NMR Spectra. An expansion of the 2D HMQC spectrum in Figure 10(a) shows different correlations corresponding to directly bonded ^{13}C-^{1}H nuclei, for which the proton chemical shift is given on the F2 axis and that for carbon on the F1 axis. The correlation assignments made according to a previous study (*36*) rely on a comparison between the chemical shifts of the observed connectivities to those of the protons and carbons in the model compounds presented in Figure 9. Correlations A and S in Figure 10(a) are assigned without ambiguity respectively to ^{13}Cα-^{1}Hα and to ^{13}Cβ-^{1}Hβ in the sidechain of structure **8**; the B and R correlations correspond respectively to the same C-H couples in structure **9**, and T is a Cβ/Hβ correlation in **10**. But some other assignments are not so clear; for instance, the correlation Cα/Hα in **10** is overlapped by other correlations in signal G, and it is not possible to make individual correlations for the sidechain in structure **12.** An expansion of the 2D HOHAHA spectrum of the same enriched MWL sample in Figure 10(b) give correlations for coupled ^{1}H-^{1}H systems: for instance the protons at 5.42 ppm have a HOHAHA correlation I with those of 4.55 ppm, but there is a partial

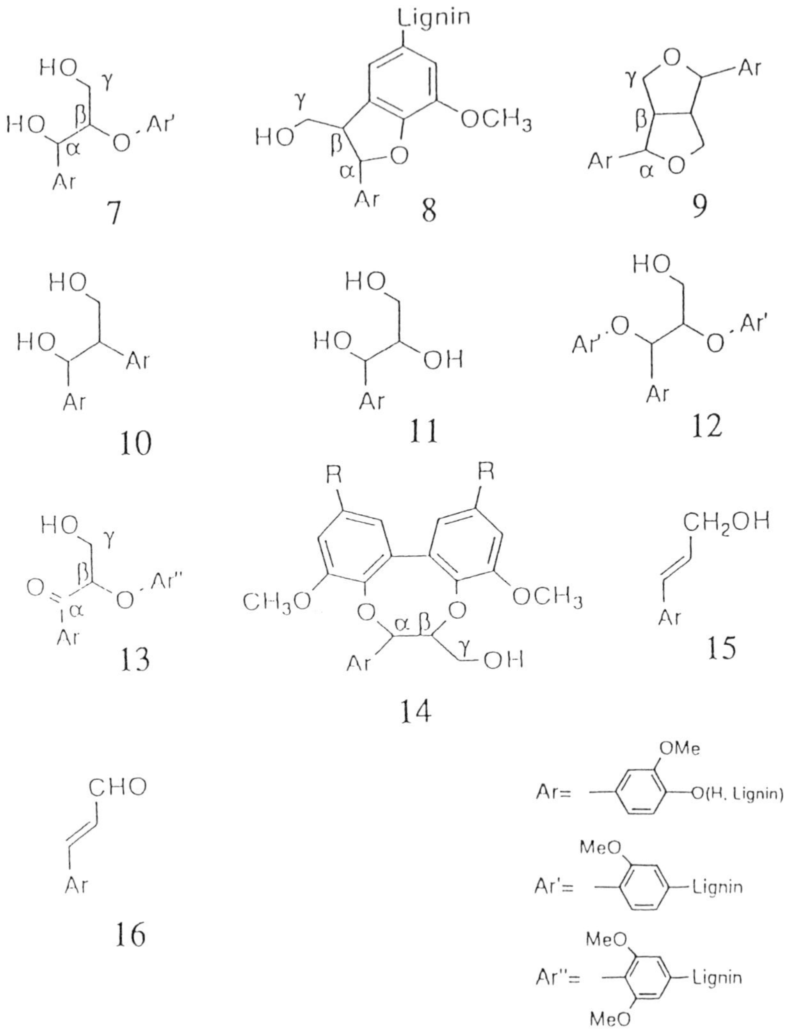

Figure 9. Major and minor bonding patterns in lignin structures observed through 2D-HMQC, 2D-HOHAHA and 3D HMQC-HOHAHA NMR pulse sequences. The lignin sample used was an acetylated ^{13}C-enriched MWL isolated from *Populus tremula* grown in a $^{13}CO_2$ enriched atmosphere.

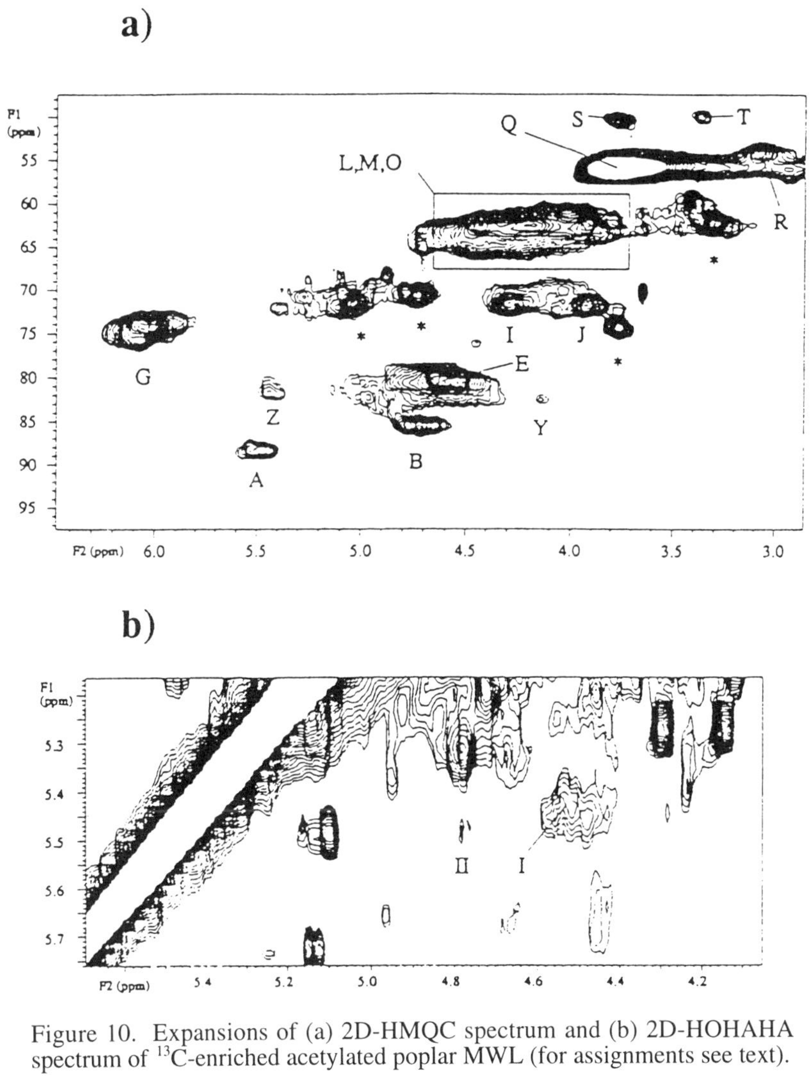

Figure 10. Expansions of (a) 2D-HMQC spectrum and (b) 2D-HOHAHA spectrum of ^{13}C-enriched acetylated poplar MWL (for assignments see text).

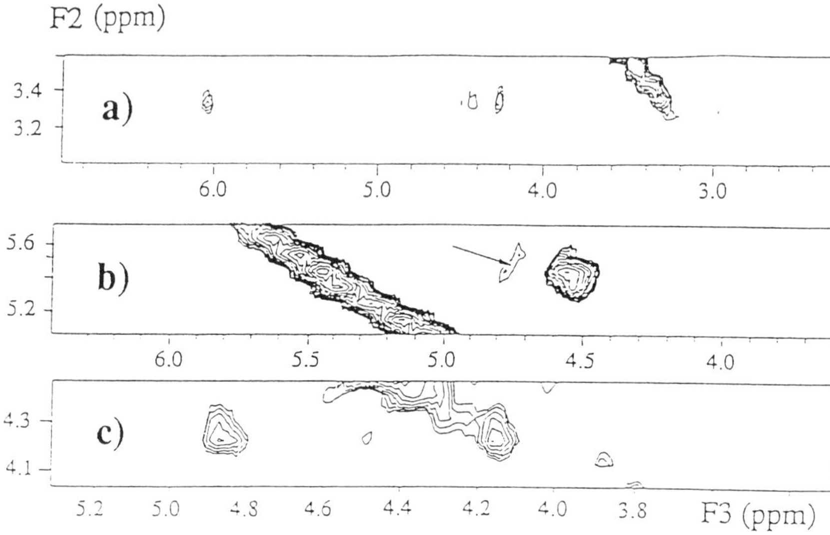

Figure 11. 3D HMQC-HOHAHA spectrum of ^{13}C-enriched acetylated poplar MWL: selected F2F3 slices in HOHAHA spectrum for different F1 values, giving H/C correlation (HMQC) and H//H correlation (HOHAHA). (a) Hβ/Cβ//Hα (3.35/50.4//6.03 ppm) and Hβ/Cβ//HγHγ′ (3.35/50.4//4.2-4.5 ppm) in **10**; (b) Hα/Cα//Hβ (5.48/80.5//4.7 ppm) in **12**; (c) Hβ/Cβ//Hα (4.14/82.5//4.85 ppm) in **14**.

overlap with the Hα/Hγ correlation of structure 2; there is also a very small correlation II at 5.45-5.53/4.75-4.80 ppm which cannot be clearly assigned.

3D HMQC/HOHAHA NMR Spectra. The 3D assignment illustrated in Figure 11 provides a good example of the resolving power of the 3D HMQC-HOHAHA experiment in the case of the complete assignments for the sidechain spin-system of β-1 structures **10** (*37, 38*). In a previous study (*36*) it was not possible to assign the chemical shifts of Cα and Cγ in **10,** as the expected 2D HMQC Hα/Cα and Hγ/Cγ correlations were overlapped by the strong correlations from structures **7** and **8**. For different δ^{13}C values selected on the F1 axis of the 2D HMQC spectrum in Figure 10(a), we obtained the F2F3 slices of HOHAHA spectra shown in Figure 11 at (a) F1 = 50.4 ppm, (b) F1 = 80.5 ppm and (c) F1 = 82.5 ppm. The F2F3 slice selected at the 50.4 ppm F1 value shows the HOHAHA correlation Hβ//Hα between Hβ at 3.35 ppm (F2 axis) and Hα at 6.03 ppm, and correlation Hβ//Hgγ,Hγ′ between Hβ and Hgγ,Hγ′ at 4.2-4.5 ppm (F3 axis). From the correlation T in the HMQC spectrum (Figure 10(a), this proton Hβ belongs to a Hβ/Cβ (3.35/50.4 ppm) group, and thus we finally conclude that the connectivities are the Hβ/Cβ//Hα and Hβ/Cβ//Hγ,Hγ′ in structure **10**. Hereby the 3D HMQC-HOHAHA experiment has provided unambiguous assignments for all the protons and carbons belonging to this β-1 spin system. From Figure 11(b) one deduces as well the Hα/Cα//Hβ connectivity (5.48/80.5//4.7) corresponding to Hα-Cα-Cβ-Hβ in compound **12** and from Figure 11(c) the Hβ/Cβ//Hα (4.144/82.5//4.85) connectivity corresponding to Hβ-Cβ-Cα-Hα in compound **14.**

Structural Information Obtained from the 3D NMR Study. The 3D HMQC/HOHAHA NMR spectra have provided some complete and unambiguous ^{1}H and ^{13}C assignments for the sidechains in lignin structures such as **10,** which could not be achieved with 2D pulse sequences. They provide as well direct evidence for the presence of structures which, although present in very low amounts, are of great structural importance, namely structures **10**, 1,2-diarylpropane, **12**, noncyclic benzyl aryl ether, and **14**, dibenzodioxocin, the last two units representing structures that can accommodate branch-points in the polymeric lignin network. The existence of β-1 structures, **10**, in lignin was for a long time a matter of debate, because all the 1D and 2D NMR experiments gave rise to spectra with overlapping signals. Only the 3D HMQC-HOHAHA experiment led to an unambiguous detection of this structure which possesses the unique feature in lignin structure of linking a phenylpropane unit to a benzene ring without a three-carbon sidechain. Trilignol **12** is difficult to observe not only because of its low frequency but also because the labile α–*O*–4′ linkage is very easily cleaved as a result of mild chemical treatment. Last but not least, it was shown here that the 6,7-dihydrodibenzo-(*e,g*)(1,4)-dioxocin, **14**, a recently discovered eight-membered ring unit in softwood lignin (*35*), also occurs in hardwood lignin.

Literature Cited

1. Nimz, H. *Ber.* **1966**, *99*, 469-474.
2. Lundquist, K. *Acta Chem. Scand. B* **1979**, *33*, 418-420.
3. Lundquist, K.; von Unge, S. *Acta Chem. Scand. B* **1986**, *40*, 791-797.
4. Hauteville, M.; Lundquist, K.; von Unge, S. *Acta Chem. Scand. B* **1986**, *40*, 31-35.
5. Brezny, R. J.; Schraml, M.; Kvicalva, J.; Chvalovski, V. *Holzforschung* **1985**, *39*, 297-303.
6. Nimz, H. H.; Lüdemann, H.-D. *Holzforschung* **1976**, *30*, 33-40.
7. Nimz, H. H.; Tschirner, U.; Stälhe, M.; Lehmann, R.; Schlosser, M. *J. Wood Chem. Technol.* **1984**, *4*, 265-284.
8. Bardet, M.; Gagnaire, D.; Nardin, R.; Robert, D.; Vincendon, M. *Holzforschung* **1986**, *40 (Suppl.)*, 17-24.
9. Barelle, M.; Fernandes, J. C.; Froment, P.; Lachenal, D., *J. Wood Chem. Technol.* **1992**, *12*, 413-424.
10. Granata, A.; Argyropoulos, D. *J. Agric. Food Chem.* **1995**, *43*, 1538-1544.
11. Brunow, G.; Sipilä, J.; Lundquist, K.; von Unge, S. *Cellulose Chem. Technol.* **1988**, *22*, 191-199.
12. Ede, R.; Brunow, G.; Simola, L. K.; Lemmetynen, J. *Holzforschung* **1990**, *44*, 95-101.
13. Ralph, J.; Helm, R. *J. Agric. Food Chem.* **1991**, *39*, 705-709.
14. Ralph, J. *Magn. Reson. Chem.* **1993**, *31*, 357-363.
15. Robert, D.; Bardet, M.; Lundquist, K.; von Unge, S. *Proc. 5th Internat. Symp. Wood Pulp. Chem.* **1989**, *1*, 217-219.
16. Guittet, E.; Lallemand, J. Y.; Lapierre, C.; Monties, B. *Tetrahedron Lett.* **1985**, *26*, 2671-2674.
17. Stomberg, R.; Lundquist, K. *Nord. Pulp Pap. Res. J.* **1994**, *9*, 37-43.
18. Xie, Y.; Terashima, N. *Mokuzai Gakkaishi* **1991**, *37*, 935-941..
19. Terashima, N.; Seguchi, Y.; Robert, D. *Holzforschung* **1991**, *45 (Suppl.)*, 35-39.
20. Xie, Y.; Robert, D.; Terashima, N. *Plant Physiol. Biochem.* **1994**, *32*, 243-249.
21. Robert, D. In *Methods in Lignin Chemistry*; Lin, S. Y., Dence, C. W., Eds.; Springer-Verlag: Berlin, 1992; pp 250-271.
22. Terashima, N.; Fukushima, K. *Wood Sci. Technol.* **1988**, *22*, 259-270.
23. Nose, M.; Bernards, M.; Furlan, M.; Zajicek, J.; Eberhardt, T. L.; Lewis, N. G. *Phytochemistry* **1995**, *39*, 71-79.

24. Gellerstedt, G. In *Methods in Lignin Chemistry*; Lin, S. Y., Dence, C. W., Eds.; Springer-Verlag: Berlin, 1992; pp 322-333.
25. Lapierre, C.; Pollet, B.; Monties, B.; Rolando, C. *Holzforschung* **1991**, *45*, 61-68.
26. Neirinck,V. *PhD Thesis*; Université J. Fourier: Grenoble, 1994; 266 pp.
27. Westermark, U.; Lidbrandt, O.; Eriksson, I. *Wood Sci. Technol.* **1988**, *22*, 243-250.
28. Taylor, R. *Electrophilic Aromatic Substitution*; Wiley: Chichester, 1990, 513 pp.
29. Larock, R. C. *Organomercury Compounds in Organic Synthesis*; Springer-Verlag: Berlin, 1985; 423 pp.
30. Granger, P. In *Transition Metal Nuclear Magnetic Resonance*; Pregosin, P., Ed.; Elsevier: Amsterdam, 1991; p 265.
31. Maudsley, A. A.; Ernst, R. R. *Chem. Phys. Lett.* **1977**, *50*, 368-372.
32. Bax, A.; Griffey, R. H.; Hawkins, B. L. *J. Am. Chem. Soc.* **1983**, *105*, 7188-7190.
33. Neirinck, V.; Robert, D.; Nardin, R. *Magn. Reson. Chem.* **1993,** *31,* 815-822.
34. Neirinck, V.; Robert, D.; Nardin, R. *J. Chim. Phys. Phys.-Chim. Biol.* **1995**, *92*, 1819-1822.
35. Karhunen, P.; Rummakko, P.; Sipilä J.; Kilpeläinen I.; Brunow, G. *Tetrahedron Lett.* **1994**, *36*, 169-170.
36. Kilpeläinen, I.; Sipilä, J.; Brunow, G.; Lundquist, K.; Ede, R. *J. Agric. Food Chem.* **1994**, *42*, 2790-2794.
37. Kilpeläinen, I.; Ämmälahti, E.; Brunow, G.; Robert, D. *Tetrahedron Lett.* **1994**, *35,* 9267-9270.
38. Kilpeläinen, I.; Ämmälahti, E.; Brunow, G.; Robert, D.; Bardet, M. submitted to *J. Org. Chem.*
39. Wijmenga, S. S.; Hallenga, K.; Hilbers, C. W. *J. Magn. Reson.* **1989**, *84*, 634-642.

Chapter 18

Characterization of Milled Wood Lignins and Dehydrogenative Polymerisates from Monolignols by Carbon-13 NMR Spectroscopy

Chen-Loung Chen

Department of Wood and Paper Science, North Carolina State University, Raleigh, NC 27695–8005

^{13}C NMR spectroscopy was used to characterize milled wood lignin (MWL) from spruce (*Picea glauca*) and Zhong-Yan Mu (*Bischofia polycarpa)* as well as dehydrogenative polymerisates (DHPs) from monolignols. The analysis of a quantitative ^{13}C NMR spectrum of spruce MWL showed that this guaiacyl type lignin had 8–8′, 8–5′, 5–5′ and 8–*O*–4′ linkages which were interpreted as approximating 2, 9, 24 and 53 per 100 C_9-units, respectively, whereas the frequency of 7–*O*–4′ linkages was less than 3 per 100 C_9-units. The frequencies of both 5–5′ and 8–*O*–4′ linkages were much higher than values previously estimated. The spectrum of the Zhong-Yan Mu MWL showed that it is a guaiacyl-syringyl lignin with a guaiacylpropane-syringylpropane molar ratio of approximately 86:14. The dehydrogenative polymerization of monolignols is discussed on the basis of the components identified among the reaction products.

Earlier work on the characterization of organic compounds by nuclear magnetic resonance (NMR) spectroscopy mainly relied on ^{1}H NMR spectroscopy. The major reason for this is that the ^{1}H nucleus is the most abundant among the nuclei that can be detected by NMR techniques. However, the advent of Fourier transform NMR (FT-NMR) spectroscopic techniques heralded rapid developments in the feasibility of obtaining natural abundance ^{13}C NMR spectra within reasonable periods of time. Thus, in spite of the low natural abundance of ^{13}C (1.1 atom %) compared to ^{1}H (99.98%), ^{13}C NMR spectroscopy has become as useful as ^{1}H NMR spectroscopy for the structural analyses of organic substances.

Among the various physical and chemical methods available for characterizing lignin, natural abundance ^{13}C NMR spectroscopy has been shown to be among the most reliable and comprehensive of the techniques (*1-11*). It furnishes comprehensive data about the structures of all carbons in lignin, in contrast to other physical and chemical analytical methods which may only supply partial structural information.

Advantages of ^{13}C NMR over ^{1}H NMR Spectroscopy

There are several advantages of ^{13}C NMR over ^{1}H NMR spectroscopy for the structural determination of organic compounds. In a ^{13}C NMR spectrum, spectral data are obtained from the 'backbone' of the molecule that provides information about the nature of all carbons in the molecule, whether carbonyl, nitrile, quaternary groups, and so forth. The second advantage is that ^{13}C NMR spectra are not complicated by spin-spin coupling between ^{13}C nuclei. This is because the probability of having two such nuclei adjacent to each other in the same molecule is low enough that the possibility of ^{13}C-^{13}C coupling can be ignored. Moreover, ^{13}C NMR spectra are usually obtained with proton decoupling (*12*), so that only single carbon resonances are observed. The third advantage is that the ^{13}C NMR chemical shift range for the majority of diamagnetic organic compounds is approximately 240 ppm compared to about 12 ppm for ^{1}H NMR spectroscopy. Finally, the signals from samples in solution are normally relatively sharp. All these factors imply that there is better resolution and less overlap of the signals in the ^{13}C NMR spectra of organic compounds. This is of particular value for polymeric natural products such as lignin.

Disadvantages of ^{13}C NMR over ^{1}H NMR Spectroscopy

One of the disadvantages of ^{13}C NMR over ^{1}H NMR spectroscopy is that the peak areas in a ^{13}C NMR spectrum, obtained with continuous proton decoupling, may not be proportional to the number of the respective ^{13}C NMR nuclei giving rise to these signals because of the nuclear Overhauser enhancement (nOe) effect and different relaxation times of the various carbons. In order to obtain a quantitative ^{13}C NMR spectrum for an organic compound, an inverse gated decoupling pulse sequence (where the proton irradiation is switched off during the relaxation delay before the ^{13}C excitation pulse) must be used to minimize nOe effects while the relaxation delay between successive $\pi/2$ pulses must be at least 5 times the longest ^{13}C longitudinal relaxation time. The other disadvantage is that the ^{13}C nucleus is approximately 5700 times less sensitive to NMR detection than the ^{1}H nucleus because of its relatively low natural isotopic abundance and smaller magnetogyric ratio.

Experimental Procedures

Preparation of Lignin Samples. To obtain ^{13}C NMR spectra of naturally occurring lignin, the lignin must be isolated free from other components present in the plant tissues (*13*). For this purpose, the procedure of Björkman (*14*) is used, in general, to prepare milled wood lignin samples (MWL) from the corresponding woody plant tissues. The milled wood lignin usually contains up to 5% of carbohydrates. Technical lignins obtained from pulping spent liquors also contain appreciable amounts of carbohydrates and fatty acids. Therefore, care must be taken to remove these contaminants as much as possible when purifying both MWL and technical lignin preparations. After purification, the lignin preparations must be dried in a drying pistol at 50°C *in vacuo* for 24 h using P_2O_5 as drying agent. The dried, purified lignin preparations are then often subjected to elemental (C, H, and O), methoxyl and carbohydrate analyses prior to obtaining their ^{13}C NMR spectra.

Preparation of Sample Solution. About 300-400 mg of the dried, purified lignin preparation is dissolved in 2 mL of a suitable solvent, with tetramethylsilane (TMS) being added as an internal reference. Insoluble materials, if any, are filtered off with a sintered glass funnel (filter disk 10 mm, i.d., porosity 25-50 mm). The concentration of the sample is typically *ca.* 15-20% (w/v). The sample solution is then transferred to a 5 mm or 10 mm diameter NMR sample tube, depending upon the probe of the instrument. In general, a ^{13}C NMR spectrum of a lignin preparation in a suitable solution is obtained with a ^{13}C NMR spectrometer operating in a Fourier transform (FT) mode at a frequency of more than 50 MHz for the ^{13}C nucleus. A solution of the lignin is placed in the probe of the NMR spectrometer, and the ^{2}H nucleus of the solvent, *i.e.* either dimethylsulfoxide-d_6 (DMSO-d_6) or acetone-d_6, is used as an internal lock for the spectrometer radio-frequency field. The optimum parameters for the operation of the spectrometer are determined in part by the experimental data obtained.

Routine ^{13}C NMR Spectra. In general, a routine ^{13}C NMR spectrum of a lignin preparation is recorded at 25-50°C with a pulse width corresponding to a flip angle in the range of 30-60°, a data acquisition time of *ca.* 0.5-1.0 sec and a pulse delay of *ca.* 0.5-2 sec, with the number of transient acquisitions being *ca.* 10,000-20,000.

Quantitative ^{13}C NMR Spectra. To facilitate effective analysis of a lignin ^{13}C NMR spectrum, the spectrum must be quantitative as far as peak areas are concerned (*7, 8*). In this regard the following factor must be considered: the excitation pulse width; the time required to reestablish the Boltzmann equilibrium state after the rf pulse; elimination of nuclear Overhauser enhancement due to proton decoupling; and the number of transient acquisitions to obtain a reasonable signal-to-noise ratio. In general, a quantitative ^{13}C NMR spectrum of lignin can be obtained by an inverse gated decoupling (IGD) pulse sequence using a 90° pulse with a data acquisition time of *ca.* 0.6 sec and pulse delay of *ca.* 12 sec employing *ca.* 20,000 transient acquisitions at an operational frequency of more than 50.3 MHz. The quantitative accuracy of the spectrum must be verified, as will be discussed in the following section.

Results and Discussion

Analysis of the ^{13}C NMR Spectrum of Milled Wood Lignin (MWL) from Spruce (*Picea glauca*). Figure 1 shows a quantitative ^{13}C NMR spectrum of milled wood lignin (MWL) from spruce (*Picea glauca*). It was recorded with a Bruker WM 250 NMR spectrometer operating at 62.9 MHz for the frequency of the ^{13}C nuclei and using an IGD pulse sequence and quadrature detection (*15*). Deuterated dimethylsulfoxide (DMSO-d_6) was used as solvent. The spectrum was acquired from a 10 mm o.d. sample tube with a sample concentration of approximately 17.5% (*w/v*) at 50°C using a 90° pulse corresponding to a 22 msec pulse width, 14,285 Hz sweep width and 11 sec pulse delay with approximately 15,000 transient acquisitions. TMS was used as the internal chemical shift reference. DEPT or APT spectra of the samples may be useful for discriminating between primary and tertiary carbon signals, on the one hand, and secondary and quaternary carbon signals, on the other. The absence of signals in the 110-103 ppm chemical shift range of the spectrum shows clearly that the lignin is a guaiacyl lignin characteristic of those found in softwoods (gymnosperms).

Quantitative Nature of the Spectrum. The spectrum shows that the signal at 55.6 ppm, an aromatic methoxyl carbon, integrates at approximately 0.98 carbon per aromatic ring (Tables I and II). This corresponds to a frequency of approximately 0.98 OCH_3 per C_9-unit for the MWL. The elemental analysis of the MWL gave a methoxyl content of 0.94 per C_9-unit [$C_9H_{7.66}O_2(H_2O)_{0.44}(OCH_3)_{0.94}$] MWL (*15, 16*). Thus, the methoxyl content estimated from the ^{13}C NMR spectrum deviates by approximately 4% from that obtained by elemental analysis. This is well within the ±5% deviation that can be expected for the ^{13}C NMR spectroscopic estimate, and thus the ^{13}C NMR spectrum has been confirmed to be quantitative.

Selection of the 160-103 ppm Chemical Shift Range of the Predominantly Aromatic Region as an Internal Standard. In general, the predominantly aromatic region (160-103 ppm) contains an insignificant number of signals originating from contaminants if the MWL preparation is properly purified (*16*). It is also well established that spruce MWL contains approximately three 4–*O*-alkylated coniferyl

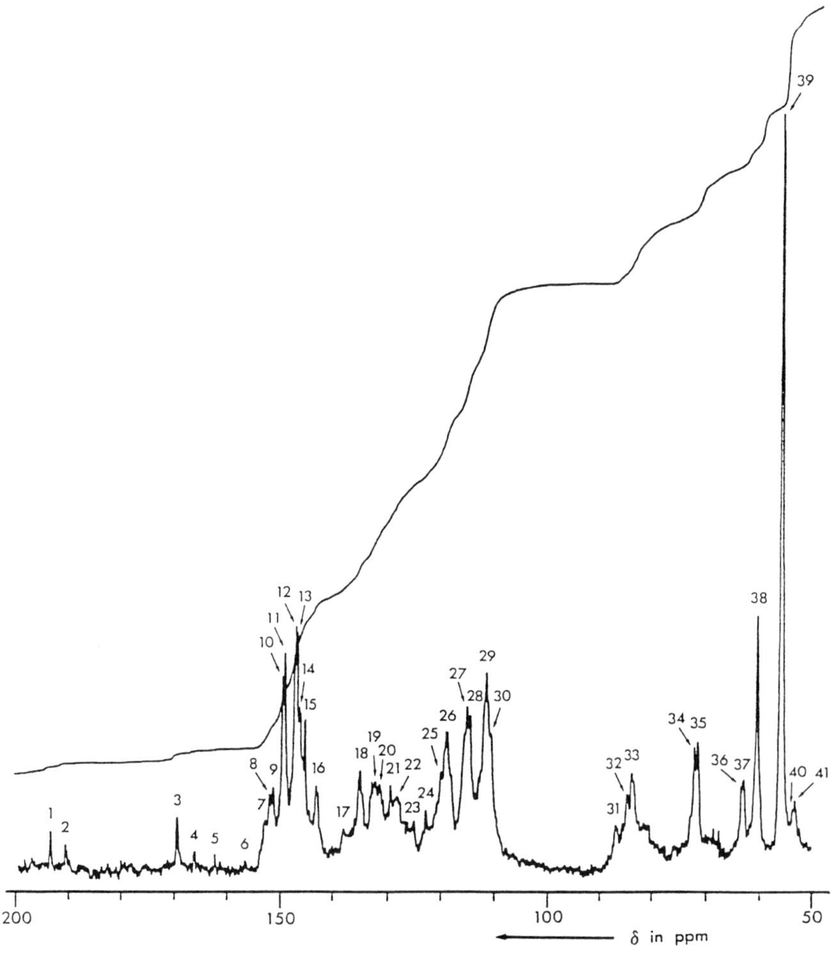

Figure 1. Quantitative ^{13}C NMR spectrum of milled wood lignin (MWL) from spruce (*Picea glauca*) obtained by inverse gated proton decoupling. Solvent: DMSO-d_6. (Adapted from ref. *15*.)

Table I. Number of Carbons Corresponding to Chemical Shift Regions for the ^{13}C NMR Spectrum of Spruce (*Picea glauca*) Milled Wood Lignin (MWL).[a]

Chemical Shift Range (ppm)	Integral[b]	Number of Carbons[c] per Aromatic Ring
194-193	1.1	0.06
192-191	0.7	0.04
158-154	0.8	0.04
154-150.5	6.8	0.36
150.5-148.5	6.2	0.33
148.5-145.5	19.8	1.04
145.5-140	6.1	0.32
140-124	27.9	1.47
124-103	48.4	2.56
90-57	44.0	2.32
57-54.5	18.6	0.98
54.5-52.5	2.4	0.13

[a] Adapted from reference *15*.
[b] Total integral for 160-100 ppm chemical shift range = 116. Thus, the integral for one aromatic carbon = 116/6.12 = 18.95, assuming that spruce MWL contains 3 units of Ar–CH=CH–CHO and 3 units of Ar–CH=CH–CH_2OH substructures per 100 C_9-units.
[c] For type of carbon see Table II.
Ar = 3-methoxy-4-hydroxyphenyl.

Table II. ^{13}C Chemical Shifts and Signal Assignments for Spruce (*Picea glauca*) Milled Wood Lignin (MWL).[a] Solvent: DMSO-d_6.

Signal Number	Chemical Shift (ppm)	Signal Intensity[b]	Assignment[c]
1	193.4	w	C=O in Ar–CH=CH–CHO C=O in Ar–CO–CH(–OAr) –C–
2	191.6	vw	C=O in Ar–CHO
3	169.4	w	ester C=O in R–O–CO–CH_3
4	166.2	vw	C=O in Ar–COOH; ester C=O in Ar-CO-OR
5	162.3	vw	unknown
6	156.4	vw	C-4 in H-units
7	152.9	w	C-3/C-3′ in etherified 5–5′ units; C-7 in Ar–CH=CH–CHO units
8	152.1	w	C-3/C-5 in etherified S-units and B-ring of 4–*O*–5′ units
9	151.3	w	C-4 in etherified G-units with 7-C=O group
10	149.4	s	C-3 in etherified G-units
11	149.1	s	C-3 in etherified G-units (8–*O*–4′ type)
12	146.8	s	C-4 in etherified G-units
13	146.6	s	C-3 in nonetherified G-units (8–*O*–4′ type)
14	145.8	m	C-4 in nonetherified G-units C-4′ in ring B of 8–5′ units
15	145.0	m	C-4/C-4′ of etherified 5–5′ units
16	143.3	m	C-3′ in ring B of 8–5′ units C-4/C-4′ of nonetherified 5–5′ units
17	138.0	w	unknown
18	134.6	w	C-1 in etherified G-units
19	132.4	w	C-5/C-5′ in etherified 5–5′ units
20	131.1	m	C-1 in nonetherified G-units
21	129.3	w	C-8 in Ar–CH=CH–CHO
22	128.0	w	C-7 and C-8 in Ar–CH=CH–CH_2OH
23	125.9	w	C-5/C-5′ in nonetherified 5–5′ units
24	122.6	vw	C-1 and C-6 in Ar–CO–C–C units
25	119.9	m	C-6 in G-units
26	118.4	s	C-6 in G-units
27	115.1	s	C-5 in G-units
28	114.7	s	C-5 in G-units
29	111.2	s	C-2 in G-units
30	110.4	m	C-2 in G-units
31	86.6	vw	C-8 in S type 8–*O*–4′ units (*erythro*)
32	84.6	w	C-8 in G type 8–*O*–4′ units (*threo*)
33	83.8	m	C-8 in G type 8–*O*–4′ units (*erythro*)
34	71.8	m	C-7 in G type 8–*O*–4′ units (*erythro*)
35	71.2	m	C-7 in G type 8–*O*–4′ units (*threo*) C-9 in G type 8–8′ units
36	63.2	w	C-9 in G type 8–*O*–4′ units w/ 7-C=O
37	62.8	w	C-9 in G type 8–5′, 8–1′ units
38	60.2	s	C-9 in G type 8–*O*–4′ units
39	55.6	vs	C in Ar–OCH_3
40	53.9	vw	C-8/C-8′ in 8–8′ units
41	53.4	vw	C-8 in 8–5′ units

[a] Adapted from reference *15*.
[b] vw = very weak; w = weak; m = moderate; s = strong; vs = very strong.
[c] H = *p*-hydroxyphenylpropane; G = guaiacylpropane; S = syringylpropane.

aldehyde units and three 4–*O*-alkylated coniferyl alcohol units per 100 C_9-units. Thus, the signals in this range embody those from 6.12 aromatic and alkenyl carbons. It follows that the total integral over this range divided by 6.12 is equivalent to one carbon atom.

Estimation of Phenylcoumaran (8–5′) Units in the Spruce MWL. The 54.5-52.5 ppm chemical shift range embodies 0.13 carbons/aromatic ring. Consequently the total frequency of 8–5′ and 8–8′ linkages is approximately 12-14 per 100 C_9-units. Since the frequency of 8–8′ linkages in spruce lignin was estimated to be 2 per 100 C_9-units by potassium permanganate oxidation (*17, 18*), the total carbons in the 8–8′ units is viewed to approximate 4 per 100 C_9-units. It follows, therefore, that the total frequency of 8–5′ linkages is approximately 9 per 100 C_9-units and that the total C-8 carbons in the 8–5′ units is approximately 8-10 per 100 C_9-units. This carbon gives rise to the signal at 53.4 ppm (*19*). Erickson *et al.* (*17*) previously gave a frequency of approximately 12 such units per 100 C_9-units for spruce MWL, on the basis of potassium permanganate oxidation studies.

Nature of Signals in the 194-191 ppm Chemical Shift Region. The chemical shift range around 194 ppm embodies approximately 0.06 carbons per aromatic ring, and may include the -CHO of 4–*O*-alkylated coniferyl aldehyde groups, and the -C=O of 4–*O*-alkylated uncondensed guaiacylpropane units with a 7-carbonyl group. The former were estimated as 3 units per 100 C_9-units on the basis of the Wiesner reaction (*18*), and consequently the 4–*O*-alkylated guaiacylpropane units with a 7-carbonyl group may also approximate 3 in number per 100 C_9-units. In addition, the signal at 191 ppm approximates 0.04 carbons per aromatic ring and corresponds to the -CHO of 4–*O*-alkylated vanillin substructures. Thus, the total number of 4–*O*-alkylated uncondensed guaiacyl residues with a 7-carbonyl is approximately 7 per 100 C_9-units.

Estimation of the Total Biphenyl (5–5′) Units in the Spruce MWL. The signals in the 154-150.5 ppm chemical shift range embody approximately 0.36 carbons/aromatic ring. The carbons that could give rise to the resonances in this region are: C-3/C-3′ of etherified biphenyl (5–5′) substructures and C-3′ of half-etherified 5–5′ units (*15*); C-3′/C-5′ in ring B of diphenyl ether (4–*O*–5′) substructures; C-4 of 4–*O*-alkylated uncondensed guaiacyl residues with a 7-carbonyl group (*4*); C-3/C-5 of 4–*O*-alkylated, uncondensed syringylpropane units (*1, 20*); and C-7 of cinnamaldehyde (Ar–CH=CH–CHO) units (*7*).

In general, spruce MWL contains approximately two 4–*O*–5′ substructures (*18*), seven 4–*O*-alkylated uncondensed guaiacyl residues with a 7-carbonyl group (see Nature of Signals in the 194-191 ppm Chemical Shift Region), three 4–*O*-alkylated coniferyl aldehyde substructures, one 4–*O*-alkylated uncondensed syringylpropane unit and three cinnamaldehyde (Ar–CH=CH–CHO) units per 100 C_9-units. The total number of the C-3/C-3′ in etherified 5–5′ units and C-3′ in half-etherified 5–5′ units is therefore approximately 0.20/aromatic ring [= 0.36 – 0.04 – 0.07 – 0.02 – 0.03]. This implies that approximately 19-21 guaiacylpropane units per 100 units are involved in etherified and half-etherified 5–5′ units.

Furthermore, the 145.5-140 ppm chemical shift range embodies approximately 0.32 carbons/aromatic ring. The carbons giving rise to these resonances are the C-4/C-4′ in the 5–5′ units, regardless of whether etherified or not (*15*), and the C-3′ in 8–5′ units (*21*). As discussed previously (see Estimation of Phenylcoumaran (8–5′) Units in the spruce MWL), the total C-8 of 8–5′ units is estimated from the 54.5-52.5 ppm chemical shift range to be approximately 8-10 per 100 C_9-units. Thus, even considering the possible extent of experimental error, the spruce MWL contains approximately 22-24 guaiacylpropane residues per 100 C_9-units that are involved in 5–5′ linkages, *i.e.* the total frequency of 5–5′ linkages is approximately 12 per 100 C_9-units. This value is slightly higher than the 10 per 100 C_9-unit estimated using the potassium permanganate oxidation method (*17, 18*).

A further inspection of the ^{13}C NMR spectrum of spruce MWL reveals that fully etherified 5–5′ units predominate over those which are partially (half) or non-etherified. This is evidenced by the presence of signals in the 145-144 ppm range and the absence of any resonances between 143 and 142 ppm (*15*). The former correspond to C-4/C-4′ in etherified 5–5′ substructures and C-4 in the etherified ring of partially etherified 5–5′ linkages, while the latter would correspond to C-4 in the phenolic rings of half- or non-etherified 5–5′ units. The C-3′ in 8-5′ units would also give rise to a signal in the range between 144 and 143 ppm (*21*). Thus, of the 23-25 guaiacylpropanes per 100 C_9-units involved in 5–5′ linkages, approximately 19-21 are present as fully or half- etherified units. The remainder, approximately 4-6, are half or non-phenolic 5–5′ units. It is of interest to note that the fully etherified 5–5′ units may occur in lignin in the form of dibenzo-2H,3H-1,4-dioxicin structures (*22, 23*) that may undergo mechanolysis during the ball milling of spruce wood in the process of milled wood lignin isolation to give partially half etherified 4′–*O*–8″ linked 5–5′ units (*24*).

Estimation of 8-Aryl Ether (8–*O*–4′) Units in Spruce MWL. The 150.5-148.5 ppm chemical shift range embodies 0.33 carbons/aromatic ring. The signals in this range correspond to the C-3 of etherified guaiacyl groups (*20*). The frequency of 7–*O*–4′ and 8–*O*–4′ linkages can be estimated as approximately 32-34 per 100 C_9-units. This does not include 7–*O*–4′ and 8–*O*–4′ units having a 5–5′ linkage as an aryl ether moiety, such as the dibenzodioxicin and half-etherified 4′–*O*–8″ linked 5–5′ units. Assuming that the fully and half-etherified 5–5′ units contributing to 7–*O*–4′ and 8–*O*–4′ linkages are in dibenzodioxicin and (half-etherified) 4′–*O*–8′ linked 5–5′ substructures, then the total frequency of 7–*O*–4′ and 8–*O*–4′ linkages would be approximately 53 [= 33 + 20] per 100 C_9-units. Since the frequency of 7–*O*–4′ units is estimated to be approximately 8 per 100 C_9-units (*17*, *18*, *25*), the frequency of 8–*O*–4′ linkages is estimated at 44-46 per 100 C_9-units.

Very recently, Ede and Kilpeläinen searched for 7–*O*–4′ units in the MWL's from a number of softwood and hardwood species employing 2D NMR spectroscopic methods such as the HOHAHA and HMQC techniques (*26*). The results indicate that, if 7–*O*–4′ units are present in MWL's, they are present at a level below the detection limit of the techniques, that is less than 0.3 per 100 C_9-units. The implications of this finding are twofold: the frequency of the 8–*O*–4′ linkages in softwood lignin is approximately 52-54 per 100 C_9-units, which is much higher than the values previously estimated; and the number of dibenzo-2H,3H-1,4-dioxicin structures present in softwood lignin is insignificant.

Estimation of Degree of 'Condensation'. Softwood lignins contain several 'condensed' aromatic structures such as the biphenyl (5–5′) and phenylcoumaran (8–5′) linkages. If they did not contain 'condensed' aromatic structures, then the signals in the 124-103 ppm chemical shift range should embody 3 carbons per aromatic ring. The deficiency in the number of carbons in this chemical shift range must then be due to the presence of 'condensed' aromatic structures. Thus, the total number of 'condensed' aromatic units and the degree of 'condensation', can be estimated by subtracting from the total number of carbons per aromatic ring estimated from the 124-103 ppm chemical shift range. Thus, the degree of 'condensation' for the spruce MWL was approximately 0.44 [= 3 - 2.56] per aromatic ring, *i.e.* 44 per 100 C_9-units.

Analysis of the ^{13}C NMR Spectrum of Milled Wood Lignin (MWL) from Zhong-Yang Mu (*Bischofia polycarpa*). Zhong-Yang Mu (*Bischofia polycarpa*; Euphorbiaceae) is a hardwood-timber species widely distributed in southeastern China. Figure 2 shows the quantitative ^{13}C NMR spectrum of the MWL from this wood species acquired using inverse gated proton decoupling (*27*). The quantitative nature of the spectrum is confirmed by the fact that signal 29 (aromatic methoxyl carbons) in the spectrum integrates to approximately 1.16 carbons per phenyl group

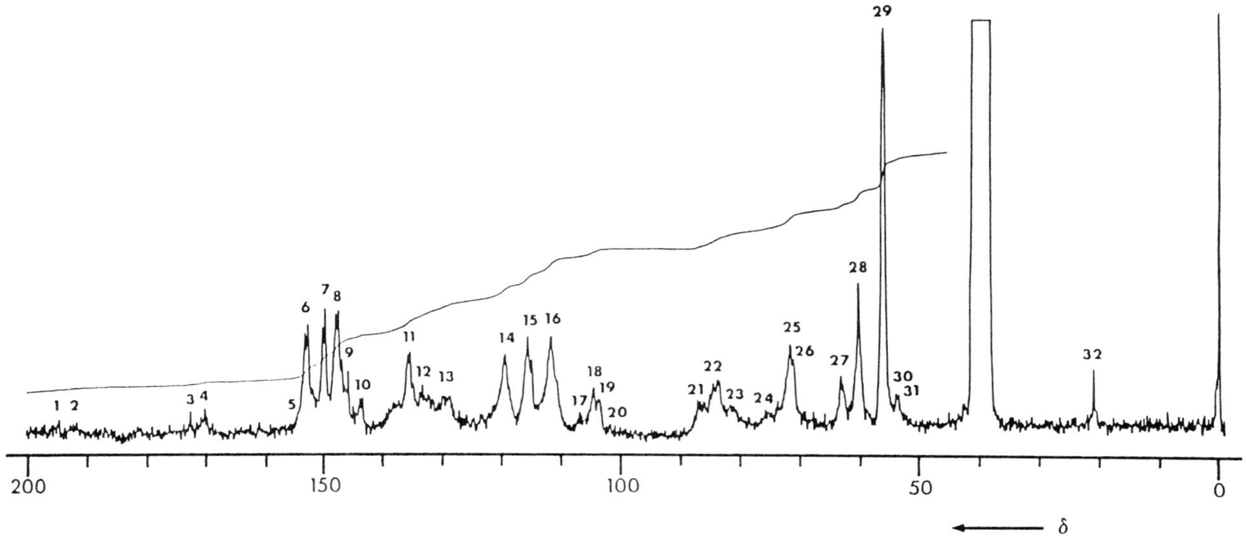

Figure 2. Quantitative ^{13}C NMR spectrum of milled wood lignin (MWL) from Zhong-Yang Mu (*Bischofia polycarpa*) obtained by inverse gated proton decoupling. Solvent: DMSO-d_6. (Adapted from ref. *27*.)

(Tables III and IV). This corresponds to a methoxyl content of approximately 1.16 per C_9-unit. The elemental analysis of the MWL gave a methoxyl content of 1.13 per C_9-unit [$C_9H_{6.49}O_2(H_2O)_{0.92}(OCH_3)_{1.13}$]. Thus, the methoxyl content of the MWL estimated from the ^{13}C NMR spectrum differs by less than 3% from the value obtained by elemental analysis. This is within a 5% deviation from the latter value, the limit of error that can be expected for the ^{13}C NMR spectroscopic estimation.

CH, CH_2 and CH_3 ^{13}C NMR Subspectra for *B. polycarpa* MWL. Figure 3 shows the CH, CH_2 and CH_3 ^{13}C NMR subspectra of the *B. polycarpa* MWL, edited by the DEPT sequence (*28-31*), while Figure 4 depicts the quaternary ^{13}C NMR subspectrum of the MWL obtained by subtracting the DEPT edited CH, CH_2 and CH_3 subspectra from the full inverse gated decoupled ^{13}C NMR spectrum (*27*). In Table III, the quantitative ^{13}C NMR spectrum of the MWL is divided into several spectral regions on the basis of the DEPT edited subspectra. Table IV shows the assignments of signals which were deduced from previously published data (*7-9, 15, 16, 19, 20, 21*).

Verification of the Presence of Carbohydrates. Carbohydrate analysis of the *B. polycarpa* MWL showed that it possessed a carbohydrate content of approximately 6.2%. Indeed, the inversed gated decoupled ^{13}C NMR spectrum of the MWL (Figure 2 and Table IV) reveals that the MWL contains a substantial amount of carbohydrates as evidenced by the presence of signals 4, 20, 24 and 32 at 169.6, 101.6, 76-73 and 20.9 ppm corresponding to *O*-acetyl C=O, C-1 and C-2 of xylan, C-3 and C-4 of xylan, and *O*-acetyl CH_3, respectively. Thus, the ^{13}C NMR spectral data support the results of the carbohydrate analysis of the MWL (*27*).

Guaiacyl-Syringyl Nature of *B. polycarpa* MWL. The quantitative ^{13}C NMR spectrum of the *B. polycarpa* MWL (Figure 2) further shows that its lignin is of a guaiacyl-syringyl type. The presence of guaiacyl-propane units is evidenced by signals 14, 15 and 16 at 119.6, 115.8 and 112.2 ppm corresponding to C-6, C-5 and C-2, respectively, of the guaiacyl aromatic ring. The presence of syringylpropane units is revealed in signals 18 and 19 at 104.9 and 103.7 ppm, both corresponding to C-2/C-6 of the syringyl aromatic ring, in addition to signal 6 at 152.6-152.3 ppm corresponding to C-3/C-5 of 4–*O*-alkylated syringyl residues (*20*).

Predominance of 8-Aryl Ether (8–*O*–4′) Linkages. In contrast to the rather weak signal 9 at 145.6 ppm corresponding to C-4 of the guaiacyl ring, signals 7 and 8 at 149.9-149.4 ppm and 147.7-147.2 ppm are very strong (Figure 2). Since signals 7 and 8 correspond to C-3 and C-4 of 4–*O*-alkylated guaiacyl rings (*20*), these moieties must be present in the lignins predominantly as 4–*O*-alkyl ethers. Moreover, they seem to be present in the lignin in the form of 8–*O*–4′ units because of the presence of relatively strong signals 22, 25 and 28 at 84.6, 71.8 and 60.2 ppm, respectively. These signals correspond to C-8, C-7 and C-9/8–*O*–4′ linkages, respectively (*19*). The presence of syringylpropane structures is also revealed in signals 18 and 19 at 104.9 and 103.7 ppm, both corresponding to the C-2/C-6 of syringyl rings, in addition to signal 6 at 152.6-152.3 corresponding to C-3/C-5 of 4–*O*-alkylated syringyl residues (*20*). In addition to signals 25 and 26, the presence of signal 21 at 87.0 ppm further indicates that 4–*O*-alkylated syringylpropane units are also present in the lignins mostly in the form of 8–*O*–4′ linkages. Signals 30 and 31 at 53.8 and 53.4 ppm indicate the presence of 8–8′ and 8–5′ substructures in the lignin sample.

Selection of Aromatic Region (160-103 ppm) as an Internal Standard. The aromatic region (160-103 ppm) in the inversed proton decoupled ^{13}C NMR spectrum, was again chosen as an internal standard to analyze the spectra quantitatively, since there are no signals from carbohydrate contaminants. The DEPT edited CH subspectrum (Figure 3a) of the *B. polycarpa* MWL indicates the presence of cinnamaldehyde- and cinnamyl-alcohol-type structures as revealed in signals 1, 5 and

Table III. Number of Carbons Corresponding to Chemical Shift Ranges for ^{13}C NMR Spectrum of Zhong-Yang Mu (*Bischofia polycarpa*) Milled Wood Lignin (MWL).[a]

Spectral Region	Chemical Shift Range (ppm)	Integral[b]	Number of Carbons[c] Per Aromatic Ring
Aromatic quaternary C[d]	156-128	33.0	3.61
Aromatic tertiary C	128-103	23.0	2.51
Syringyl C-2/C-6	108-103	3.3	0.36
Oxygenated aliphatic C[e]	90.0-57.5	28.6	3.13
Aromatic methoxyl C	57.5-54.5	10.6	1.16
C-8 in 8–8′ & 8–5′	54.5-52.5	0.8	0.09

[a] Adapted from reference *27*.

[b] Total integral for 160-103 ppm chemical shift range = 56; thus, the integral for one aromatic carbon = 56/6.12 = 9.15, assuming that the Zhong-Yang Mu MWL contains 3 Ar–CH=CH–CHO and 3 Ar–CH=CH–CH_2OH substructures per per 100 C_9-units.

[c] For type of carbon see Table IV.

[d] Including 0.12 vinyl carbons/aromatic ring.

[e] Consisting of carbons in lignin sidechains and carbohydrates.

Table IV. ^{13}C Chemical Shifts and Signal Assignments for Zhong-Yang Mu (*Bischofia polycarpa*) Milled Wood Lignin (MWL).[a] Solvent: DMSO-d_6.

Signal Number	Chemical Shift (δ)	Intensity[b]	Assignment[c]
1	194.0	vw	C=O in Ar–CH=CH–CHO units; C=O in Ar–CO–CH(–OAr) –C– units
2	191.6	vw	C=O in Ar–CHO units
3	172.0	vw	C=O in -COOH of aliphatic acid units
4	169.6	w	C=O in *O*-acetyl groups of xylan units
5	152.9	vw	C-7 in Ar–CH=CH–CHO units
6	152.9-152.3	s	C-3/C-3′ in etherified 5–5′ units; C-3/C-5 in etherified S-units and B-ring of 4–*O*–5′ units
7	149.4-149.2	s	C-3 in etherified G-units
8	147.7-147.2	s	C-3 in etherified G-units (8–*O*–4′ type); C-3 in nonetherified G-units; C-4 in etherified G-units; C-3/C-5 in nonetherified S-units
9	145.6	w	C-4 in etherified G-units
10	143.2	m	C-4′ in B-ring of 8–5′ units; C-4/C-4′ in nonetherified 5–5′ units; C-3′ in B-ring of 8–5′ units; C-4/C-4′ in nonetherified 5–5′ units;
11	135.6	s	C-1 in etherified G-units; C-4 in etherified S-units
12	135.0-131.0	m	C-1 in nonetherified G- and S-units; C-5/C-5′ in etherified 5–5′ units; C-1 in nonetherified G-units
13	130.0-129.0	m	C-8 in Ar–CH=CH–CHO units; C-7 and C-8 in Ar–CH=CH–CH_2OH units
14	119.0	s	C-6 in G-units
15	115.8	s	C-5 in G-units
16	112.2	s	C-2 in G-units
17	106.5	w	C-2/C-6 in etherified and nonetherified S-units with 7-C=O group
18	104.7	m	C-2/C-6 in etherified and nonetherified S-units
19	103.7	m	C-2/C-6 in etherified and nonetherified S-units and S-type 8–8′ units
20	101.6	vw	C-1 in xylan
21	87.0	m	C-8 in S-type 8–*O*–4′ (*erythro*)
22	84.6	s	C-8 in G-type 8–*O*–4′ units (*erythro*)
23	81.2	m	unknown
24	76.0-73.0	w	C-2/C-3/C-4 in xylan units
25	72.2	s	C-7 in G-type 8–*O*–4′units (*erythro*)
26	71.6	w	C-9 in G- and S-type 8–*O*–4′units (both *threo*); C-9 in G- and S-type 8–8′ units
27	62.9	s	C-9 in G- and S-type 8–5′ and 8–1′ units C-9 in G- and S-type 8–*O*–4′ units with a 7-C=O; C-5 in xylan
28	60.2	s	C-9 in G- and S-type 8–*O*–4′ units
29	55.8	vs	C in Ar-OCH_3 units
30	53.8	w	C-8/C-8′ in G- and S-type 8–8′ units
31	53.4	vw	C-8 in G- and S-type 8–5′ units
32	20.4	s	C in CH_3 of *O*-acetyl groups of xylan

[a] Adapted from reference 27.
[b] vw = very weak; w = weak; m = moderate; s = strong; vs = very strong.
[c] G = guaiacylpropane; S = syringylpropane.

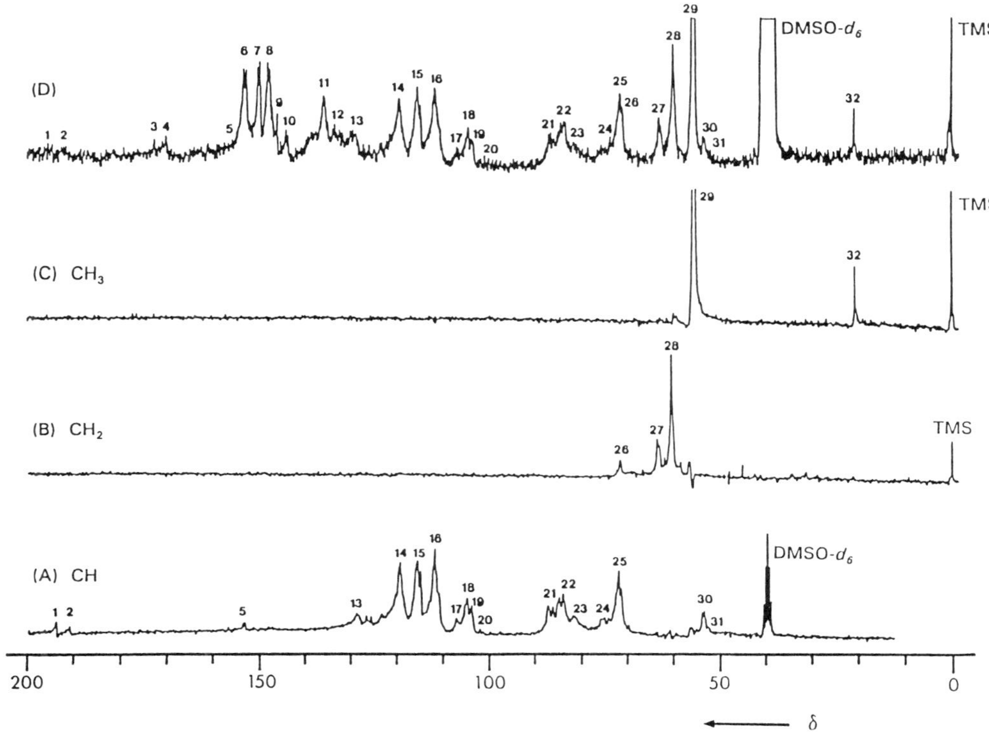

Figure 3. DEPT-edited ^{13}C NMR subspectra of milled wood lignin (MWL) from Zhong-Yang Mu (*Bischofia polycarpa*): (A) CH, (B) CH_2, (C) CH_3 subspectra, and (D) overall spectrum obtained by inverse gated proton decoupling. Solvent: DMSO-d_6. (Adapted from ref. *27*.)

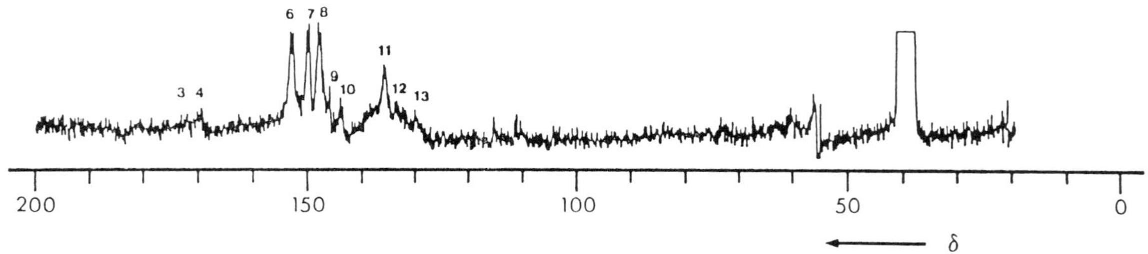

Figure 4. DEPT-edited ^{13}C NMR quaternary carbon subspectrum of milled wood lignin (MWL) from Zhong-Yang Mu (*Bischofia polycarpa*): difference spectrum produced by subtracting the DEPT-edited CH, CH_2 and CH_3 subspectra from the overall spectrum obtained with inverse gated proton decoupling. (Adapted from ref. 27.)

13 at 194.0, 152.9 and 129.7-129.4 ppm, respectively. From the integration of these signals, the cinnamaldehyde- and cinnamyl-alcohol-type units in the lignin were each estimated to be approximately 0.03 per aromatic ring. Consequently, the integral over the aromatic region in the inverse gated proton decoupled ^{13}C NMR spectrum should again correspond to 6.12 carbons per aromatic ring.

Estimation of the Ratio of Guaiacylpropane to Syringylpropane Units, Methoxyl Content and Frequency of Diphenyl Ether (4–*O*–5′) Linkages. Since an uncondensed syringylpropane unit has two equivalent C-2/C-6 carbons, an approximate frequency of syringylpropane moieties per C_9-unit in the *B. polycarpa* MWL can be estimated from the total number of C-2/C-6 carbons. (Table III), *i.e.* 0.18 per C_9-unit [= 0.36/2]. Because 4-hydroxyphenylpropane units in MWL are negligible, as indicated by nitrobenzene oxidation studies (*27*), the MWL consists mostly of guaiacylpropane and syringylpropane units in an approximate ratio of 82:18. Assuming this is correct, then the methoxyl content of the MWL would be 1.18 per C_9-unit. The methoxyl content of the MWL is thus estimated somewhat higher than the elemental analytical value of 1.13 per C_9-unit and the value of 1.16 per C_9-unit estimated from the aromatic methoxyl spectral region (Table III). The excess is probably due to the presence of diphenyl ether (4–*O*–5′) units. The chemical shift of C-2 and C-6 in 5-aroxyguaiacylpropane is in the 110-103 ppm range. The frequency of 4–*O*–5′ linkages can be estimated to be approximately 0.03-0.04 per C_9-unit [= 1.18 - (1.13 - 1.16)/2] for the MWL. Thus, it follows that the methoxyl content of the MWL is approximately 1.14 moles per C_9-unit. The MWL would then consist of guaiacyl-propane and syringylpropane units in an approximate ratio of 86:14.

Estimation of Degree of Condensation. If the *B. polycarpa* MWL did not contain units condensed in the aromatic ring, the number of aromatic tertiary carbons would be 2.86 carbons per C_9-unit [= 0.86 x 3 + 0.14 x 2]. From the aromatic tertiary carbon spectral region (Table III), a value of 2.51 carbons per C_9-unit is obtained. The degree of condensation for the aromatic moieties in the MWL then approximates 0.35 [= 2.86 - 2.51] per C_9-unit, *i.e.* 35 per 100 C_9-units. The *B. polycarpa* MWL is, therefore, a rare guaiacyl-syringyl lignin (hardwood lignin), in which the guaiacyl-propane residues predominate over syringylpropane moieties.

Dehydrogenative Polymerisates (DHPs) from Monolignols. When coniferyl alcohol was originally oxidized by dioxygen using mushroom press-sap as catalyst, it underwent dehydrogenative polymerization to form a dehydropolymerisate (DHP), which appeared to be lignin-like (*32*). Eventually, the major oxidizing enzyme in the mushroom press-sap was identified as laccase (EC 1.10.3.2). In the 1950's, Freudenberg *et al.* (*33-36*) reinvestigated the dehydrogenative polymerization of coniferyl alcohol and its homologs (monolignols) by dioxygen using fungal laccase. Freudenberg *et al.* (*37*) also found laccase-like and peroxidase activities in the differentiating xylem of spruce (*Picea abies*). Furthermore, Higuchi and Ito found that coniferyl alcohol underwent dehydrogenative polymerization to give a DHP by hydrogen peroxide with peroxidase (EC 1.11.1.7) (*38*). In addition, the monolignol substrate was also oxidized to a DHP by dioxygen when laccase from the Japanese lacquer tree (*Rhus vernicifera* Stokes) and phenoloxidase were used as catalysts. Peroxidase was further implicated in the biosynthesis of lignin by experiments using crude enzyme preparations that showed an increase in the rate of DHP formation from coniferyl alcohol on addition of hydrogen peroxide (*39*). Based on these results, Freudenberg (*40*) postulated that laccase and peroxidase could be involved in the biosynthesis of lignin *in vivo*.

Several years later, Nakamura (*41*) showed that a purified *Rhus* laccase (EC 1.10.3.2) could catalyze the oxidation of pyrocatechol, hydroquinone and ascorbic acid by dioxygen, but could not catalyze the oxidation (dehydrogenative polymerization) of coniferyl alcohol by dioxygen to form a DHP, at least on the basis of turbidity measurements and changes in absorbance at 280 nm up to a reaction time of 16 h. The enzyme also did not catalyze the oxidation of creosol or guaiacol. These results led Nakamura to conclude that laccase does not participate in the *in vivo* biosynthesis of lignin. From hindsight, this conclusion was premature since Nakamura did not investigate the kinetics of the reactions, particularly the rates of oxidation and the nature of the oxidation products. Very recently, Okusa *et al.* (*42*) showed that *Rhus* laccase does catalyze the oxidation of coniferyl alcohol by dioxygen, but very slowly. After a reaction time of 144 h, most of the substrate had disappeared with the formation of *ca.* 20 mol% pinoresinol, 25 mol% dehydrodiconiferyl alcohol, and 2 mol% guaiacylglycerol-8-coniferyl ether, but no DHP.

Recently, a laccase was purified from a sycamore (*Acer pseudoplatanus* L.) cambial cell suspension culture medium (*43*). This *Acer* cell laccase was further studied and characterized (*44, 45*). Sterjiades *et al.* (*46*) further showed that the enzyme could oxidize monolignols, suggesting that it is involved in lignin biosynthesis. Savidge and Udagama-Randeniya (*47*) found cell wall-bound coniferyl alcohol oxidase activity in the differentiating xylem of five conifer species (gymnosperms) using a *p*-anisidine assay. The authors postulated that this activity is responsible for the polymerization of monolignols to lignins. Similarly, an acetone-insoluble cell wall enriched fraction in *Forsythia* species catalyzed the preferential formation of (+)-pinoresinol, over its (–)-antipode, from *trans*-coniferyl alcohol with no cofactor other than oxygen (*48*). Very recently, a laccase has been isolated and characterized from the cell walls of differentiating xylem of loblolly pine (*Pinus taeda*) (*49*). This laccase coincided in time and place with lignin formation in pine. In addition, it catalyzed the dehydrogenative polymerization of coniferyl alcohol by dioxygen to produce the corresponding DHP *in vitro*.

Monolignol Oxidation Catalyzed by Different Laccases. Fungal laccases from *Coriolus versicolor* (*50*) and *Pycnoporus coccineus* (*51*) readily catalyze the oxidation of coniferyl alcohol by dioxygen to produce the corresponding DHP (*42*), although *Pycnoporus* laccase catalyzes the dehydrogenative polymerization of the substrate much faster than both *Coriolus* and *Rhus* laccases. The latter are very ineffective in catalyzing this oxidation. In these oxidations, the reaction time required for the disappearance of all the substrate (in 20 mL 0.14 M solution) was 12 h for the *Pycnoporus* laccase, 43 h for the *Coriolus* laccase, and 145 h for *Rhus* laccase. In the *Pycnoporus* laccase-catalyzed oxidation, a DHP was produced in approximately 96 mol% yield. In the *Coriolus* laccase-catalyzed oxidation, dehydrodiconiferyl alcohol, an oligomeric mixture and DHP were produced in approximately 7 mol%, 60 mol% and 28 mol% yields, respectively. The constituents of the oligomeric mixture were not identified. In the *Rhus* laccase-catalyzed oxidation, by contrast, pinoresinol, dehydrodiconiferyl alcohol, guaiacylglycerol-8-coniferyl ether and an oligomeric mixture were produced in approximately 21 mol%, 26 mol%, 2 mol% and 40 mol% yields, respectively, but no DHP was formed. In these laccase-catalyzed oxidations, the major dimeric products were pinoresinol and dehydrodiconiferyl alcohol while guaiacylglycerol-8-coniferyl ether was a minor dimeric product.

Monolignol Oxidation Catalyzed by Peroxidases. Both peroxidases, Type I and II, from horseradish *Armoracia rusticana* readily catalyze the oxidation of coniferyl alcohol by hydrogen peroxide to produce the corresponding DHP. In general, the rates of the peroxidase-catalyzed oxidation are much faster than those of the corresponding laccase-catalyzed oxidations. In these oxidation reactions, the reaction time required for the disappearance of all the substrate (in 20 mL 0.14 M solution) was approximately 3 h for both peroxidases. As with the laccase-catalyzed oxidations, pinoresinol and dehydrodiconiferyl alcohol were the major dimeric products, and guaiacylglycerol-8-coniferyl ether was a minor dimeric product.

Oxidation of Aromatic Compounds Catalyzed by Laccases in the Presence of a Mediator. In the presence of a mediator, such as 2,2′-azino-bis(3-ethylbenzthiazoline-6-sulfonic acid) [diammonium salt, ABTS-$(NH_4)_2$], both *Coriolus* and *Pycnoporus* laccases catalyze the oxidation of 3,4-dimethoxybenzyl alcohol in the presence of dioxygen to 3,4-dimethoxybenzaldehyde (*52, 53*). Under the same reaction conditions, *Coriolus* laccase also catalyzes the oxidation of benzyl alcohol by dioxygen to benzaldehyde, but *Pycnoporus* laccase does not. In addition, *Coriolus* laccase catalyzes the oxidation of nonphenolic toluene derivatives by dioxygen to give the corresponding benzaldehyde derivatives in the presence of a mediator (*54*), but *Pycnoporus* laccase again does not (Miyakoshi, T., personal communication, 1996). Since white rot fungi such as *Coriolus versicolor* and *Pycnoporus coccineus* decompose woody tissues using the resulting carbohydrate and lignin fragments as nutrients for their growth, one may assume that there are naturally occurring mediators present in these fungal systems that in effect activate appropriate enzymes such as laccases with respect to lignin degradation. However, the nature of these putative mediators has not been established.

Characterization of DHPs and Related Dehydrogenatively Produced Intermediates by ^{13}C NMR Spectroscopy. The characterization of DHPs and related dehydrogenatively formed intermediates by ^{13}C NMR spectroscopy is carried out in a similar manner to that described for the characterization of lignin preparations. If the intermediates of the dehydrogenative polymerization of monolignols are to be characterized, they must be isolated from the corresponding reaction mixtures. For their characterization in turn, the DHPs must themselves be free of dimeric and oligomeric components.

Isolation of Dimeric Intermediates and DHP (Dehydrogenatively Produced) from Monolignols. An aqueous reaction mixture from dehydrogenative polymerization of coniferyl alcohol was extracted by shaking with $ClCH_2CH_2Cl$. The $ClCH_2CH_2Cl/H_2O$-insoluble material (DHP) was filtered and purified. The $ClCH_2CH_2Cl$-soluble part was chromatographed on a silica gel column (50 cm x 1.5 cm i.d.) using $CHCl_3$-CH_3OH (20:1, v/v) as eluent to isolate the following individual dimeric products; pinoresinol, tlc R_f 0.57 ($CHCl_3$-CH_3OH, 9:1 v/v), mp 119-120°C (from EtOH), lit. 120-121°C (*55*); dehydrodiconiferyl alcohol, tlc R_f 0.41 ($CHCl_3$-CH_3OH, 9:1 v/v), mp 154-155° (from CH_2Cl_2-petroleum ether 1:1, v/v), lit. 156-157°C (*34*); guaiacylglycerol-8-coniferyl ether, tlc R_f 0.38 ($CHCl_3$-CH_3OH, 9:1 v/v), colorless oil, lit. colorless amorphous powder (*35*). Although the 8–*O*–4′ dimer gave a single spot by *tlc* analysis, the ^{13}C NMR spectrum of the product showed that it consisted of *erythro*- and *threo*-diastereomers in a molar ratio of approximately 2:1. The ^{13}C NMR spectral data for these three dimeric products are summarized in Table V (*42*).

DHPs Formed in the Presence of Fungal Laccases and Type II Horseradish Peroxidase. Figure 5 shows the ^{13}C NMR spectra of DHPs obtained from the dehydrogenative polymerization of coniferyl alcohol under the catalytic influence of *Coriolus* and *Pycnoporus* laccases and Type II horseradish peroxidase at pH 7. Table VI summarizes the ^{13}C chemical shifts of the important signals in these DHPs and their assignments based on published data (*7-11, 15, 19, 20, 21, 56*). The spectra of all three DHPs are rather similar. The major substructures are linkages of the types 8–5′, 8–8′ and 8–*O*–4′, while minor substructures are 5–5′ units; the completely etherified 5–5′ units may include dibenzo-2H,3H-1,4-dioxicin structures.

Table V. ^{13}C Chemical Shifts for Pinoresinol, Dehydrodiconiferyl Alcohol and Guaiacylglycerol-8-Coniferyl Ether Produced by Dehydrogenation of Coniferyl Alcohol Catalyzed by Laccases and Peroxidases.[a] Solvent: DMSO-d_6.

Carbons	^{13}C Chemical Shifts of Dimeric Products (ppm)			
	Pinoresinol	Dehydrodiconiferyl Alcohol	Guaiacylglycerol-8-Coniferyl Ether *erythro*	*threo*
C-7	84.9	87.3	71.6	70.9
C-8	53.8	53.2	83.7	84.4
C-9	71.0	63.2	60.2	60.2
C-1	132.2	132.3	133.1	133.3
C-2	110.4	110.6	111.2	111.4
C-3	147.5	147.1	147.0	147.0
C-4	145.9	146.3	145.5	145.5
C-5	115.1	114.8	114.6	114.5
C-6	118.6	118.4	119.0	119.5
OCH_3	55.6	55.6[b]	55.6[b]	55.6[b]
C-7′	84.9	129.4	128.7	128.7
C-8′	53.8	127.9	128.7	128.7
C-9′	71.0	61.8	61.6	61.6
C-1′	132.2	130.4	130.1	130.0
C-2′	110.4	110.6	109.8	109.9
C-3′	147.5	143.5	149.8	149.8
C-4′	145.9	147.4	147.9	147.7
C-5′	115.1	128.9	115.6	115.6
C-6′	118.6	115.3	119.0	119.0
OCH_3	55.6	55.5[b]	55.5[b]	55.5[b]

[a] Adapted from reference *42*.
[b] These values are interchangeable

Table VI. Resonance Assignments in the ^{13}C NMR Spectra of Dehydropolymerisates (DHPs) from Coniferyl Alcohol.[a] Solvent: DMSO-d_6

Resonance Number	Chemical Shift (ppm)	Signal Intensity[b]	Assignment
1	193.5	w	C=O in Ar–CH=CH–CHO; Ar–(C=O)–CH–C units
2	191.2	w	C=O in Ar–CHO units
3	152.8	w	C-3/C-3′ in etherified 5–5′ units
4	149.9	w	C-3 in etherified G (8–*O*–4′) units
5	147.6	m	C-3 in nonetherified G-units; C-4 in etherified G-units
6	146.1	w	C-4 in nonetherified G-units
7	143.8	w	C-3 in cyclic 8–5′ units
8	132.5	w	C-1 in nonetherified G-units
9	130.7	w	C-1 in etherified G-units
10	129.2	m	C-7/C-8 in Ar–CH=CH–CH_2OH units
11	128.7	m	C-7/C-8 in Ar–CH=CH–CH_2OH units
12	125.7	w	C-5/C-5′ in nonetherified 5–5′ units
13	120.4	w	C-6 in etherified and nonetherified G-units
14	118.8	w	C-6 in etherified and nonetherified G-units
15	115.4	m	C-5 in etherified and nonetherified G-units
16	110.7	m	C-2 in etherified and nonetherified G-units
17	87.3	w	C-7 in 7–*O*–4′ and cyclic 8–5′ units
18	85.3	w	C-7 in 8–8′ units; C-8 in 8–*O*–4′ units
19	71.1	m	C-9 in 8–8′ units; C-7 in 8–*O*– 4′ units
20	63.1	m	C-9 in 7–*O*–4′ and cyclic 8–5′ units
21	61.8	s	C-9 in Ar–CH=CH– CH_2–OH units
22	60.3	m	C-9 in 8–*O*– 4′ units
23	55.8	s	C in Ar–OCH_3 units
24	53.7	m	C-8/C-8′ in 8–8′ units
25	53.1	w	C-8 in 8–5′ units

[a] Adapted from reference *42*.
[b] w = weak; m = moderate; s = strong.

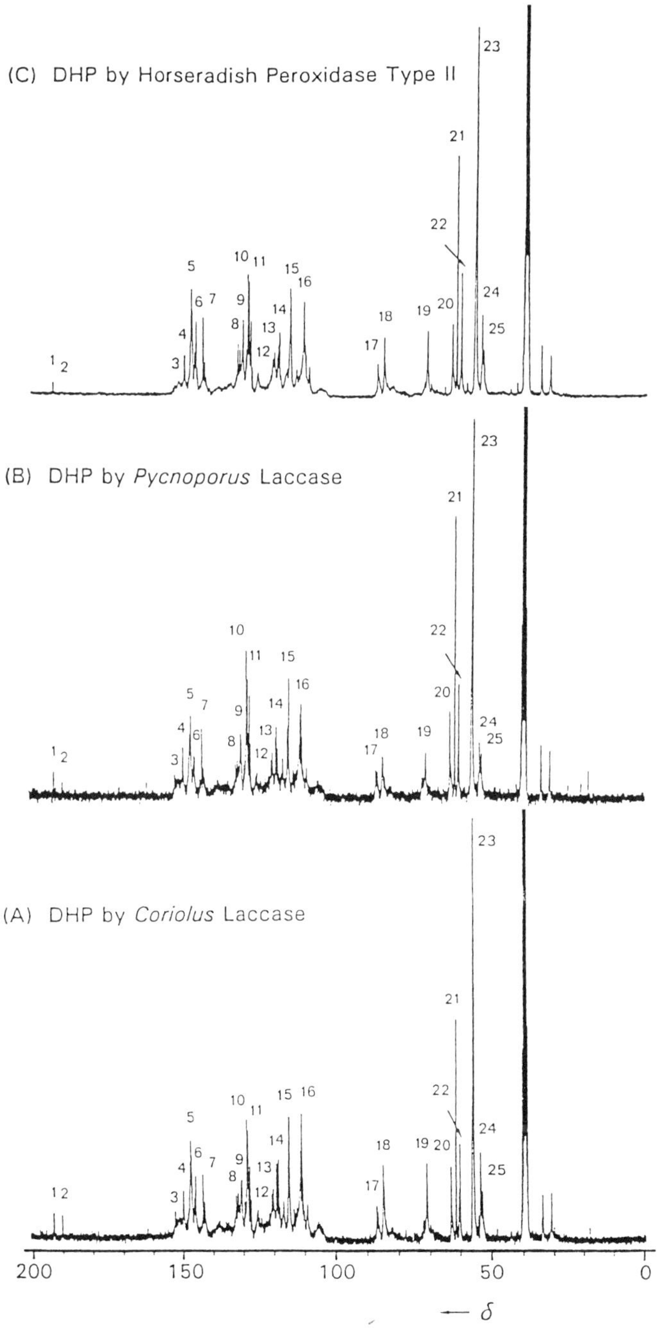

Figure 5. ^{13}C NMR spectra of dehydropolymerisates (DHPs) prepared from coniferyl alcohol using (A) *Coriolus* laccase, (B) *Pycnoporus* laccase, and (C) horseradish peroxidase as catalyst. (Adapted from ref. *42*.)

Literature Cited

1. Lüdemann, H.-D.; Nimz, H. *Makromol. Chem.* **1974**, *175*, 2409.
2. Nimz, H; Mogharab, I.; Lüdemann, H.-D. *Makromol. Chem.* **1974**, *175*, 2563.
3. Nimz, H; Robert, D.; Faix, O.; Nemer, M. *Holzforschung* **1981**, *35*, 16.
4. Chen, C.-L.; Chua, M. G. S.; Evans, J.; Chang, H-M. *Holzforschung* **1982**, *36*, 239.
5. Lapierre, C.; Lallemand, J. Y.; Monties, B. *Holzforschung* **1982**, *36*, 275.
6. Kringstad, K. P.; Mörck, R. *Holzforschung* **1983**, *37*, 237.
7. Bardet, M.; Foray, M. F.; Robert, D. *Makromol. Chem.* **1985**, *186*, 1495.
8. Landucci, L. L. *Holzforschung* **1985**, *39*, 355.
9. Chen, C.-L.; Robert, D. In *Methods in Enzymology, Biomass*; Wood, W. A., Kellogg, S. T., Eds.; Academic Press: San Diego, CA, 1988; Vol. 161B, pp 137-174.
10. Landucci, L. L.; Deka, G. C.; Roy, D. N. *Holzforschung* **1992**, *46*, 505.
11. Robert, D. In *Methods of Lignin Chemistry*; Lin, S. Y., Dence, C. W., Eds.; Springer-Verlag: Heidelberg/Berlin, 1992; pp 250-273.
12. Ernst, R. R. *J. Chem. Phys.* **1966**, *45*, 3845.
13. Björkman, A. *Svensk Papperstidn.* **1956**, *59*, 477.
14. Lundquist, K. In *Methods of Lignin Chemistry*; Lin, S. Y., Dence, C.W., Eds.; Springer-Verlag: Heidelberg/Berlin, 1992; pp 65-75.
15. Drumond, M.; Aoyama, M.; Chen, C.-L.; Robert, D. *J. Wood Chem. Technol.* **1989**, *9*, 421.
16. Robert. D.; Chen, C.-L. *Holzforschung* **1989**, *43*, 323.
17. Erickson, M.; Larsson, S.; Miksche, G. *Acta. Chem. Scand.* **1973**, *27*, 903.
18. Adler, E. *Wood Sci. Technol.* **1977**, *11*, 168.
19. Bardet, M.; Gagnaire, D.; Nardin, R.; Robert, D.; Vincendon, M. *Holzforschung* **1986**, *40 (Suppl.)*, 17.
20. Hassi, H. Y.; Aoyama, M.; Tai, D.; Chen, C.-L.; Gratzl, J. S. *J. Wood Chem. Technol.* **1987**, *7*, 555.
21. Miyakoshi, T.; Chen, C.-L. *Holzforschung* **1992**, *46*, 39.
22. Karhunen, P.; Rummakko, P.; Sipilä, J.; Brunow, G.; Kilpeläinen, I. *Tetrahedron Lett.* **1995**, 169.
23. Karhunen, P.; Rummakko, P.; Pajunen, A.; Brunow, G. *J. Chem. Soc., Perkin Trans. I.* **1996**, 2303.
24. Wu, Z.-H.; Sumimoto, M.; Tanaka, H. *Holzforschung* **1994**, *48*, 400.
25. Freudenberg, K; Harkin, J. M; Werner, H.-K. *Chem. Ber.* **1964**, *97*, 909.
26. Ede, R. M.; Kilpeläinen, I. *Res. Chem. Intermed.* **1995**, *21*, 313.
27. Pan, D.; Tai, D.; Chen, C.-L.; Robert, D. *Holzforschung* **1990**, *44*, 7.
28. Bardet, M.; Foray, M-F.; Robert, D. *Makromol. Chem.* **1985**, *186*, 1495.
29. Bendall, M. R.; Pegg, D. T.; Doddrell, D. M.; Hull, W. E. *DEPT Brüker Information*; Brüker Analytische Messtechnik: Karlsruhe, **1982**.
30. Doddrell, D. M.; Pegg, D. T.; Bendall, M. R. *J. Mag. Reson.* **1982**, *48*, 323.
31. Benn, R.; Gunther, G. *Angew. Chem. Internat. Ed.* **1983**, *22*, 350.
32. Freudenberg, K.; Richtzenhain, H. *Chem. Ber.* **1943**, *76*, 997.
33. Freudenberg, K.; Reznik, H; Boesenberg, H; Rasenack, D. *Chem. Ber.* **1952**, *85*, 641.
34. Freudenberg, K.; Hübner, H. H. *Chem. Ber.* **1952**, *85*, 1181.
35 Freudenberg, K.; Schlüter, H. *Chem. Ber.* **1955**, *88*, 617.
36. Freudenberg, K.; Neish, A. C. *Constitution and Biosynthesis of Lignin*; Springer-Verlag: New York, NY, 1968.
37. Freudenberg, K.; Harkin, J.; Reichert, M.; Fukuzumi, T. *Chem. Ber.* **1958**, *91*, 581.
38. Higuchi, T; Ito, Y. *J. Biochem.* **1958**, *45*, 575.
39. Higuchi, T. In *Biochemistry of Wood*; Kratzl, K., Billek, G., Eds.; Pergamon Press: New York, NY, 1959; pp 161-188.
40. Freudenberg, K. *Nature* **1959**, *183*, 1152.

41. Nakamura, W. *J. Biochem.* **1967**, *62*, 54.
42. Okusa, K.; Miyakoshi, T.; Chen, C.-L. *Holzforschung* **1996**, *50*, 15.
43. Bligny, R.; Douce, R. *Biochem. J.* **1983**, *209*, 489.
44. Driouich, A.; Lainé, A. C.; Vian, B.; Faye, L. *Plant J.* **1992**, *2*, 13.
45. Sterjiades, R.; Dean, J. E. D.; Eriksson, K.-E. L. *Plant Physiol.* **1992**, *99*, 1162.
46. Sterjiades, R.; Dean, J. E. D.; Gamble, G.; Himmelsbach, D.; Eriksson, K.-E. L. *Planta* **1993**, *190*, 75.
47. Savidge, R.; Udagama-Randeniya, P. *Phytochemistry* **1992**, *31*, 2959.
48. Davin, L. B., Bedgar, D. L.; Katayama, T.; Lewis, N. G. *Phytochemistry* **1992**, *31*, 3869.
49. Bao, W.; O'Malley, D. M.; Whetten, R. W.; Sederoff, R. R. *Science* **1993**, *260*, 672.
50. Fåfraeus, G.; Reinhammar, B. *Acta Chem. Scand.* **1967**, *21*, 2367.
51. Oda, Y.; Adachi, K.; Aita, I.; Ito, M.; Aso, Y.; Igarashi, H. *Agric. Biol. Chem.* **1991**, *55*, 1394.
52. Bourbonnais, R.; Paice, M. G. *FEBS Lett.* **1990**, *267*, 99.
53. Potthast, A.; Rosenau, T.; Chen, C.-L.; Gratzl, J. S. *J. Mol. Cat. A: Chemical* **1996**, *108*, 5.
54. Potthast, A.; Rosenau, T.; Chen, C.-L.; Gratzl, J.S. *J. Org. Chem.* **1995**, *60*, 4320.
55. Freudenberg, K.; Knof, L. *Chem. Ber.* **1957**, *90*, 2857.
56. Ellwardt, von P.-Chr.; Haider, K.; Ernst, L. *Holzforschung* **1981**, *35*, 103.

Chapter 19

Experimenting with Virtual Lignins

Lubo Jurasek

Paprican, 570 St. John's Boulevard, Pointe Claire, Quebec H9R 3J9, Canada

A three dimensional model of lignin was assembled from phenylpropane subunits which were represented as simplified space-filling structures connected by any of the 6 most common bonds found in natural softwood lignin. Computer simulations of lignin biosynthesis showed that, if lignin is synthesized in thin lamellae, the degree of crosslinking is lower than if the 'synthesis' takes place in more open space. This agrees with experimental data on the crosslinking of lignin in the fine lamellae of the secondary wall compared to the more bulky lignin in the cell corners and middle lamella. The biosynthesis of lignin within the polysaccharide network of the secondary wall was also modeled confirming the expected low degree of crosslinking. A tendency toward parallel orientation of the subunits with respect to the microfibril direction was detected. The alignment was spontaneous, due to the elongated shapes of the spaces within which the lignin was biosynthesized. Computer 'experiments' with the pulping of middle lamella and secondary wall lignins demonstrated that the highly crosslinked lignin of the middle lamella is likely to be more resistant to degradation. Simulation of the synthesis of an unrestricted dehydropolymerisate (DHP) from a monolignol resulted in components with apparent fractal properties.

Lignin chemistry is a well developed field and a great deal is known about this natural polymer (*1*), including its reactivity with a variety of chemicals used in pulping and bleaching (*2*), its physical properties as they relate, for example, to mechanical pulping (*3*), its biodegradation by fungi (*4, 5*), and its distribution in the cell wall as revealed by microscopy (*6-8*). It is therefore somewhat surprising that lignin is still being represented only by a two-dimensional structure in the plane of the printed page while little attention is paid to its three-dimensional character. Just like the properties of other biological macromolecules, such as proteins and nucleic acids, where the shapes determine a substantial part of their function, the properties of lignin are also better understood in three dimensions. However, the relationship between structure and function has not yet been seriously contemplated, either with lignin or with the whole lignocellulosic complex found in wood fibers.

The modeling of lignin at a level of complexity sufficient, ostensibly, to warrant the use of a computer was pioneered by Glasser and Glasser (*9*) more than 2 decades ago. However efforts to model the three-dimensional structure of lignin have started only recently. Since, unlike other biological macromolecules, lignin does not seem to contain repeating motifs which would be amenable to unambiguous structure determination by crystallography or NMR spectroscopy, and since microscopy has not yet provided us with lignin images at the atomic level of resolution, computer simulation of lignin formation is the only available means of producing a three-dimensional molecular model. One line of research has opted for building representations of low molecular weight lignin model compounds (usually dimers of phenylpropane subunits) and using geometry optimization (energy minimization) algorithms to arrive at the most probable structures. Simon and Eriksson (*10*) have recently published the likely conformations of lignin dimers and oligomers linked by β–*O*–4′ ether bonds. A crystal structure of a lignin dimer was published earlier by Stromberg and Lundquist (*11*). Lignin oligomers have been modeled by Faulon *et al.* (*12*) and their interaction with cell wall carbohydrates has also been examined (*13-15*). Another approach was to build very large coarse macromolecular models with no particular regard for the details of their structure at the atomic level of resolution. Such models (often built in rectangular matrices) allow simulation of bond frequencies, polymerization dynamics, and degradation patterns of the polymer, giving excellent agreement with actual experimental data (*16-18*).

An effort to bridge the gap between the electron microscopy of lignin and its known chemical structure led to the development of a macromolecular modeling approach described in detail by Jurasek (*19, 20*). Only a general description of the methods is given here, enough for the reader to appreciate the applications. The objective of the project was to build a *macromolecular* model of the lignocellulose structure at the atomic level of resolution. The size of such a model has to be large enough to capture a representative piece of the structure, including at least parts of two cellulose microfibrils and the region between them occupied by hemicellulose and lignin. Thus the molecular weight of such structure must be rather large (about 10^6 Da). Therefore the modeling technique has to be simple enough not to overwhelm the computer.

The lignin molecular modeling approach used in this paper is based on the monolignol radical polymerization theory of lignin biosynthesis (*1*) but, rather than attempting to simulate the chemical coupling mechanisms, the program is designed to conform with the final result of the polymerization process, *i.e.* with the frequencies of various bonds in softwood lignin as reported in the literature. The biosynthetic space was initially seeded by a low concentration of precursor monomers, and it was then assumed that the polymerization occurred by random coupling of progressively more monomers to molecular chains gradually growing from these fixed seed points, and also by polymer-to-polymer coupling. The coupling process was modeled on a coarse level by docking the structures in space at locations where they fit stereochemically. The model building algorithm continuously examines the space for opportunities to add more subunits in positions and orientations such that formation of the particular bonds is facilitated. No molecular orbital computations were applied during the coarse modeling. Only after the coarse model was built were the atomic coordinates fitted by another computer algorithm and finally commercial software (HyperChem, HyperCube, Inc.) used for geometry optimization and energy minimization. The model of the lignin macromolecule is flexible and the coarse positioning of the subunits is accurate enough to allow geometry optimization without substantial dislocations of the monomeric subunits and without changing the overall shape of the macromolecule. Thus the lignin macromolecule is not depicted as a collection of subunits linked together in predefined geometries but rather it accommodates a continuum of dihedral bond angles. Whether this is true in reality would be a matter of dispute, but it appears reasonable to assume that the real lignin structure is not a collection of predefined dimer geometries but rather a structure originating from random encounters of precursor radicals, resulting in a variety of bond geometries which as a whole undergo changes as the size of the molecule increases and its shape alters to fit in the allocated space. An imposition of rigidity upon the unit-to-unit bonding geometry would limit this model to growth in free space only (such as in simulations of dehydrogenative polymerization producing monolignol dehydropolymerisates or DHPs). The growth of a rigid model in the restricted space provided by the cell wall structure would cease long before the lignin content could reach the values accepted for lignocellulosic materials.

This approach, applied to the simulation of lignin biosynthesis and its degradation by chemical pulping and by enzymes, is the basis of the following 'experiments'. *It should be noted that, unless stated otherwise, all data and images shown in this article have been generated by computer software.*

'Experiment' One: the Polymerization of Monolignols to Form Lignin

Building a molecule with a size of 10^6 Da is not a minor task, even for the fastest of today's computers. To simplify the problem, the softwood lignin structure has been first represented by a collection of ball and stick entities, where the ball represents the aromatic ring of the phenylpropane subunit, including the methoxyl group, and the stick represents the aliphatic three-carbon moiety at one end and the phenolic hydroxyl at the other (*19*). Figure 1 illustrates how this structural entity corresponds to the chemical formula. This coarse three-dimensional layout of monomers in the lignin macromolecule is called *guaraná* representation since it resembles the shape and color of the fruit of a Brazilian shrub (*Paullinia cupana*) with mind stimulating properties.

At the beginning of simulated lignin polymerization the biosynthetic space was seeded with monomers usually corresponding in number to a 50 mM concentration (normally just a few molecules). The polymerization process was governed by a set of simple rules ensuring that the most common bond types, at least, existing between the subunits (*21*) would be used, that the polymerization would be random, and, of course, that each time a new subunit be added it could not overlap or clash with other structures already in place. The growth gradually filled the available space and slowed down considerably as the density approached 1.35 g/cm^3 (close to the values reported for lignin by Ramiah and Goring (*22*) and Stamm (*23*)), at which point the polymerization was stopped. As the polymer expanded through both linear growth and branching, some parts of its structure grew close to one another. These parts were then allowed to crosslink if their configurations were favorable. Crosslinks between two adjacent polymer chains result in larger branched molecules, but crosslinks *within* a single chain have special topological significance since they result in the formation of intramolecular crosslinks, each of them corresponding to a closed loop (ring). Some of these rings were small and the dibenzodioxocin rings discovered earlier by Karhunen *et al.* (*24*) were amongst them.

Six bond types, deemed to be the most common in softwood lignin (*21*), were incorporated into the polymerized lignin model. Figure 2 shows these bond types in three dimensions. The lignin modeling described in this article was limited to the use of guaiacyl (softwood) units. Incorporation of syringyl units, as found in hardwood, would alter the model extensively. A comparison between models of the two lignin types would be rather interesting and is currently under investigation. More detailed information on the polymerization algorithm can be found in previous publications (*19, 20*) and in the computer program listing available from the author upon request.

Perhaps it should be emphasized that, once the rules are set, the polymerization proceeds spontaneously with no further input from the human model builder, thus precluding any subjective bias. The rules remain the same in all experiments described below, unless otherwise explicitly stated. Despite the invariant rules, the final results of the modeling can be quite different depending on the shape of the biosynthetic space. Very large structures with degree of polymerization (DP) in excess of 6,000 (molecular weight about 10^6 Da) can be built this way. The way the

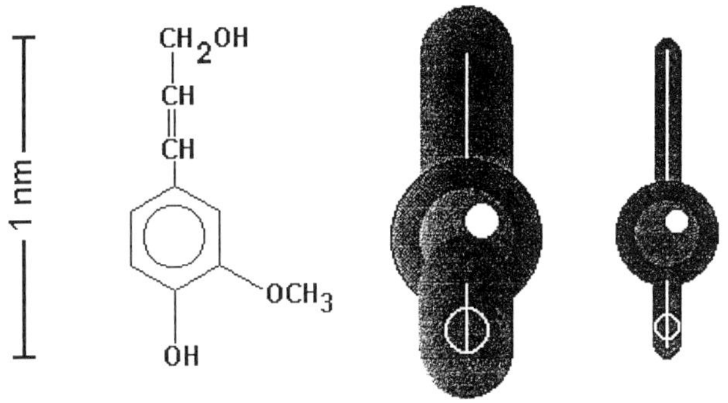

Figure 1. The structural formula of coniferyl alcohol and its *guaraná* representation. The ball includes both the aromatic ring and the methoxyl group. The white circle signifies the phenolic hydroxyl. The bulkier shape in the middle shows the monomer size used in the coarse model building. The thinner shape on the right is often used for easier visualization of the structures.

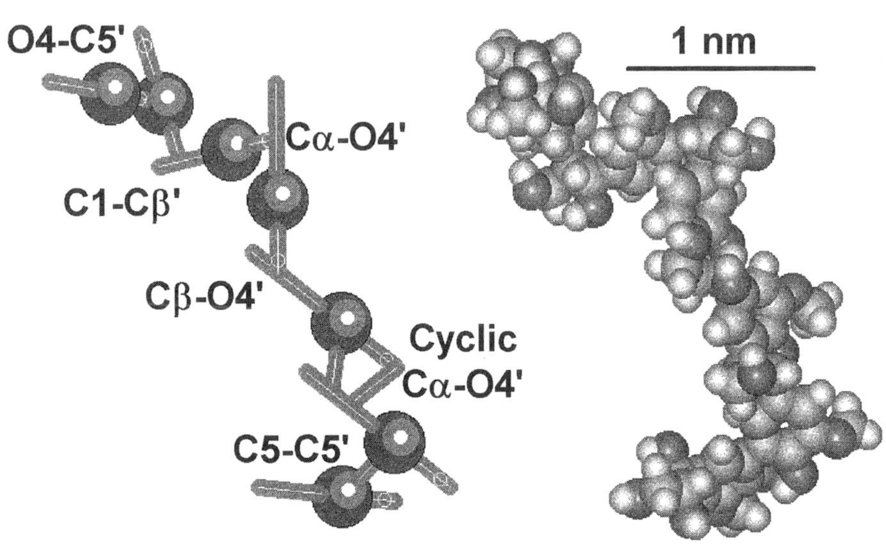

Figure 2. A lignin heptamer showing the six representative bonds used in the model building. The coarse *guaraná* representation of the heptamer is shown on the left and a refined molecular model derived from the coarse structure is on the right.

structure is laid down in space is influenced by its proximity to surfaces. Since, for example, internal crosslinking occurs less frequently at the surface (where only about one half of the space is accessible), thin films of lignin are less crosslinked than bulkier structures (*19*). To eliminate this effect, the model can be built in a boundless periodic cell (Figure 3). Although the lignin appears to be rather compact, a closer examination of its molecular structure revealed a degree of porosity sufficient to allow some diffusion of small molecules (less than 1 nm in diameter (*20*)).

Once the model is built in the defined space, the *guaraná* representation can be translated into atomic coordinates. The *guaraná* model provides the connectivities of all subunits as well as the three-dimensional coordinates of the ends of each monomer (Cγ and O4, respectively), and ensures that there is room for the aromatic ring with its methoxyl group, and for the aliphatic sidechain. Precise coordinates for all carbon and oxygen atoms (derived from HyperChem-built molecular models of guaiacylglycerol monomers) are then written in by the program. The S and R configurations of the α and β carbons are chosen at random, thus always giving an approximately equal number of *erythro-* and *threo-*configurations in the aliphatic sidechain. The appropriate covalent bonds between the neighboring monomeric subunits are also generated by the program. All this produces a complete input file (in HyperChem*.HIN format) which can be further processed by HyperChem and, if desired, transported into other commercial computational chemistry software. Although care was taken to ensure that the *guaraná* subunits are positioned so as to respect the spatial requirements for formation of the appropriate bonds and, during translation into atomic coordinates, the aromatic rings with their methoxyl groups were rotated to minimize any strain on the structure, the coordinates are ultimately accurate only to within about 0.3 nm. For more detailed work with the model, the approximate coordinates must be refined as can be done with any of the geometry optimization algorithms offered by a variety of commercial computational chemistry and molecular modeling packages. The example shown in Figure 4 is a small piece of simulated lignin built in the *guaraná* representation, translated into atomic coordinates, and then optimized by MM+ (HyperChem, Hypercube, Inc.) to give a more precise geometry. It is satisfying to observe that the lignin model structure is flexible enough to allow it easily to accommodate unstrained interatomic distances and bond angles. All computations were done *in vacuo*. Ignoring the presence of water in the system, in order to simplify computation, usually leads to minor conformational errors. A detailed description of all the rules included in the coordinate-generation algorithms is included in the computer program listing available from the author.

Lignin models built to fill a large and unbounded space [*e.g.* a periodic cell of $(8.2\ nm)^3$] are believed by the author to represent middle lamella lignins. Figure 5 shows how such a model grows to reach its target density and how some of its properties develop during the simulated growth. The cell was first seeded with 17 monomers and the growth was then simulated as continuous coupling of additional monomers to chains growing from these seed points. The process of growth and intermolecular crosslinking leading to the coalescence of the 17 molecules originating from the seed points was rapid, the final degree of intramolecular crosslinking (ring formation) was high, and the phenolic hydroxyl content decreased gradually as more of the groups underwent reaction in the polymerization process. It is interesting to note in this context Donaldson's (*8*) electron microscopic observations of middle lamella lignin growth from discrete initiation sites followed by their coalescence into more continuous space-filling structures.

The spatial orientation of the subunits in the model of the middle lamella lignin is entirely random, and the overall shape of the branched molecular backbone is that of a random coil, because the bond types are randomly distributed throughout the molecule and the angles at which the subunits are connected are variable. This agrees with the random nature of natural lignin, at least as it is found in the middle lamella. It should be noted that the middle lamella is too thick to be considered a film of lignin from a molecular perspective. The surface effects on intra-molecular crosslinking

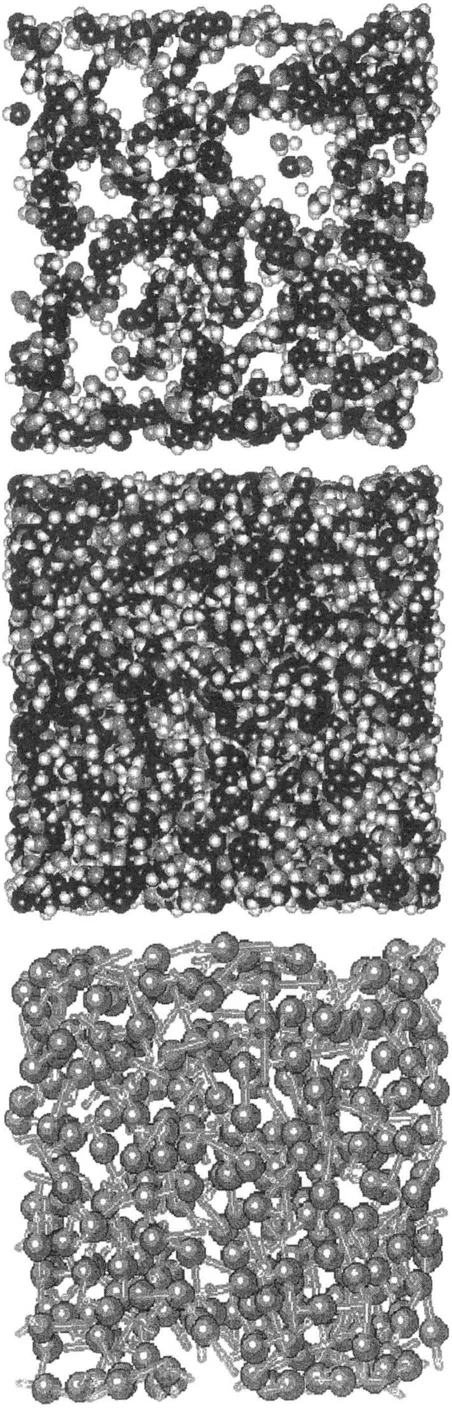

Figure 3. A model of middle lamella lignin in a boundless periodic cell (4.4 nm)^3, shown as a coarse structure on the left, at an atomic level of structural resolution in the middle, and as a 1 nm section on the right.

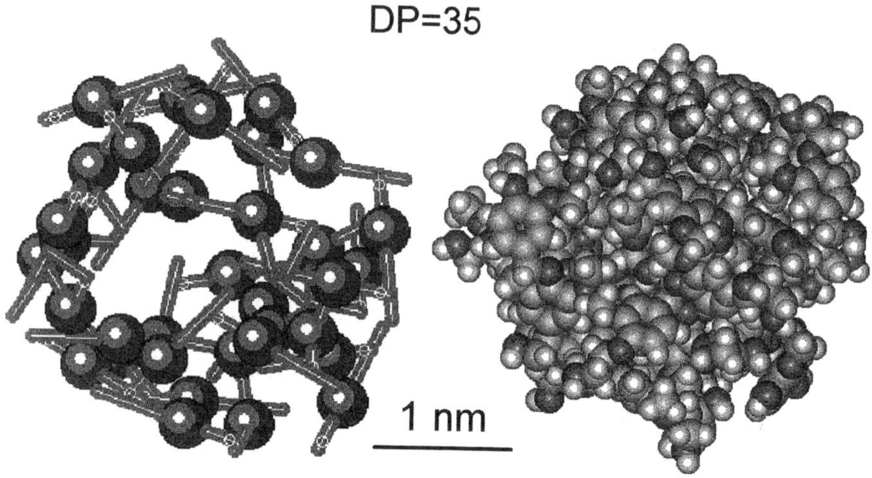

Figure 4. A model of a lignin oligomer grown in a sphere up to a density of 1.35 g/cm^3. The coarse *guaraná* representation is on the left. A geometry-optimized molecular model of the same structure is shown on the right.

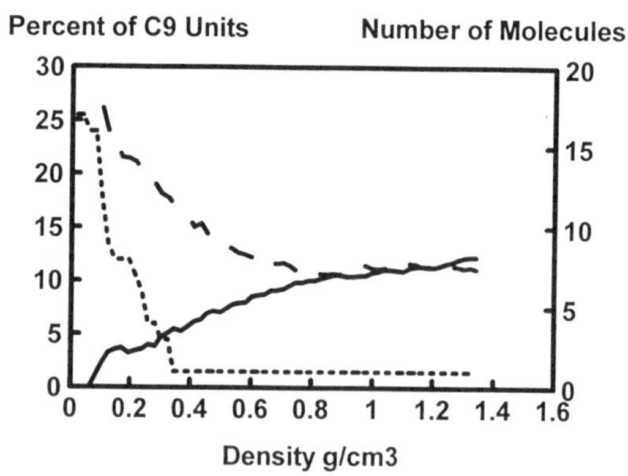

Figure 5. The progress of lignification in a small model region (cube with 8.2 nm side) of the middle lamella; plotted as a function of the increasing density of lignin in the lignifying tissue: (top line) number of lignin molecules, (- -) number of phenolic hydroxyls per 100 C_9 units of lignin, (—) number of intramolecular cross-links (rings) per 100 C_9 units.

observed in models of lignin grown as thin films diminish rapidly with increasing film thickness and become very small when the latter exceeds 3 nm (*19*); the middle lamella is about 100 times thicker than this.

'Experiment' Two: the Lignification of the Secondary Wall

The construction of a model of the secondary wall was based on a high-resolution micrograph of a cross-section of spruce secondary wall (*6*) into which a layout of cellulose microfibrils, hemicellulose molecules, and lignin was built in the spirit of the model proposed by Page (*25*) and Kerr and Goring (*26*). It was assumed that lignification was preceded by an almost complete biosynthesis of the cellulose and hemicellulose framework in the secondary wall, as microscopic evidence suggests (*27-29*). Thus the computer program first constructed two rows of cellulose microfibrils (Figure 6) and then inserted hemicellulose molecules in parallel with the microfibrils. As many hemicellulose molecules as possible were placed to form a slightly disorientated monomolecular layer on the tangential (with respect to fiber center) faces of the microfibrils as proposed by Kerr and Goring (*30*). The remaining hemicellulose molecules (there was not enough room on the tangential surfaces to accommodate them all) were distributed randomly in the remaining space. This model at that point represented a space where 50% was occupied by cellulose, 25% by hemicellulose, and 25% remained void. Such structure loosely resembles a system of steel reinforcing rods (hemicellulose) enclosed in wooden casing (cellulose) just before the cement (lignin) is poured in. The computer model of lignin, built using precisely the same polymerization rules as those used to synthesize the middle lamella lignin, then filled the void volume—sparsely at first, and finally at full lignin density, 1.35 g/cm^3

Figure 6 illustrates the progress of the simulated lignification in models of secondary wall sections. The simulation suggests initial formation of lignin clusters which gradually coalesce into a more or less compact lignin-hemicellulose composite. As lignification proceeded (Figure 7), the degree of crosslinking increased initially as fast as in the middle lamella (Figure 5) but then it grew only very slowly reaching a final value much lower than that of the middle lamella lignin. Why should the secondary wall lignin be so much less crosslinked? It is mainly because of its large interfacial surface areas. At the lignin interface with the carbohydrate matrix, the intramolecular crosslinking is limited by the fact that on the outside there is no structure with which to crosslink (crosslinking of lignin with carbohydrates has not yet been incorporated into the program). This modeling result is in agreement with published experimental evidence suggesting that the lignin in the middle lamella (and cell corners) is indeed more crosslinked than in the secondary wall (*31*).

Another feature of *secondary wall* lignification is that the process of coalescence of the oligomers originating from the seeds is much slower. In the example shown in Figure 7 the final crosslinking between the last two lignin molecules occurred just before lignification was completed. In some simulation runs it was even found that, at the end of lignification, the coalescence of lignin molecules was incomplete. This suggests that the molecular weight of lignin in the secondary wall might be lower than that in the middle lamella. The reason for slower (and perhaps incomplete) coalescence of secondary wall lignin is its limited potential to form crosslinks because of the large interfacial surface area mentioned above. Figure 7 also shows that the content of phenolic hydroxyls in the model dropped earlier and more rapidly than in the case of middle lamella lignin, but near the end of lignification it leveled off at approximately the same value. This value is close to the content of phenolic hydroxyl groups (0.13 OH/C_9 unit, *i.e.* 13%) reported for the secondary wall lignin by Whiting and Goring (*32*).

The coarse model of the secondary wall can be translated into atomic coordinates as illustrated in Figure 8. The coordinates for the lignin were generated by the program already mentioned in this article. The model of cellulose was built from the coordinates for *Valonia* cellulose I of Sarko and Muggli (*33*) and the hemicellulose

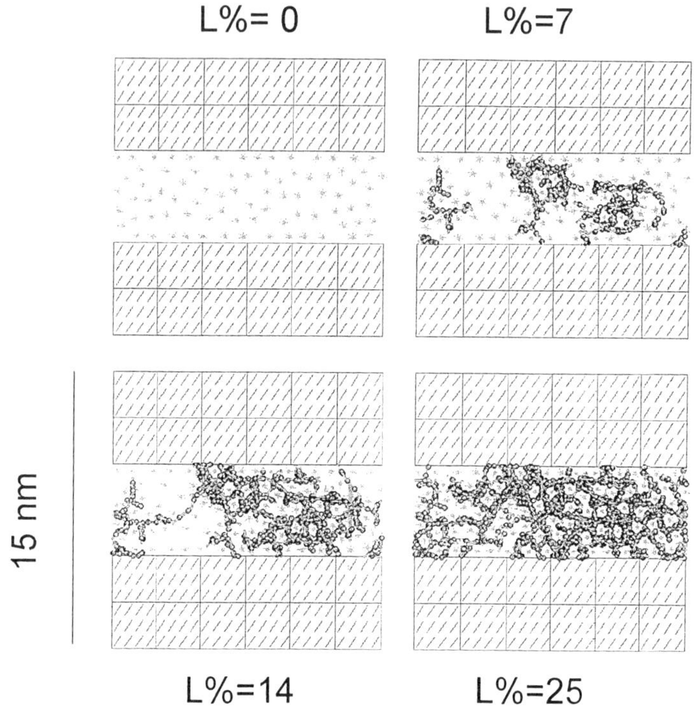

Figure 6. Models of 7 nm thick sections of the secondary wall at various stages in the lignification process. L% refers to lignin content. Oblique lines are cellulose molecules, stars are hemicellulose molecules and the dark globular structure is lignin.

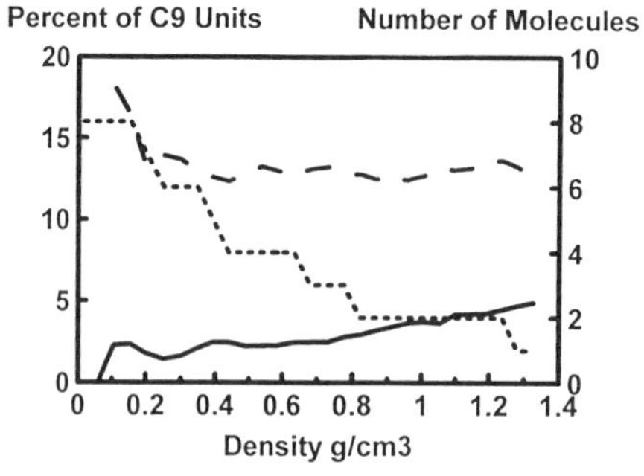

Figure 7. The progress of lignification for a small model (box 15x15x7 nm) of the secondary wall plotted as a function of the increasing density of lignin in the lignifying tissue. [See Figure 5 for explanation of symbols.]

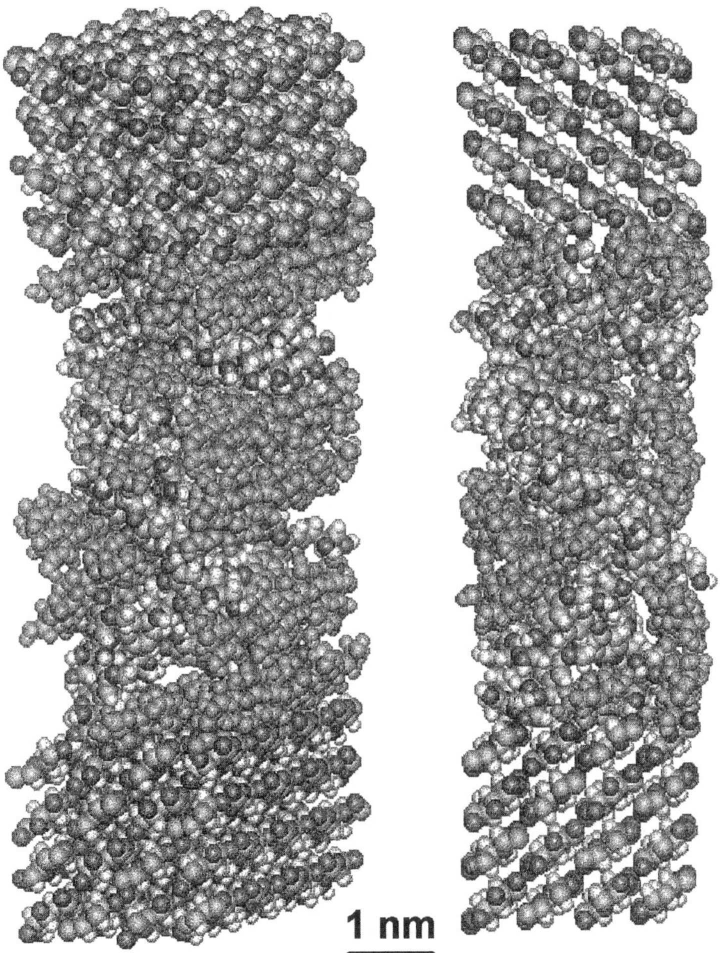

Figure 8. A molecular model of the secondary wall produced by simulated lignification and then translating part of the coarse structure shown in Figure 6 into atomic coordinates. The front view showing the cross-sectional plane is on the right; a side view is on the left.

molecules, fitted into the channels left in the lignin structure, are molecular models of a linear glucomannan built using HyperChem (Hypercube Inc.). HyperChem was also used to assemble the whole molecular model from the atomic coordinates. No geometry optimization of the secondary wall model was carried out at this stage.

An analysis of a model of the secondary wall lignin (*20*) revealed that, unlike in the middle lamella lignin, the orientation of the subunits was not entirely random. Computation of the coordinates of the *guaraná* subunits revealed a certain degree of preferential alignment with the longitudinal axis of the carbohydrate molecules (the approximate fiber direction). The degree of the ordering can be described by the order parameter used in liquid crystal theory (*34*), which was determined to be 0.232 for the model of the secondary wall lignin (a perfectly aligned parallel structure would have the parameter equal to 1, while in a perfectly random structure the parameter would be 0). This orientation implies that the aromatic rings of the secondary wall lignin are preferentially oriented in directions parallel to the fiber axis.

Why did this happen? No intentional lignin orientation was written into the lignification algorithm. The alignment occurred spontaneously as the new monomers were randomly fitted into the spaces between the carbohydrate molecules. Thus the axial orientation of the carbohydrate molecules was *imprinted* upon the lignin molecule. A degree of longitudinal alignment of the aromatic rings in the secondary wall was experimentally determined by Raman microprobe infrared spectroscopy (*35*). Goring (*36*) proposed that the lignin orientation was a consequence of the axial alignment of wood carbohydrates. More recently Donaldson's (*8*) remarkable micrographs have suggested a degree of parallel alignment of the secondary wall lignin with the carbohydrate matrix. Also Shevchenko (*37*) suggests that the nanometer scale shape of the support on which DHP lignins are deposited can affect their orientation. In agreement with Goring's hypothesis (*36*) and the microscopic evidence (*8*), the molecular modeling of lignin demonstrates and explains how the axial alignment of lignin might arise.

An alternative explanation to the Raman spectroscopic data (*35*) was offered by Faulon *et al.* (*12*). The authors conclude that there is helical order in the lignin structure because the minimum energy conformation of a lignin oligomer consisting of a stretch of 10 subunits (all connected by Cβ–*O*–4′ linkages) is a helix. The helices are believed to be aligned coaxially with the fiber and this would result in an axial alignment of the aromatic rings. While such helical structures might be found in artificial lignins, the presence of lignin helices in the secondary wall seems unlikely for a number of reasons: (1) the probability of finding stretches of 4 or more Cβ–*O*–4′ linked lignin monomers is very low in spite of the high frequency of these bonds—in the lignin model only about 4% of the whole consists of such stretches; (2) the helices do not pack tightly enough to account for the density of lignin; (3) the optical and x–ray diffraction properties of lignin (*38*) do not support the notion of such a helical arrangement.

'Experiment' Three: Pulping of the Lignin Models

The models of middle lamella and secondary wall lignins were subjected to simulated depolymerization. To approximate the conditions of kraft pulping, it was assumed that only the alkyl ether bonds of the lignin were broken and that the hemicellulose was partially removed during the process. No condensation reactions were assumed to take place during the degradation. It was also assumed that all lignin fragments with DP less than 10 would be small enough to be leached from the fiber as soon as they were formed, while larger fragments would remain in place without changing their position or molecular configuration. This computer 'experiment' has been described in more detail by Archibald *et al.* (*39*). McNaughton *et al.* (*40*) showed that lignin fragments much larger than those of DP 10 are leached from the cell wall during kraft (and even more so during sulfite) pulping. These results cannot be adequately simulated with a model representing only a small part of the cell wall but, being that as it may, the simple rules resulted in the following degradation patterns.

Figure 9 describes the characteristics of the simulated kraft pulps as they appeared from 'experiments' with a model of the secondary wall. The lignin content in the fiber decreased during the degradation and reached nearly zero when about 50% of the lignin bonds had been broken. The content of aliphatic ether bonds in the residual lignin decreased steadily during the degradation (because these were the bonds preferentially broken during kraft pulping) while the content of condensed bonds (resistant to kraft pulping) concomitantly increased, reaching about 75% of all remaining bonds at the extent of delignification corresponding to an oxygen-delignified pulp (lignin content 2% dry pulp).

It is interesting to consider some of the implications of this study in more detail. Due to the random nature of the polymerization process the distribution of the condensed and aliphatic ether linkages is irregular. Thus there are regions in the lignin structure with a higher content of condensed linkages than elsewhere. During kraft pulping these condensed domains resist depolymerization and stay intact while other domains dissolve. Therefore this kind of residual kraft lignin cannot be effectively removed by extended pulping and a different chemical basis is required to break it down. Figure 10 shows the morphological consequences of the simulated kraft pulping. It is not just a reversal of biosynthesis. A comparison of two images at the same lignin content (7%) in Figures 6 and 10 shows that, while during biosynthesis a few rather large fragments are present at this lignin content, during kraft pulping a *larger* number of *smaller* fragments is evident. One such fragment rich in condensed linkages is shown in Figure 11, with its structure translated to atomic coordinates. It would be interesting to analyze this structure in some detail and try to design a strategy to break it down effectively. The structure may contain clues to obtaining higher brightness in pulps. Because of differences in their degrees of crosslinking, it is to be expected that the breakdown patterns between the middle lamella and the secondary wall lignins will differ. The presence of intramolecular crosslinks (rings) in lignin tends to protect the structure against depolymerization. In a linear or branched polymer structure, every time a bond is broken depolymerization occurs. In an intramolecularly crosslinked structure, every time a bond within a ring is broken, no depolymerization occurs; just the degree of crosslinking is reduced. If the rings are large (composed of many subunits), then the probability of a bond being broken nonproductively (without net depolymerization) is rather high. This seems to be the case in the lignin model since the rings appear to be broken more often at the outset of the degradation process. Mainly because of their different degrees of crosslinking, there are substantial differences in the degradation patterns of the middle lamella and secondary wall lignin models. There is always a lag in the depolymerization of middle lamella lignin, when the weight average molecular weights of both the total lignin (*19*) and the small lignin fragments (when the largest lignin fragment is excluded from the molecular weight computation—see Figure 12) are examined. Indeed experimental data (*41*) have shown that the middle lamella lignin is less reactive toward chlorine than the secondary wall lignin.

'Experiment' Four: Biodegradation

The molecular model of, for example, unbleached kraft pulp secondary wall can be used to illustrate its interaction with enzymes involved in lignin degradation. As pointed out earlier (*42*) and shown in Figure 13, the molecules of the redox enzymes manganese peroxidase (*43*) and laccase seem to be too large to be able to diffuse freely into the structure of the secondary wall. The residual hemicellulose and lignin provide obstacles to diffusion and the pore sizes are just too small to accept the enzyme macromolecules. However Mn-complexes (*e.g.* Mn-oxalate) and laccase mediators [*e.g.* 2,2′-azino-bis(3-ethylbenzthiazoline-6-sulfonate); ABTS] are small enough to diffuse readily through the pores of the secondary wall. In mechanical pulps where the porosity of the cell wall is smaller (pore sizes only extend up to about 1 nm), the diffusion even of the laccase mediator could be restricted. Smaller

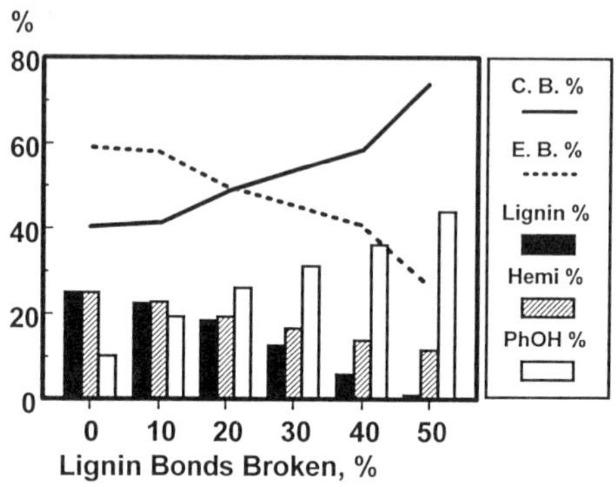

Figure 9. Simulated kraft pulping of a model of the secondary wall. C.B.% and E.B.% are the condensed and aliphatic ether bonds, respectively, expressed as percentages of all bonds. Lig% and Hemi% are the contents of lignin and hemicellulose, respectively, in the fiber. PhOH% is phenolic hydroxyl content per 100 C_9 units.

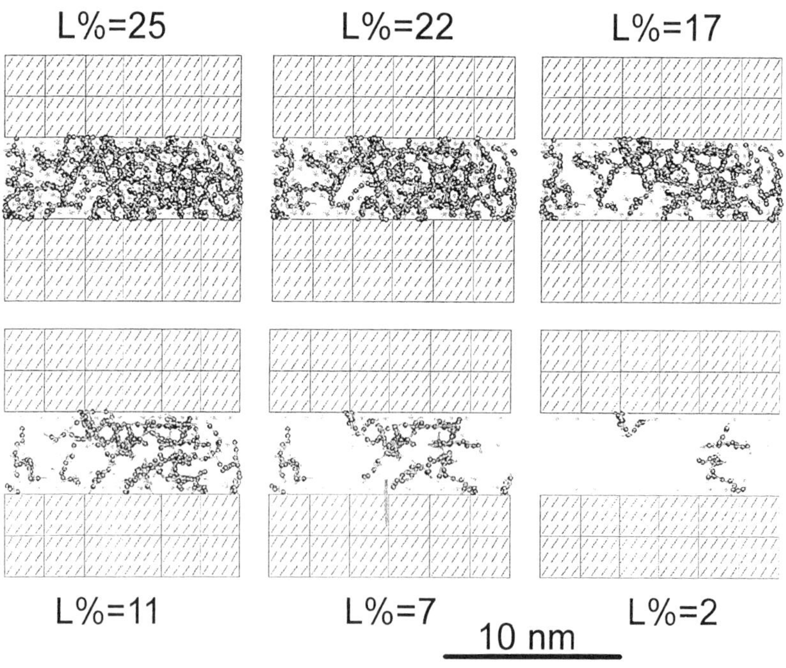

Figure 10. Delignification of the secondary wall model shown in Figure 6 during simulated kraft pulping (employing the symbols as in Figure 6).

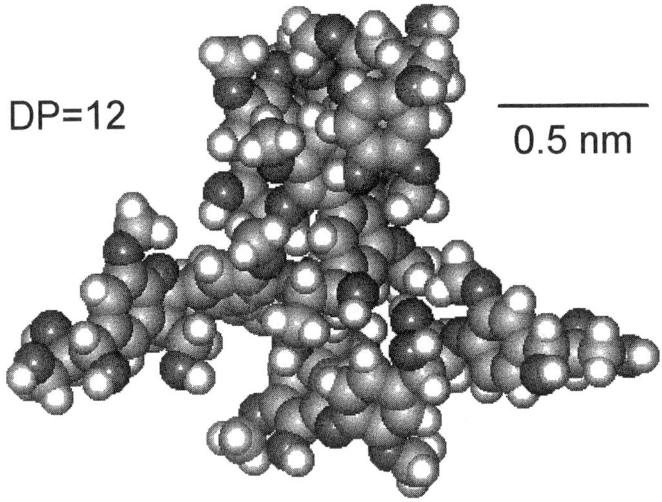

Figure 11. A fragment of residual kraft lignin at an atomic level of resolution after geometry optimization. The model consists of 12 subunits connected mostly by linkages resistant to kraft pulping. The oligomer is bonded by the following linkages: 4 β-*O*-4′, 2 5–5′, 2 β–1′, 2 β-5′ (phenylcoumaran), and a 4–*O*–5′ bond.

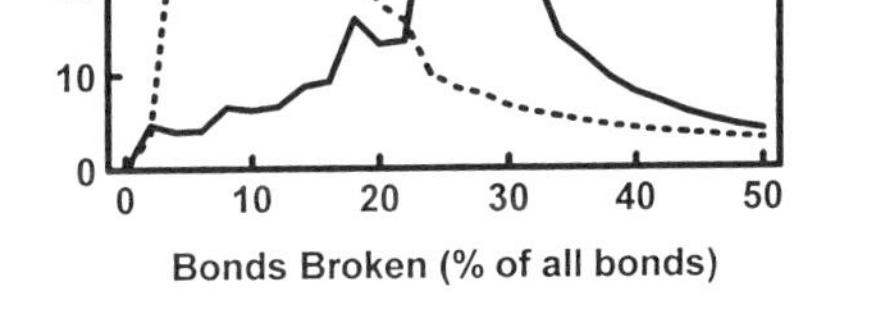

Figure 12. Changes in the weight-average molecular weight of small fragments (*i.e.* fragments other than the largest one) produced during simulated random depolymerization of middle lamella (ML) and secondary wall (SW) lignins. The release of the bigger of the 'small' fragments from the ML occurs later than from the SW.

molecules such as H_2O_2 or the Mn-oxalate complex appear to be able to diffuse more freely.

'Experiment' Five: Dehydrogenative Polymerisates (DHPs) of Monolignols

Simulated polymerization of lignin in unrestricted space yields components with various shapes depending on the polymerization conditions (*20*). This type of polymerization is meant to simulate the formation of DHP lignins in solution. The shapes of the components can be described quantitatively in terms of their fractal dimensions which can be computed by the Hausdorff scaling and box counting method (*44*). The fractal dimensions of various objects are measures of their general shapes. For example, a fractal dimension close to 3 suggests a compact particle, while a fractal dimension 2 or less usually suggests a particle with very loose, stringy, conformation. Examples of the two are illustrated in Figure 14. Fractal dimensions of real particles can be determined from physical measurements (*31, 45*) and the results compared with the models. The polymerization algorithm ('Experiment' One) normally chooses at random the points where additions of new monomers are attempted. When the normal polymerization algorithm was modified to allow monomer additions only at the chain termini (end-wise growth with no branching) and to limit the random choice of unit-to-unit axial angles (frozen geometry), the resulting structures resembled the properties of DHP lignins grown under 'Zutropf' conditions (*46*). On the other hand, when branching was encouraged by introducing a 90% bias in favor of adding new monomers at the non-terminal positions, the properties of the resulting particles resembled 'Zulauf' or 'bulk' DHP polymers (Table I). The experiments with free-growing lignin models corroborate the idea of the fractal nature of lignin (*47*). However, lignin is an imperfect fractal: it only behaves as a fractal on the submicroscopic scale.

Table 1. Fractal Dimensions of Lignins

Lignin Type	Fractal Dimensions	
	Experimental Data	Model Data (*20*)
Endwise DHP	1.66 (*45*)	1.75 ('frozen linear DHP')
Bulk DHP	2.62 (*45*)	2.53 ('maximized branching DHP')
Secondary Wall	2.27 (*31*)	2.30 ('rigid DHP')

Conclusions

Molecular modeling of lignocellulosic structures yields structures the various properties of which can be examined computationally and the results compared with experiment. The lignin model should not be considered a final version but rather it must be continuously modified to achieve the best agreement with current experimental data. It is encouraging that, even with its many inadequacies, the model reasonably approximates the natural polymer. The model of the secondary wall reflects its appearance under the microscope and explains the orientation of aromatic rings revealed in physical measurements. The predicted degrees of internal crosslinking in both middle lamella and secondary wall lignins are compatible with the actual experimental data and the effect of crosslinking on lignin reactivity is

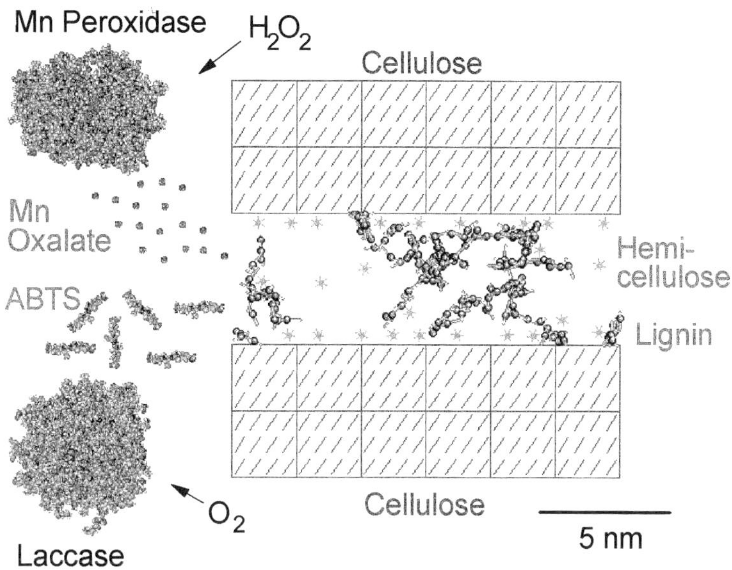

Figure 13. Comparison of the sizes of some components in various putative lignin biodegradation systems with the pores in the secondary wall model. From Jurasek *et al.* (*42*).

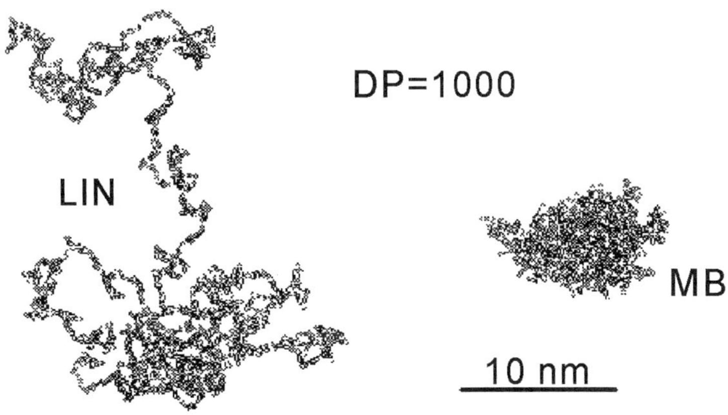

Figure 14. Examples of two different models of DHP lignins. LIN is a very loose linear lignin model structure and MB is a more compact highly branched lignin structure. The degree of polymerization of each polymer molecule was 1000.

demonstrated. A simplified model of kraft pulping suggests that the assumption of secondary condensation is not necessary to explain the high degree of condensation in the residual kraft lignin in the pulp. Modeling of DHP lignin agrees with the notion of its fractal structure. Translation of the coarse lignin model to an atomic level of resolution yields stereochemically reasonable structures that can be geometry-optimized without major rearrangements. The use of the model at present lies mainly to *explain* various properties of lignocelluloses. However, it is to be hoped that in the future the model could be applied to *predicting* lignin reactivity, submicroscopic morphology, and various physical properties.

Acknowledgments

The author gratefully acknowledges many helpful comments provided by Drs. D. Argyropoulos, J.-L. Faulon, D.G. Gray, C. Lim, M.G. Paice, and I.D. Reid and thanks Drs. N.G. Lewis and S. Sarkanen for encouragement. The project was financed by Paprican-maintaining members and, in part, by Genencor International, and by the Canadian government. The project was also a part of the program of the Protein Engineering Network of Centers of Excellence.

Literature Cited

1. Glasser, W. G. In *Pulp and Paper: Chemistry and Chemical Technology*; Casey, J. P., Ed.; Wiley: New York, NY, 1980, 3rd edn.; pp 39-111.
2. *Pulp and Paper Manufacture—Vol. 5: Alkaline Pulping*; Kocurek, M. J., Ed.; TAPPI-CPPA: Atlanta, GA, 1989.
3. Kubo, S.; Uraki, Y.; Sano, Y. *Holzforschung* **1996**, *50*, 144-150.
4. Eriksson, K.-E. L.; Blanchette, R.; Ander, P. *Microbial and Enzymatic Degradation of Wood and Wood Components*; Springer-Verlag: Berlin, 1990.
5. Reid, I. D. *Can. J. Bot.* **1995**, *73(Suppl. 1),* S1011-1018.
6. Ruel, K.; Barnoud, F.; Goring, D. A. I. *Wood Sci. Technol.* **1978**, *12*, 287-291.
7. Terashima, N.; Fukushima, K.; He, L.-F.; Takabe, K. In *Forage Cell Wall Structure and Digestibility*; Jung, H. G., Buxton, D. R., Hatfield, R. D., Ralph, J., Eds.; ASA-CSSA-SSSA: Madison, WI, 1993; pp 247-270.
8. Donaldson, L. A. *Wood Sci. Technol.* **1994**, *28*, 111-118.
9. Glasser, W. G.; Glasser, H. R.. *Macromolecules* **1974**, *7*, 17-27.
10. Simon, P.; Eriksson, K.-E. L. *Holzforschung* **1995**, *49*, 429-438.
11. Stromberg, R.; Lundquist, K. *J. Crystallogr. Spectr. Res.* **1989**, *19*, 331-339.
12. Faulon, J.-L.; Carlson, G. A.; Hatcher, P. G. *Org. Geochem.* **1994**, *21*, 1169-1179.
13. Faulon, J.-L.; Hatcher, P. *Energy and Fuels* **1994**, *8*, 402-407.
14. Bízik, F.; Tvaroska, I.; Remko, M. *Carbohydrate Res.* **1994**, *261*, 91-102.
15. Houtman, C. J.; Atalla, R. H. *Plant Physiol.* **1995**, *107*, 977-984
16. Lange, H.; Wagner, B.; Yan, J. F.; Kaler, E. W.; McCarthy, J. L. *In Proc. 7th Internatl. Symp. Wood Pulp. Chem*. **1993**, *1*, 111-122.
17. Roussel, M. R.; Lim, C. *Macromolecules* **1995**, *28*, 370-376.
18. Roussel, M. R.; Lim, C. *J. Comput. Chem.* **1995**, *16*, 1181-1191.
19. Jurasek, L. *J. Pulp Paper Sci.* **1995**, *21*, J274-279.
20. Jurasek, L. *J. Pulp Paper Sci*. **1996**, *22*, J376-380.
21. Chen, C.-L. In *Wood Structure and Composition*; Lewin, M., Goldstein, I. S., Eds.; M. Dekker: New York, NY, 1991; pp 183-261.
22. Ramiah, M. V.; Goring, D. A. I. *J. Polym. Sci. C* **1965**, *11*, 27-43.
23. Stamm, A. J. *Tappi J.* **1969**, *52*, 1498-1502.
24. Karhunen, P.; Rummakko, P.; Sipilä, J.; Brunow, G.; Kilpeläinen, I. *Tetrahedron Lett.* **1995**, *36*, 4501-4504.
25. Page, D. H. *Wood and Fiber,* **1976**, *7,* 246-248.
26. Kerr, A. J.; Goring, D. A. I. *Cellulose Chem. Technol.* **1975**, *9*, 563-573.
27. Wardrop, A. B. *Tappi J.* **1957**, *40*, 225- 243.

28. Freudenberg, K. In *The Formation of Wood in Forest Trees*; Zimmermann M. H., Ed.; Academic Press: New York, NY, 1964.
29. Terashima, N.; Fukushima, K.; Sano, Y.; Takabe, K. *Holzforschung* **1988**, *42*, 347-350.
30. Kerr, A. J.; Goring, D. A. I. *Wood Sci.* **1977**, *9*, 136-139.
31, Leclerc, D. F.; Olson, J. A. *Macromolecules* **1992**, *25*, 1667-1675.
32. Whiting, P. H.; Goring, D. A. I. *Pap. Puu* **1982**, *64*, 592-595.
33. Sarko, A.; Muggli, R. *Macromolecules* **1974**, *7*, 486-494.
34. Collings, P. J. *Liquid Crystals*; Princeton University Press: New Jersey, 1990.
35. Atalla, R. H.; Agarwal, U. P. *Science* **1985**, *227*, 636-639.
36. Goring, D. A. I. *ACS Symp. Ser.* **1989**, *397*, 2-10.
37. Shevchenko, S. M.; Bailey, G. W.; Yu, Y. S.; Akim, L. G. *Tappi J.* **1996**, *79*, 227-237.
38. Schultze, B.; Theden, G.; Vaupel, D. *Holz Roh-Werkstoff* **1937**, *1*, 75-80.
39. Archibald, F. S.; Bourbonnais, R.; Jurasek, L.; Paice, M. G.; Reid, I. D. *J. Biotechnol*. **1997**, in press.
40. McNaughton, J. G.; Yean, W. Q.; Goring, D. A. I. *Tappi J.* **1967**, *50*, 548-553.
41. Whiting, P. H.; Goring, D. A. I. *Holzforschung* **1982**, *36*, 303-306.
42. Jurasek, L.; Archibald, F.; Bourbonnais, R.; Paice, M. G.; Reid, I. D. In *Proceedings of the 1994 TAPPI Biological Sciences Symposium and 2nd International Symposium on the Applications of Biotechnology to Tree Culture, Protection, and Utilization*; TAPPI Press: Atlanta, GA, 1994; pp 1-6.
43. Sundaramoorthy, M.; Kishi, K.; Gold, M. H.; Poulos, T. L. *J. Biol. Chem.* **1994**, *269*, 32759-32767.
44. Schroeder, M. *Fractals, Chaos, Power Laws*; W. H. Freeman & Co.: New York, NY, 1991.
45. Karmanov, A. P.; Monakov, Y. B. *Khim. Drev.* **1994**, *No. 2*, 34-38.
46. Freudenberg, K. *Angew. Chem.* **1956**, *68*, 508-512.
47. Gravitis, J.; Erins, P. *Appl. Polym. Symp.* **1983**, *37*, 421-440.

Chapter 20

Efficient Ether Cleavage in Lignins: The Derivatization Followed by Reductive Cleavage Procedure as a Basis for New Analytical Methods

Fachuang Lu and John Ralph[1]

U.S. Dairy Forage Research Center, U.S. Department of Agriculture–Agricultural Research Service, 1925 Linden Drive West, Madison, WI 53706

A new method is introduced for cleaving α- and β-alkyl ethers in lignins. Acetyl bromide treatment dissolves the lignin or whole cell wall material, cleaving benzyl aryl ethers, acetylating primary hydroxyl groups, and brominating benzylic positions. The key step is reductive cleavage of the resulting α-bromo-β-ethers to give cinnamyl acetates. Finally, the product is acetylated to minimize the number of components and to facilitate GC separation. The method has been named the '**DFRC**' procedure, an acronym for the methodology involved (**D**erivatization **F**ollowed by **R**eductive **C**leavage). Low molecular mass products can be extracted from the crude mixture; GC analysis gives data similar to those from analytical thioacidolysis. The basic method itself can be adapted for numerous other analytical determinations of value to lignin researchers.

Lignin is a complex polymer derived from free-radical coupling reactions of hydroxycinnamyl alcohols (*1, 2*). Although lignification is not random, it is still widely held that there are no regularly repeating structures of any significant length. Certainly there is enormous stereochemical heterogeneity (*3*). The β-ether interunit linkage predominates in most natural lignins and there is significant evidence for sequences of β-ether linked units of three or more along the polymer chain (*4, 5*). It is well recognized that cleaving ether linkages in lignin effects a dramatic degree of depolymerization. Cleaving ethers is the basis of chemical pulping and various lignin analytical methods. If lignin were a truly random infinite linear polymer with randomly incorporated β-ether linkages (it is not!), Figure 1 shows the 'mer' proportions, from simple probability theory, resulting from cleavage of all β-ethers. For example, with 40% randomly distributed β-ethers, some 16% monomers would be expected, and 84% of the product would be hexamers or smaller. Similarly, if 50% of the interunit linkages are β-ethers, 25% monomers and 94% hexamers or smaller would result from cleaving all β-ethers. Although natural lignins are not correctly modeled by such an infinite, linear, random polymer, these numbers give crude estimates of the expected amounts and distributions of products. Conversely, if the values obtained in practice vary widely from these numbers, the actual distribution might give some clues to the grouping of lignin interunit linkages.

[1]Corresponding author.

Analytical thioacidolysis (*6, 7*) is today one of the most effective methods for efficiently cleaving ether bonds in lignin to produce identifiable monomeric and dimeric degradation products which can be quantitated. In this chapter we describe an even more efficient method for cleaving lignin α- and β-ethers. The similar release of fragments obtained by this approach also lends itself to development of an analytical method or, as we suggest below, a series of methods. It has already become so important in our own work that a description of the basis of the methodologies is presented here with the aim of hastening their further development. Analytical methodologies such as those described here that identify aspects of composition, structure, and regiochemistry are valuable aids in understanding lignin biosynthesis as well as degradation pathways.

The Reaction of Lignins with Acetyl Bromide

Acetyl bromide (AcBr) is well-known for its ability to facilitate dissolution of lignocellulosic materials in acetic acid. A rapid and simple method for lignin determination using AcBr (*8*) has been widely applied (*9*) and modified for various purposes (*11-13*). Iiyama and Wallis (*14*), after observing the changes in UV-spectra and UV-specific absorption coefficients of lignin model compounds after AcBr treatment, proposed that AcBr acted as an acylating reagent similar to acetyl chloride, acylating hydroxyls (both aliphatic and phenolic) present in lignin to produce acetylated derivatives. Nimz (*15*) had previously demonstrated aryl ether cleavage, bromination and acetylation of a phenylcoumaran model compound using AcBr. It was also suspected that AcBr might acylate aromatic rings of lignin structural units in some cases in the presence of perchloric acid (*14*). Later, they examined the reactions of AcBr (with and without perchloric acid) with lignin model compounds and saccharides (*16*). They found that the rates of the various reactions were in the order *O*-acetylation > bromine substitution >> β-ether cleavage ≈ C-acetylation > demethylation. Their reactions of veratrylglycerol-β-guaiacyl ether **GG-1b** (Figure 2), a 4-*O*-etherified β-ether model compound, gave rather complex arrays of products as determined by HPLC (*16*). The reactions included full acetylation, acetylation/α-bromination, and β-ether cleavage. We have discovered (*17*) that remarkable efficiency and selectivity can be obtained under milder conditions than those described previously (*16*).

Acetyl bromide can effect three types of reactions of value in the study of lignins: acetylation, benzylic bromination, and α-ether cleavage (Figure 3) (*17*). With the correct choice of solvents and conditions, these reactions are more selective and higher yielding than those that result under the conditions used in lignin determination methods based on AcBr (*16*).

Acetylation of Hydroxyl Groups. Acylation of alcohols and phenols with acyl halides using basic or acidic catalysts is well known, although AcBr is only rarely used for preparative purposes. Results (*17*) from a series of model compounds **1** (Figure 2) showed that primary γ-OHs were readily acetylated by AcBr treatment. Although secondary benzylic hydroxyls, when treated with AcBr in AcOH or 1,4-dioxane, were also acetylated, they subsequently formed α-bromo derivatives **7** in almost quantitative yield (Figure 3) as discussed below. Phenolic OHs were acetylated more slowly than primary (γ-) or secondary benzylic (α-) OHs. These reactions are probably acid-catalyzed since small amounts of water in the samples will produce HBr.

Bromination of Benzyl Alcohols. Zawadowski (*18*) observed the bromination of benzylic hydroxyl groups with AcBr, and Iiyama and Wallis (*16*) proposed that bromides were formed among the products of their reactions. Bromotrimethylsilane (TMSBr) and HBr are known to effectively form benzyl bromides from benzyl alcohols, and these bromides have been utilized in schemes to generate lignin model

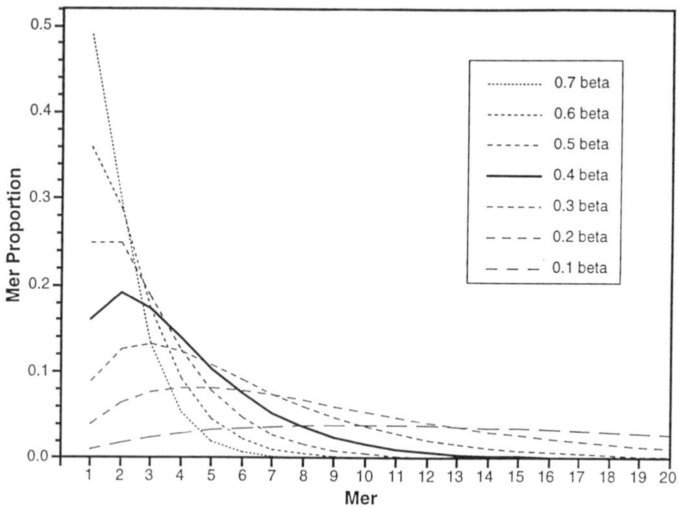

Figure 1. Proportion of monomers, dimers and higher mers expected from β-ether cleavage of an infinite linear polymer with randomly disposed linkages having proportions of β-ether linkages ranging from 10% to 70%.

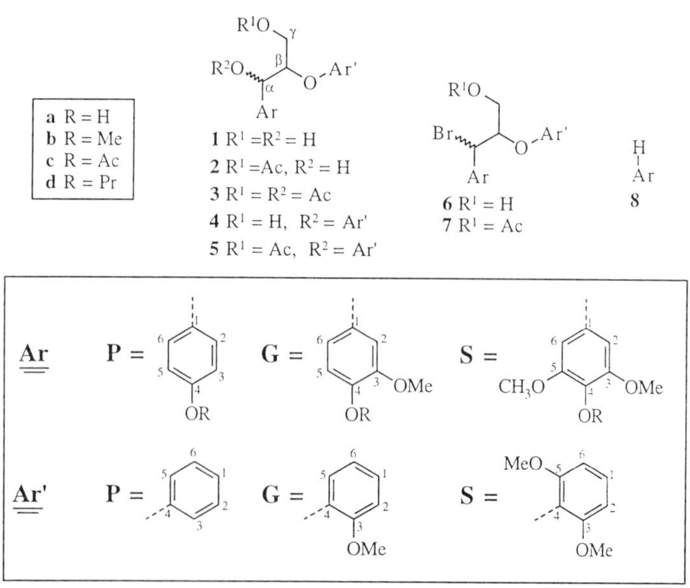

Figure 2. Shorthand notation for structures. Compounds are named using P, G and S (to identify *p*-coumaryl, guaiacyl and syringyl moieties) using the Ar ring followed by the Ar′ ring (if there is one), then the compound number and the phenolic hydroxyl derivatization indicator **a-d**. For example, **GG-1a** denotes free phenolic guaiacylglycerol-β-guaiacyl ether. Later in the manuscript, trimers follow the same format specifying Ar, Ar′, and the third Ar″ ring (not shown), *e.g.* **GSS-1a**.

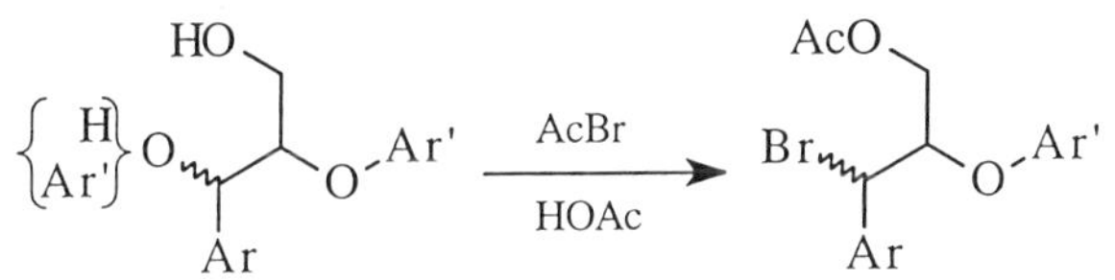

1-5
(a,b)

7
(c,b)

Figure 3. Reaction of β-aryl ether model compounds with AcBr gives acetylated α-bromo-compounds **7**; α-aryl ethers are cleanly cleaved.

quinone methides (*19, 20*). Model compounds **1a** and **1b**, which are often used as representatives of β–*O*–4′ linked structural elements predominant in softwood lignins, were completely converted to acetylated α-bromo derivatives **7b** and **7c** after AcBr treatment in acetic acid or 1,4-dioxane (Figure 3) (*17*). This observation confirms the nature of the products proposed by Iiyama and Wallis (*16*). The cleanliness of AcBr reactions is illustrated in Figure 4, which depicts ^{1}H NMR spectra of *crude* total products from AcBr treatment of various model compound of the type **1** (see Figure 2). AcBr is therefore synthetically useful for the selective bromination of benzylic hydroxyl groups in the presence of other primary or secondary hydroxyls. Additionally, if stereochemical integrity is not an issue, AcBr may also be used to selectively acetylate primary and/or phenolic hydroxyls in the presence of benzylic hydroxyls because the resulting benzylic bromide can be readily hydrolyzed (acetone/water) back to the benzylic hydroxyl groups without affecting the acetates.

It was previously found (*17*) that acetylation of primary γ-hydroxyl groups proceeded faster than that of benzylic hydroxyls during the AcBr treatment in acetic acid. The α-bromo product **7** is formed *via* the α-acetate intermediate **3** (*e.g.* **3b** from **1b**). Therefore, Iiyama and Wallis' observation (*16*) of diacetates is an indication of either too short a reaction time or that the bromides **7** were partially hydrolyzed in subsequent work-up. Treatment of model compound *threo*-**GG-1a** with AcBr in AcOH or 1,4-dioxane resulted in a mixture of 2 diastereomers (about 1:1) of the α-bromo derivatives **GG-7c**, suggesting an S_N1 reaction *via* the benzyl carbocation ion. Conversion of the benzyl acetate to the bromide will occur in free-phenolic or 4-etherified compounds but is very slow with 4-*O*-acetates. No bromide formation was observed when the fully acetylated compound **GG-3c** was used. This could be because little HBr is produced when compounds with no hydroxyls react; much of the reactivity is presumably attributable to HBr. The completeness of reactions in dioxane or acetic acid implies that phenolic acetylation by AcBr is much slower than benzylic alcohol acetylation.

Cleavage of Non-cyclic Benzyl Aryl Ethers. The presence and quantity of non-cyclic benzyl aryl ether (α–*O*–4′) substructures in lignin have been controversial (*21-23*). Methods for distinguishing α- from β-ethers are becoming more important because of the fundamentally different mechanisms for their formation (nucleophilic addition to quinone methides *vs.* radical coupling) and the consequent biochemical implications (*24*). There are several methods used to cleave and quantitate benzyl aryl ether substructures. Perhaps the simplest is a mild acid hydrolysis (*25*). The increase in phenolic content is measured by employing reaction conditions which cleave β–*O*–4′ ethers more slowly than α–*O*–4′ ethers. Recently, we developed a bromotrimethylsilane (TMSBr) method which works effectively on lignin model compounds (*26*) but was not suitable for quantitative analysis of α-aryl ether linkages in lignin. This is because the reagent rapidly reacts with water while lignin has very low solubility in totally anhydrous solvents that are compatible with TMSBr (*e.g.* dioxane). We assumed that AcBr would efficiently cleave non-cyclic benzyl aryl ethers although this was not specifically mentioned by Iiyama and Wallis (*16*). Cyclic benzyl aryl ether lignin model compounds are efficiently cleaved (*15*). AcBr has been used for the cleavage of cyclic ethers under more drastic conditions. For example, refluxing simple cyclic ethers with AcBr and Lewis acid catalysts yielded acetoxy-bromoalkanes (*27*).

We found that the benzyl aryl ether model compound **GG-4b** was cleaved rapidly and cleanly, without affecting the β-ether after >12 h (*17*). NMR spectra of the products showed that the α-bromo acetylated product **GG-7b** and 2-methoxyphenyl acetate **G-8c** were formed essentially quantitatively in AcOH or 1,4-dioxane. The mechanism of α-ether cleavage by AcBr presumably follows an S_N1 process resulting in a mixture of α-bromo diastereomers **GG-7b**. The possibility that the cleavage reaction follows an S_N2-like mechanism with fast isomerization cannot be ruled out; anchimerically assisted reactions followed by bromide isomerization have recently been observed in this laboratory (*26*).

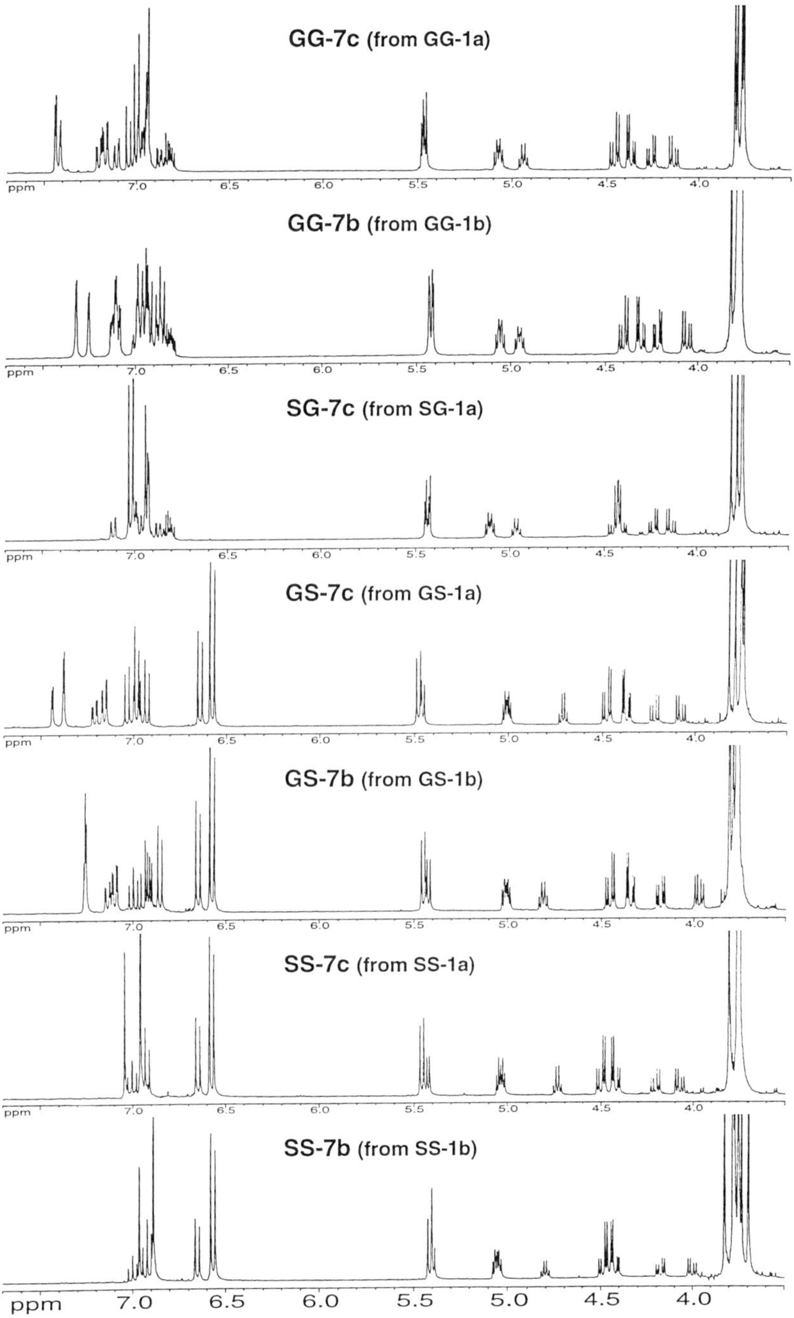

Figure 4. Partial ^{1}H NMR spectra of *crude* AcBr reaction products from a variety of dimeric β-aryl ether model compounds.

Summary of Lignin Reactions with AcBr

Lignin dissolution by AcBr is an extremely effective process (*8, 10, 12, 14, 17*). The NMR spectra shown in Figure 4 are for *crude* products from AcBr reactions with various syringyl/guaiacyl lignin model β-ethers. NMR spectra of milled tissue lignins following AcBr treatment (Figure 5) show that the bromides are also produced as major products from polymeric lignins. Solubilization with AcBr provides a promising means for characterizing whole lignins by NMR spectroscopy. The relative rates of reactions observed between AcBr and β-aryl ether lignin model compounds under mild conditions are: primary (γ-) OH acetylation > benzyl aryl ether cleavage/ bromination > secondary benzylic (α-) OH acetylation > benzyl acetate to benzyl bromide substitution >> phenolic OH acetylation. Under these conditions, no β-aryl ether cleavage or demethylation (reported previously under harsher conditions (*16*)) were detectable. The cleanliness and reproducibility of these reactions suggest that lower reaction temperatures should be examined for AcBr lignin determination methods. The selective bromination of benzyl alcohols by AcBr is synthetically useful. Selective cleavage of benzyl aryl ethers by AcBr may provide a new technique for the analysis of α-aryl ether linkages and for quantitating hydroxycinnamic acids that are passively incorporated into lignin (*24, 28*). Such applications are being evaluated in our laboratory.

Reductive Cleavage of β-Bromo Ethers

The major contribution here (*29, 30*) is the recognition that reaction with AcBr has produced compounds that are almost ideally susceptible to reductive cleavage. The β-bromo ethers **7** that are produced are amenable to formal two-electron transformations (Figure 6). Related ethers have been cleaved using Zn dust (*31*) or Cr(II)en (*32, 33*), for example. Reaction of the lignin or lignin model products with Zn^0 in dioxane/acetic acid/water at room temperature produces the cleaved products within minutes. NMR spectra of the *crude* products resulting from AcBr treatment followed by reaction with Zn^0 are shown in Figure 7.

It should be noted that the 'adduct' mechanism (*34, 35*) for the cleavage of free phenolic β-ethers by reduced anthraquinone species is a formal two-electron process with considerable similarity when viewed in general terms from the perspective of Figure 6b. The adduct **11** between AHQ^{2-} and a lignin quinone methide undergoes Grob fragmentation to cleave the β-ether bond. The reaction with lignin α-ketones **12** also occurs readily (Figure 6b) but is not as convenient for lignin reactions since lignins oxidized at the benzylic position are particularly insoluble and intractable (*29*). The actual mechanism presumably involves single electron transfer steps rather than a concerted two-electron reduction (Figure 6c) although precedence for the latter is known (*36*). Similar reactions with 1,2-dibromides where the bromines can be oriented in *anti* (180°) or *syn* (or 0°) periplanar positions show complete stereo-selectivity, implicating concerted two-electron reduction (*36-38*). In our case, a stereoselective reaction (Figure 6d) should lead to ~50:50 *trans:cis* products since the starting bromides equilibrate to a 50:50 *syn:anti* (*threo:erythro*) mixture. However, since the products are ~95% *trans*, the mechanism cannot be concerted and must rather involve carbanion intermediates (Figure 6c).

It is possible to run the Zn^0 reaction under anhydrous conditions to produce aryl propenyl derivatives **10** (*e.g.* eugenol **G-10** in Figure 6a). However, we were not able to find conditions that would cleanly give such products without some of the cinnamyl acetates **9** also being formed in larger amount. That limitation, the increased sensitivity of the reaction to the exact conditions, and the realization that the hydroxycinnamyl acetates **9** were more diagnostic of the lignin structure, led to our exclusive development of the method based on the production of hydroxycinnamyl acetates.

Thus the two steps involving AcBr treatment followed by reductive cleavage cleanly produces *cis-* and *trans*-hydroxycinnamyl acetates **9** (Figure 8). These could

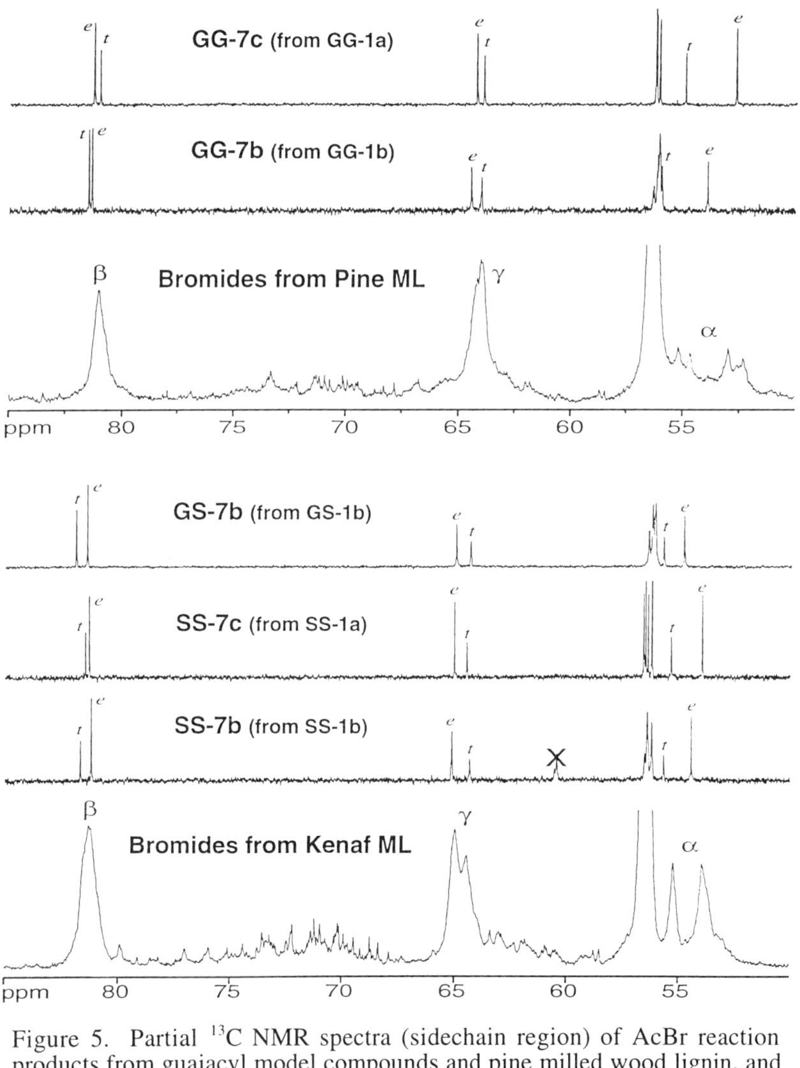

Figure 5. Partial ^{13}C NMR spectra (sidechain region) of AcBr reaction products from guaiacyl model compounds and pine milled wood lignin, and syringyl models and a kenaf milled lignin. Analogous bromides are clearly formed in lignins and model compounds.

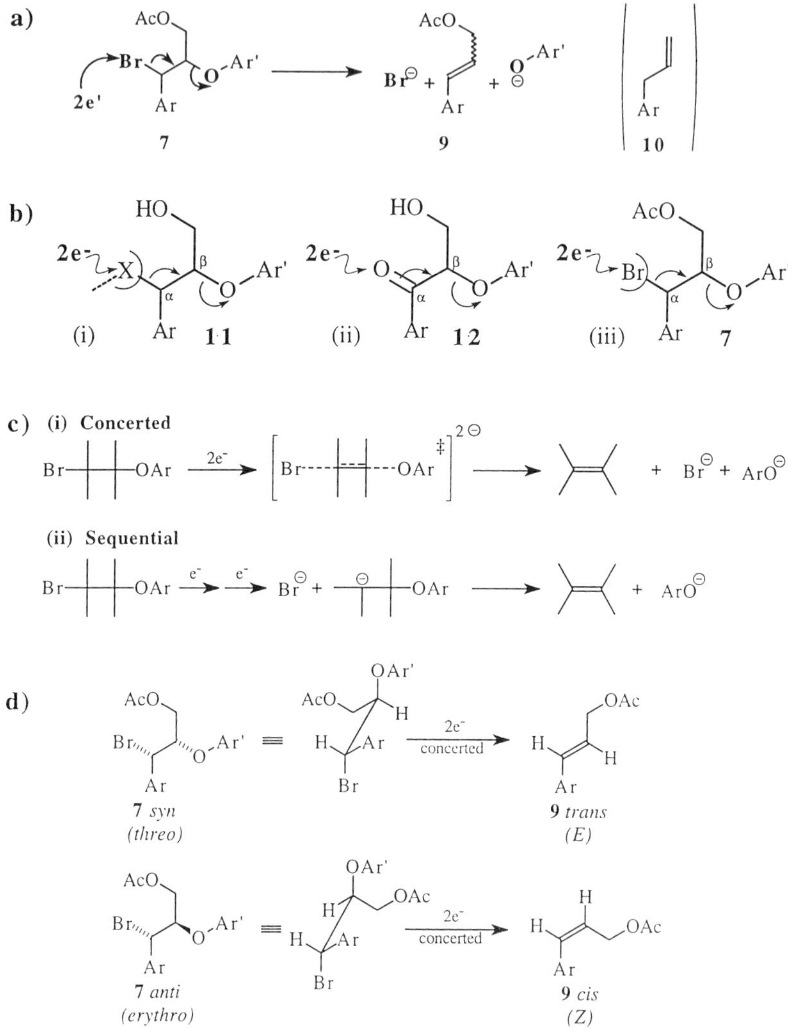

Figure 6 (a). Reductive cleavage of β-bromoethers **7** to give hydroxycinnamyl acetates **9**; compounds **10** can be formed in stronger acid. (b) Three examples of conceptually similar formally two-electron processes: (i) cleavage of ethers by anthraquinone *via* adduct mechanism (see text); (ii) reduction of α-keto-β-ethers with Zn/H^+ efficiently cleaves the ether; (iii) bromides proced by AcBr treatment are cleavable through formally two-electron processes. (c) Two alternative reaction mechanisms for reductive cleavage: (i) concerted two-electron reduction; (ii) two sequential single-electron steps *via* an intermediate carbanion. (d) Stereochemistries expected from concerted two-electron reduction of the two bromide isomers of **7**; since products from a 50:50 *syn:anti*-**7** favor *trans*-**9** highly over the *cis*-isomer, the mechanism must be sequential.

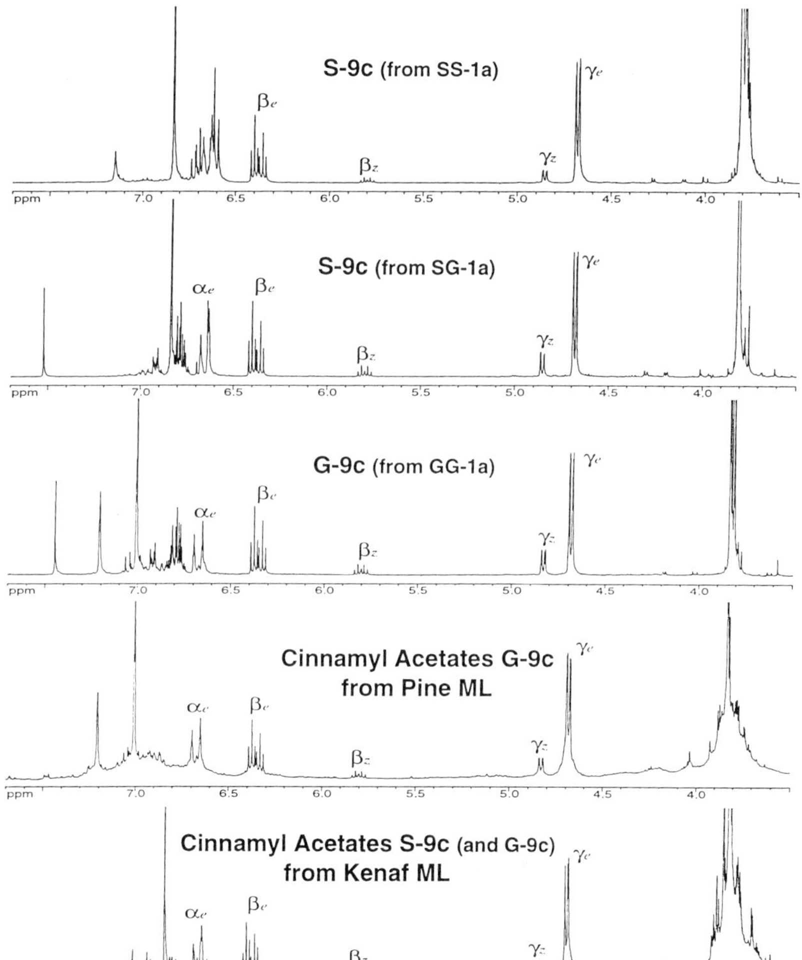

Figure 7. Partial ^{1}H NMR spectra of crude products from full DFRC treatments of model dimers **1** and milled tissue lignins from pine and kenaf. The production of cinnamyl acetates **9** (Figure 6a) is clearly evident, even in the lignins.

be separated by GC and quantitated directly at this stage. Indeed there are some advantages to doing so as we shall see later. However, such an approach necessitates separately quantitating free phenolic and phenol-acetylated compounds for each monomer aromatic ring (S, G, or P); free phenolic components arise in the reduction step from originally etherified structures, whereas phenolic acetates are produced from initially free-phenolic or α-etherified units in the AcBr step (Figure 8). In order to obtain the simplest and most easily quantifiable mixture, the obvious solution was to simply acetylate the product before GC quantitation. Thus the third step, acetylation, produces fully acetylated hydroxycinnamyl acetates **9c**, which are stable and ideal for GC quantitation (and GC-MS identification).

Attempts to Convert *cis/trans* Isomers to Single Products

Although the *trans:cis* ratio is high, it would be advantageous to convert the isomers to a single product thereby allowing even easier quantitation. The simplest way to remove isomers from the products **9** would seem to be hydrogenation (Figure 9). Regrettably this apparently simple step is capricious and difficult to realize in practice in our hands. The cinnamyl esters are readily hydrogenated to arylpropane derivatives **14** in competition with the desired double-bond saturation reaction to give propyl acetates **13**. Hydrogenolysis is normally considered to be less significant with Pt than Pd catalysts, but this was not borne out on these substrates. Hydrogenation without first acetylating any free phenols is a particular problem. Hydrogenation with Rh/C in EtOAc was the best condition we found; the hydrogenolysis products from acetylated phenolic models comprised <1% of the mixture, but with lignin where free phenolic moieties are generated (prior to acetylation) ~5% hydrogenolysis products are typical. The cheaper Pd/C catalyst produces <8% hydrogenolysis products. Finally, hydrogenation masks the origin of compounds arising from minor units in lignin. Although hydrogenation may find eventual application in the DFRC method, its use now has been abandoned in favor of the simpler 3-step method depicted in Figure 8.

Mass Spectra

One of the most appealing aspects of using acetate derivatives is the utility of their mass spectra (Figure 10). Although fragmentation to the acylium ion (m/z 43) occurs, the spectra observed above mass 45 are usually more informative than those of TMS derivatives. Thus, the molecular ion is often significant, and even-massed ion-radicals arising from various acetate fragmentation/rearrangements are particularly diagnostic and structurally revealing. For example, aromatic acetates are lost as neutral ketene (mass 42) so M-42 peaks signify aromatic acetates (Figure 10). By contrast, unsaturated sidechain γ-acetates also lose $CH_3CO^{\bullet}$ (mass 43), and fully saturated sidechains fragment by a McLafferty rearrangement (*39*) eliminating neutral acetic acid (mass 60). As an example, this made identification of the novel 5-hydroxyconiferyl alcohol product **15** particularly simple; from the molecular ion, it loses 42 (a phenolic acetate), 42 again (another phenolic acetate), and then 42/43 (unsaturated sidechain acetate). This product, identified first by thioacidolysis (*40*), is significant particularly in certain brown midrib mutants of various grasses which have been shown to be OMT-deficient (*41*). Obviously, the kenaf used for this analysis also has 5-hydroxyconiferyl alcohol incorporated into its lignin. In fact, all hardwoods, grasses and legumes that we have subjected to the DFRC procedure release this unit. We have confirmed that there is no possibility of methoxyl group demethylation under these conditions. Since thioacidolysis does not reveal this structure in normal hardwood lignins, we conclude that the DFRC method simply releases them with greater efficiency.

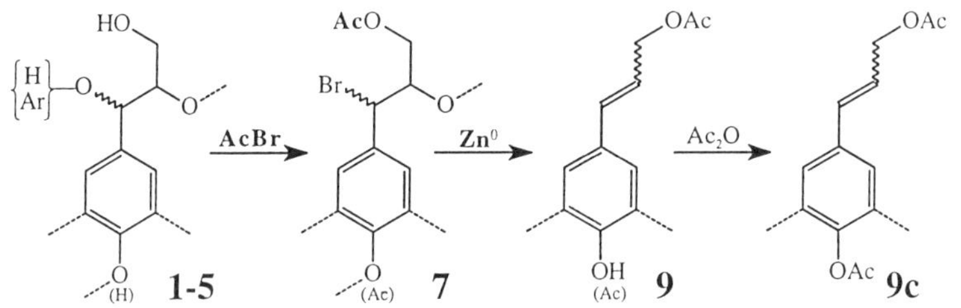

Figure 8. The full DFRC scheme: cell wall dissolution and derivatization using AcBr in acetic acid, reductive cleavage using zinc, and final acetylation to give hydroxycinnamyl acetates **9c** from β-ether linkages.

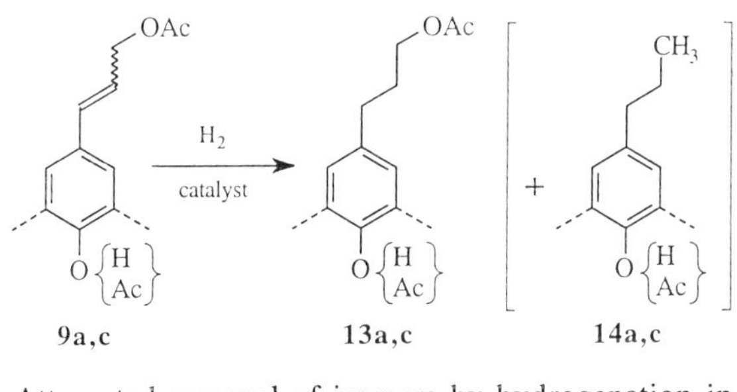

Figure 9. Attempted removal of isomers by hydrogenation in our hands produces hydrogenolysis side-products **14** in addition to the desired propyl acetates **13**.

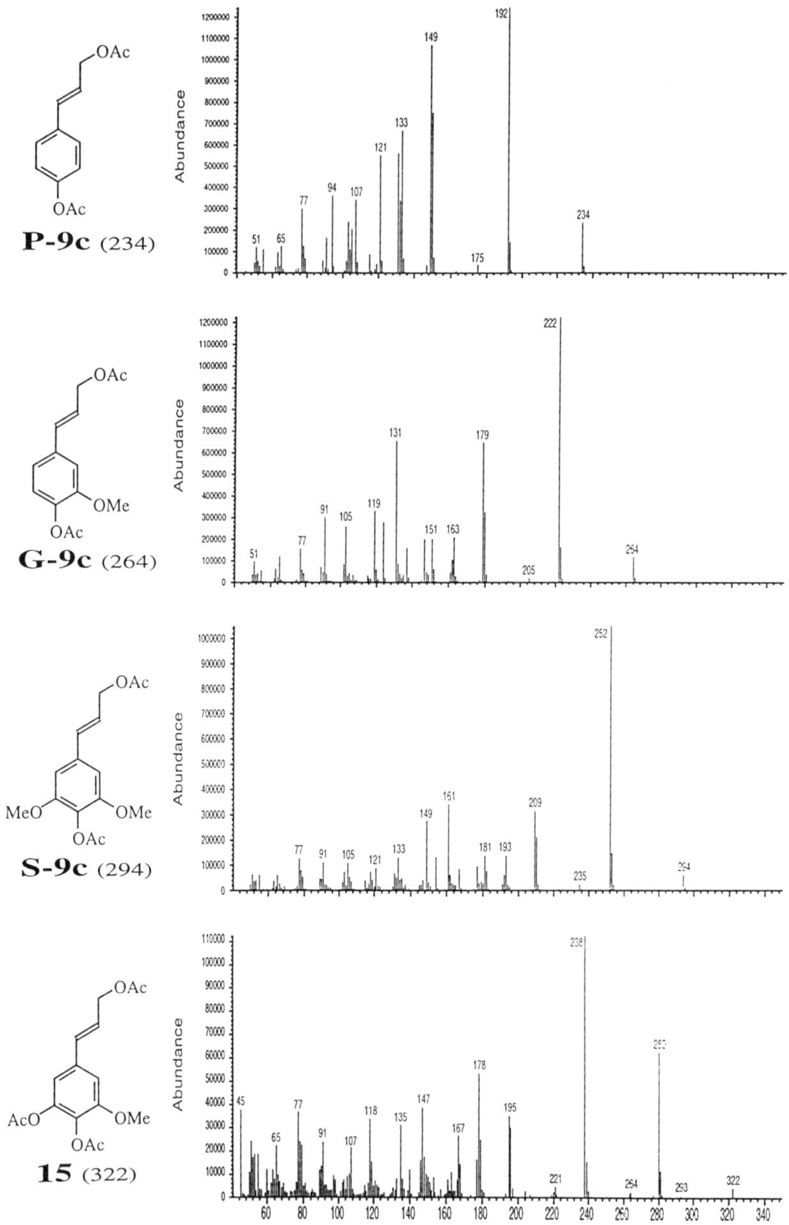

Figure 10. Mass spectra of the monomer acetates. Aromatic acetates readily lose 42 (ketene) whereas the acetylated unsaturated sidechains lose 42/43. The mass spectra have significant molecular ions and their fragmentations are quite predictable. The lower spectrum (of **15**), with two 42 losses and a 42/43 loss, is clearly from the 5-hydroxyconiferyl alcohol-derived units in lignin.

Experimental Section

The DFRC Procedure—Analysis of Monomers

The scheme and conditions for producing analyzable monomers to give data analogous to that from the original (one-step) thioacidolysis method are specified in the following subsections. The full scheme is shown in Figure 8.

Materials and Reagents

Acetyl bromide, dioxane, acetic acid and zinc dust were purchased from Aldrich Chemical Co. and used as supplied. Commercial analytical reagent grade solvents were used without further purification.

AcBr stock solution: AcBr and acetic acid, 8:92 or 20:80 by volume (Table I).
Acidic reduction medium: dioxane/acetic acid/water (5:4:1, v/v/v).
Acetylation reagent: pyridine:acetic anhydride (1:1, v/v).
Internal standard for GC quantitation: tetracosane.

Table I summarizes the conditions for various sample types.

Detailed Experimental Methods—Model Compounds

(For lignins or cell wall samples, see Table I for times, conditions, and amounts)

AcBr step. To a 10 mL round bottom flask containing about 5 mg lignin model compound was added 2.5 mL of AcBr stock solution. The mixture was kept at room temperature with gentle stirring overnight. Finally, the solvent was either removed by rotary evaporation below 50°C or under a stream of air.

Reductive cleavage step. The above residue was dissolved in 2.5 mL dioxane/acetic acid/water (5:4:1, v/v/v) solution. Zinc dust (50 mg) was added to the well-stirred solution. Stirring was continued for 30 min. The mixture was completely transferred into a separatory funnel with CH_2Cl_2 (10 mL) and saturated NH_4Cl (10 mL), and an internal standard (2.5 mg tetracosane in methylene chloride) added. The pH of the aqueous phase was adjusted to less than 3 by adding 3% HCl, the contents of the separatory funnel vigorously mixed, and the organic layer separated. The water phase was extracted twice more with CH_2Cl_2 (2 x 5 mL). The combined CH_2Cl_2 portions were dried ($MgSO_4$) and the solvent evaporated to dryness under reduced pressure.

Acetylation step. The residue was acetylated in 1.5 mL dichloromethane containing 0.2 mL acetic anhydride and 0.2 mL of pyridine for 40 min. All volatile components were removed completely by co-evaporation with ethanol under reduced pressure. The residue was used for NMR, GC and GC-MS characterization.

GC determination and GC-MS. The resulting reaction mixture products were dissolved in 2 mL of methylene chloride and 1 µL of this solution was used for GC analysis. They were quantitatively determined by GLC (Hewlett Packard 5980): column 0.20 µm x 30 m SPB-5 (Supelco); He carrier gas, 1 mL/min; 30:1 split ratio; injector 220°C; initial column temperature 150°C, ramped 10°C/min to 310°C, held 10 min; FID detector, 310°C. The amounts of individual monomers **9c** were determined using response factors (RFs) derived from pure standards using tetracosane as internal standard. Relative retention times (RRTs) and GC response factors relative to the tetracosane internal standard are given in Table II. EI-MS data were collected on a Hewlett Packard HP5970 mass selective detector attached to the same GC; the column and flow conditions differed slightly between the GC-FID and GC-MS configurations, so absolute retention times differ between the two.

Representative GC (FID) or GC-MS (TIC) traces that illustrate the nature of the products for a variety of samples are depicted in the following sections on model compounds, isolated lignins, and whole cell wall samples. Clearly the major peaks arise from guaiacyl and syringyl units involved in ether linkages. Smaller amounts of

Table I. Summary of Reaction Conditions for the DFRC Procedure with Various Substrates.

Step	Models	Lignins	Cell Walls
Wt of material	5 mg	10 mg	20 mg*
AcBr step			
AcBr reagent	8:92	8:92	20:80
temperature	RT/50°C	RT/50°C	50°C
time	16 h/2 h	16 h/2 h	3 h
Reductive cleavage step			
Acidic solution	2.5 mL	2.5 mL	2.5 ml
Zn^0	50 mg	50 mg	50 mg
time	0.5 h	0.5 h	0.5 h
time for α-keto β-ethers	16 h	16 h	16 h
internal standard (tetracosane)	2.5 mg	0.5 mg	50-300 μg**
Acetylation step			
acetylation solvent (CH_2Cl_2)	1.5 mL	1.5 mL	1.5 mL
pyridine/Ac_2O	0.2 mL ea	0.2 mL ea	0.2 mL ea
acetylation time	40 min	40 min	40 min
GC quantitation			
volume with CH_2Cl_2	2 mL	0.2 mL	0.2 mL
volume injected	1 μL	1 μL	2 μL
initial temperature (time)	150 (1)	150 (1)	140 (1)
intermed. temp. (ramp rate)	—	—	240 (3/min)
final temp. (ramp rate, °C/min)	310 (10/min)	310 (10/min)	310 (30/min)
final temp. (hold time)	310 (10)	310 (10)	310 (10)
total run time	27 min	27 min	47 min

* The amount of sample required depends on the amount of lignin in the sample and the cleavable β-ether frequency. 20 mg is a comfortable level for most cell wall samples. It is possible to use lower quantities for the three sample types listed here.

** As with any analysis, the amount of standard should be reasonably matched to the amount of the compounds being quantitated. Using the conditions described in the rest of the table, the following may help in selecting the appropriate level for analysis: bass (hardwood), 300 μg; pine (softwood), 200 μg; alfalfa (legume), 50 μg; corn (grass), 50 μg.

Note: For lignins, using the longer GC temperature program that is employed for cell walls may improve resolution and separate out minor peaks.

Table II. Mass Spectral and GC Data for Products 9, from DFRC Procedure

Cpd	RRT*	RF**	m/z (relative intensity)
P-9c$_{cis}$	0.58	1.74	234(19, M^+), 192(100), 149(85), 133(54), 121(44), 107(28), 94(29), 77(24)
P-9c$_{trans}$	0.63	1.74	
G-9c$_{cis}$	0.69	1.85	264(11, M^+), 222(100), 179(37), 163(9), 151(12), 131(27), 119(15), 91(14)
G-9c$_{trans}$	0.76	1.85	
S-9c$_{cis}$	0.81	2.06	294(8, M^+), 252(100), 209(24), 193(10), 181(8), 161(17), 149(13), 133(6)
S-9c$_{trans}$	0.88	2.06	

* Relative retention time. ** Response factor.

p-hydroxyphenyl and 5-hydroxyconiferyl alcohol-derived units are also readily apparent. More minor components are considered later. It should be noted that retaining the functionality at the γ-position makes these products more diagnostically related to lignin than if we had chosen to target the arylpropane products (see above).

Quantitative Aspects

Lignin Model Compounds. Model compound studies are useful to test reactions for their selectivity and yields. Conditions which fail to give near-quantitative yields of desired products from appropriate model compounds may be less effective with lignins. As asserted above, the AcBr reaction is very selective (if the specified conditions are adhered to), and the Zn^0/H^+ reductive cleavage is also extremely efficient. The final acetylation step is also virtually quantitative. Consequently the DFRC procedure produces hydroxycinnamyl acetates from β-ether dimers and trimers in very high yields (Table III) and produces exceptionally well-resolved chromatograms (Figure 11). Such yields are currently unattainable by thioacidolysis, which typically produces 70-75% of the primary tri-thioethyl derivatives from such models (*42, 43*).

Isolated Lignins. Isolated 'milled wood lignins' produce well-resolved chromatograms due to the high lignin concentration (and correspondingly low polysaccharide content) and because dissolution in AcBr is rapid and complete. Figure 12 shows a series of isolated lignin materials representing hardwoods (willow), softwoods (pine), grasses (bamboo) and dicots (kenaf). In all chromatograms, peaks of interest are well resolved and quantitation is straightforward. Table IV gives product yields for a few lignins and makes a comparison with results from thioacidolysis determined in our laboratory. [We utilize the thioacidolysis method as described, but have never done a full optimization and cannot guarantee that we have optimal values. However, the molar values are close to those in the literature for similar materials when expressed on a Klason lignin basis (*7, 44*). It appears from the kenaf lignin that the DFRC method may give a better recovery of syringyl products, but extensive comparisons have not been made.]

Whole Cell Walls. As with other degradative procedures, low molecular weight polysaccharide degradation products are produced by the DFRC method. Initial procedures on whole cell walls attempted to achieve complete dissolution of the walls during the AcBr step. However, this was subsequently found to be unnecessary. The lignin component is quite completely solubilized in 2-3 h even though some insoluble residue remains; product yields do not go up after that period and quantitation becomes more difficult due to the presence of higher amounts of polysaccharide degradation products. Figure 13 shows well-resolved chromatograms from a representative softwood (pine) and hardwood (bass). The GC temperature program is extended (Table 1) to provide better resolution from the polysaccharide components. We have not run a sufficient variety of samples to determine whether these spectra are typical. Very recent trials with a maize sample having a high *p*-coumarate content have polysaccharide peaks that interfere with the lignin monomer determinations. This sample, however, reveals the *p*-coumarate esters, as described later. Time will tell if the DFRC procedure can be applied to the same range of materials as thioacidolysis — there may be instances where thioacidolysis is superior.

Minor Monomers Released from Isolated Lignins and Cell Wall Samples

Other monomeric lignin-derived products are seen in chromatograms from plant materials. Given the utility of the reactions with β-ethers, there is reasonable certainty that many of these products arise from minor structures in lignin. As such they are poised to provide more detailed characterization of lignin components. Examination of the reactivities and products of minor lignin units such as

Table III. Comparison of Monomer Percent Yields from Thioacidolysis, Acidolysis and DFRC Methods.

Compound	PG-1a	GG-1a	GS-1a	SG-1a	SS-1a	GGG-1a	GSS-1a G%	GSS-1a S%
DFRC method	93.2	94.5	95.0	93.5	96.4	92.6	97.0	93.4
thioacidolysis (*42, 43*)	---	70/75	---	75	52-57	---	---	---
acidolysis (*42*)	---	69	---	---	32	---	---	---

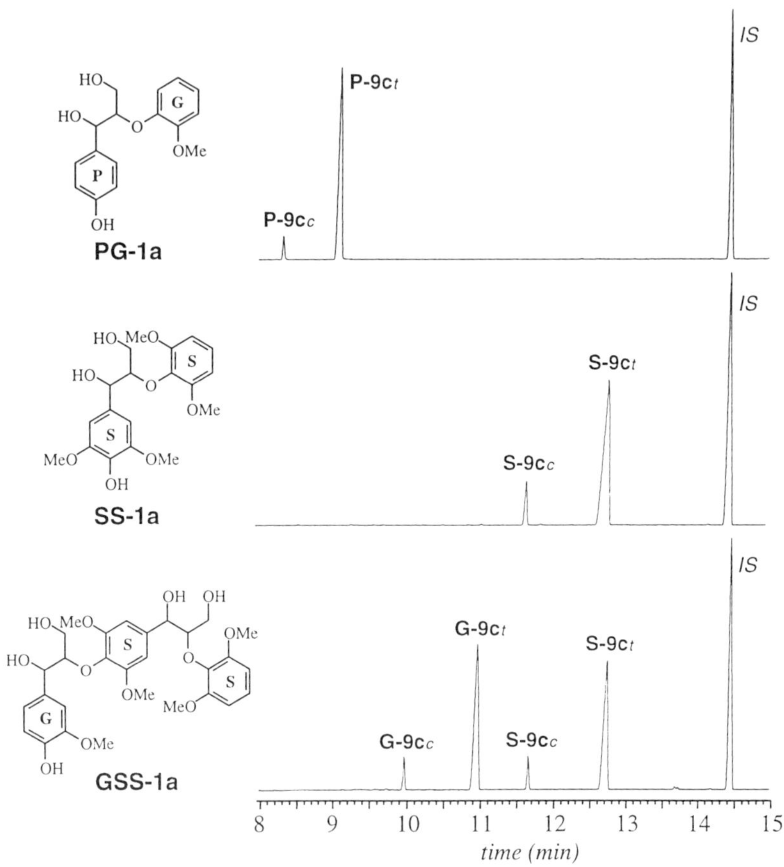

Figure 11. GCs of DFRC products **9c** from three model compounds showing the cleanliness of the products following the three reaction steps. P, G, and S represent *p*-hydroxyphenyl, guaiacyl, and sinapyl products, as per Figure 2. The end unit on each model has no sidechain and is not present as such in natural lignins; its products appear early in the chromatogram (not shown). *t* = *trans*, *c* = *cis* isomers.

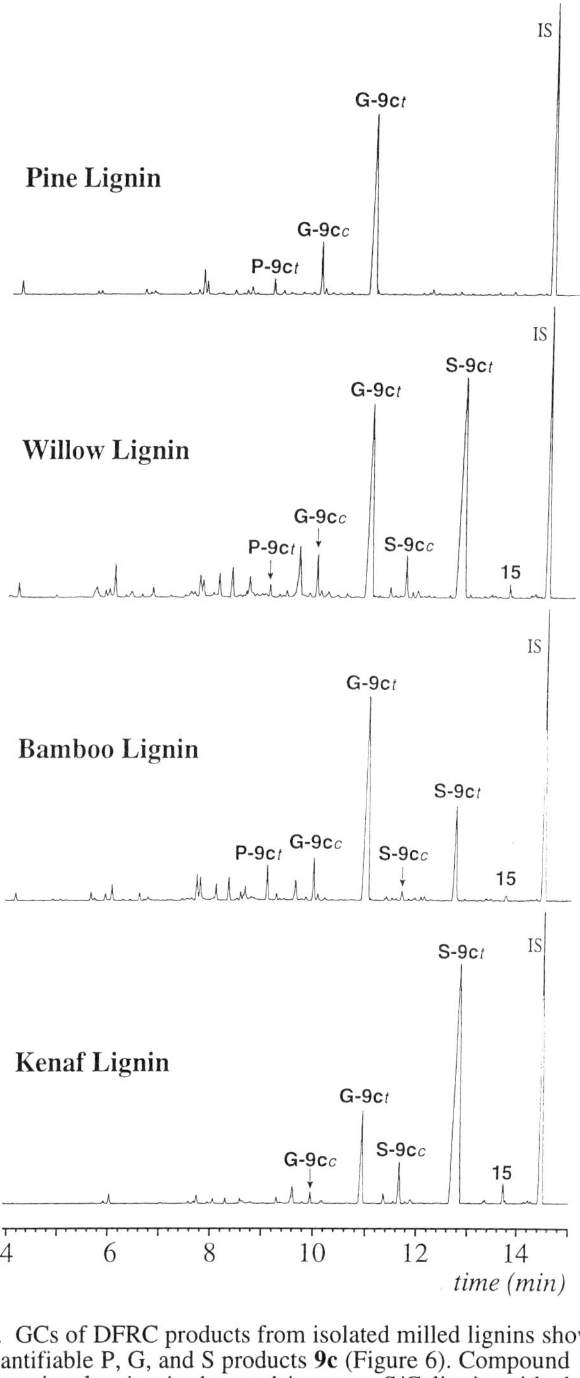

Figure 12. GCs of DFRC products from isolated milled lignins showing the readily quantifiable P, G, and S products **9c** (Figure 6). Compound **15**, from 5-hydroxyguaiacyl units, is detected in every S/G lignin with the DFRC method.

Table IV. DFRC Yields of the Chief H, G and S Monomers from DFRC Degradation and Thioacidolysis of Various Milled Tissue Lignins. (Corrections for *ca.* 85-95% lignin content are not made.)

Lignin	Yield (%)		Molar Yield (μmoles/g)		Relative Distribution H : G : S	
	DFRC	Thio*	DFRC	Thio	DFRC	Thio
Pine	17.1	–	651	–‡	0.03:1.0:--	0.02:1.0:0.01†
Kenaf	36.4	–	1258	967	t:1.0:5.70	--:1.0:5.08
Willow	20.6	–	733	597	0.03:1.0:1.39	--:1.0:1.54

* Data are available but do not allow for relevant comparison—molar yields are directly comparable between the two methods.

‡ Pine lignin yields are reported at ~800 μmoles/g (*42*); adjusting our figures to a Klason lignin (~85%) basis gives comparable yields.

† The thioacidolysis detection of S-units in pine cannot be substantiated.

t = trace.

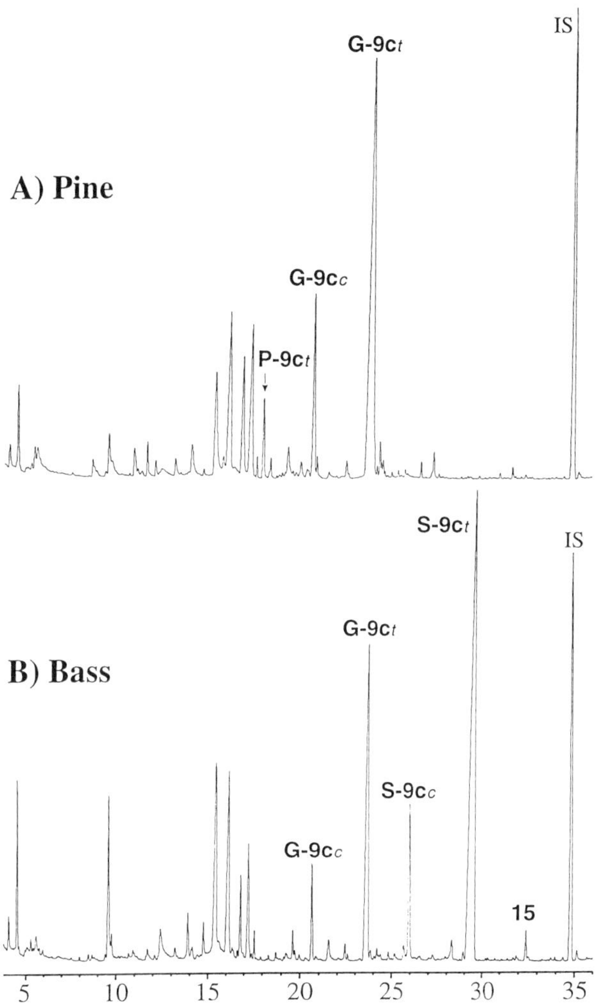

Figure 13. GC-FID chromatograms of DFRC monomers from extractive-free wood samples. A) Pine (softwood) and B) Bass (hardwood). The normal P, G and S peaks **9c** are shown as well as the 5-hydroxy-product **15**. Early peaks in the chromatogram are largely polysaccharide-derived.

cinnamaldehyde, benzaldehyde, hydroxycinnamyl alcohol, and dihydrocinnamyl alcohol end-groups, 5–hydroxyconiferyl alcohol units, and α–carbonyl structures, are currently underway in our laboratories.

An example of the range of products obtained is seen in Figure 14 and accompanying Table V from a bamboo milled wood lignin. Some of the products are currently unidentified and their sources are unknown, whereas the identities of others are quite obvious from their mass spectra. Full identification and authentication of these minor products, along with their likely sources, will be forthcoming.

Advantages over Thioacidolysis

In our opinion, the following advantages are seen in the DFRC procedure.

1. Reactions are more selective and more quantitative. As noted above, the AcBr and the Zn^0 reactions are extremely clean. Reactions of model compounds through the entire process produce ~95% yields of the targeted products (Table III). Thioacidolysis yields in our laboratories are typically 75% (*43*); this seems to agree with those reported by Lapierre on the recoverability of products that have been subjected to thioacidolysis conditions (*42*). Thioacidolysis produces some chain-shortened products (aryl ethyl derivatives rather than aryl propyl) from side reactions. The product yields that we note here (Table IV) attest to the higher yielding reactions in lignins as well.
2. The reactions are perhaps simpler and occur under milder conditions. Reactions at above 100°C under acidic conditions, as used in acidolysis and thioacidolysis, are not typically used in organic syntheses because their rates are more difficult to predict and control. The reactions described here are sufficiently high yielding to be synthetically as well as analytically useful.
3. Malodorous chemicals are not required. No discussion of thioacidolysis is complete without mention of the foul stench of the required reagent (ethane thiol), although its impact can be minimized by careful work, utilizing bleach for trapping, etc. In some laboratories, it is simply not possible to perform thioacidolyses because of restrictions on tolerable levels of odor.
4. Esters are not cleaved during the DFRC procedure. In thioacidolysis, esters are partially cleaved (*43, 45*) and this causes problems with grasses (for example) where *p*-coumarate esters at the lignin γ-position (*46*) can be quite prevalent. γ-Esters remain completely intact after the DFRC procedure and this can be used to tremendous advantage in identifying and characterizing esterified lignin components. (See also 'Other Applications' below.)
5. The scheme is amenable to targeting products and to a rather wide variety of future applications. Some of these are discussed below ('Other Applications').

Disadvantages with respect to Thioacidolysis

There are also disadvantages to the DFRC procedure. Among them are the following.

1. Interference from carbohydrate degradation products is encountered. Polysaccharides are also degraded under DFRC conditions. At present, the identities of the most labile polysaccharides and their reaction products have not been established. However, GC peaks from acetylated non-lignin products are clearly evident (GC-MS). Products from isolated lignins are clean and such contamination is not a problem. With whole plant materials, the impact can also be minimized (Figure 13). It appears, from the limited comparisons that we have made that woods and legumes are eminently compatible with the method but recent trials with maize causes some concern for grasses. The ability to find and quantitate the lignin peaks of interest in a wide range of samples will determine the range of usefulness of this method. Silica solid-phase adsorption of saccharide components (prior to the acetylation step) may be a partial solution; the lignin products are more solvent soluble and will easily elute from SiO_2.

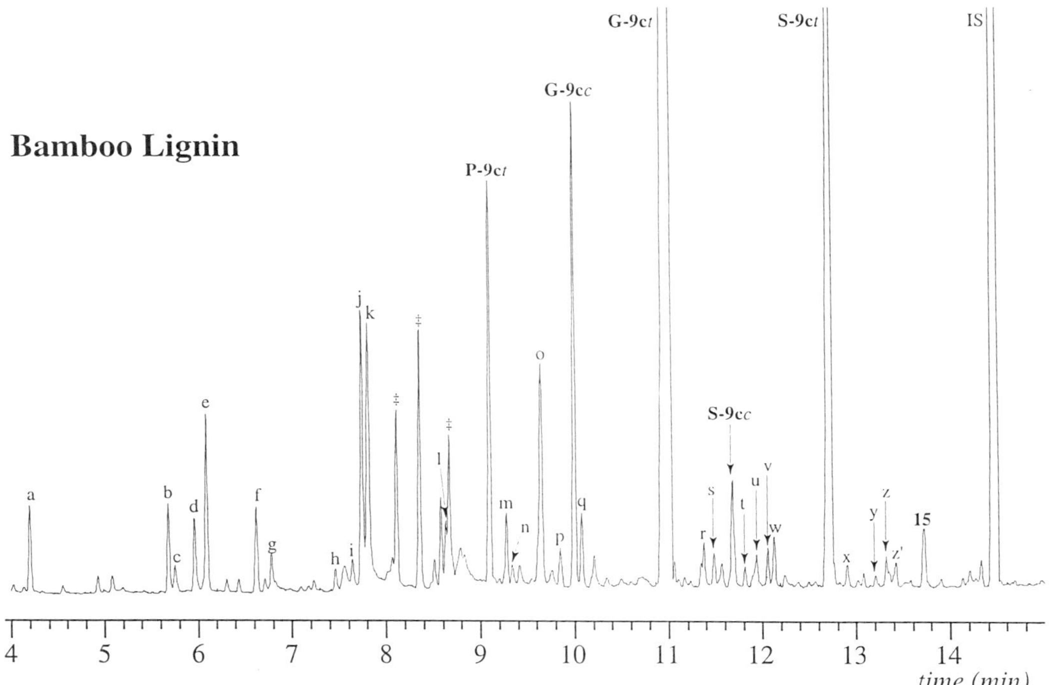

Figure 14. Vertically expanded GC FID of DFRC products from a bamboo lignin showing minor peaks that are identifiable in lignins and cell wall samples. Table V lists peak identities (where known) and MS data for them. ‡ = carbohydrate-derived peaks.

Table V. Structural and Mass Spectral Characteristics of DFRC Monomers from Isolated Lignins and Lignocellulosic Materials.

GC* Peak	Structure** assignment	Molecular ion and prominent fragments m/z (relative intensity)
a	G-CH$_3$	180(8), 138(91), 123(100), 95(27), 77(31)
b	G-CH$_2$CH=CH$_2$	206(9), 164(100), 149(26), 133(24), 131(23), 103(15), 91(25)
c	G-CHO	194(6), 152(100), 151(96), 123(10), 109(19), 79(17)
d	P-CH$_2$OAc	208(17), 166(100), 124(77), 107(100), 106(80), 95(26), 84(36)
e	S-CH$_3$	210(6), 168(100), 153(33), 125(12), 107(10), 79(8)
f	G-?	206(11), 164(100), 149(23), 133(16), 131(16), 103(14), 91(19)
g	G-CO-CH$_3$	208(5), 166(42), 151(100), 123(14), 79(16)
h	G-CH$_2$CO-CH$_3$	222(5), 180(28), 137(100), 122(6), 94(8), 77(10)
i	S-CH$_2$CH=CH$_2$	236(5), 194(100), 179(11), 163(8), 147(8), 131(12), 119(13), 91(14)
j	G-CH$_2$OAc	238(6), 196(100), 154(65), 137(82), 122(17), 107(10), 93(14)
k	G-CO-CH$_2$CH$_3$	222(4), 180(26), 151(100), 123(10), 108(5), 79(6)
l	S-?	236(14), 194(100), 179(19), 163(35), 151(13), 131(16), 120(25), 91(22)
m	G-?	264(8), 222(46), 179(100), 162(12), 151(48, 135(10), 119(53), 107(10), 91(32)
n	S-CH$_2$CO-CH$_3$	252(5), 210(48), 167(100), 149(16), 137(16), 121(11), 103(10), 91(9), 84(17)
o	S-CH$_2$OAc	268(5), 226(100), 184(67), 167(66), 152(6), 123(26)
o′	S-CO-CH$_2$CH$_3$	252(3), 210(36), 181(100), 153(10), 138(6), 109(5)
o″	G-CH$_2$CH$_2$CH$_2$Br	288(3), 286(3), 246(26), 244(28), 137(100), 122(5), 107(6)
p	G-CHOAcCOCH$_3$	280(4), 236(8), 195(27), 153(100), 125(8), 107(4), 93(12), 65(8)
q	G-?	264(8), 222(48), 179(100), 162(22), 151(46), 137(12), 119(56), 107(8), 91(37)
r	S-CHOAcCOCH$_3$	310(4), 268(14), 225(26), 183(100), 155(12), 123(14), 95(7)
s	S-CH$_2$CH$_2$CH$_2$Br	316(6), 374(60), 194(5), 167(100), 151(11), 137(6), 107(6)
t	S-?	294(8), 252(41), 209(100), 192(28), 181(40), 149(58), 137(12), 121(20), 103(15)
u	G-CHOAcCH$_2$CH$_2$Br	346(5), 302(36), 245(7), 224(3), 195(12), 153(100), 131(16), 91(9)
v	G-CHOAcCH$_2$CH$_2$OAc	324(6), 282(56), 222(23), 179(54), 163(20), 153(100), 131(36), 107(12), 93(22), 73(17)
w	G′-CH=CHCH$_2$OAc	292(10), 250(21), 208(72), 165(21), 148(100), 137(19), 120(22), 103(12), 91(20)
x	G-CO-CHOAcCH$_2$OAc	338(4), 296(24), 210(20), 193(4), 178(6), 151(100), 137(3), 123(7), 93(4)
y	G-CHOAc-CHOAc-CH$_2$OAc	382(6), 340(10), 322(3), 280(52), 238(8), 195(18), 178(56), 153(100), 137(6), 103(7), 93(9)
z	S-CHOAcCH$_2$CH$_2$Br	376(3), 332(46), 330(46), 292(11), 290(10), 273(8), 183(100), 167(8), 155(20), 123(13), 91(10)
z′	G-CHOAc-CHOAc-CH$_2$OAc	382(2), 340(7), 280(16), 238(5), 195(22), 178(30), 153(100), 137(5), 103(6), 73(9)
15	G″-CH=CHCH$_2$OAc	322(6), 280(63), 238(100), 195(35), 178(44), 167(20), 147(25), 135(22), 118(19)

* See Figure 14.

**

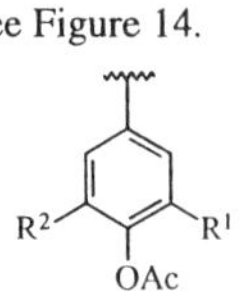

P R^1 = R^2 = H
G R^1 = OMe, R^2 = H
S R^1 = R^2 = OMe

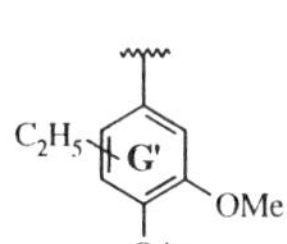

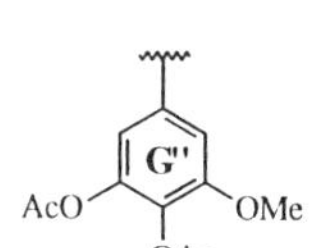

Alternatively, crude reversed-phase separations may be applicable. These matters are currently being investigated in our laboratories.

2. The extent to which the DFRC procedure has been developed is still quite limited. At this point, we do not know what happens to a number of structural units that have already been investigated by thioacidolysis. For example, the ability to estimate vinyl ethers is useful in pulping studies; we suspect that vinyl ethers will be cleaved in our process to produce two-carbon-sidechain products but this needs to be confirmed. We do not yet know what happens to aldehydes that are naturally present and may be at elevated levels in certain brown midrib grasses and CAD-deficient mutants or transgenic plants. We have already identified dimers without any methodological change (except to run the GC chromatograms for a longer time); these are potentially quantifiable, as they are in the two-step thioacidolysis/Raney nickel method, but this work remains to be done. We know that β-1 models behave as expected, but do not yet know what happens to phenylcoumarans, resinols, dibenzodioxicins (*47*) and other structures in lignin. Obviously, these deficiencies will be the focus of forthcoming research.

Other Applications

As has been noted above, the DFRC method is a fairly convenient equivalent of the thioacidolysis one-step (monomer) method. It should also provide dimers without further modification; we will be concentrating on dimer identification and analysis subsequently. The method is also amenable to the identification of novel units in lignin such as those produced from 5-hydroxyconiferyl alcohol incorporation into lignins. But these just touch on the possibilities. The following is a selection of applications/modifications that are currently being explored in our laboratory.

Free Phenolic *vs*. Etherified β-Ethers. The ability to determine the proportion of free phenolic *vs.* etherified units that can be released is of considerable importance. This has been elegantly achieved in thioacidolysis by pre-methylating the samples (*48, 49*) and the same procedure would provide equivalent information here. The only drawback is that complete methylation is difficult to achieve and is a lengthy process. The DFRC method apparently allows the distinction between free phenolic and etherified β-ethers with only slightly modified steps.

This distinction comes from the observation that phenolic groups in the original lignin are acetylated in the AcBr step. In fact, released α-ethers will also become acetylated and this must be borne in mind. In contrast, providing the AcBr is completely removed after the first step, the new phenols that are produced by cleavage of β-ethers in the Zn^0 step, remain largely free phenolic. A small amount of acetylation does appear to accompany the Zn^0 reduction step. However, the solution is simple—propionic acid instead of acetic acid can be used for the reduction step. Propionylation instead of the final acetylation would be a trivial modification. Providing acetate/propionate exchange does not occur (it does not in preliminary trials), analysis by GC or GC-MS would immediately reveal separate propionylated (therefore originally etherified) monomers and acetylated (originally free phenolic) monomers. A more elegant possibility may be to utilize an acidic catalyst that does not acetylate. HCl is too strong, but we have had success with other mild acids that do not derivatize the released phenol. Then (monomeric) products that are free phenolic are those that were originally etherified and cleaved in the Zn^0 step, whereas those from originally free phenolic units (or α-ethers) become acetylated in the first AcBr step.

The process using propionic anhydride is illustrated with a trimeric model **GSS-1a** in Figure 15. Although, as with thioacidolysis, it must be remembered that the information provided is solely from units that can be released, the ability to distinguish between such free phenolic and etherified units is most valuable.

Ester Determinations—Quantitation and Regiochemistry. Various plants have esterified lignins. Grass lignins are acylated by *p*-coumarate; we have recently shown (*46*) that this is solely at the γ-position in maize and ryegrass and suspect that the regiochemistry holds true for all grasses. Kenaf has recently be shown to possess a most unusual lignin (*50*); it is ~50% acetylated. In this case the regiochemistry is less strict, with about 95% at γ- and 5% at the α-position, but acetyl groups (unlike *p*-coumaroyl groups) are known to migrate (*51*). What had not been determined was whether the *p*-coumarate in grass lignins was on guaiacyl or syringyl units. Terashima *et al.* (*52*) had shown by microautoradiography that *p*-coumarate 'appeared' in lignin later in the lignification process, at the same time that syringyl units were predominantly incorporated. Although thioacidolysis cleaved esters substantially, we were able to separate and identify coniferyl and sinapyl *p*-coumarate derived products from maize (*43*). Approximately an 8-fold excess of the sinapyl *p*-coumarate product was released, strongly suggesting that sinapyl units were the main sites for *p*-coumarate on this mature maize. Thioacidolysis cleaved about 45% of the esters in a single model study we carried out (*43*); thus acylated components become under-quantitated while monomers are also under-estimated if the intact esters are not taken into account. Whether the *p*-coumarate is first esterified by sinapyl alcohol and the dimer co-polymerized with monolignols to form a lignin with the *p*-coumarate at the γ-position, or whether the *p*-coumaroylation occurs subsequent to lignification is still debated. In fact there are a number of remaining mysteries about *p*-coumarate in lignin (*24, 43, 46*). The DFRC method is vastly superior to thioacidolysis for identifying and characterizing such esters on lignin since esters are not cleaved under these conditions. Employing this method (Figure 16) on the same maize lignin as was used for extensive NMR studies (*46*), a lignin which contained about 18% *p*-coumarate, produced large peaks for the sinapyl *p*-coumarate derived product **S-17c** and for the coniferyl *p*-coumarate derived product **G-17c** (Figure 16). Figure 17 shows a GC-MS total ion chromatogram in which these intact esters are readily identified. Small amounts of doubly unsaturated analogs of **17** are also present later in the chromatogram (not shown). Significant simplification can be obtained with a hydrogenation step for these esters. A modification to the reaction may also be revealing for lignins such as kenaf where acetate groups are on the sidechain. The equivalent reaction with propionyl bromide has been shown to be reasonably effective (but not as good as AcBr). If (as we suspect) transesterification will not occur, carrying out such a reaction should produce γ-acetylated monomers from those that originally bore an acetate, but γ-propionates from those that originally possessed a γ-hydroxyl group. Such investigations applied to shed light on the interesting class of naturally acetylated lignins will begin shortly in our laboratory. Acetate groups on lignins have been noted (*53, 54*) but are generally missed in NMR spectroscopic analyses due to the common practice of acetylating lignins prior to NMR analysis (to improve their solubility and spectral dispersion properties).

Determination of Hydroxycinnamic Acids in Lignins. In grasses, and possibly legumes and some woody species, hydroxycinnamic acids (esterified to cell wall polysaccharides) are intimately incorporated into lignins (*24, 55*). Simple α-ethers arise from the reaction of the phenol with intermediate quinone methides produced during lignification whereas a whole range of products are produced by direct radical cross-coupling reactions between the hydroxycinnamates and lignin monomers/oligomers. As reviewed in this book (*28*), such cross-coupling products between ferulate and coniferyl/sinapyl alcohol have not only been found (in a ryegrass) but there is strong evidence that ferulate is acting as a nucleation site for lignification—a site in the cell wall at which lignification begins (*55*). Analytically, it is possible to measure hydroxycinnamate esters, of course, by simple room temperature saponification. Of the radically cross-coupled products, it is only possible to release those that are ether-linked. Determination of ethers has been successfully accomplished by acidolysis and high temperature base treatment. A clever scheme even allows the determination of the amount of hydroxycinnamates that are both

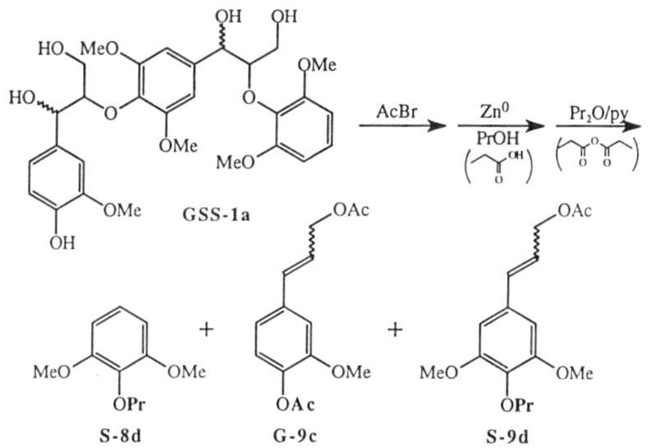

Figure 15. Distinguishing free-phenolic from etherified β-ether units *via* a modification to the DFRC method; the acidic solvent for the Zn^0 step does not contain acetic acid, and propionic anhydride is used instead of acetic anhydride for the final acylation step. As an example, a trimeric **GSS** model produces acetylated and free-phenolic products. The acetylated product **G-9c** arises from the initially free-phenolic G-unit, whereas the propionylated products **S-8d** and **S-9d** are those that were initially etherified. Consequently, those released units that were initially free-phenolic can be quantitated against those that were originally etherified.

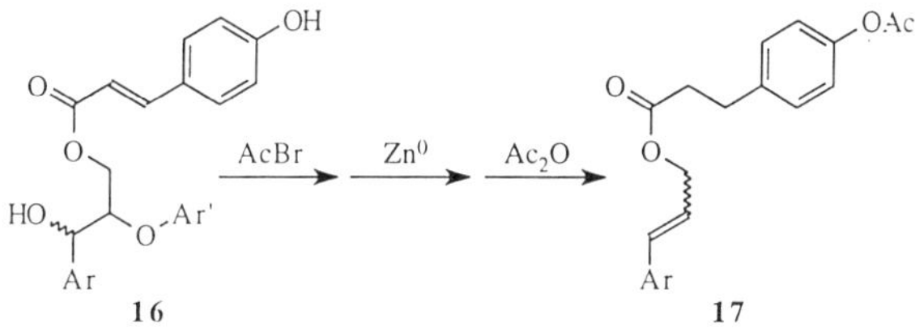

Figure 16. *p*-Coumaroylated β-ether units **16** in grass lignins will release intact ester products **17**, coniferyl and sinapyl dihydro-*p*-coumarates. Such products arise since esters are not cleaved in the procedure and prove that *p*-coumarate groups are esterified to syringyl and guaiacyl components of grass lignins.

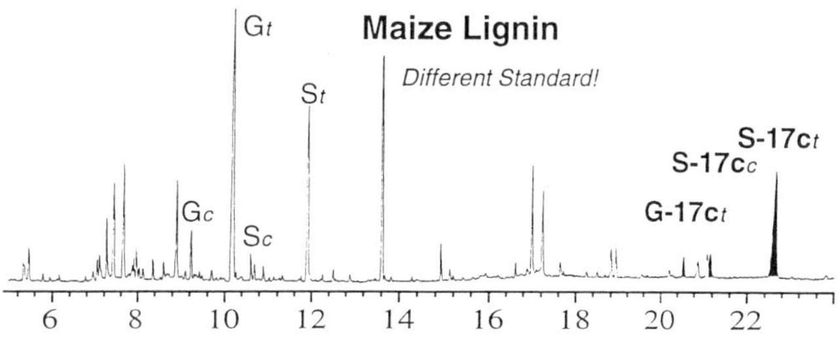

Figure 17. Total ion chromatogram (TIC) from (old version of, non-optimal) DFRC method on a maize lignin isolate. The hydroxycinnamyl *p*-coumarate derivative peaks **17** are readily observed; the sinapyl *p*-coumarate derived product **S-17c** is particularly abundant. Normal sinapyl monomers (**S-9c**, labeled S) are significantly lowered. *t* = *trans*, *c* = *cis*.

etherified and esterified (*56*). Such valuable data reveal a lot about the biochemistry and importance to the plant of hydroxycinnamates in the cell wall.

The DFRC procedure stands poised to implement further methods. In its simplest form it may provide another means for releasing of etherified hydroxycinnamates without the need for high-temperature reactions. A saponification step will still be required, however, since primary esters are not cleaved during the AcBr or Zn^0 steps. A variety of information may become available by applying the saponification, AcBr and Zn^0 steps in different orders.

Summary

A combined application of the effective cell wall dissolving reactions that are attained by acetyl bromide under mild conditions in suitable solvents, followed by reductive cleavage reactions, provides a powerful new methodology for cleaving ethers in lignin. In its most basic form, such a methodology provides an attractive new chemical basis for lignin analysis, giving results similar to those of thioacidolysis. This DFRC method is already proving valuable in identifying novel units in various genetic mutants (*57*), but the method has other facets that convey considerable potential for certain structural and regiochemical studies. Although we are working intensively to bring some of these methods to fruition, the range of possibilities is enormous and we feel that it is more important to make the basic methods available now for researchers to exploit rather than to try and hold back this information until we can perfect more individual protocols ourselves. We are actively working on assessing the potential for characterizing dimers and higher oligomers, determining the potential to distinguish etherified from free phenolic (releasable) units in lignins, identifying esterified lignin components, and determining how the methods can be used to detail hydroxycinnamate incorporation into lignins. Obviously, with this work now reported, we may not be the first to develop viable methodologies in these areas. We also expect that the methodologies will be expanded and refined. We trust that others will engage in the same spirit of scientific freedom and alert us and all interested parties to the new possibilities as they arise.

Acknowledgments

The authors are grateful to John Grabber for performing the thioacidolytic analyses, to Ron Hatfield for his work on applying the AcBr method to difficult samples and for valuable discussions, and for a USDA-National Research Initiative competitive grant (# 95-03675) from the Improved Utilization of Wood and Wood Fiber program.

Literature Cited

1. Harkin, J. M. In *Oxidative Coupling of Phenols*; Taylor, W. I., Battersby, A. R., Eds.; Marcel Dekker: New York, NY, 1967; pp 243-321.
2. Sarkanen, K. V.; Ludwig, C. H. *Lignins—Occurrence, Formation, Structure and Reactions*; Wiley-Interscience: New York, NY, 1971.
3. Ralph, J. *Magn. Reson. Chem.* **1993**, *31*, 357-363.
4. Ralph, J.; Grabber, J. H. *Holzforschung* **1996**, *50*, 425-428.
5. Lundquist, K. *Svensk Papperstidn.* **1974**, *76*, 704-710.
6. Lapierre, C. In *Forage Cell Wall Structure and Digestibility*; Jung, H. G., Buxton, D. R., Hatfield, R. D., Ralph, J., Eds.; ASA-CSSA-SSSA: Madison, WI 1993; pp 133-166.
7. Rolando, C.; Monties, B.; Lapierre, C. In *Methods in Lignin Chemistry*; Dence, C. W., Lin, S. Y., Eds.; Springer-Verlag: Berlin-Heidelberg, 1992; pp 334-349.
8. Johnson, D. B.; Moore, W. E.; Zank, L. C. *Tappi* **1961**, *44*, 793-789.
9. Dence, C. W. In *Methods in Lignin Chemistry*; Lin, S. Y., Dence, C. W., Eds.; Springer-Verlag: Heidelberg, 1992; pp 33-61.

10. Morrison, I. M. *J. Sci. Food Agr.* **1972**, *23*, 1463-1469.
11. Bagby, M. O.; Cunningham, R. L.; Maloney, R. L. *Tappi* **1973**, *56*, 162-163.
12. Iiyama, K.; Wallis, A. F. A. *Wood Sci. Technol.* **1988**, *22*, 271-280.
13. Marton, J. *Tappi* **1967**, *50*, 335-337.
14. Iiyama, K.; Wallis, A. F. A. *Holzforschung* **1989**, *43*, 309-316.
15. Nimz, H. *Liebigs Ann. Chem.* **1966**, *691*, 126-133.
16. Iiyama, K.; Wallis, A. F. A. *J. Wood Chem. Technol.* **1990**, *10*, 39-58.
17. Lu, F.; Ralph, J. *Holzforschung* **1996**, *50*, 360-364.
18. Zawadowski, T. *Rocz. Chem.* **1968**, *42*, 297-307.
19. Ralph, J.; Young, R. A. *J. Wood Chem. Technol.* **1983**, *3*, 161-81.
20. Johansson, B.; Miksche, G. E. *Acta Chem. Scand.* **1972**, *26*, 289-308.
21. Nimz, H. *Wood Sci. Technol.* **1981**, *15*, 311-316.
22. Leary, G. J. *Wood Sci. Technol.* **1982**, *16*, 67-70.
23. Ede, R. M.; Kilpeläinen, I. *Res. Chem. Intermediates* **1995**, *21*, 313-328.
24. Ralph, J.; Helm, R. F. In *Forage Cell Wall Structure and Digestibility*; Jung, H. G., Buxton, D. R., Hatfield, R. D., Ralph, J., Eds.; ASA-CSSA-SSSA: Madison, WI, 1993; pp 201-246.
25. Lai, Y.; Guo, X. *Holzforschung* **1992**, *46*, 311-314.
26. Ralph, J.; Helm, R. F.; Fort, R. C.; Elder, T. J. *J. Chem. Soc., Perkin Trans. I* **1994**, 2117-2121.
27. Veejendra, K. Y.; Alex, G. F. *J. Org. Chem.* **1986**, *51*, 3372-3374.
28. Ralph, J.; Hatfield, R. D.; Grabber, J. H.; Jung, H. G.; Quideau, S.; Helm, R. F. *ACS Symp. Ser.* **1997**, this volume.
29. Lu, F.; Ralph, J. *211th Amer. Chem. Soc. Natl. Mtg. Abstr.* **1996,** *Cell*-110.
30. Lu, F.; Ralph, J. *J. Nat. Prod.* **1997**, submitted.
31. House, H. O.; Ro, R. S. *J. Amer. Chem. Soc.* **1958**, *80*, 182.
32. Kochi, J. K.; Singleton, D. M.; Andrews, L. J. *Tetrahedron* **1968**, *24*, 3503-3515.
33. Greene, A. E.; Charbonnier, F.; Luche, M. J.; Moyano, A. *J. Amer. Chem. Soc.* **1987**, *109*, 4752-4753.
34. Landucci, L. L. *Tappi* **1980**, *63*, 95-99.
35. Landucci, L. L.; Ralph, J. *J. Org. Chem.* **1982**, *47*, 3486-3495.
36. Evans, D. H.; O'Connell, K. M. In *Electroanalytical Chemistry—A Series of Advances*; Marcel Dekker: New York, NY, 1986; Vol. 14, pp 113-207.
37. Fry, A. J. In *Organic Electrochemistry*; Springer-Verlag: New York, NY, 1972; Vol. 34, pp 1-44.
38. Cassanova, J.; Rogers, H. H. *J. Org. Chem.* **1974**, *39*, 2408-2410.
39. McLafferty, F. W. *Interpretation of Mass Spectra,* 2nd ed.; The Benjamin/Cummings Publishing Company: Reading, MA, 1973; pp 58.
40. Lapierre, C.; Tollier, M. T.; Monties, B. *C. R. Acad. Sci., Ser. 3* **1988**, *307*, 723-728.
41. Chabbert, B.; Tollier, M. T.; Monties, B.; Barriere, Y.; Argiller, O. *J. Sci. Food Agric.* **1994**, *64*, 349-355.
42. Lapierre, C.; Monties, B.; Rolando, C. *J. Wood Chem. Technol.* **1985**, *5*, 277-292.
43. Grabber, J. H.; Quideau, S.; Ralph, J. *Phytochemistry* **1996**, *43*, 1189-1194.
44. Lapierre, C. In *Forage Cell Wall Structure and Digestibility*; Jung, H. G., Buxton, D. R., Hatfield, R. D., Ralph, J., Eds.; ASA-CSSA-SSSA: Madison, WI, 1993; pp 133-166.
45. Chesson, A.; Provan, G. J.; Russell, W.; Scobbie, L.; Chabbert, B.; Monties, B. *J. Sci. Food Agr.* **1996**, in press.
46. Ralph, J.; Hatfield, R. D.; Quideau, S.; Helm, R. F.; Grabber, J. H.; Jung, H.-J. G. *J. Amer. Chem. Soc.* **1994**, *116*, 9448-9456.
47. Karhunen, P.; Rummakko, P.; Sipilä, J.; Brunow, G.; Kilpeläinen, I. *Tetrahedron Lett.* **1995**, *36*, 169-170.
48. Lapierre, C.; Rolando, C. *Holzforschung* **1988**, *42*, 1-4.
49. Lapierre, C.; Monties, B.; Rolando, C. *Holzforschung* **1988**, *42*, 409-11.
50. Ralph, J. *J. Nat. Prod.* **1996**, *59*, 341-342.

51. Helm, R. F.; Ralph, J. *J. Wood Chem. Technol.* **1993**, *13*, 593-601.
52. Terashima, N.; Fukushima, K.; He, L.-F.; Takabe, K. In *Forage Cell Wall Structure and Digestibility*; Jung, H. G., Buxton, D. R., Hatfield, R. D., Ralph, J., Eds.; ASA-CSSA-SSSA: Madison, 1993; pp 247-270.
53. Das, N. N.; Das, S. C.; Sarkar, A. K.; Mukherjee, A. K. *Carbohydr. Res.* **1984**, *129*, 197-207.
54. Sarkanen, K. V.; Chang, H.-M.; Allan, G. G. *Tappi* **1967**, *50*, 587-590.
55. Ralph, J.; Grabber, J. H.; Hatfield, R. D. *Carbohydr. Res.* **1995**, *275*, 167-178.
56. Lam, T. B. T.; Iiyama, K.; Stone, B. A. *Phytochemistry* **1992**, *31*, 1179-1183.
57. Ralph, J.; MacKay, J. J.; Hatfield, R. D.; O'Malley, D. M.; Whetten, R. W.; Sederoff, R. R. *Science* **1997**, *277*, 235-239.

Chapter 21

Structural Diversity in Lignans and Neolignans

Adrian F. A. Wallis

CSIRO Forestry and Forest Products, Private Bag 10, South Clayton MDC, Victoria 3169, Australia

Lignans are a diverse group of optically active plant phenols formed by the union of two phenylpropane units through the center carbon atoms (C-8 /C-8′) of their sidechains. Phenylpropane dimers which are linked through atomic centers other than the 8,8′ carbons are frequently termed neolignans. The compounds can contain free phenolic hydroxyl groups or aromatic alkoxy groups. The range of most of the naturally occurring lignan and neolignan structures are outlined in this chapter, and methods for their structural elucidation are discussed. Some pitfalls in the determination of structures by NMR spectroscopy are given. Approaches to the biomimetic synthesis of lignans and neolignans are illustrated by the oxidative coupling reactions of isoeugenol and 4-propenylsyringol. In some cases the laboratory preparation of lignans by oxidative coupling has occurred in advance of their isolation from plant materials. The remarkable stereo-specificity noted in the oxidative coupling of *p*-hydroxyphenyl-propenes through the C-8/C-8′ carbon atoms is rationalized in terms of mechanisms passing through favored intermediate complexes. However, this stereospecificity is not observed in every case, so that more study is required for a more complete understanding of the reaction mechanisms involved.

The term 'lignan' was given by Haworth in 1936 to a group of optically active phenylpropanoid dimeric extractives linked through the central (C-8/C-8′)carbon atoms of their propane sidechains (**1**) (*1*). The family of compounds was extended in 1972 by Gottlieb to include 'neolignans', phenylpropanoid dimers linked through atoms other than the C-8/C-8′ carbon atoms of the sidechains (*2*). However, Gottlieb later revised the definition to cover only those compounds formed by the oxidative dimerization of allyl or propenylphenols (*3*). The revised classification has not gained general acceptance, and the original definition of neolignans is more frequently used (*4, 5*). Related compounds occur in the form of lignans and neolignans such as sesquilignans, dilignans and trilignans with 3, 4 and 6 phenylpropane units, respectively. Hybrid lignans are exemplified by flavonolignans, and norlignans have one or more carbon atoms from their propane sidechains missing (*6*). Examples of lignans with different skeletons include one acyclic structure (**2**), three monocyclic structures (**3-5**), two bicyclic structures (**6** and **7**) and two tricyclic structures (**8** and **9**). Neolignans have more varied skeletons, and examples include a 3,3′-linked biphenyl (**10**), 8,1′- and 8,3′-linked 1,2-diarylpropanes (**11** and **12**) and the 3-*O*-4′- and 8-*O*-4′-linked ethers (**13** and **14**).

Lignans were originally thought to be low molecular weight lignin fragments, although unlike lignin they are optically active, which implies enzymic control of the final steps in their biosynthesis. Both lignans and neolignans are understood to be formed by enzyme-catalyzed oxidative coupling of phenolic precursors, and the resulting dimers can subsequently undergo oxidation, reduction or alkylation to give the final product (see Chapters 22 and 25, this volume). They are widely distributed throughout the plant kingdom, occurring in up to 70 plant families in roots, rhizomes, wood, stems, leaves, fruits and seeds (*5*).

Lignans have been the subject of several reviews, notably in two books entirely devoted to them (*5, 7*), in chapters of monographs (*8, 9*), and several journal reviews (*6, 10-18*). They have become topical in recent times because increasing numbers of them have been described, from 14 in Haworth's 1936 review (*1*) to 440 in 1987 (*5*) (exclusive of neolignans), and because of their strong and varied biological activities. Notably, derivatives of podophyllotoxin (**15**) are used clinically in cancer chemotherapy (*19*), kadsurenone (**16**) is a potent antagonist of a mediator of inflammatory diseases (*20*), enterolactone (**17**) has digitalis-like activity (*21*) and masoprocol (**18**) is useful for treating various skin disorders (*22*).

Structural Elucidation of Lignans and Neolignans

Structure determination for the lignans and neolignans has followed similar lines to that for other classes of natural products. Various forms of chromatography have been used for the separation and tentative identification of unknown compounds—for instance capillary gas chromatography (GC) for separation of lignan glucosides (*23*), pyrolysis-GC (*24*) and high performance liquid chromatography (*25*) for lignans. Once separated, the components can be identified by spectroscopic comparison with authentic compounds, or by synthesis. GC itself has been used as a tool for the classification of lignans (*26*).

Ultraviolet spectrophotometry has also been a useful technique for establishing the class to which an unknown lignan belongs. All lignans show a basic UV absorption pattern typical of aromatic compounds, with three bands in the regions of 210, 230 and 280 nm (*5*). The infrared spectrum is also a useful tool in structure determination for lignans. However, lignans often occlude small guest molecules such as the solvent from which they are crystallized, that can distort the information obtained from their infrared spectra (*27*).

Mass spectrometry applied to lignans leads to extensive fragmentation and, although the spectra do not show many daughter ions of diagnostic significance, they do allow the characterization of aromatic rings and basic features of dimeric structures (*5*). Direct vapor injection is not an appropriate technique for obtaining mass spectra of lignan glycosides, because they are generally too involatile. For such compounds, chemical ionization (*28*), field desorption and fast atom bombardment (*29*) are important anciliary mass spectral techniques.

The most useful spectral technique for structural elucidation of lignans and neolignans is nuclear magnetic resonance (NMR) spectroscopy. The technique is particularly sensitive to structural variations (*30*), and there are substantial differences between the spectra of different classes of lignans. Both ^{1}H and ^{13}C (*31*) NMR spectroscopy are useful and complementary methods; indeed the ^{13}C NMR spectroscopy of lignans has been the subject of extensive reviews (*32, 33*).

Although NMR spectroscopy has been a very powerful technique for elucidating the structures of lignans and neolignans, some errors in assignment have occurred through incorrect interpretation of the spectra. For instance for thomasic acid, a

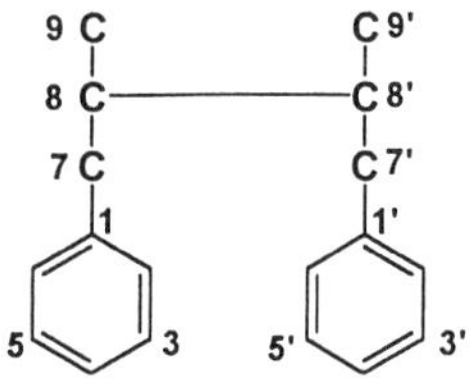

1

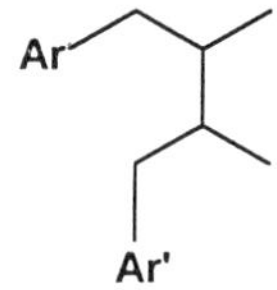

2

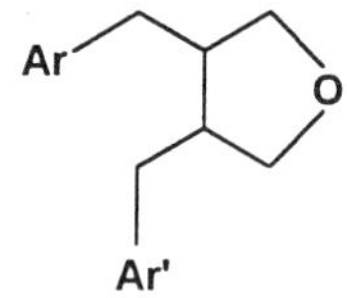

3

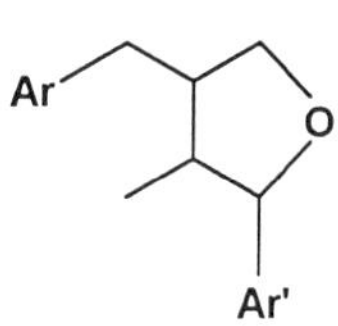

4

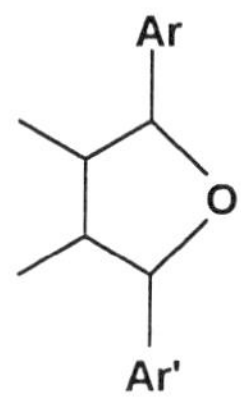

5

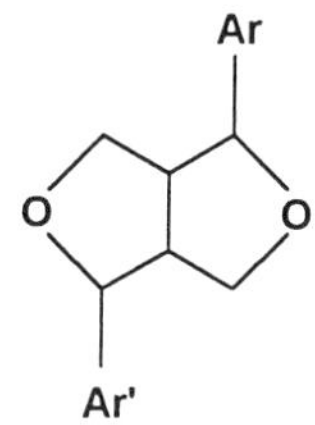

6

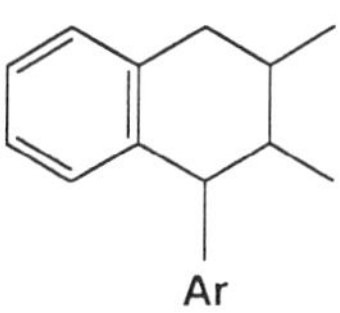

7

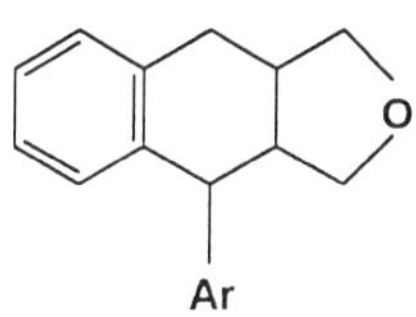

8

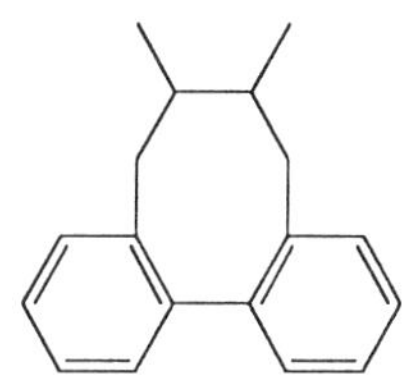

9

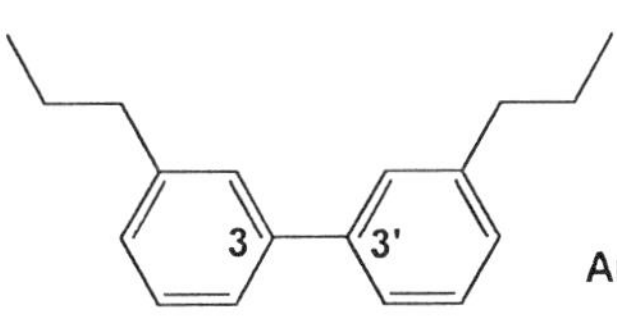

10

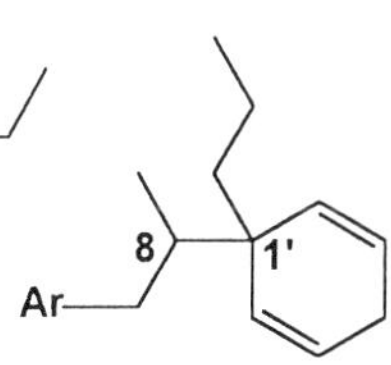

11

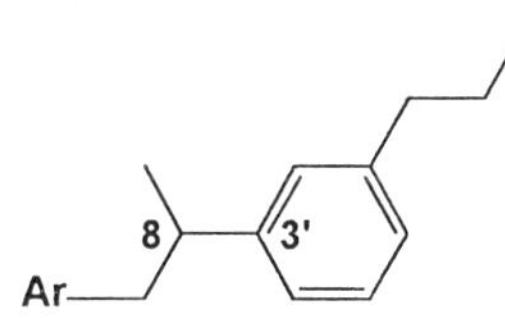

12

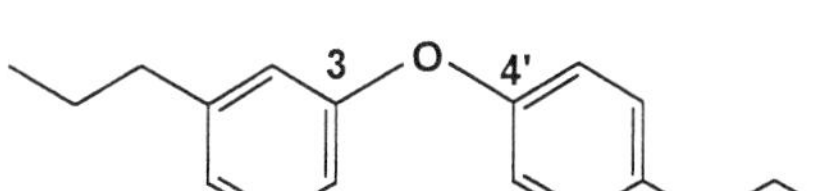

13

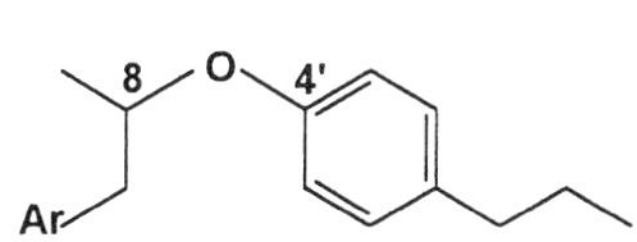

14

lignan isolated from the heartwood of the American rock elm *Ulmus thomasii*, a *cis* configuration was assigned to the C-7′ and C-8′ substituents in **19a** on the basis of the small coupling constant $J_{7',8'}$ in its ^{1}H NMR spectrum, indicative of a small dihedral angle between the C-H bonds (*34*). This was later revised to the *trans* configuration **19b**, when it was recognized that the compound assumed a *trans*-diaxial conformation (*35*). This conformation has subsequently been assigned to other dihydronaphthalene lignans (*36, 37*). A similar correction has been made to the assignment of a guaiacyl-dihydronaphthalene (*38*).

In spite of the current sophistication of NMR spectroscopic techniques, errors still occur in the assignment of configuration to neolignans on the basis of ^{1}H and ^{13}C NMR spectra. Green *et al.* (*39*) originally proposed the unusual 8–*O*–3′ structure for a lignan from the roots of *Piper capense*, with 1,3,5-substitution of the etherified ring, as a result of NMR spectral analysis. The structure was later confirmed to be that of the conventional 8–*O*–4′ neolignan **20** (*40*). Corrections have been made to the assignment of configuration for two similar structures, the 8–*O*–4′ linked isomers of the neolignan dehydrodiconiferyl alcohol (*41*) isolated from *Arum italicum* (*42*). The original errors of assignment were made on the basis of the deceptively simple pattern for the three aromatic proton signals (two singlets) in their ^{1}H NMR spectra, caused by coincident chemical shifts of the two protons at C-5′ and C-6′.

NMR spectroscopy has been used to ascertain the conformation of lignans and neolignans in solution. The probable conformations of isomeric tetrahydrofuranoid lignans were obtained from their ^{1}H (*43*) and ^{13}C (*44*) NMR spectra. Those of 8–*O*–4′ neolignan isomers were also determined by ^{1}H (*45, 46*) and ^{13}C (*47*) NMR spectroscopic examination, and it was concluded that hydrogen bonds between the benzylic hydroxyls and adjacent ether oxygen atoms influenced the conformation. This was confirmed in the solid state by x-ray crystal structure determinations (*48, 49*).

Synthesis of Lignans and Neolignans

The biosynthesis of lignans and neolignans has received attention in various reviews (*5, 7, 8*), and is discussed in more detail in Chapters 22 and 25 of this volume. Both classical organic syntheses and biomimetic syntheses involving phenolic oxidative coupling have been employed as approaches to the synthesis of lignans and neolignans in the laboratory (*5, 17, 50*). These syntheses have value for structural confirmation of the compounds, and for possible pharmaceutical applications. The absolute configurations of lignans have been assigned through optical rotatory dispersion, circular dichroism and x-ray crystallography (*5*). Several enantioselective syntheses for lignans and neolignans have been outlined (*51-53*).

Biomimetic syntheses of lignans are illustrated in the lignans derived from the oxidative dimerization of isoeugenol and 4-propenylsyringol. Oxidation of isoeugenol (**21**) with one-electron oxidants gives rise to a mesomeric free radical where the unpaired electron resides on the oxygen, C-5 and C-8 atoms, and the coupling of the various radical mesomers gives rise to a variety of products.

Erdtman (*54*) found that the arylcoumaran **22a** (dehydrodiisoeugenol) was the product of the oxidative dimerization of isoeugenol in the presence of ferric chloride. This observation led him to propose that similar radical intermediates were involved in the biosynthesis of lignin. Formation of the arylcoumaran arises from radical coupling between the O and C-5 atomic centers to give an intermediate quinone-methide and subsequent cyclization leads to the arylcoumaran product. Forty years after Erdtman's paper describing the structure of dehydrodiisoeugenol, the compound was reported as a 8-3′-neolignan in the wood of *Licaria aritu* (*55*) and in nutmeg (*Myristica fragrans*) (*56*), and since that time it has been isolated from other plant materials also (*57-61*).

Oxidation of *Z*-isoeugenol (**21b**) with hydrogen peroxide catalyzed by a peroxidase enzyme gave, as one of the products, the arylcoumaran **22b** with the sidechain retaining its Z stereochemistry (*45, 62*). However, the *trans* arrangement of

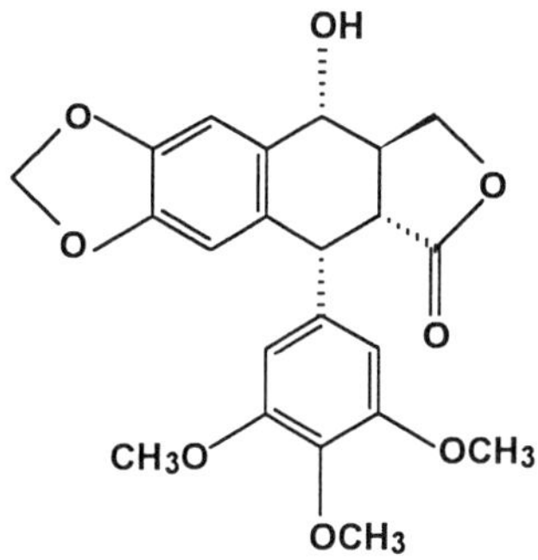

15

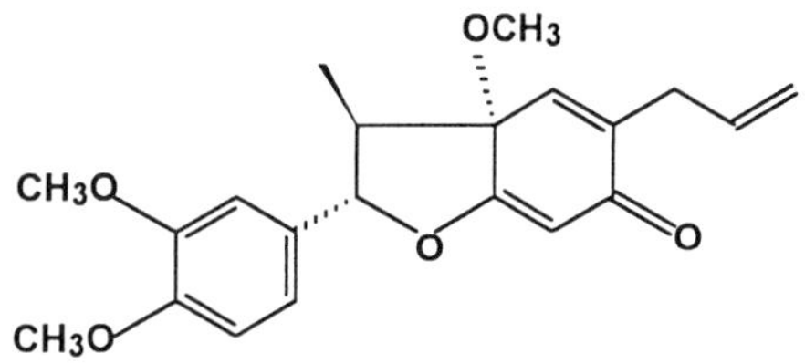

16

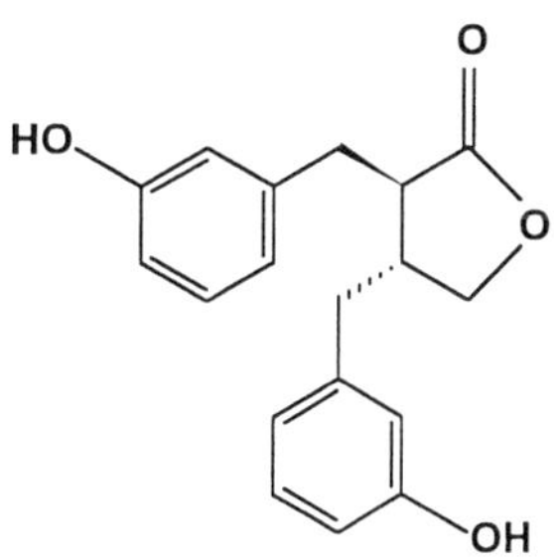

17

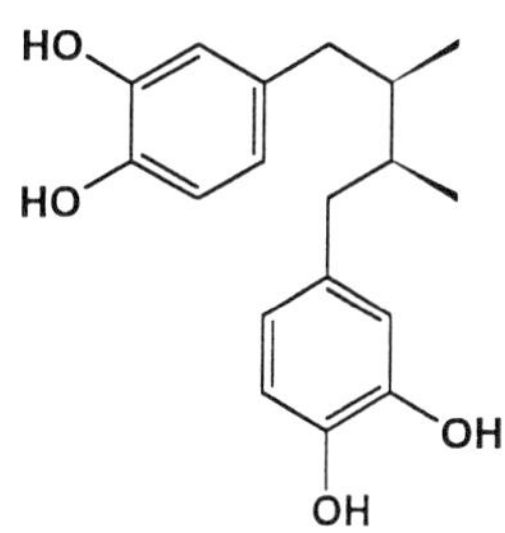

18

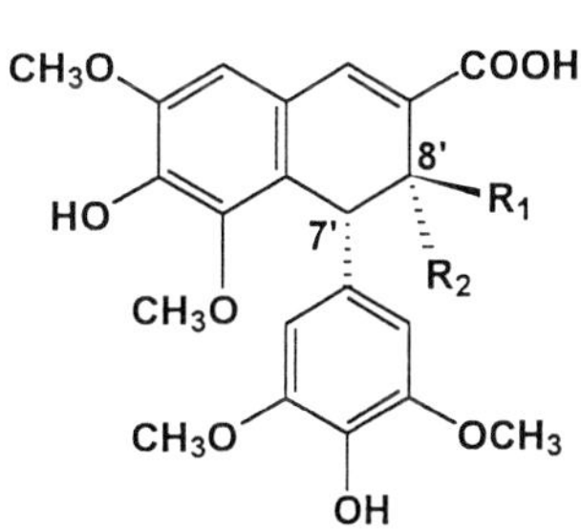

	R_1	R_2
19 a	H	CH_2OH
b	CH_2OH	H

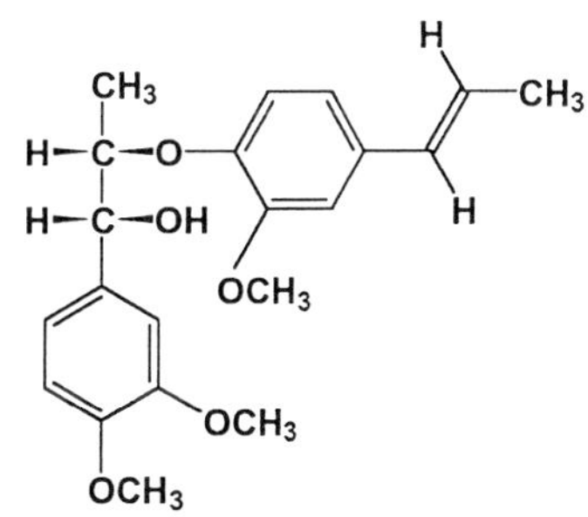

20

substituents on the coumaran ring implies rotation about the C-7–C-8 bond of the quinonemethide intermediate prior to cyclization (*62*). There have been no reports of the isolation of naturally occurring neolignans with *Z*-propenyl sidechains, although arylcoumarans analogous to **22** with *cis* arrangements of the substituents on their coumaran rings have been proposed as structures for neolignans from the needles of *Picea abies* (*63*) and the inner bark of *Betula pendula* (*64*).

The products of the oxidation of *E*-isoeugenol (**21a**) by hydrogen peroxide with peroxidase included a mixture of the *threo* and *erythro* isomers of the 8–*O*–4′ neolignans **23a** and **24a** (*45*). The *Z*-isomer of isoeugenol **21b** treated similarly gave rise to neolignans **23b** and **24b** with the *Z*-propenyl sidechains retained (*45*). These products arise from radical coupling between the O and C-8 atomic centers followed by addition of water to the resulting quinonemethide. Both *threo* and *erythro* neolignans **23a** and **24a** have been identified in the bark and root of *Machilus thunbergii* (*60*) and in mace (from *Myristica fragrans*) (*65*), while the *erythro* isomer has been found in the roots of *Nardostachys jatamansi* (*66*). The methyl ethers of *threo* **23a** (*66, 67*) and *erythro* **24a** (*39, 40, 66*) have also been found in plant extracts.

Both *E*- and *Z*-isoeugenol gave on oxidation by hydrogen peroxide with peroxidase, the tetrahydrofuranoid lignans **25a** and **26a**. These lignans are derived from stereospecific *threo* radical coupling between the two C-8 atomic centers to give a bisquinonemethide; water adds to one quinonemethide to give a benzyl alcohol, which subsequently attacks the second quinonemethide with the formation of the tetrahydrofuran ring. The stereospecific formation of the *threo* bisquinonemethide has been rationalized in terms of an intermediate complex involving overlapping aromatic rings. The complex **27** which leads to *threo* coupling is more sterically favored than that (**28**) which would give *erythro* coupling. For 8–8′ coupling of radicals from a phenol with bulky *tert*-butyl substituents, both *threo* and *erythro* coupling products have been found (*68*); it was postulated that here the bulky substituents hindered the formation of the intermediate complex.

Both tetrahydrofuranoid lignans **25a** and **26a** have been isolated from mace (from *Myristica fragrans*) (*69*), and compound **26a** has also been described in the extracts of *Jatropha grossidentata* (*70*). The dimethyl ethers of **25** and **26** are known as galbelgin (**25b**) and veraguensin (**26b**), and were first isolated from *Himantandra belgraveana* (*71*) and *Ocotea veraguensis* (*72*), respectively.

Reaction of *E*-4-propenylsyringol (**29a**) with ferric chloride in aqueous acetone has afforded the 8–*O*–4′-coupled *threo* and *erythro* dehydrodimers **30a** and **31a**, respectively, as the major products (*46*). Similar reaction of *Z*-4-propenylsyringol (**29b**) with ferric chloride has yielded the corresponding products with *Z*-propenyl sidechains, **30b** and **31b**. The yields of the compounds were improved when silver oxide was used as the oxidant (*73*). The *erythro* isomer **31a** has been isolated from the fruit of *Virola carinata* (*74*), whereas the *threo* isomer **30a** has been obtained as its aryl methyl ether from *Piper polysyphorum* (*75*).

Whereas the 8–*O*–4′ coupled products predominated in the oxidation mixture from *E*-4-propenylsyringol (**29a**) in the acidic ferric chloride medium, reaction of **29a** with hydrogen peroxide in the presence of peroxidase yielded a mixture of the *threo*-coupled tetrahydrofurans **32a** and **33a** as the sole products (*45*). However, reaction of *Z*-4-propenylsyringol with hydrogen peroxide in the presence of peroxidase gave the four tetrahydrofurans **32a**-**35a**, resulting from both *threo* and *erythro* coupling (*45*). Three of the isomers **32a**-**34a** have been found as components of mace (from *Myristica fragrans*) (*69*), while **32a** has been isolated from the fruit of *Licaria aurea* (*76*) and the leaves of *Virola pavonis* (*77*). The dimethyl ethers, **32b** (*76, 78*) and **34b** (*79*), have also been found in plant materials.

The absence of stereospecificity in coupling between the two C-8 atomic centers of radicals derived from *Z*-4-propenylsyringol is surprising in view of the stereospecificity in 8–8′ coupling between the radicals from isoeugenol and *E*-4-propenylsyringol. Stereospecific *threo* coupling has been noted in other oxidative coupling reactions of 1-arylpropenes (*80-87*). More study is required for a clearer understanding of the factors influencing the stereochemistry of oxidative coupling

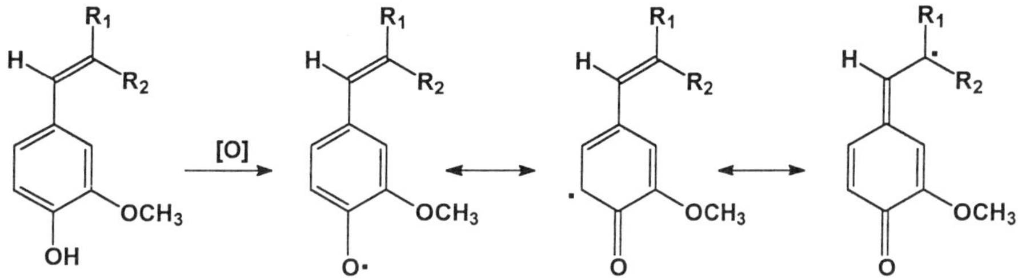

21 M_O M_5 M_8

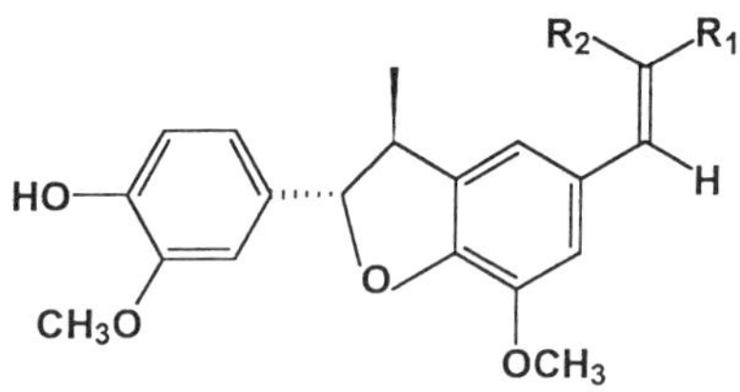

22

a $R_1 = CH_3$ $R_2 = H$

b $R_1 = H$ $R_2 = CH_3$

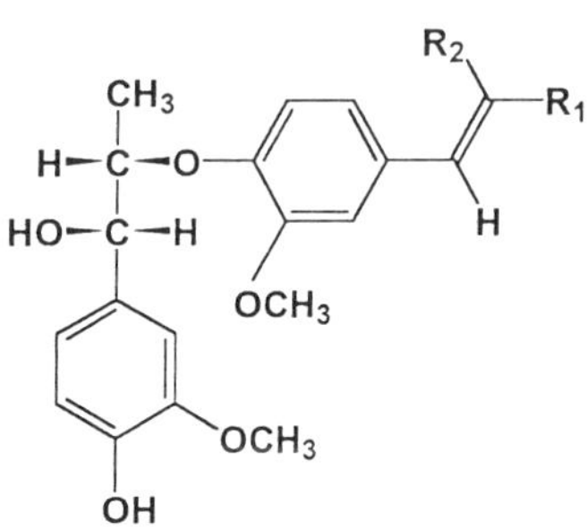

23

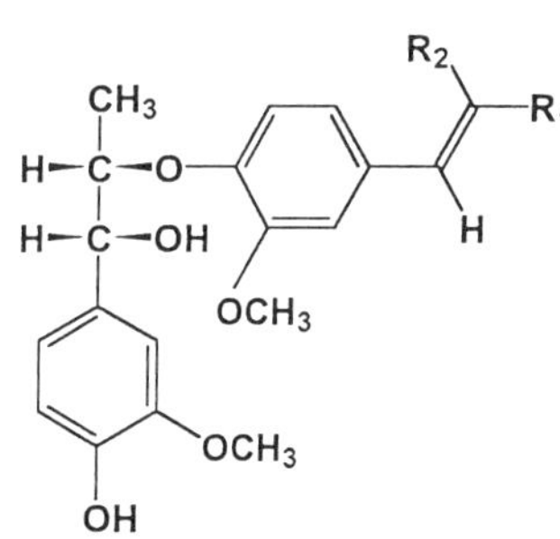

24

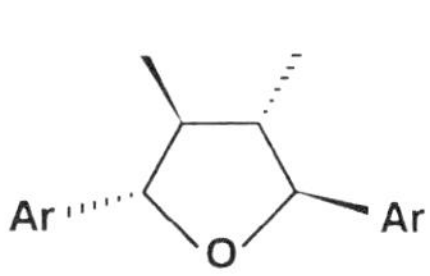

25

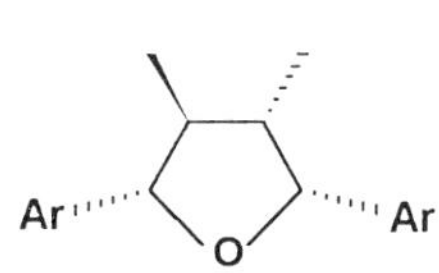

26

a Ar = 4-hydroxy-3-methoxyphenyl

b Ar = 3,4-dimethoxyphenyl

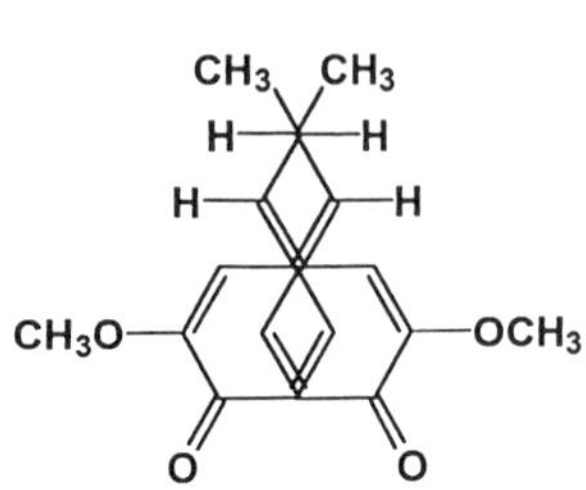

27

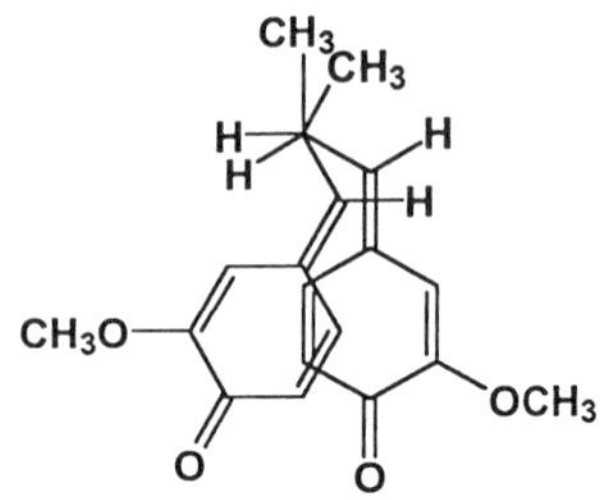

28

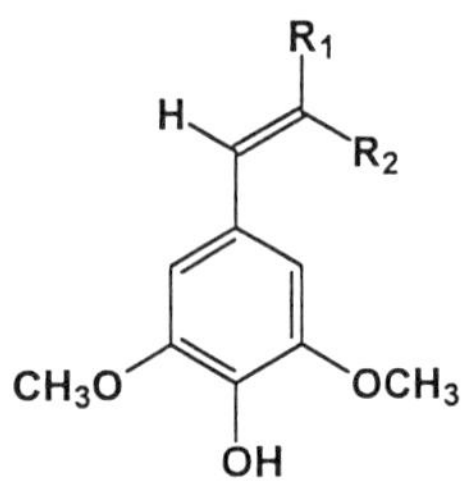

29

a $R_1 = CH_3$ $R_2 = H$

b $R_1 = H$ $R_2 = CH_3$

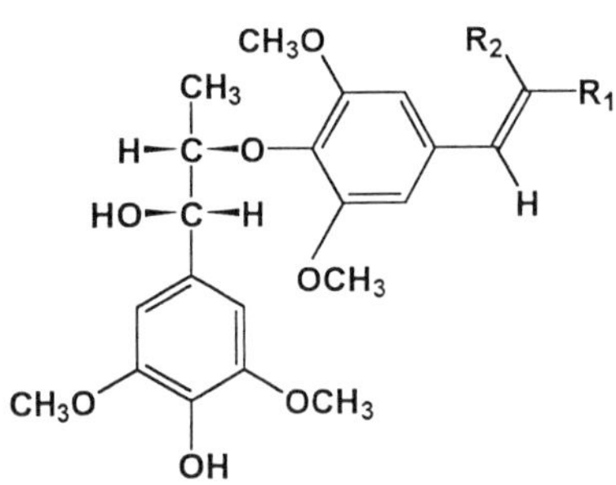

30

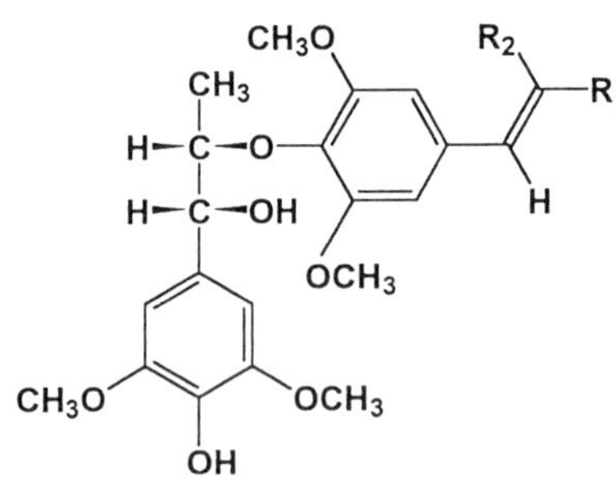

31

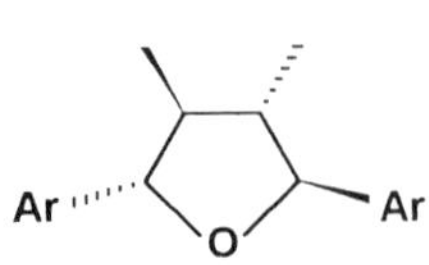

32

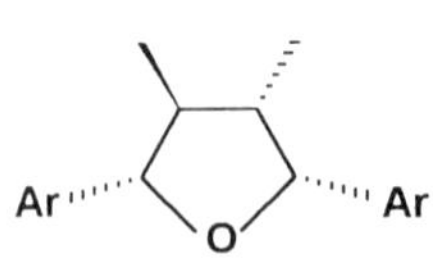

33

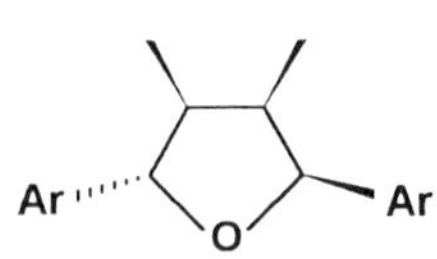

34

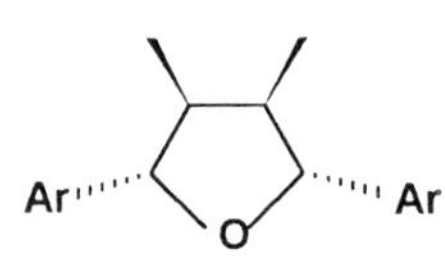

35

a Ar = 4-hydroxy-3,5-dimethoxyphenyl

b Ar = 3,4,5-trimethoxyphenyl

processes leading to the formation of lignans, although a recent study showed some *erythro*-coupled product from the oxidation of methyl sinapate (*88*). The number and type of structures found in this remarkable group of natural products will undoubtedly increase apace, particularly because of the intense interest in their biological properties.

Literature Cited

1. Haworth, R. D. *Ann. Rep. Progr. Chem.* **1936**, *33*, 266-279.
2. Gottlieb, O. R. *Phytochemistry* **1972**, *11*, 1537-1570.
3. Gottlieb, O. R. *Fortschr. Org. Chem. Naturst.* **1978**, *35*, 1-72.
4. Moss, G. P. *Lignan and Neolignan Nomenclature*; IUPAC-IUB Joint Commission on Biochemical Nomenclature, **1987**, *25.3*, 1.
5. Ayres, D. C.; Loike, J. D. *Lignans—Chemical, Biological and Clinical Properties*; Cambridge University Press: Cambridge, 1990.
6. Ward, R. S. *Nat. Prod. Rep.* **1995**, *12*, 183-205.
7. Rao, C. B. S. *Chemistry of Lignans*; Andhra University Press: Waltair, India, 1978.
8. Davin, L. B.; Lewis, N. G. In *Phenolic Metabolism in Plants*; Stafford, H. A., Ibrahim, R. K., Eds.; Plenum Press: New York, NY, 1992; pp 325-375.
9. Gottlieb, O. R.; Yoshida, M. In *Natural Products of Woody Plants, I*; Rowe, J. W., Ed.; Springer-Verlag: Berlin, 1989, pp 439-511.
10. Castro, M. A.; Gordaliza, M.; Del Corral, J. M. M.; San Feliciano, A. *Phytochemistry* **1996**, *41*, 995-1011.
11. Ward, R. S. *Nat. Prod. Rep.* **1993**, *10*, 1-28.
12. Jensen. S.; Hansen, J.; Boll, P. M. *Phytochemistry* **1993**, *33*, 523-530.
13. Whiting, D. A. *Nat. Prod. Rep.* **1990**, *7*, 349-364.
14. Massanet, G. M.; Pando, E.; Rodriguez-Luis, F.; Zubia, E. *Fitoterapia* **1989**, *60*, 3-35.
15. Whiting, D. A. *Nat. Prod. Rep.* **1987**, *4*, 499-525.
16. Whiting, D. A. *Nat. Prod. Rep.* **1985**, *2*, 191-211.
17. Pelter, A. *Recent Advances in Phytochemistry* **1986**, *20*, 201-241.
18. Chatterjee, A.; Banerji, A.; Banerji, J.; Pal, S. C.; and Bhosal, T. *Proc. Indian Acad. Sci. (Chem. Sci.)* **1984**, *93,* 1031-1057.
19. Anon. *Chem. Brit.* **1981**, *17*, 457.
20. Chang, M. N.; Han, G.-Q.; Arison, B. H.; Springer, J. P.; Hwang, S.-B.; Shen, T. Y. *Phytochemistry* **1985**, *24*, 2079-2082.
21. Braquet, P.; Senn, N.; Robin, J.-P.; Esanu, A.; Garay, R. P. *J. Hypertension*, **1986**, *4* (suppl. 5), S461-S464.
22. Olsen, E. A.; Abernethy, M. L.; Kulp-Shorten, C.; Callen, J. P.; Glazer, S. D.; Huntley, A.; McCray, M.; Monroe, A. B.; Tschen, E.; Wolf, J. E. *J. Am. Acad. Dermatol.* **1991**, *24*, 738.
23. Kraus, R.; Spiteller, G. *Liebigs Ann. Chem.* **1990**, 1205-1213.
24. Galletti, G. C.; Ward, R. S.; Pelter, A. *J. Anal. Appl. Pyrol.* **1991**, *21,* 281-293.
25. Betto, P.; Gabriele, R.; Galeffi, C. *J. Chromatogr.* **1992**, *594*, 131-135.
26. Ayres, D. C.; Chater, R. B. *Tetrahedron* **1969**, *25*, 4093-4098.
27. Hartwell, J. L.; Schrecker, A. W. In *Progr. Chem. Org. Nat. Prod.*; Zechmeister, L., Ed.; Springer: Vienna, 1958; Vol. 15, pp 83-166.
28. Milne, G. W. A.; Fales, H. M.; Axelrod, T. *Anal. Chem.* **1971**, *43*, 1815-1820.
29. Chapman, J. R. *Practical Organic Mass Spectrometry—A Guide for Chemical and Biochemical Analysis*; 2nd Edition; Wiley: Chichester, 1993.
30. Günther, H. *NMR Spectroscopy—Basic Principles, Concepts, and Applications in Chemistry*; Wiley: New York, NY, 1994.
31. Breitmaier, E.; Voelter, W. *Carbon-13 NMR Spectroscopy;* 3rd edition; VCH Verlag: Weinheim, 1987.
32. Agarwal, P. K.; Thakur, R. S. *Magn. Reson. Chem.* **1985**, *23*, 389-418.

33. Agarwal, P. K.; Pathak, A. K. *Magn. Reson. Chem.* **1994**, *32*, 753-773.
34. Seikel, M. K.; Hostettler, F. D.; Johnson, D. B. *Tetrahedron* **1968**, *24*, 1475-1488.
35. Wallis, A. F. A. *Tetrahedron Lett.* **1968**, 5287-5288.
36. Ayres, D. C.; Harris, J. A. *Chem. Commun.* **1969**, 1135-1136.
37. Agata, I.; Hatano, T.; Nishibe, S.; Okuda, T. *Phytochemistry* **1989**, *28*, 2447-2450.
38. Wallis, A. F. A. *Chem. Ind.* **1974**, 259.
39. Green, T. P.; Galinis, D. L.; Wiemer, D. F. *Phytochemistry* **1991**, *30*, 1649-1652.
40. Reilly, M. J.; Wallis, A. F. A.; Lundquist, K.; Stomberg, R. *J. Wood Chem. Technol.* **1992**, *12*, 471-483.
41. Gellerstedt, G.; Lundquist, K.; Wallis, A. F. A.; Zhang, L. *Phytochemistry* **1995**, *40*, 263-265.
42. Greca, M. D.; Molinaro, A.; Monaco, P.; Previtera, L. *Phytochemistry* **1994**, *35*, 777-779.
43. Sarkanen, K. V.; Wallis, A. F. A. *J. Heterocyclic Chem.* **1973**, *10*, 1025-1027.
44. Fonseca, S. F.; Barata, L. E. S.; Rúvera, E. A.; Baker, P. M. *Can. J. Chem.* **1979**, *57*, 441-443.
45. Sarkanen, K. V.; Wallis, A. F. A. *J. Chem. Soc. Perkin Trans. I* **1973**, 1869-1878.
46. Wallis, A. F. A. *Aust. J. Chem.* **1973**, 1571-1576.
47. Braga, A. C. H.; Zacchino, S.; Badano, H.; Sierra, M. G.; Rúveda, E. A. *Phytochemistry* **1984**, *23*, 2025-2028.
48. Wallis, A. F. A.; Lundquist, K.; Stomberg, R. *Acta Chem. Scand.* **1991**, *45*, 508-516.
49. Stomberg, R.; Lundquist, K.; Wallis, A. F. A. *J. Crystallogr. Spectrosc. Res.* **1982**, *23*, 319-333.
50. Ward, R. S. *Chem. Soc. Rev.* **1982**, *11*, 75-125.
51. Zacchino, S. A.; Badano, H. *J. Nat. Prod.* **1988**, *51*, 1261-1265.
52. Zacchino, S. A.; Badano, H. *J. Nat. Prod.* **1991**, *54*, 155-160.
53. Zacchino, S. A. *J. Nat. Prod.* **1994**, *57*, 446-451.
54. Erdtman, H. *Liebigs Ann. Chem.* **1933**, *503*, 283-294.
55. Aiba, C. J.; Corrêa, R. G. C.; Gottlieb, O. R. *Phytochemistry* **1973**, *12*, 1163-1164.
56. Forrest, T. P.; Forrest, J. E.; Heacock, R. A. *Naturwissenschaften* **1973**, *60*, 257-258.
57. Forrest, J. E.; Heacock, R. A.; Forrest, T. P. *J. Chem. Soc. Perkin Trans. I* **1974**, 205-209.
58. Hattori, M.; Hada, S.; Watahiki, A.; Ihara, H.; Shu, Y.-Z.; Kakiuchi, N.; Mizuno, T.; Namba, T. *Chem. Pharm. Bull. Japan* **1986**, *34*, 3885-3893.
59. Achenbach, H.; Groß, J.; Dominguez, X. A.; Cano, G.; Star, J. V.; Brussolo, L. C.; Muñoz, G.; Salgado, F.; López, L. *Phytochemistry* **1987**, *26*, 1159-1166.
60. Shimomura, H.; Sashida, Y.; Oohara, M. *Phytochemistry* **1987**, *26*, 1513-1515.
61. Yoshida, M. *ACS Symp. Ser.* **1995**, *588*, 85-92.
62. Sarkanen, K. V.; Wallis, A. F. A. *Chem. Commun.* **1969**, 298-299.
63. Lundgren, L. N.; Popoff, T.; Theander, O. *Phytochemistry*, **1981**, *20*, 1967-1969.
64. Smite, E.; Pan, H.; Lundgren, L. N. *Phytochemistry*, **1995**, *40*, 341-343.
65. Hada, S.; Hattori, M.; Tezuka, Y.; Kikuchi, T.; Namba, T. *Phytochemistry*, **1988**, *27*, 563-568.
66. Bagchi, A.; Oshima, Y.; Hikino, H. *Planta Med.* **1991**, *57*, 96-97.
67. Barata, L. E. S.; Baker, P. M.; Gottlieb, O. R.; Rùveda, E. A. *Phytochemistry* **1978**, *17*, 783-786.
68. Sarkanen, K. V.; Wallis, A. F. A. *J. Chem. Soc. Perkin Trans. I* **1973**, 1878-1881.
69. Hattori, M.; Hada, S.; Kawata, Y.; Tezeka, Y.; Kikuchi, T.; Namba, T. *Chem. Pharm. Bull. Japan* **1987**, *35*, 3315-3322.

70 Schmeda-Hirschmann, G.; Tsichritzis, F.; Jakupovic, J. *Phytochemistry* **1992**, *31*, 1731-1735.
71. Birch, A. J.; Milligan, B.; Smith, E.; Speake, R. N. *J. Chem. Soc.* **1958**, 4471-4476.
72. Crossley, N. S.; Djerassi, C. *J. Chem. Soc.* **1962**, 1459-1462.
73. Zanarotti, A. *J. Chem. Res. (M)* **1983**, 2625-2637.
74. Cavalcante, S. H.; Yoshida, M.; Gottlieb, O. R. *Phytochemistry* **1985**, *24*, 1051-1055.
75. Ma, Y.; Han, G. Q.; Li, C. L.; Cheng, J. R.; Arison, B. H.; Hwang, S. B. *Yaoxue Xuebao* **1991**, *26,* 345-350; *Chem. Abstr.* **1991**, *115*, 247446.
76. Bárbosa-Filho, J. M.; da Silva, M. S.; Yoshida, M.; Gottlieb, O. R. *Phytochemistry* **1989**, *28*, 2209-2211.
77. Ferri, P. H.; Barata, L. E. S. *Phytochemistry* **1991**, *30*, 4204-4205.
78. Holloway, D.; Scheinmann, F. *Phytochemistry* **1974**, *13*, 1233-1236.
79. Conserva, L. M.; da Silva, M. S.; Filho, R. B. *Phytochemistry* **1990**, *29*, 257-260.
80. Erdtman, H. *Svensk Kem. Tidskr.* **1935**, *47*, 223-758.
81. Cartwright, N. J.; Haworth, R. D. *J. Chem. Soc.* **1944**, 535-537.
82. Freudenberg, K.; Rasenak, D. *Chem. Ber.* **1953**, *86*, 755-758.
83. Freudenberg, K.; Schraube, H. *Chem. Ber.* **1955**, *88*, 16-23.
84. Freudenberg, K.; Harkin, J. M.; Reichert, M.; Fukuzumi, T. *Chem. Ber.* **1958**, *91*, 581-590.
85. Freudenberg, K.; Lehmann, B. *Chem. Ber.* **1960**, *93*, 1354-1366.
86. Chapman, O. L.; Engel, M. R.; Springer, J. P.; Clardy, J. C. *J. Am. Chem. Soc.* **1971**, *93*, 6696-6698.
87. Wallis, A. F. A. *Aust. J. Chem.* **1973**, *26*, 1571-1576.
88. Setälä, H.; Pajunen, A.; Kilpeläinen, I.; Brunow, G. *J. Chem. Soc. Perkin Trans. I* **1994**, 1163-1165.

Chapter 22

The Biochemical Control of Monolignol Coupling and Structure During Lignan and Lignin Biosynthesis

Norman G. Lewis and Laurence B. Davin

Institute of Biological Chemistry, Washington State University, Pullman, WA 99164–6340

Over the last six decades or so, various theories and views have emerged to account for lignin and lignan assembly mechanisms. None have yet been able to satisfactorily describe all of the biochemical, structural and anatomical observations associated with formation of either *distinct* metabolic class *in vivo*. In this chapter a quite different perspective in terms of substance, approach and conclusion is given which now appears to account for all of the observations hitherto made. It is based primarily on results from the *definitive* elucidation of monolignol biochemical reactions, established at both the protein and gene levels, which reveals that monomer coupling processes do not occur haphazardly as often depicted. In this context, the *first example* of regio- and stereochemical control of monolignol bimolecular phenoxy radical coupling was that which gave the dimeric 8–8′ linked lignan, (+)-pinoresinol, through the participation of a dirigent protein. The gene encoding this protein has *no* sequence homology to that of any other of known function, perhaps helping to explain further why stereoselective bimolecular phenoxy radical coupling, as controlled by this new class of proteins, had previously gone undiscovered. It is now contemplated that distinct arrays of dirigent protein sites (or some equivalent) stipulate macromolecular assembly mechanisms in the cell wall leading to the various lignins encountered. That is, to specifically account for the well-defined heterogeneous nature of lignin biopolymers within particular cell-wall types, it is proposed that arrays of dirigent protein sites in the developing wall specify the nature (monomer composition and bond frequency) of the primary lignin chain. There is growing evidence to show that specific lignin-related proteins are translocated to predetermined areas within the *pre-lignified* cell wall. An array of lignin monomers are then targeted to *presumed* specific sites on these proteins within the wall at about the same time. Although the precise mechanistic details need to be elucidated, lignification is then envisaged to occur *via* end-wise polymerization of the aligned monomers thereby generating the initial primary lignin chain(s), with chain replication occurring *via* a template polymerization effect as proposed by Sarkanen. Such a mechanism for lignin biosynthesis does not, however, readily account for formation of minor (reduced) substructures (*e.g.* dihydrodehydrodiconiferyl alcohol), *supposedly* present in the lignin macromolecule, which would suggest pre- or post-coupling modifications. As discussed elsewhere, these are viewed instead to be the products of a distinct pathway to the lignans.

Until now, it has been generally felt that Nature's second most abundant macromolecule, lignin, is randomly assembled in the cell walls of vascular plants. How likely indeed is such a supposition to be correct? A review of the evidence would indicate that this hypothesis appears somewhat incongruous, particularly in the light of recent results. To address this issue, it is best to begin with some discussion of the initial hypothesis for lignin formation. In that view, transient free-radical entities, derived from up to three monolignols (each varying by degree of aromatic ring methoxylation) were postulated as undergoing random coupling in the plant cell wall to ultimately afford the lignin biopolymers (*1-3*). That is, any enzymatic involvement was viewed to cease following the action of *at least* two types of oxidases (laccases and/or peroxidases), which generated the free-radicals from the corresponding phenolic substrates. Subsequent coupling processes (*i.e.* lignin assembly) were then considered to occur non-enzymatically (see Figure 1).

Are the arguments in support of this biochemical mechanism as compelling today as they apparently were some forty years ago or so? The answer to that question would seem to be no. There are now at least ten important observations which require a re-appraisal of the random coupling theory, and its replacement with a more all-encompassing hypothesis which takes into account other facts and/or observations. Each of these ten points is presented in summary form below.

(1) The 'lignin-forming' oxidase system used for the first preparations of artificial lignin dehydropolymerisates (so-called 'lignin *in vitro*') was not obtained from actively lignifying tissues. It was instead a crude extract from mushrooms possessing oxidative capacity (*4, 5*), and only later were laccases and peroxidases employed (*1-3, 6*). What other biopolymer can trace its biosynthetic provenance back to enzymes that were isolated from an organism not even involved in its formation?

(2) It has been repeatedly reported that artificial lignins and natural lignins are identical. These include statements (with emphasis added) such as:

- "Natural lignin and lignin formed *in vitro* are *identical*" (*1*);
- "In den zwei letzten Jahren wurde es zur Gewißheit, daß das biosynthetische Lignin aus Coniferylalkohol und das Coniferenlignin, abgesehen von einigen Prozenten *p*-Cumar- und Sinapinalkohol-Komponente im Naturstoff, *identisch* sind und daß auch ihr Bildungsmechanismus übereinstimmt" (*7*); and
- "La comparaison physique et chimique mène à la conclusion que les préparations naturelles et artificielles sont *identiques*" (*2*).

The conclusion of being *identical* serves as the *linchpin* of the random coupling theory. It was based on results from comparison of artificial and presumed natural lignin-derived preparations. These studies used a variety of spectroscopic techniques and chemical degradation procedures that were available at the time, *e.g.* the infrared spectral comparison of 'lignin *in vitro*' (*2, 7, 8*) to that of 'Björkman' (*9*) and 'Brauns' (*10*) lignin preparations, with the latter two being *presumed* to closely represent native lignins.

The *identical* nature of both 'artificial' and lignins *in vivo* has, however, long been questioned. For example, Holgar Erdtman in 1957 (*11*), who is generally viewed as the father of the dehydrogenation theory leading to lignin formation, emphasized that "In some respects, they ['artificial' lignins] are similar to lignins, but in others they are *distinctly* different". Additionally, in 1971, Herbert Hergert (*12*) noted that "although the artificial lignin is asserted to contain the same building units as spruce wood lignin, the infrared spectra (*2, 8, 13*) show important differences in carbonyl content (1715 cm^{-1}), ethylenic double bonds (975 cm^{-1}), content of ether and hydroxyl groups (shape of 1140 cm^{-1} band and intensity ratios of 1140 and 1035 cm^{-1} bands) and aromatic substitution patterns (peak ratios in 900-800 cm^{-1} region)". It is now *very well documented*, using more modern approaches, *e.g.* by ^{13}C NMR spectral analysis and thioacidolysis (*14, 15*), that artificial and natural lignins *are not identical*, *i.e.* in terms of bond types, bond frequencies and polymer size.

(3) The predominant interunit linkage in natural lignins is the 8–*O*–4′ aryl ether bond (see Figure 2). For example, it is viewed to correspond to ~50% of the interunit linkages in the lignin of softwoods, whose constitution is primarily coniferyl alcohol-

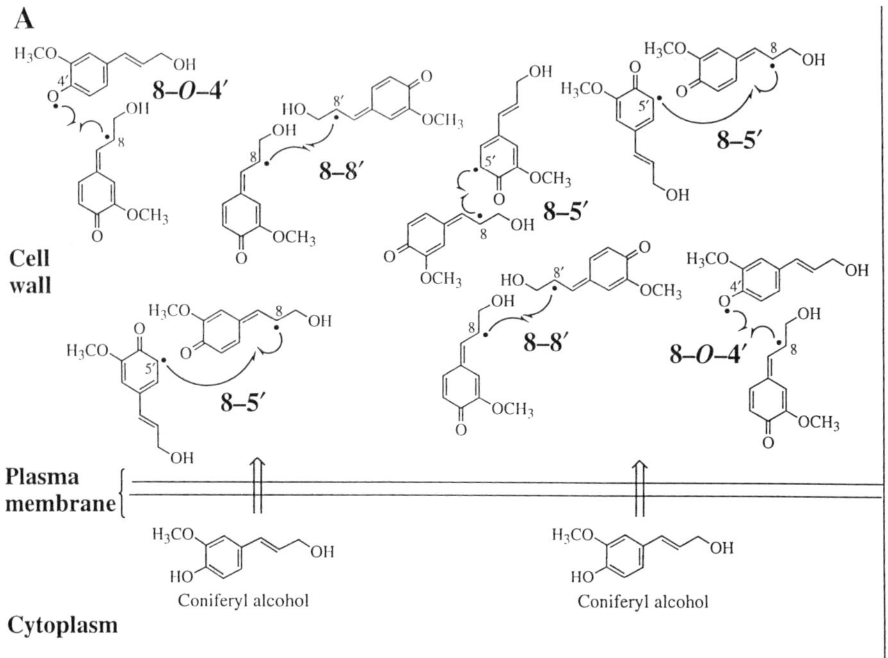

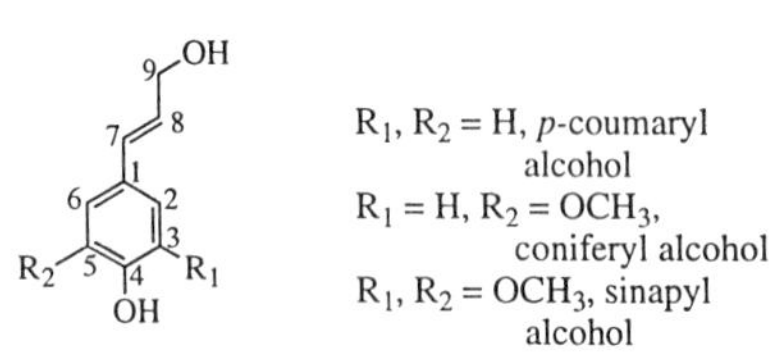

Figure 1. Representation of the Freudenberg hypothesis for random monolignol coupling during softwood lignin biosynthesis. **A,** coniferyl alcohol is transported from the cytoplasm into the lignifying cell wall with subsequent one electron oxidation and random coupling of the free-radical forms. Note: For simplicity, only three modes of random coupling (8–5′, 8–8′ and 8–*O*–4′) are shown. Additionally, only coniferyl alcohol, rather than its putative monolignol glucoside, is shown as the species being transported into the cell wall, and **B,** the three monolignols involved in lignin biosynthesis.

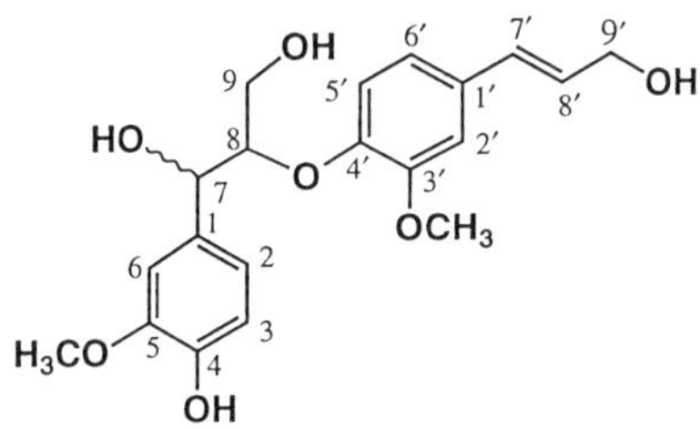

8–*O*–4′-dilignol
(±)-Erythro/threo guaiacylglycerol
8–*O*–4′-coniferyl alcohol ethers

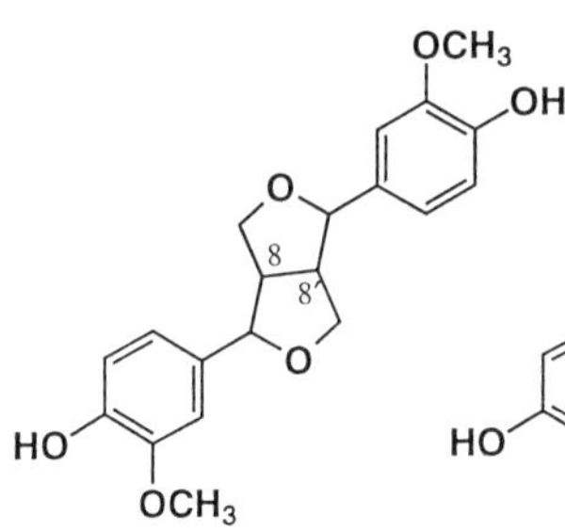

8–8′-dilignol
(±)-Pinoresinol

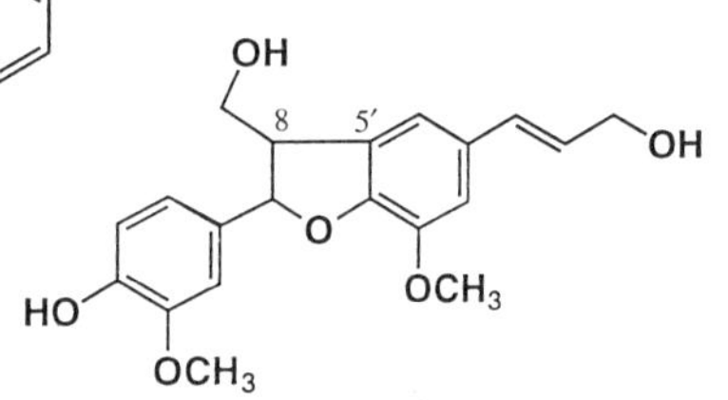

8–5′-dilignol
(±)-Dehydrodiconiferyl alcohol

Figure 2. Major racemic dimeric products obtained by random free-radical coupling of coniferyl alcohol *in vitro* with subsequent intramolecular cyclization or nucleophilic attack of the corresponding quinone methide with hydroxide anion.

derived. The other substructures are present in lower amount, *e.g.* 8–5′ (~10%). The remaining linkages are of a miscellaneous type, but, interestingly, they apparently contain few 8–8′ (≤2%) linkages (*16*). These bonding frequencies arose from detailed chemical analyses of lignin degradation products, primarily achieved by Adler and co-workers from the late 1940's onwards.

In vitro coupling reactions, on the other hand, using coniferyl alcohol and O_2/laccase were examined by the Freudenberg school. These initially gave both an artificial lignin, and various 8–5′, 8–8′ and 8–*O*–4′ racemic dimeric lignan products, in a ratio of 26:13:9 (see Table I and Figure 2), *i.e.* where 8–5′ coupling *predominated* (*5*). It was suggested, however, that if a slow release of coniferyl alcohol (by enzymatic hydrolysis of coniferin by β-glucosidase) was instead effected, a different type of artificial lignin was obtained. Although this was not characterized further, it was proposed that coupling now occurred in such a way so that the 8–*O*–4′ dilignol dimers were thought to prevail (~60%) (*5*): however, this bonding frequency was apparently never truly quantified.

Over the next decade, the ratios of the dimeric products obtained *in vitro* were *continually revised* in various reviews (*3, 17*). For example, Erich Adler (*18*) referred to the 8–*O*–4′ dimer as being the dominant product *in vitro* (~70%), as reported in Freudenberg reviews (*19-21*). Yet, even as late as in 1968, there was still uncertainty about whether it prevailed or not *in vitro*, since it was noted that the "guaiacylglycerol-β-coniferyl ether (8–*O*–4′ dilignol) *probably* surpasses in yield even pinoresinol and dehydrodiconiferyl alcohol" (emphasis added) (*3*). The revisions of the 8–*O*–4′ frequency of the dimers were also noted and commented upon by Kyosti V. Sarkanen in 1971 (*22*), see footnote * below in Table I. Whatever reasons actually held for the continual shifting of 8–*O*–4′ bond frequencies *in vitro*, artificial and natural lignins *were not identical*.

Table I. Relative Amounts of Direct Coupling and Disproportionation Products in the Dehydrogenation of Coniferyl Alcohol by Laccase and Oxygen.* [Table modified from K. V. Sarkanen in ref. (*22*)]

Product	% of total	% of dimers	References
Unchanged coniferyl alcohol	3	–	(*5*)
Polymer (DHP)	45	–	(*5*)
8–5′ Dilignol	26[1]	54	(*5*)
8–8′ Racemic pinoresinol	13	27	(*5*)
8–8′ Racemic epipinoresinol	trace	–	(*7*)
8–*O*–4′ Dilignol	9[2]	19	(*5*)
5–5′ Dilignol	trace (?)	–	(*23*)
Coniferyl aldehyde	0.2	–	(*5*)

1. An approximate yield of 40% also reported (*24*)
2. Incomplete recovery

* In his reviews Freudenberg (*3*) has quoted 15% yield for 8–5′ dilignol, and approximately equal yield for racemic pinoresinol and the highest yield for 8–*O*–4′ dilignol, contradicting the data published in the original papers.

(4) At least six proteins, peroxidase (*1-3, 6, 25-27*), laccase (*1-3, 6, 28-31*), peroxidase and laccase (*1-3, 25, 32*), coniferyl alcohol oxidase (*33-35*), (poly)phenol oxidase (*36*) and cytochrome oxidase (*37*) have been implicated in the final stage of lignin assembly. Even today there is no general consensus as to which is most important. To our knowledge, there is no other biopolymer existing in Nature that is viewed as employing *any one of a number of enzymes* to facilitate its formation. That is, there is no other example where such a plethora of enzymes have been proposed,

with each having identical roles for the same biochemical step, in this case the one-electron oxidation of the substrate.

(5) Plant laccases have long been postulated as involved in lignification (*1-3, 6, 28-32*). In contrast to the other monolignols, *p*-coumaryl alcohol is oxidized at a much lower rate by all of the plant laccases examined thus far (*30-32*). This is something of a contradiction if laccases and *p*-coumaryl alcohols are targeted, as postulated, to the early stages of lignin assembly. It is also seldom appreciated that these *plant* enzymes do not convert coniferyl and sinapyl alcohols into lignins *in vitro*, but rather primarily give various racemic lignan dimers which can, upon exhaustion of the monomer supply, be converted into tetramers and so forth (*2*).

(6) The random coupling hypothesis ignores the fact that lignin monomer composition and lignin interunit linkages *in vivo* are *not* products of chance. In contrast, *they are fully predictable for a specific cell type.* For example, individual monolignols, such as *p*-coumaryl and coniferyl alcohols, are *selectively* targeted to distinct regions within the cell walls (*e.g.* middle lamella *versus* secondary wall) and tissue (*e.g.* vessel *versus* fiber lignins) where they undergo coupling, as shown by UV-microscopy (*38-41*), radiolabeling/autoradiography (*42-45*) and immunolabeling (*46, 47*). It is further believed that *p*-coumaryl alcohol is primarily involved in the initial stages of lignin assembly in the middle lamella/cell corners, whereas coniferyl alcohol is mainly targeted to the secondary wall. It can, therefore, be concluded that there is a *very high level of order* in the regional deposition and inter-unit linking of individual monomeric components in specific areas of the cell wall during lignin formation *in vivo*. This is neither predicted nor observed from *in vitro* experiments.

(7) Results from various lignin analyses only rarely describe the type of tissue being examined, *i.e.* whether it is from sapwood, heartwood, diseased wood, reaction (*e.g.* compression) wood and so forth. This is of utmost importance, given that lignification is essentially only associated with sapwood formation and that most woody samples analyzed are composed of >95% heartwood. Tissues, such as heartwood, diseased and 'stressed' woods, compression wood, *etc.*, often accumulate (in addition to the lignins) large amounts of colored *non-structural* phenolics, these being primarily complex mixtures of oligomeric lignans and flavonoids (see Chapter 25) (*12, 48, 49*). They are primarily laid down as insoluble deposits into the fully developed pre-lignified xylem sapwood *via* infusion from specialized conducting cells, *e.g.* ray parenchyma (*50*) during heartwood formation. A substantial proportion is *only solubilized* under conditions approaching lignin dissolution, such as in those employed for milled wood lignin, dioxane-HCl and Björkman lignin preparations. Hergert has long cautioned the need for carefully specifying the tissue type used for lignin analyses, given that such *non-structural* deposits can interfere substantially with the lignin analysis being conducted (*12, 49*).

Indeed, because of the structural similarities of *non-lignin, non-structural* phenolic substances to lignin itself, these can often be misidentified as either 'abnormal' lignins (*51*) or even as lignin 'aging' products (*52*), *if sufficient care is not taken.* Accordingly, such contaminants can extensively contribute to the supposed proportion of 8–5′ and 8–8′ linked 'lignin substructures' depending upon the extent of such infusions (see Chapter 25).

(8) Other than lignin biosynthesis, there are additional metabolic fates for monolignols. For example, lignans are also formed from coniferyl alcohol (*53-56*), and can often occur in optically active forms, although the antipode present can vary with the plant in question (*57-60*). Most lignans are biosynthesized whereby bimolecular phenoxy radical coupling is highly controlled in terms of the regio- and stereochemistry of the products obtained (*56*). Given the fact that lignins and lignans are often derived from the same precursors, it is curious, therefore, that little attention has been given to either any possible biochemical relationships in the enzymology of their formation, or how their biochemical pathways might differ.

Lignans can exist as a wide variety of structures ranging from simple dimers (*61, 62*) to higher (insoluble) oligomeric deposits (*12, 48, 49*), such as in western red cedar (*Thuja plicata*) heartwood, where they account for up to 20% by weight (*63*).

In the latter case, they are essentially only deposited in large amount *during* heartwood formation, as a *post-lignification* response.

Because of a lack of suitable methodologies to readily distinguish between the lignins and oligomeric lignans, this has often resulted in the latter being trivially described as 'lignin-like', 'abnormal lignins', Brauns' lignins and so forth—even though detailed analyses reveal profound distinguishing features. Indeed, failure in certain studies to distinguish between lignins and (oligo)lignans would be akin to liking celluloses and hemicelluloses as being the same (since they are all monosaccharide-derived), and then ignoring the differences in their monomeric building blocks and the interunit linkages in their polymers [*e.g.* such as cellulose and callose which are β(1-4) and β(1-3) linked polymers, respectively].

(9) There are no *known* intermediates formed either during, or remaining after, the polymerization of monolignols to give the high molecular weight, structural, lignins *in vivo*. In contrast, the so-called 'soluble' lignins (*e.g.* Brauns' lignins) isolated from various species were instead oligomeric lignans, derived from a completely different biochemical pathway (see Chapter 25).

(10) There is not a single well-defined example of macromolecular cell wall assembly (biopolymer formation) that has adequately been duplicated *in vitro*. This is true not only for the lignins but also for all of the carbohydrate-derived polymers, including that of cellulose. This should serve as a cautionary reminder that elucidation of these biochemical processes will undoubtedly reveal important subtleties (in terms of macromolecular assembly) and help clarify issues such as how lignification is actually orchestrated *in vivo*.

Thus, in summary, early attempts to delineate the actual biochemical mechanisms attendant to bi- and macromolecular phenoxy radical coupling, such as those leading to the lignans and lignins, resulted in a hypothesis that does not account for the formation of either class of metabolites. Its basis lay in the study of enzyme-catalyzed reactions of monolignols *in vitro* (so-called oxidative coupling) to give preparations which were originally described at that time to represent lignin structure *in vivo* (*1-3, 7*), but which, in fact, do not. According to this original perception, lignin macromolecular assembly only required oxidative enzymes, of no specific type, to generate the corresponding free-radical entities, and these then underwent random coupling to afford the polymeric lignins.

The importance of systematically delineating the *actual* biochemical processes involved in *control* of phenoxy radical coupling *in vivo* cannot be over-emphasized, given its central importance to both vascular plant cell wall macromolecular assembly and plant defense. Indeed, the already recognized differences between artificial and natural lignins strongly suggest that subtle biochemical processes controlling coupling *in vivo* have gone undetected.

In order to provide the necessary background for the development of a new hypothesis for control of phenoxy radical coupling and lignin/lignan assemblies, this contribution is divided into four sections:

Early theories of lignification and lignan formation.
Biochemical pathways to the monolignols: Lignin and lignan branchpoint pathways.
A turning point in plant phenol coupling: Dirigent proteins.
Macromolecular assemblies: Lignin formation and arrays of dirigent protein sites.

Early Theories of Lignification and Lignan Formation

The path to our current knowledge of monolignol coupling processes, leading to the lignans and lignins, spans well over a century. As with most scientific endeavors, it was punctuated by a relatively small number of seminal observations and insights along the way. This section discusses several critical junctures, together with other perspectives which vied for favor prior to ultimately being discarded.

The beginnings of the discovery of the pathway leading to the monolignols can be traced back to observations made by De Candolle in 1819. As described by Bernard Monties in 1989 (*64*), "While wood was still considered to be a uniform chemical compound, A. De Candolle (1819) (*65, 66*) coined the word '*la lignine*' to describe the fraction of wood insoluble after treatment with solvents and mild acid. In 1838, Anselme Payen (*67, 68*) established that this compound, also named '*le ligneux*', was a mixture of two distinct products: '*la cellulose*' and '*l'incrustation ligneuse*,' *i.e.* lignin incrustation..."

In 1874, the isolation and structural characterization of a substance from *Abies excelsa*, *A. pectinata*, *Pinus strobus*, *P. cembra* and *Larix europaea* was reported by Ferdinand Tiemann and Wilhelm Haarmann (*69*) and named coniferin. It was subsequently shown to release glucose and coniferyl alcohol upon action of emulsin (*70*). Peter Klason later developed, near the turn of the century, a process to quantify lignin in secondary xylem (wood) *via* measurement of the insoluble residue remaining after 70-72% sulfuric acid digestion (*71, 72*). By 1908, Klason commented that lignin was a condensation product of coniferyl alcohol and *oxy*coniferyl alcohol (*73*), and later K. Kürschner, in 1925, speculated lignin to be a polymer of coniferin (*74, 75*).

By 1933-34, there was considerable disagreement and confusion about the very constitution of lignin, with various investigators proposing that it was derived either from cellulose (*76-79*), pentoses/pentosans (*80, 81*) or pectin (*82*), as summarized by Max Phillips (*83*). On the other hand, Karl Freudenberg had proposed that lignin was a linear polymer derived from coniferyl alcohol, linked *via* 9–*O*–4′ linkages (Figure 3), based upon perceived dehydration reactions occurring *in vivo* (*84, 85*).

This viewpoint was challenged later by Holgar Erdtman (*86*) who suggested that coniferyl alcohol-like molecules were linked instead *via* dehydrogenative coupling to give the polymeric lignins. This concept arose from the conversion of isoeugenol into the dimeric phenylcoumaran (8–5′ linked lignan), trivially named dehydrodiiso-eugenol (*86, 87*), using a crude mushroom extract for catalysis. Interestingly, this same dehydrogenative conversion had previously been described by H. Cousin and H. Hérissey in 1908 (*88*), except that the product was incorrectly identified. Just a few years earlier, in 1900 at Ann Arbor, M. Gomberg had marvelously discovered the existence of free-radicals (*89*). This signaled the beginnings of the mechanistic basis for one electron oxidative (dehydrogenative) processes, and thus the means to ultimately afford both the lignans and lignins. Gomberg's seminal contribution ends with the immortalized statement that "This work will be continued as I wish to reserve the field for myself."

Following Erdtman's seminal insight into dehydrogenative coupling, almost two decades elapsed before Freudenberg (*1-3, 20, 90*), Karl Kratzl (*91-94*), Arthur Neish (*95-99*) and their co-workers established (a) that various substances from phenylalanine/shikimic acid onwards to the monolignols, *p*-coumaryl, coniferyl and sinapyl alcohols (Figure 4), were intermediates in lignin biosynthesis, and (b) that the monolignols could undergo dehydrogenative coupling/polymerization *in vitro* (*3*). As discussed earlier, the *in vitro* studies culminated in the identification of (±)-dehydrodiconiferyl alcohols, (±)-pinoresinols and (±)-erythro/threo guaiacyl-glycerol 8–*O*–4′-coniferyl alcohol ethers (see Figure 2)]. These findings were important, given that a means for forming the possible coupling products had now been identified, in spite of the serious concerns previously discussed about the initial enzyme source and the wide fluctuations in 8–*O*–4′ bonding frequencies.

Two major factors had emerged by the end of the nineteen sixties: first, uncontrolled random coupling of free-radical species *in vitro* in open solution, could occur at different positions in the monolignol-derived molecules, to initially afford various racemic lignan dimers; second, these were considered to undergo further oxidative coupling to generate the lignin skeleton (*3*). Thus, in contrast to any other known biological process, the formation of Nature's second most abundant natural product was *envisaged to be satisfactorily duplicated by the random encounter of its free-radical precursors in a test tube*. A comparable sequence of events was

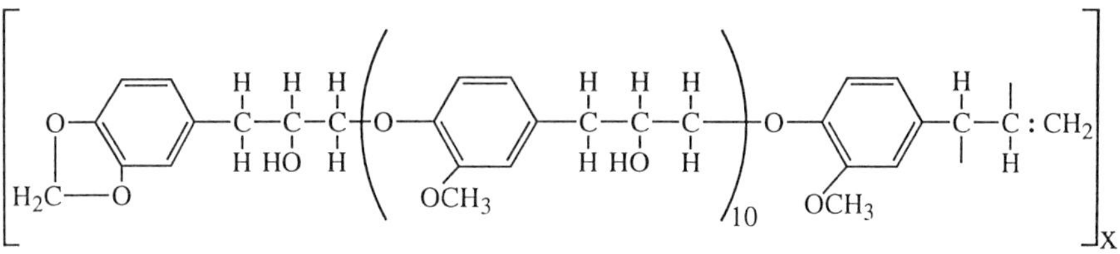

Figure 3. An earlier model for lignin structure reported in 1929 and 1933 by Freudenberg (*84, 85*).

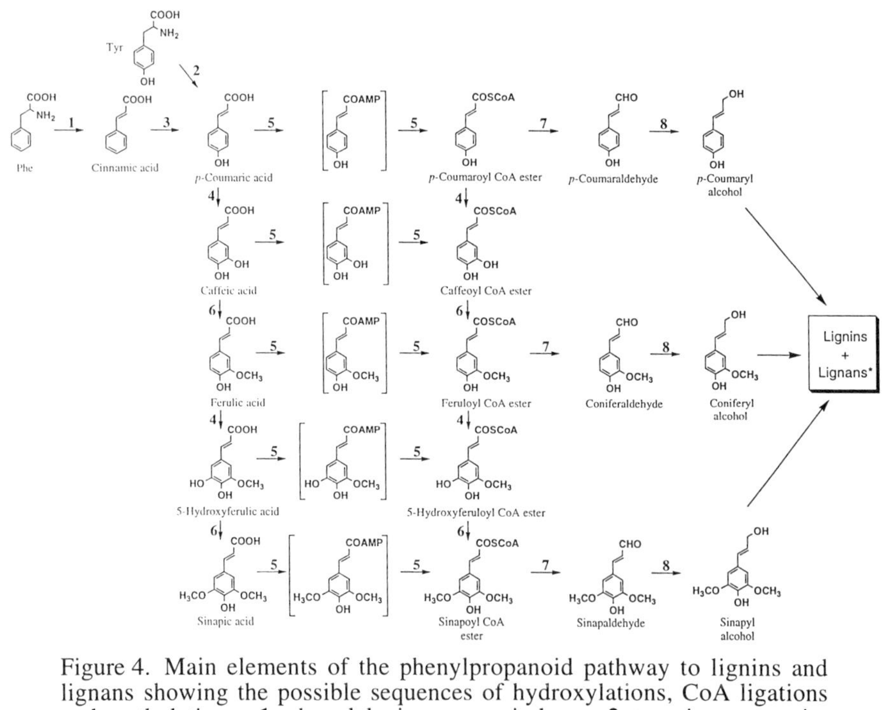

Figure 4. Main elements of the phenylpropanoid pathway to lignins and lignans showing the possible sequences of hydroxylations, CoA ligations and methylations. **1,** phenylalanine ammonia-lyase, **2,** tyrosine ammonia-lyase (mainly in grasses), **3,** cinnamate-4-hydroxylase, **4,** hydroxylases, **5,** CoA ligases involving AMP and CoA ligation, respectively, **6,** *O*-methyltransferases, **7**, cinnamoyl-CoA:NADP oxidoreductases (CCR), **8,** cinnamyl alcohol dehydrogenases (CAD). [Note *: This only shows the lignans derived from monolignols, and not those from allylphenols nor other modifications. The bulk of the lignans seem to be monolignol-derived.]

proposed to occur *in vivo*: free radical species were thought to combine in the lignifying cell walls, *via* diffusion-driven encounters, to first generate the aforementioned dimers and, eventually, the lignin polymer.

Shortly after Erdtman's 1933 proposal for lignin formation, Haworth in 1937 in England described an abundant series of natural products found in vascular plants as lignans (*100*). These were viewed at that time to only represent dimeric substances linked *via* their 8–8′ bonds, *e.g.* (+)-pinoresinol (Figure 5). Importantly, and unlike the lignins, they were primarily found in an *optically active* form, suggesting perhaps that some other mode of assembly than that described for lignin was in place. Erdtman subsequently spent a part of the remainder of his life in attempting to determine how regiospecific and stereoselective control of phenolic coupling giving the lignans occurred, but was repeatedly thwarted in his attempts (*101*). Nevertheless, recognition was given to the presumed fact that phenolic coupling was, at least for the lignans, under some sort of biological control.

The following years until the present took very different, but rewarding, approaches to the study of the lignins and lignans. As indicated earlier, Adler (as well as Mischke) and co-workers had attempted to systematically apply chemical degradation procedures to various tissues from selected plant species, especially lignified woody gymnosperms, in order to *estimate* the overall relative frequencies of interunit linkages in lignins [see (*16*) for a review]. Although constrained by the methodologies available at that time (such as potassium permanganate degradation, acidolysis and so forth), salient features of the native lignin polymer began to gradually emerge (Table II). As can be seen, these studies indicated that the 8–*O*–4′ linkage predominated, with the 8–5′ substructure next most abundant (~10%). In comparison, the contribution of the 8–8′ linked moiety was very low (<2%) in all plant species examined. Moreover, the latter was almost *never found* as pinoresinol, but instead in other forms (*e.g.* as a divanillyltetrahydrofuran derivative) (*16*). Nevertheless, this and other 'minor' reduced substructures were also considered to be *possibly* part of lignin. It was recognized at that time, however, that formation of such modified substructures could not readily be rationalized simply on the basis of direct oxidative polymerization of monolignols. (See Chapter 25 for an explanation and clarification for formation of oligolign*a*ns).

Table II. Proportions of different types of bonds connecting phenylpropane units in Björkman lignin from spruce (*Picea abies*). [Reconstructed from E. Adler (*16*)]

Bond Type	Proportions (%)
8–*O*–4′	50
8–5′	9-12
5–5′	9.5-11
4–*O*–5′	3.5-4
8–1′	7
8–8′	2

Subsequently, following application of a variety of different ^{13}C NMR spectroscopic approaches, in both the solid and solution state, it became further evident that the structures of lignins *in situ*, as well as that of isolated lignin-derived preparations (*14, 15, 102-105*), were not duplicated by artificial lignin preparations; the introduction of thioacidolysis as a means to assess lignin structures reinforced and extended these observations (*15*). A significant addition to these studies came even more recently in work from Gösta Brunow and co-workers, where it was established that the cyclic dibenzodioxicin substructure was abundant in softwood lignins (see chapter 10) (*106, 107*). This, at long last, provided clarification to the disputed existence of various 7–*O*–aryl ethers in lignin macromolecular structure.

Given the long-held random coupling hypothesis for lignification, it is instructive at this point to reflect upon the biological observations associated with its formation *in vivo*, *i.e.* in terms of how lignification is carefully and reproducibly orchestrated throughout all vascular plant species examined. During 'normal' plant growth and development, cell differentiation and morphogenesis are programmed to yield the familiar tissues and organs of all vascular plants. In sapwood formation in trees, for example, generation of the lignified secondary xylem is a gradual process differentially extending from the cambial zone to the fully developed vascular tissue (*108*). It begins with cells in the cambial region undergoing radial expansion when in their primary wall phase. At this stage the primary walls are composed mainly of cellulose microfibrils, in no particular orientation, together with non-cellulosic polysaccharides and other proteins of no truly defined function [*e.g.* extensin (*109*)]. When this phase of development is completed, the remaining polysaccharide components of the secondary wall (S_1, S_2 and S_3 layers) are sequentially laid down, but where the cellulose microfibrils are now orientated (Figure 6A) (*110*).

Importantly, the bulk of lignin biosynthesis *follows* polysaccharide deposition after the stage where the overall architecture of the thickened cell wall has been established. Moreover, it is initiated *at precise points* in the cell-corners or just inside the primary wall, prior to extending along the middle lamella and into the secondary wall (*111-115*). These lignin initiation points appear, based on immunolabeling studies, to be coincident with the deposition of proposed proline-rich proteins secreted from the Golgi apparatus and transported across the plasma membrane and into *specific* regions within the maturing cell wall (*46, 116*). Although no biochemical clarification has ever been given to what scaffolding might mean mechanistically, these proteins have been suggested as 'lignin scaffolds'.

That lignin biosynthesis is initiated at specific sites far removed from the plasma membrane is a key observation. It is key because the monomeric precursors transported from the cytosol to the plasma membrane not only have to negotiate their way into the inner confines of the wall to sites where lignin biosynthesis is initiated, but more importantly it reveals a very highly controlled process.

It is also significant that the various monolignols are deposited in a *non-uniform* manner throughout the wall. Recognition of this heterogeneity in lignin composition was first established by the work of David Goring and co-workers, who made several important observations (*38-41*). In gymnosperm tracheids, for example, it was reported that the middle lamella/cell-corner regions had a higher *p*-coumaryl alcohol-derived lignin content that that of the secondary wall, which was predominantly of coniferyl alcohol origin (*40*). With the angiosperms, the fiber and vessel walls also had lignins which were heterogeneous throughout the confines of the wall. That is, for instance, in white birch (*Betula papyrifera*), the middle lamella region is rich in coniferyl alcohol, whereas the vessels and fiber walls were primarily derived from coniferyl and sinapyl alcohols, respectively (*38, 39, 41*).

These findings were subsequently confirmed through other approaches, and the ramifications recognized. In a series of radiotracer experiments, Nori Terashima (*42-45*) and others reported that there was a precise temporal and spatial deposition of *specific* monolignols during tracheary element, vessel and fiber formation. Their data also revealed that in gymnosperms, such as in Japanese black pine (*Pinus thunbergii* Parl.) (*43, 44*), *p*-coumaryl alcohol was initially laid down primarily in the middle lamella/cell-corner regions, whereas coniferyl alcohol was mainly deposited into the secondary wall. Additionally, in angiosperms, *e.g. Syringa vulgaris*, a comparable situation to gymnosperm lignin again existed for vessel development, whereas the fibers also contained sinapyl alcohol-derived components in their secondary walls (*45*).

Thus it had emerged that lignin monomer deposition was in fact occurring in a highly specific and targeted manner. Indeed, in 1989, Terashima summarized the results of a number of biological observations by different investigators regarding

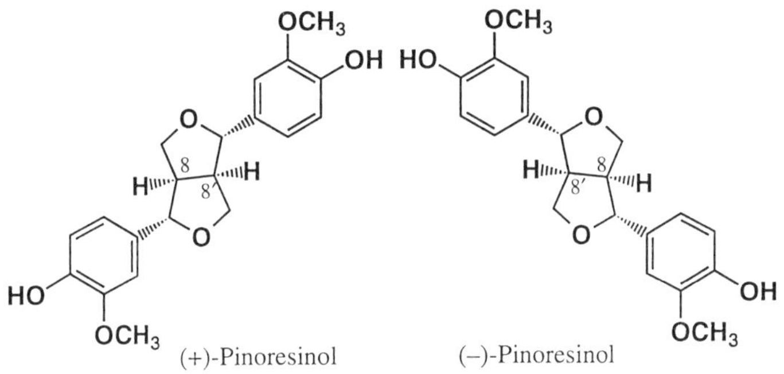

Figure 5. Schematic representation of (+)- and (–)-pinoresinol antipodes.

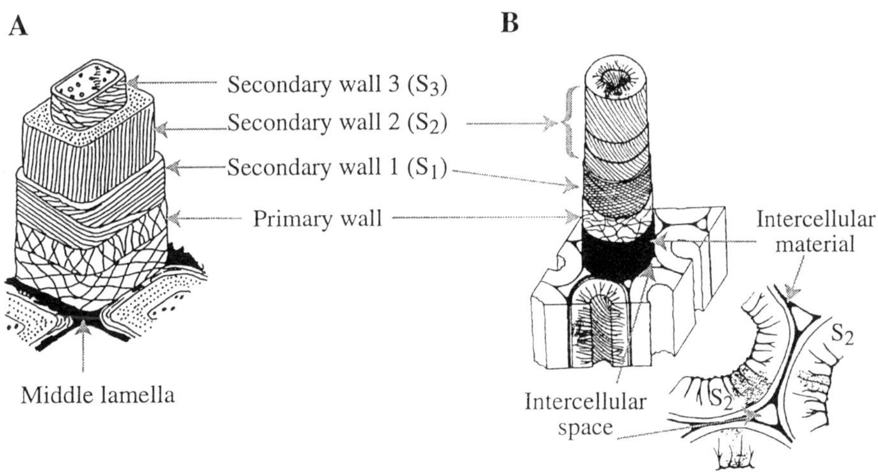

Figure 6. Schematic model of the cell wall structure of **A,** a 'normal' softwood tracheid and **B,** a typical compression wood tracheid. [Redrawn from refs. (*110*) and (*149*), respectively, in a manner to illustrate cellulose microfibril orientation in different cell wall layers.]

lignification *in vivo*, and concluded that "protolignin is not a disordered co-polymer of various monolignols. Instead, macromolecular formation occurs in a biochemically regulated manner, and the heterogeneous nature of protolignin is a natural and inevitable consequence of its *unique mechanism of biogenesis*" (*45*) (emphasis added). In short, the UV microscopy (*38-41*), radiolabeling (*42-45*), and immunolabeling (*46, 47*) studies had revealed apparent initiation sites for lignin biosynthesis, and a process of heterogeneous monolignol deposition into different parts of the cell walls. Since then, immunolabeling studies using antibodies raised against different monolignol dehydropolymerisates have further emphasized the heterogeneity of lignin structures in the cell wall (*46, 47*). Thus, from a purely biological perspective, lignification had a clearly defined order in its monolignol deposition and polymerization. However, there was still no clarification as to how lignin assembly itself was controlled.

During this hiatus, in other parts of the world, primarily in the laboratory of Otto R. Gottlieb, different but related studies were vigorously underway. In these extensive studies, which focused mainly on the Magnoliaceae and Lauraceae, a range of (typically) optically active lignan structures began to emerge possessing various interunit linkages (*117-119*) (Figure 7). Inspection of these and other diverse skeleta strongly suggested that specific coupling mechanisms must indeed be responsible for the generation not only of the 8–8′ linked lignans, but also other skeletal types. [The reader should note that the term lignan only describes 8–8′ linked products, even though there are other coupling modes (*e.g.* 8–5′, 8–1′, 8–*O*–4′) present in the plant kingdom. Although the descriptor, neolignan, was proposed for such metabolites, the use of the term lignan, where the mode of coupling is stipulated, is both preferred and used in this chapter.]

Thus all of these findings, originating from various places, began to emphasize two important trends: the first was that lignin biosynthesis *in vivo* was a much more controlled process than the early *in vitro* experiments would have suggested. The second, from the lignan chemotaxonomical studies, strongly indicated there were precise enzymatic systems conferring specificity in monomer (monolignol) coupling leading to specific skeletal forms. It, therefore, became an urgent priority for us to reconcile how these biochemical processes actually occurred, particularly since it had been definitively established (*120-122*) that monolignols could, in fact, be employed to generate both the lignans and the lignins.

Biochemical Pathways to the Monolignols: Lignin and Lignan Branchpoint Pathways

No discussion on the oxidative coupling of plant phenols would be complete without brief mention of the biochemical pathway to the monolignols and their recently established branchpoints. As can be seen in Figure 4, the monolignol pathway consists of five types of enzymatic conversions, namely deamination, aromatic ring hydroxylations, *O*-methylations, CoA ligations and NADPH-dependent reductions. [For a detailed review, see ref. (*123*).]

Elucidation of the pathway began in the early fifties where it was established, primarily in the laboratories of Neish (*95, 97-99*) and Freudenberg (*1-3*), that radiolabeled cinnamic acid and coniferin served as lignin precursors. It was not, however, until 1961, with the discovery of phenylalanine ammonia-lyase by Joan Koukol and Eric Conn (*124*), that the point of linkage between primary metabolism and so-called secondary metabolism was identified. Following that seminal discovery, a significant effort was placed on isolating and identifying the various enzymes involved in aromatic ring hydroxylation (*125-127*), CoA ligation (*128-130*) and *O*-methylation (*131-134*), as well as subsequent reductive steps catalyzed by cinnamoyl CoA reductase and cinnamyl alcohol dehydrogenase (*128-130*), respectively. The hydroxylation and methylation steps have since been extensively studied, with the conversion of cinnamic acid to *p*-coumaric acid, catalyzed by cinnamate-4-hydroxylase, perhaps being the most well understood (*135, 136*). On the

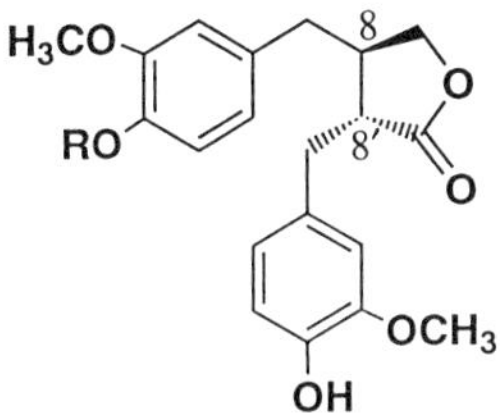

R = H, Matairesinol
R = CH_3, Arctigenin

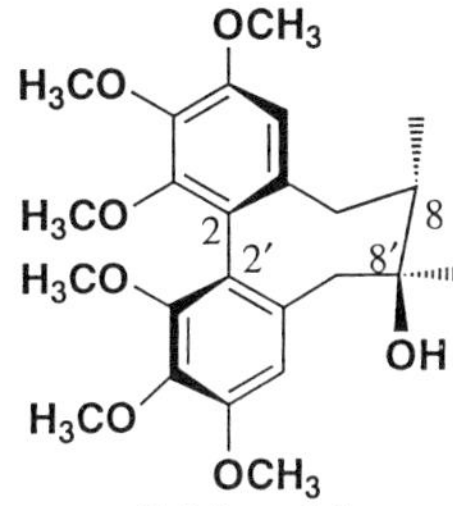

Schizandrin

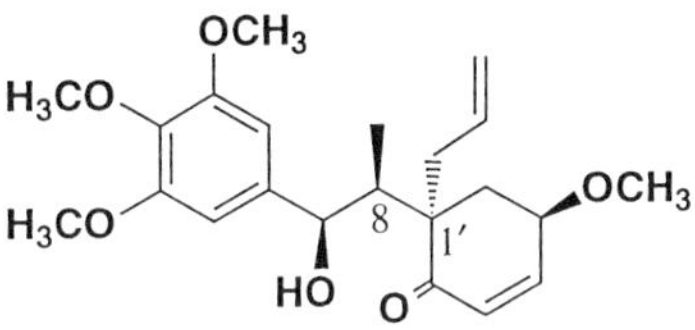

Megaphone

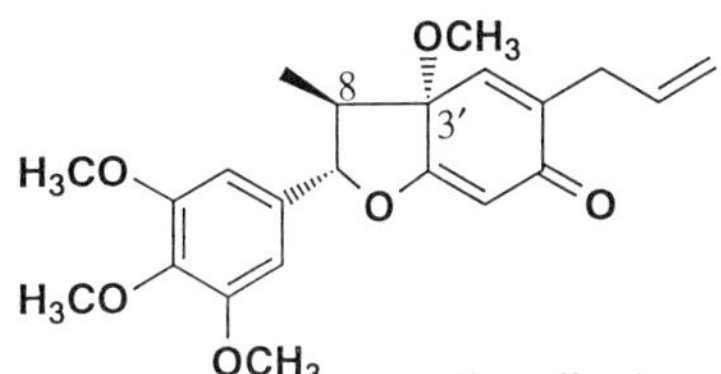

Mirandin A

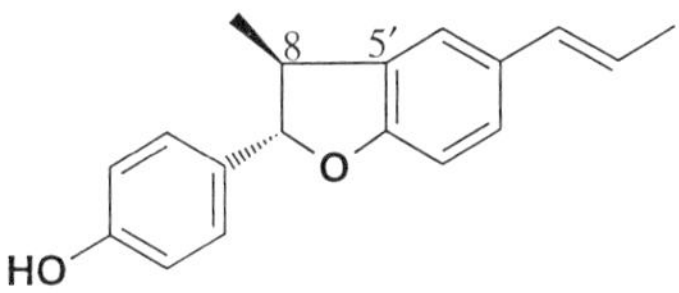

Conocarpan

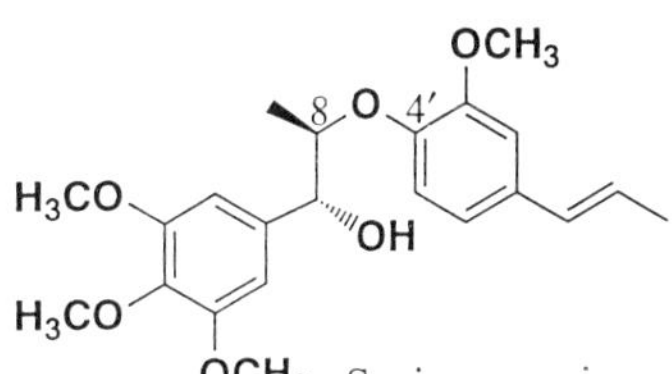

Surinamensin

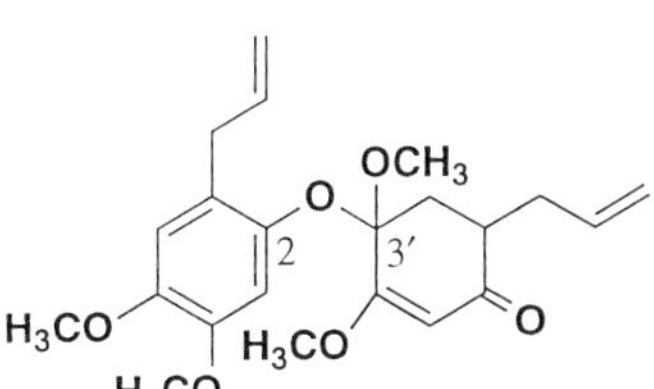

Lancilin

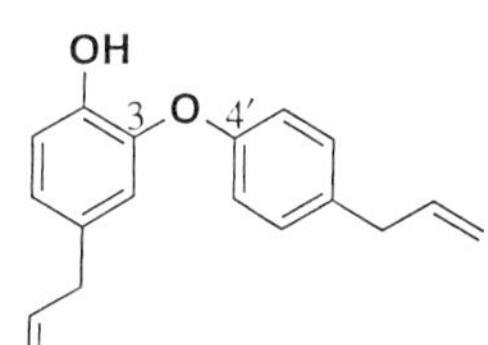

Isomagnolol

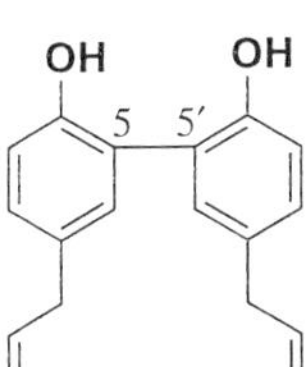

Magnolol

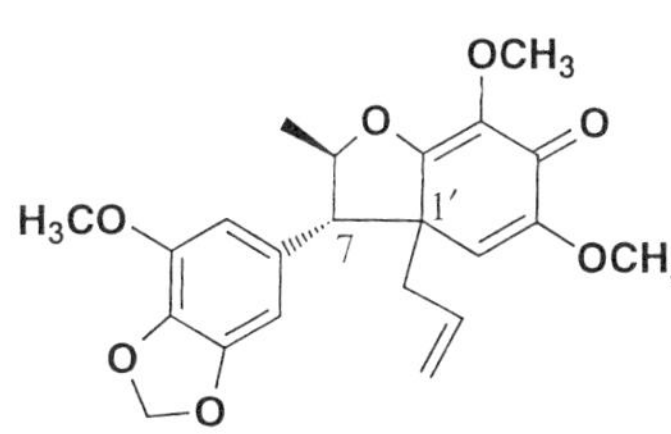

Chrysophyllon IA

Figure 7. Examples of various lignans encompassing differing interunit linkages.

other hand, the subsequent aromatic hydroxylation steps leading to caffeic and, to a lesser extent, 5-hydroxyferulic acids are still not particularly well-defined (*127, 137*). As a further complication, the sequence of methylation/CoA ligation conversions is purposefully left ambiguous, even though the various enzymes and genes have apparently been obtained; readers are directed to references (*138-141*) for a more thorough discussion of this aspect of phenylpropanoid metabolism.

Despite this work, several cautionary remarks need to be made regarding our current knowledge of phenylpropanoid metabolism: (i) metabolic control analysis (*142-144*) has never been reported for the pathway to the monolignols, and hence control or regulatory points have not yet been identified (but see Chapter 1); (ii) while Figure 4 represents the overall pathway, various branchpoints *e.g.* to the flavonoids and suberins also exist, which are frequently cell- and/or tissue-specific. Thus control mechanisms are presumably in place for either partial induction of the overall pathway or redirection of flux as needed (*e.g.* by suppression of specific enzymatic steps); (iii) it is possible that the ambiguity of the CoA ligation/methylation sequence may, in fact, be a subtle means for metabolite sorting into specific branchpoints. For example, the late Joe Varner (Washington University, personal communication) considered that the free acids and CoA-ligated forms might be diverted to vessel and fiber lignins, respectively. Another alternative, based on our own studies, is that this is possibly a means for channeling carbon into either the lignin or lignan pathways.

The major reason for briefly discussing the monolignol pathway in this chapter is again to underscore the fact that monolignols, such as coniferyl alcohol, have more than one metabolic fate, a fact which curiously is seldom alluded to. It is, accordingly, even more unfortunate that some studies still confuse lignins and lignans as being the same (*31, 35*), even though they are distinct biochemical classes!

The lignans are a very abundant, structurally diverse, class of natural products, which are near ubiquitous in all vascular plants being found in stems, roots, bark, flowers, leaves, fruits and seeds and so forth (*61*). In some instances, they can be deposited in very large amounts *and co-occur in lignified tissues* (*12, 48, 49*). As stated earlier, in western red cedar heartwood, lignans, consisting of dimers such as plicatic acid and its oligomeric and polymeric congeners, can account for up to 20% or so of the lignified wood (*63*) and are significantly responsible for the characteristic color/texture/durability of the tissue (see Chapter 25). Hergert, and others even earlier, long ago pointed out that during heartwood formation these substances are primarily infused into the lignified sapwood of certain species through specialized cells (*49, 50*).

It should be recognized, however, that it is only now that the lignin and lignan pathways are in fact being biochemically differentiated (see Chapter 25). Accordingly, more detailed studies will require clarification of how their pathways also differ in terms of: (1) subcellular locations and tissue-specificities of the lignan metabolites, relative to that of lignin; (2) whether the enzymology leading to the monolignols targeted for lignin and lignan biosynthesis differs [*e.g.* through involvement of distinct cinnamyl alcohol dehydrogenase (CAD) isoforms in different subcellular locations or tissues (see Figure 4)]. In this context it may be significant that putative catalytically distinct CAD isoforms have been identified *i.e.* CAD 1P and CAD 2P in *Eucalyptus gunnii* Hook (*145*); and (3) the sites of lignan post-coupling modifications including both subcellular locations and tissue-specificities.

A Turning Point in Plant Phenol Coupling: Dirigent Proteins

In order to gain the necessary insight into how phenolic coupling was actually controlled *in planta*, attention was, therefore, first directed towards formation of the simplest phenol coupling products, namely that affording the lignans. As discussed elsewhere (*146*), chemotaxonomic data suggested that there is only a relatively limited number of different lignan bimolecular coupling modes present in the plant kingdom, with that encompassing the 8–8′ linkage being the most prevalent. (See Figure 7 for this and other examples of different lignan skeletal types). From their optical activities, these were considered to arise *via* bimolecular phenoxy radical coupling, but where the stereoselectivity must be *explicitly* controlled.

Initial experiments employed *Forsythia* sp. as a plant material, since this was an abundant source of the 8–8′ linked lignans, pinoresinol, matairesinol and arctigenin (see Figures 5 and 7), and, which more importantly, only accumulated them in optically pure form. The readily soluble protein fraction from the stem tissue was first assayed for coniferyl alcohol coupling ability. This only gave tritium labeled (±)-dehydrodiconiferyl alcohols, (±)-pinoresinols and (±)-erythro/threo guaiacylglycerol 8–*O*–4′-coniferyl alcohol ethers (upon incubation with [9-^{3}H]coniferyl alcohol), but only when H_2O_2 was added as a cofactor, due to the action of non-specific peroxidases.

Attention was next focused on the *Forsythia* 'insoluble' residue remaining after removal of the soluble proteins. Upon incubation with [9-^{3}H]coniferyl alcohol, in the absence of any exogenously provided cofactor, this crude enzyme preparation catalyzed the preferential formation of (+)-pinoresinol over its (–)-antipode (and other lignan forms) (*53*). This was the first indication that a stereoselective coupling mode had been detected enzymatically. As shown in Figure 8, subsequent purification of the (+)-pinoresinol synthesizing machinery was not an easy task, given the fact that essentially *every* fraction from the cation exchange (Mono S) column was capable of oxidizing *E*-coniferyl alcohol due to the presence of contaminating non-specific oxidase(s). Significantly, however, fractions eluting at high salt concentrations (see chromatographic profile) were able to engender (+)-pinoresinol synthesis.

Subsequent chromatographic separation of the 333 mM Na_2SO_4 fraction, using a BioCad Perfusion Chromatography system, enabled the (+)-pinoresinol catalyzing system (final elution peak in Figure 8) to be partially separated into four components (Figure 9A). Of these, Fraction I (Figure 9B) was purified to apparent homogeneity, and was established to be an ~78 kDa protein, on the basis of both gel permeation chromatography and analytical ultracentrifugation (*56*). However, it did not possess any (oxidative) catalytic activity by itself. Fraction III (Figure 9C), on the other hand, had laccase-like activity, and electron spin resonance analysis further suggested that it was a typical plant laccase. Fraction III also displayed typical laccase properties, catalyzing conversion of coniferyl alcohol into formation of racemic (±)-dehydrodiconiferyl alcohols, (±)-pinoresinols and (±)-erythro/threo guaiacyl-glycerol-8–*O*–4′-coniferyl alcohol ethers, in relative ratios of 4:2:1, *i.e.* where the 8–5′ linkage again predominated. On the other hand, when both fractions I and III were combined together in judicious amounts, phenolic coupling was now such that only (+)-pinoresinol formation occurred [see ref. (*56*)].

Thus, in order to confer control of the long sought after regio- and stereospecificity in lignan formation, a 78 kDa protein coupling agent was necessary. This finding was first made with *Forsythia* sp. (*56, 147*), but has since been extended to other plant systems such as sesame (*Sesamum indicum*) and *Linum usitatissimum*, where in the latter case only the (–)-antipode is formed (J. L. Ford, Washington Sate University, 1997, unpublished observations). In terms of a mechanism of action, the protein apparently only serves to bind and orientate the substrate molecules (presumed to be the free-radical entities), since it has no detectable active (oxidative) site (*56*); the term *dirigent* (Latin = to guide or align) protein was coined to describe its unique function. Stereoselective coupling only occurs with the dirigent protein when auxiliary oxidative capacity was provided, through addition of either a non-specific oxidase (*e.g.* laccase) or a one electron oxidant. Accordingly, the means to achieve stereoselective coupling leading to optical activity profoundly differs from that proposed by Freudenberg to afford the lignins, where only random coupling had been implicated.

The effect of the 78 kDa dirigent protein on stereoselective coupling of *E*-coniferyl alcohol is shown in Figure 10. As can be seen, when the auxiliary oxidase (*e.g.* laccase) or an one-electron oxidant (such as FMN) is incubated *alone* with *E*-coniferyl alcohol, conversion into the three racemic lignan dimers occurs (Figures 10 A and C). On the other hand, when the 78 kDa dirigent protein is then

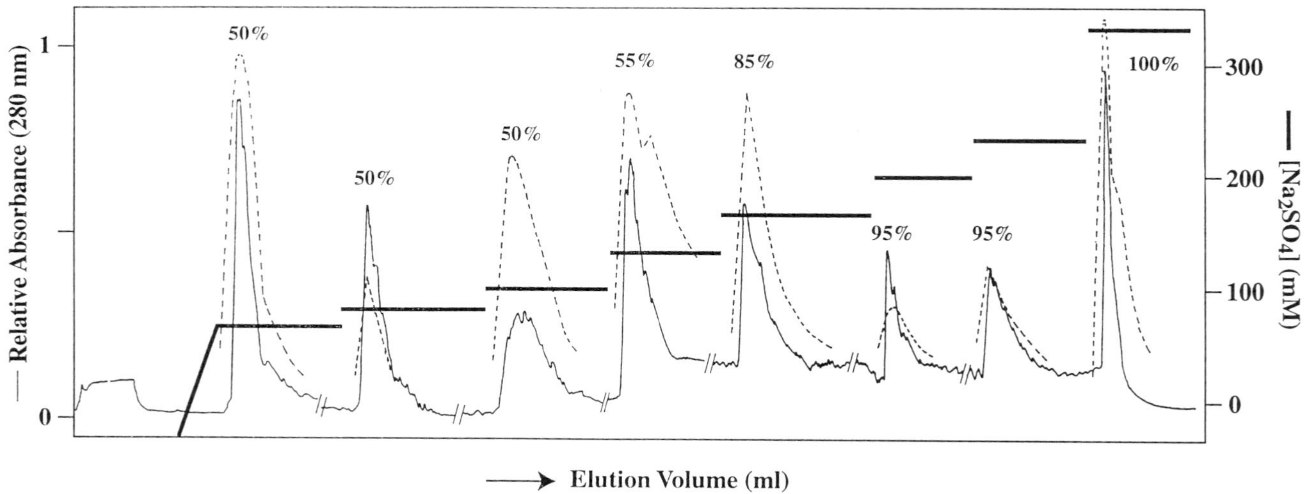

Figure 8 Fractionation of a crude *Forsythia intermedia* protein mixture catalyzing (+)-pinoresinol formation using Mono S HR5/5 (Pharmacia) column chromatography [see (*56*) for details]. Proteins were desorbed with Na_2SO_4 by implementing a series of step gradients as shown (—). Each fraction (peak) was able to catalyze formation of [9,9′-^{3}H]pinoresinol from [9-^{3}H]coniferyl alcohol in the absence of any exogenously added co-factor (-----). Numbers in parentheses on Figure refer to the optical purity of the pinoresinol obtained when that fraction was incubated with coniferyl alcohol [*e.g.* 50% = racemic, 100% = only (+)-enantiomer].

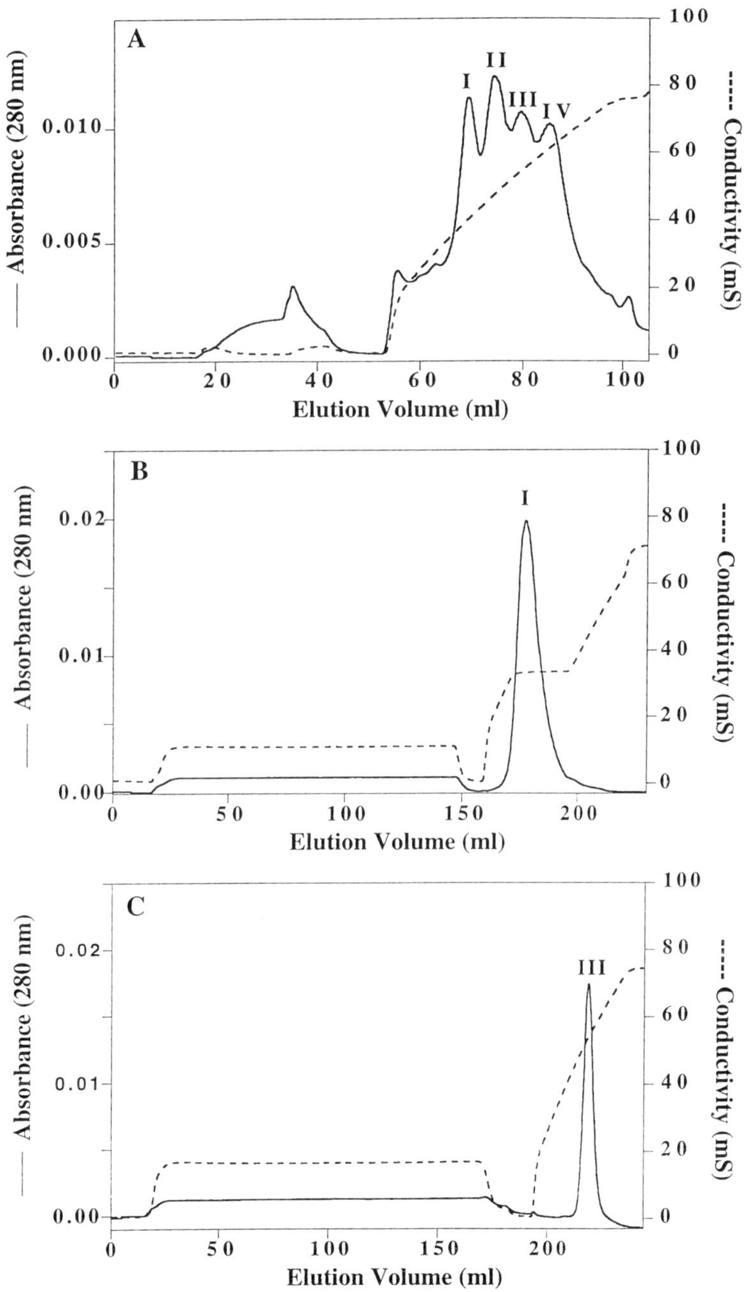

Figure 9. Fractionation of protein mixture catalyzing (+)-pinoresinol formation into components I-IV by perfusion chromatography [POROS SP-M column; see (*56*) for elution details]. **A,** separation of proteins into four overlapping fractions I to IV, **B,** purified fraction I, and **C,** purified fraction III.

added to the enzyme mixture, the coupling mode is redirected to afford essentially only (+)-pinoresinol (Figures 10 B and D). Note also that in both cases, addition of the 78 kDa dirigent protein has little, if any, effect on the rate of *E*-coniferyl alcohol depletion. This, nevertheless, represented the *first* example of stereoselective bimolecular phenoxy free-radical coupling *in vitro*, being manifested in a manner that could not have been predicted. Moreover, stereoselectivity was only observed with *E*-coniferyl alcohol as a substrate; it did not occur when *p*-coumaryl or sinapyl alcohols were used.

Given that *Forsythia* and *Linum* species catalyze distinct coupling modes [*i.e.* affording (+)- and (–)-pinoresinols, respectively], this further indicates that a new class of proteins exists, and also strongly suggests that other proteins engendering formation of other (distinct) skeletal types await discovery. Additionally, these findings also revealed important features as to how bimolecular phenoxy radical coupling reactions are actually controlled *in vivo*, and how non-specific auxiliary oxidases (*i.e.* laccase, peroxidase) can be conscripted for a precise function, albeit if only for providing general oxidative capacity.

The *Forsythia* 78 kDa dirigent protein was next shown by SDS-PAGE to possess a 25-27 kDa subunit, suggesting that the native protein perhaps existed as a trimer. N-terminal and internal amino acid analyses generated sufficient sequence data in order to synthesize degenerate primers, which in turn were used in a PCR-based approach to generate a ~400 bp probe. The latter was used to screen the corresponding *Forsythia* cDNA library and in this way, a full-length clone was obtained but whose resulting amino acid sequence only gave a MW ~18 kDa, with the discrepancy rationalized as being due to glycosylation (*148*). There are, however, several important salient features about the amino acid sequence other than its length: first, there was no homology to any other protein of known function; second there were no sites indicating a cofactor requirement (*e.g.* heme- or copper-binding site, *etc*.); third, its leader sequence revealed it as part of the secretory system, and fourth, it was a member of a multigene family.

Heterologous expression of the *Forsythia*-derived recombinant 'dirigent' protein was next achieved using a *Spodoptera*/baculovirus system, this being satisfyingly capable of engendering the conversion of two molecules of *E*-coniferyl alcohol into (+)-pinoresinol, provided a suitable oxidase or oxidant was added. Moreover, other putative cDNA's encoding presumed dirigent proteins have since been obtained from a variety of plant sources of interest, *i.e.* western red cedar (*Thuja plicata*), western hemlock (*Tsuga heterophylla*), *Eucommia ulmoides* and loblolly pine (*Pinus taeda*).

According to the kinetic data obtained to this point with the *Forsythia* dirigent protein (*56*), all evidence suggests that its role is to capture the free-radical species (released following one-electron oxidation) and to bind and orientate these in a manner whereby only stereoselective coupling can occur. It is now of considerable importance to establish the site(s) on the dirigent protein that involve substrate binding. Given that only an 18 kDa protein is involved (albeit apparently existing as a 78 kDa glycosylated trimer), it is necessary to determine whether each 18 kDa protein subunit binds one or two substrate molecules; delineating this mechanistic question is the subject of an ongoing investigation.

Thus these studies demonstrated that stereoselective coupling of plant phenols can and does occur, this having been established at the protein, recombinant protein and molecular level. Not only representing the first example of control of phenoxy bimolecular radical coupling [with a protein being encoded by a gene sequence that has no homology to any other of known function], it also demonstrated that phenoxy bimolecular radical coupling processes need not be left to chance. This, it is believed (see next section), has important ramifications for lignification.

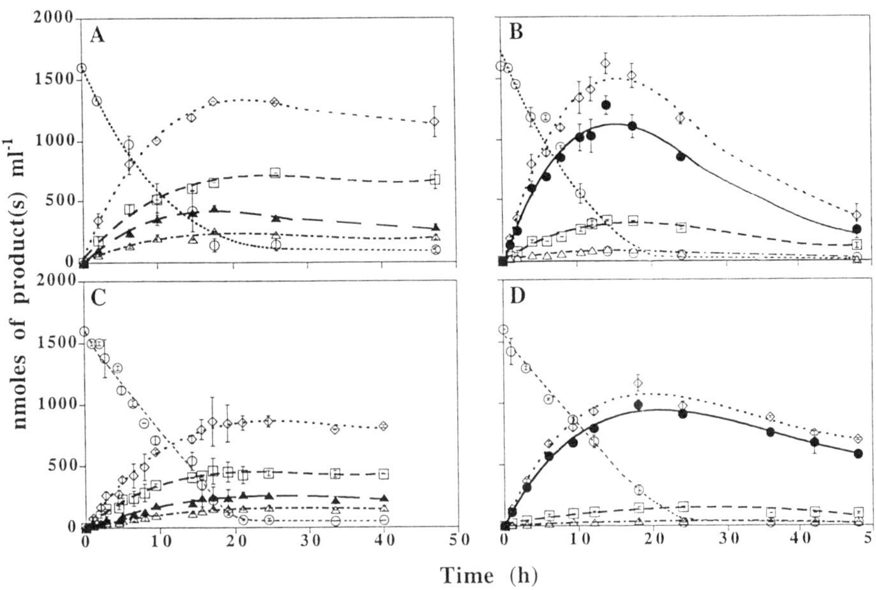

Figure 10. Time course of *E*-coniferyl alcohol depletion and formation of lignans from *E*-coniferyl alcohol when incubated in presence of: **A,** *F. intermedia* laccase (10.67 nmol l^{-1}), **B,** *F. intermedia* laccase (10.67 nmol l^{-1}) and 78 kDa protein (768 nmol l^{-1}), **C,** FMN (0.5 mmol l^{-1}) and **D,** FMN (0.5 mmol l^{-1}) and 78 kDa protein (768 nmol l^{-1}). Comparable results were also obtained with other oxidants, such as ammonium peroxydisulfate *i.e.* there is no specific requirement for FMN as an oxidant. ❍: Coniferyl alcohol (represented in phenylpropanoid dimer equivalents); ●: (+)-Pinoresinol; ▲: (±)-Pinoresinols; □: (±)-Dehydrodiconiferyl alcohols; Δ: (±)-*threo-erythro* guaiacylglycerol 8–*O*–4′-coniferyl alcohol ethers; ◊: Sum of all lignan products. [See Figures 1 and 2 for chemical structures; redrawn from ref. (*56*).]

Macromolecular Assemblies: Lignin Formation and Arrays of Dirigent Protein Sites

The previous sections summarized certain recent findings as regards bimolecular phenoxy radical coupling control *in vitro*, the involvement of dirigent proteins, and just a few of the limitations in attempting to duplicate lignin *in vivo* with artificial preparations. Consequently, it is now worthwhile reflecting upon some of the requirements and constraints for lignification *in vivo* as far as they can *currently* be gauged. Indeed, each of these factors *must* be fully considered when attempting to explain biochemical mechanisms leading to the lignin biopolymers.

From a historical perspective, perhaps two of the most intriguing aspects about lignification is the *apparent* lack of any optical activity in isolated lignin-derived biopolymers (and their corresponding fragments), as well as the inherent variability of lignin monomer composition within cell walls of different tissues, organs, cells and even calli. Indeed, this lack of optical activity and lignin variability seemed to provide convincing proof to the original contention that all lignins must be randomly assembled. Yet, while lignins are undoubtedly variable, it is often overlooked that this variability is both specific and predictable. For example, lignin present in softwood 'normal' secondary xylem tracheary elements is *predictably* different from that present in the primary xylem, phloem, needles and periderm cells, as well as those of the fibers and vessels of hardwoods—in terms of both the relative amounts and monolignol composition. Although this variability of lignins, within and between different species, is often viewed as a confounding property, it should be considered *instead* as an important clue. That is, why do particular cell types have predictable patterns of lignin structure which differ from those of others? Put in another way, perhaps a shortcoming of the random coupling hypothesis is that it essentially ignored this single most important facet of lignification, namely that of the predictability of lignin composition and amounts which are a function of *specific* cell types.

This is nicely illustrated with the example of lignins present in 'normal' and 'compression' wood xylem cells in conifers (softwood), respectively. 'Normal' lignified secondary wall xylem is formed when the tree stem is growing upright parallel to its gravitational vector. On the other hand, if the stem is displaced from its vertical axis, certain cells—originally destined for 'normal' secondary xylem development—are now *conscripted* to the formation of compression wood cells. The physiological effect of this 'plasticity' in cell wall formation is to strategically redeploy certain cells, for a reinforcing capacity, in order to attempt to buttress and realign the stem. As shown in Figure 6, these conscripted cells now differ dramatically from those of 'normal' xylem in many ways. In loblolly pine (*Pinus taeda*), for example, the effects of this *cellular reprogramming* results in the compression wood tracheary elements being rounder and shorter (10-40% less) than their 'normal' xylem counterparts, with thicker cell walls and almost double the specific gravity (*149*). Additionally, the S_2 layer becomes deeply fissured, the S_3 layer is absent, with the cellulose having a lower degree of polymerization and being less crystalline. That is, the cell wall architecture is substantially changed in biopolymer content, composition, molecular size and cellulose microfibril orientation. While the mode of lignin deposition appears to follow fairly well that of 'normal' secondary xylem development, its *p*-coumaryl alcohol content is increased quite substantially, from about 5 to 30%; comparable observations have been made in the study of *Cryptomeria japonica* compression wood formation.

Herein lies the most important key to lignification: both lignin monomer composition and lignin amount can differ dramatically during compression wood formation, relative to that in 'normal' secondary xylem formation, even though they were derived from identical cells to begin with. In short, 'normal' and 'compression' wood cell formation reveal a *pre-programmed* plasticity in cell wall development, which varies predictably depending upon the cue(s) for formation of a particular cell (wall) type.

Indeed, although seldom stated as such, numerous studies have clearly demonstrated that for the cell wall assembly of a *specific cell type*, both the pattern of lignification (monomer amount and composition) and the deposition of specific structural carbohydrate biopolymers and other proteinaceous substances are rigorously pre-programmed. Given this fact, how then can an individual cell, when conscripted for lignin formation, so precisely and reproducibly determine both the monomeric content and/or degree of lignification in a region ostensibly beyond control of the living part of the cell—namely across the plasma membrane and in the depths of the cell wall/middle lamella. In this regard, the random coupling hypothesis would ascribe no particular advantage to such a control of lignification, other than simply as a mechanism for filling up voids within the plant cell walls. If this were correct, monomer pattern deposition would solely be a function of chance, in terms of both formation and/or transport of monolignols into the lignifying cell wall.

This is not the case. Instead, the formation of each monomer type and their differential transport and deposition within the maturing cell wall, including the polymerization event itself, is a function of the particular cell type being formed. Any theory of lignification *in vivo* must, therefore, take this fact into account, including provision of an explanation *as to how lignification remains under control of the cell itself*.

Bearing these defining factors in mind, a working hypothesis is proposed below for control of lignin monomer content and assembly, based on the discovery of dirigent proteins (*56*) and the Sarkanen template theory for lignin polymerization (*150*). This new hypothesis appears to be able to provide a rational model which meets all known requirements for lignification *in vivo*. Prior to discussing it in detail, however, a brief summary of the previous model for cell wall lignification is first provided.

In earlier paradigms of lignin biosynthesis, it was proposed that monolignols were first converted into their phenolic glucosides (*e.g. E*-coniferyl alcohol → coniferin) (*1-3*). These were then envisaged to be transported *via* vesicles to the plasma membrane and released into the cell wall. Action of cell wall bound β-glucosidase(s) was implicated to regenerate the monolignols, with the latter subsequently undergoing polymerization when in contact with the appropriate (per)oxidase(s).

Although this model was readily adopted, curiously none of its shortcomings were either addressed and/or reconciled. First, it would involve an equimolar amount of glucose being released in the cell wall, relative to that of *E*-coniferyl alcohol and, hence, lignin. But no evidence for glucose accumulation has ever been shown. Second, in studies of metabolism from phenylalanine to the monolignols and formation of 'lignin' in *Pinus taeda* cell suspension cultures, no accumulation of any coniferin, or other phenolic glucosides, *within* the cells was noted; yet all other metabolites in the pathway from phenylalanine were observed (A.M. Anterola, unpublished results, Washington State University, 1997). Third, earlier work by Stefan Marcinowski and Hans Grisebach suggested that the monolignol glucoside (coniferin) was, at best, responsible only for a part of the lignin formed, based on turnover experiments (*151*). Additionally, in lignifying poplar plants (Brian Ellis, University of British Columbia, 1997, personal communication), there was no evidence for any β-glucosidase activity in its lignifying tissues, an aspect which is apparently true for many angiosperms. The provisional assumption can, therefore, be made that some other mechanism is in place for monolignol transport into the cell walls, which does not primarily involve β-glucosidase catalyzed hydrolysis of monolignol glucosides.

Regardless of the mode of monomer transport, a most important observation is the appearance of uniformly growing lignin domains at sites far removed from the plasma membrane in the maturing cell wall. In this context, L. Donaldson (*115*) reported that initiation of lignin deposition and biosynthesis occurs near or adjacent to the original cell plate, *i.e.* in the region where the middle lamella between cells ultimately develops. Additional sites of lignin biosynthesis are also observed in the primary wall. Although this evidence was based only upon potassium permanganate ($KMnO_4$) staining of lignifying cell walls of radiata pine (*Pinus radiata*), it was also observed that individual lignin lamellae (4-5 nm wide) developed at these initiation sites before they began to coalesce. While aware of the potential pitfalls of correlating lignin synthesis with histochemical staining [see (*152*)], there is, however, corroborative autoradiographic evidence using radiolabeled lignin precursors which further supports these assertions (*42-45*). It would, therefore, appear that the key to defining how lignin biosynthetic assembly is actually controlled *lies in the vicinity* of these initiation sites.

How then is precise control of the heterogeneous nature of the lignin biopolymeric assembly *absolutely* specified for a particular cell wall type, whether the lignin is in either 'normal' and 'compression' wood in softwoods, or fiber and vessel walls in hardwoods? Clearly, the formation of a specific pre-programmed cell wall developmental event first involves the precise deposition of various carbohydrate polymeric and proteinaceous substances. Once formed, these are exquisitely targeted to specific regions (sites) in the developing cell walls, and in so doing the cell establishes its overall wall architecture. Moreover, even if lignification is inhibited, as shown by the use of the phenylalanine ammonia-lyase inhibitor, AOPP, during *Zinnia* tracheary element formation, the unlignified cell walls still display their familiar architecture, as established by deposition of the carbohydrate constituents (cellulose and hemicelluloses), as well as the other proteinaceous components (*153*).

When carbohydrate deposition is completed, however, the subsequent deposition of lignins is reported to be temporally and spatially correlated with secretion from the Golgi apparatus of specific proteins. These, in turn, are thought to be proline-rich and are targeted to precise regions in the maturing developing cell wall (*46, 116, 154*). It can be provisionally concluded that these proteins represent *at least one* important component of the *pre-determined* or *pre-programmed* stages of lignin assembly *in vivo*.

As discussed earlier, a new insight into phenolic coupling *in vivo* came in 1996 when it was recognized that the dirigent protein in *Forsythia* sp. directed only the coupling of two coniferyl-alcohol derived moieties (*i.e.* as their presumed free-radical counterparts), but not that of either *p*-coumaryl or sinapyl alcohols (*56*). This revealed that considerable substrate specificity could be built into the coupling process, in this instance where only (+)-pinoresinol formation was being controlled. This led us to consider that a comparable process could also be in place for targeting and polymerizing specific monomers in the lignifying cell wall, in a precise, predictable and reproducible manner. As described in the previous section, the gene for the dirigent protein encodes for an ~18 kDa subunit, although it is currently unknown whether each subunit binds (the free-radical form) of either a single monomeric species or two monomers. Assuming the former case, this would imply that the gene encoding this basic subunit *encodes a single substrate binding site for one coniferyl-alcohol derived monomer*. If arrays of substrate binding sites were laid down, or juxtapositioned, together, this could result in formation of a primary lignin chain (*i.e.* the first strand of the lignin biopolymer).

Another insight to emerge was the concept of template polymerization for lignification and, thus, how chain replication might occur *in vivo*. This concept resulted from interpretation of findings concerning the *in vitro* polymerization of coniferyl alcohol in the presence of a lignin polymer viewed as still containing softwood 'native' lignin characteristics. In the presence of H_2O_2 and peroxidase (*150*), it was concluded that the 'native' lignin functioned as a template for monolignol polymerization and lignin assembly, *i.e.* self-replication. This study, however, yielded no new information on the nature of the interunit linkages within the artificial lignin formed *de novo*. Nevertheless, it was proposed that lignin assembly occurred *via* an end-wise polymerization—where the existing lignin macromolecule functioned as a template. (As the reader is undoubtedly aware, template polymerization, giving rise to replication of various polymeric chains, is very well-known in many aspects of biochemistry and chemistry.) Accordingly, such a mechanism could explain observations made *in situ* as to why lignin domains were 'growing' in specific areas in the cell wall (at lignin initiation sites), prior to ultimately coalescing with one another.

While this template theory might explain how and why lignin monomers can align themselves with the pre-existing lignin matrix prior to *de novo* lignin synthesis, it gave no insight whatsoever into the constitution of the *initial* proteinaceous template for lignification. One possible explanation is that specific (dirigent or dirigent-like) proteins are targeted to precise (lignin initiation) sites in the developing cell wall. It can be anticipated that the resulting arrays contain distinct subunits (of dirigent proteins) which *specifically* bind different monolignols and encode different bonding types, *i.e.* that lignification is under the control of arrays of distinct dirigent protein sites specifying formation of (particular) lignin chains (S. Sarkanen, discussion with N. G. Lewis).

If correct, this would provide an appealing mechanism for the control of all known aspects of the lignification process *in vivo, i.e.* where clearly *different* substrate binding motifs, in distinctive orientations, could encode information establishing both the monomer content, and interunit linkages required for a particular lignin chain assembly for a specific cell type. Further, once this information is processed (*i.e.* giving the primary chain), lignification would then continue by a template (self-replication) polymerization process. Indeed, such a process could also explain the perceived lack of optical activity of lignins in a number of different ways. Two examples will suffice: it could be that the primary chains, for example, form complementary 'mirror images' *via* template replication, or that there are two distinct types of proteins each encoding formation of complementary chains that effectively cancel out any measurable optical activity.

In summary, this new hypothesis provides the best explanation thus far as to how monolignol polymerization and lignin assembly is orchestrated, in both a predictable and specific manner, while still remaining under control of the cell—even in its last act prior to death. Resolving whether this occurs as such, together with determination of each and every biochemical factor specifying/controlling lignin assembly *in vivo*, will be the subject of further intense investigation. There is, however, sufficient proof that the process of lignification is not a series of random biochemical events *in vivo*. The challenge to science now resides in providing a rational explanation for lignin formation in a *convincing biochemical manner*.

Acknowledgments

The authors thank the United States Department of Energy (DE-FG03-97ER20259), the United States Department of Agriculture (9603622), the National Science Foundation (MCB09631980), the National Aeronautics and Space Administration (NAG100164), McIntire-Stennis and the Lewis B. and Dorothy Cullman and G. Thomas Hargrove Center for Land Plant Adaptation Studies for generous support of this study. Thanks are also extended to Vincent Burlat for helpful discussions.

Literature Cited

1. Freudenberg, K. *Nature* **1959**, *183*, 1152-1155.
2. Freudenberg, K. *Bull. Soc. Chim. France* **1959**, 1748-1753.
3. Freudenberg, K. In *Constitution and Biosynthesis of Lignin*; Freudenberg, K., Neish, A. C., Eds.; Springer-Verlag: New York, NY, 1968, pp 47-122.
4. Freudenberg, K.; Richtzenhain, H. *Chem. Ber.* **1943**, *76*, 997-1006.
5. Freudenberg, K.; Schlüter, H. *Chem. Ber.* **1955**, *88*, 617-625.
6. Freudenberg, K.; Harkin, J. M.; Reichert, M.; Fukuzumi, T. *Chem. Ber.* **1958**, *91*, 581-590.
7. Freudenberg, K.; Lehmann, B. *Chem. Ber.* **1960**, *93*, 1354-1366.
8. Freudenberg, K. *Croatica Chem. Acta* **1957**, *29*, 189-194.

9. Björkman, A. *Nature* **1954**, *174*, 1057-1058.
10. Brauns, F. E. *J. Amer. Chem. Soc.* **1939**, *61*, 2120-2127.
11. Erdtman, H. *Ind. Eng. Chem.* **1957**, *49*, 1386-1387.
12. Hergert, H. L. In *Lignins—Occurrence, Formation, Structure and Reactions*; Sarkanen, K. V., Ludwig, C. H., Eds.; Wiley-Interscience: New York, NY, 1971, pp 267-297.
13. Freudenberg, K.; Siebert, W.; Heimberger, W.; Kraft, R. *Chem. Ber.* **1950**, *83*, 533-539.
14. Brunow, G.; Ede, R. M.; Simola, L. K.; Lemmetyinen, J. *Phytochemistry* **1990**, *29*, 2535-2538.
15. Brunow, G.; Kilpeläinen, I.; Lapierre, C.; Lundquist, K.; Simola, L. K.; Lemmetyinen, J. *Phytochemistry* **1993**, *32*, 845-850.
16. Adler, E. *Wood Sci. Technol.* **1977**, *11*, 169-218.
17. Freudenberg, K. *Adv. Chem. Ser.* **1966**, *59*, 1-21.
18. Adler, E. *Ind. Eng. Chem.* **1957**, *49*, 1377-1383.
19. Freudenberg, K. In *Fortschritte d. Chemie Org. Naturstoffe*; Zechmeister, L., Ed.; Springer Verlag: Vienna, 1954, pp 43-82.
20. Freudenberg, K. *Angew. Chem.* **1956**, *68*, 508-512.
21. Freudenberg, K. *Angew. Chem.* **1956**, *68*, 84-92.
22. Sarkanen, K. V. In *Lignins—Occurrence, Formation, Structure and Reactions*; Sarkanen, K. V., Ludwig, C. H., Eds.; Wiley Interscience: New York, NY, 1971, pp 95-163.
23. Freudenberg, K.; Renner, K. C. *Chem. Ber.* **1965**, *98*, 1879-1892.
24. Freudenberg, K.; Rasenack, D. *Chem. Ber.* **1953**, *86*, 755-758.
25. Higuchi, T. *Physiol. Plant.* **1957**, *10*, 356-372.
26. Nose, M.; Bernards, M. A.; Furlan, M.; Zajicek, J.; Eberhardt, T. L.; Lewis, N. G. *Phytochemistry* **1995**, *39*, 71-79.
27. Barcelo, A. R. *Protoplasma* **1995**, *186*, 41-44.
28. Higuchi, T. *J. Biochem.* **1958**, *45*, 515-528.
29. Driouich, A.; Lainé, A.-C.; Vian, B.; Faye, L. *Plant J.* **1992**, *2*, 13-24.
30. Sterjiades, R.; Dean, J. F. D.; Eriksson, K.-E. L. *Plant Physiol.* **1992**, *99*, 1162-1168.
31. Bao, W.; O'Malley, D. M.; Whetten, R.; Sederoff, R. R. *Science* **1993**, *260*, 672-674.
32. Sterjiades, R.; Dean, J. F. D.; Gamble, G.; Himmelsbach, D. S.; Eriksson, K.-E. L. *Planta* **1993**, *190*, 75-87.
33. Savidge, R.; Udagama-Randeniya, P. *Phytochemistry* **1992**, *31*, 2959-2966.
34. Udagama-Randeniya, P.; Savidge, R. *Electrophoresis* **1994**, *15*, 1072-1077.
35. McDougall, G. J.; Stewart, D.; Morrison, I. M. *Planta* **1994**, *194*, 9-14.
36. Mason, H. S.; Cronyn, M. *J. Amer. Chem. Soc.* **1955**, *77*, 491.
37. Koblitz, H.; Koblitz, D. *Nature* **1964**, *204*, 199-200.
38. Fergus, B. J.; Goring, D. A. I. *Holzforschung* **1970**, *24*, 113-117.
39. Fergus, B. J.; Goring, D. A. I. *Holzforschung* **1970**, *24*, 118-124.
40. Whiting, P.; Goring, D. A. I. *Wood Sci. Technol.* **1982**, *16*, 261-267.
41. Saka, S.; Goring, D. A. I. In *Biosynthesis and Biodegradation of Wood Components*; Higuchi, T., Ed.; Academic Press: Orlando, Fl, 1985, pp 51-62.
42. Terashima, N. *Holzforschung* **1986**, *40 (Suppl.)*, 101-105.
43. Terashima, N.; Fukushima, K. *Wood Sci. Technol.* **1988**, *22*, 259-270.
44. Terashima, N. In *Plant Cell Wall Polymers—Biogenesis and Biodegradation*; Lewis, N. G., Paice, M. G., Eds.; ACS Symposium Series: Washington, DC, 1989, Vol. 399, pp 148-159.
45. Terashima, N.; Fukushima, K. In *Plant Cell Wall Polymers—Biogenesis and Biodegradation*; Lewis, N. G., Paice, M. G., Eds.; ACS Symposium Series: Washington, DC, 1989, Vol. 399, pp 160-168.
46. Müsel, G.; Schindler, T.; Bergfeld, R.; Ruel, K.; Jacquet, G.; Lapierre, C.; Speth, V.; Schopfer, P. *Planta* **1997**, *201*, 146-159.

47. Burlat, V.; Ambert, K.; Ruel, K.; Joseleau, J. P. *Plant Physiol. Biochem.* **1997**, *35*, 645-654.
48. Hergert, H. L. *J. Org. Chem.* **1960**, *25*, 405-413.
49. Hergert, H. L. In *Cellulose Chemistry and Technology*; Arthur, J. C., Jr., Ed.; ACS Symposium Series; American Chemical Society: Washington, DC, 1977, Vol. 48, pp 227-243.
50. Chattaway, M. M. *Aust. For.* **1952**, *16*, 25-34.
51. Ralph, J.; MacKay, J. J.; Hatfield, R. D.; O'Malley, D. M.; Whetten, R. W.; Sederoff, R. R. *Science* **1997**, *277*, 235-239.
52. Lai, Y. Z.; Sarkanen, K. V. In *Lignins—Occurrence, Formation, Structure and Reactions*; Sarkanen, K. V., Ludwig, C. H., Eds.; Wiley Interscience: New York, NY, 1971, pp 165-240.
53. Davin, L. B.; Bedgar, D. L.; Katayama, T.; Lewis, N. G. *Phytochemistry* **1992**, *31*, 3869-3874.
54. Chu, A.; Dinkova, A.; Davin, L. B.; Bedgar, D. L.; Lewis, N. G. *J. Biol. Chem.* **1993**, *268*, 27026-27033.
55. Dinkova-Kostova, A. T.; Gang, D. R.; Davin, L. B.; Bedgar, D. L.; Chu, A.; Lewis, N. G. *J. Biol. Chem.* **1996**, *271*, 29473-29482.
56. Davin, L. B.; Wang, H.-B.; Crowell, A. L.; Bedgar, D. L.; Martin, D. M.; Sarkanen, S.; Lewis, N. G. *Science* **1997**, *275*, 362-366.
57. Zhuang, L.-g.; Seligmann, O.; Jurcic, K.; Wagner, H. *Planta Medica* **1982**, *45*, 172-176.
58. Vaquette, J.; Cavé, A.; Waterman, P. G. *Planta Medica* **1979**, *35*, 42-47.
59. Kitagawa, S.; Nishibe, S.; Benecke, R.; Thieme, H. *Chem. Pharm. Bull.* **1988**, *36*, 3667-3670.
60. Fang, J.-M.; Hsu, K.-C.; Cheng, Y.-S. *Phytochemistry* **1989**, *28*, 3553-3555.
61. Ayres, D. C.; Loike, J. D. *Chemistry and Pharmacology of Natural Products. Lignans: Chemical, Biological and Clinical Properties*; Cambridge University Press: Cambridge, England, 1990, pp 402.
62. Gottlieb, O. R.; Yoshida, M. In *Natural Products of Woody Plants—Chemicals Extraneous to the Lignocellulosic Cell Wall*; Rowe, J. W., Kirk, C. H., Eds.; Springer Verlag: Berlin, 1989, pp 439-511.
63. MacLean, H.; Gardner, J. A. F. *For. Prod. J.* **1956**, *6*, 510-516.
64. Monties, B. In *Methods in Plant Biochemistry*; Harborne, J. B., Ed.; Academic Press: London, 1989, Vol. 1, pp 113-157.
65. De Candolle, A. P. *Theorie Elémentaire de la Botanique*; Deterville: Paris, 1819, pp 454.
66. De Candolle, A. P. In *Physiologie Végétale*; Bechet, Ed.; Deterville: Paris, 1832, Vol. 1, pp 194.
67. Payen, A. *C. R. Acad. Sci., Paris* **1838**, *7*, 1052-1056.
68. Payen, A. *C. R. Acad. Sci., Paris* **1838**, *7*, 1125.
69. Tiemann, F.; Haarmann, W. *Ber. deutsch. chem. Ges.* **1874**, *7*, 608-623.
70. Tiemann, F. *Ber. deutsch. chem. Ges.* **1875**, *8*, 1127-1136.
71. Klason, P. *Cellulosechem.* **1923**, *4*, 81.
72. Klason, P. *Bericht über die Hauptversammlung des Vereins der Zellstoff- und Papier Chemiker* **1908**, 52-53.
73. Klason, P. *Arkiv. Kemi., Mineral. Geol.* **1908**, *3*, 17.
74. Kürschner, K. In *Zur Chemie der Ligninkörper*; F. Enke: Stuttgart, 1925, pp 77.
75. Kürschner, K.; Schramek, W. *Tech. Chem. Papier-Zellstoff-Fabr* **1932**, *29*, 35.
76. Cross, C. F.; Bevan, E. J.; Beadle, C. In *Cellulose*; Longmans, Green and Co.: London, 1916, pp 177-181.
77. König, J.; Rump, E. *Chemie u. Structur der Pflanzenmembran.*; 1914.
78. Fuchs, W. *Biochem. Z.* **1927**, *180*, 30-34.
79. Fuchs, W. *Biochem. Z.* **1928**, *192*, 165-166.
80. Klason, P. *Arkiv. Kemi., Mineral. Geol.* **1917**, *6*, 21.
81. Rassow, B.; Zschenderlein, A. *Z. angew. Chem.* **1921**, *34*, 204.
82. Ehrlich, E. *Cellulosechem.* **1931**, *11*, 161.

83. Phillips, M. *Chem. Rev.* **1934**, *14*, 103-170.
84. Freudenberg, K.; Zocher, H.; Dürr, W. *Chem. Ber.* **1929**, *62*, 1814-1823.
85. Freudenberg, K. *J. Chem. Educ.* **1932**, *9*, 1171-1180.
86. Erdtman, H. *Liebigs Ann. d. Chem.* **1933**, *503*, 283-294.
87. Erdtman, H. *Biochem. Z.* **1933**, *258*, 172-180.
88. Cousin, H.; Hérissey, H. *Bull. Soc. Chim.* **1908**, 1070-1075.
89. Gomberg, M. *J. Am. Chem. Soc.* **1900**, *22*, 757-771.
90. Freudenberg, K.; Niedercorn, F. *Chem. Ber.* **1958**, *91*, 591-597.
91. Kratzl, K.; Hofbauer, G. *Monatsh. Chem.* **1956**, *87*, 617 (and references therein).
92. Kratzl, K.; Billek, G.; Klein, E.; Buchtela, K. *Monatsh. Chem.* **1957**, *88*, 721-734.
93. Kratzl, K.; Hofbauer, G. *Monatsh. Chem.* **1958**, *89*, 96-101 (and references therein).
94. Kratzl, K.; Faigle, H. *Monatsh. Chem.* **1958**, *89*, 708-715.
95. Brown, S. A.; Neish, A. C. *Can. J. Biochem. Physiol.* **1955**, *33*, 948-962.
96. Brown, S. A.; Neish, A. C. *Nature* **1955**, *175*, 488-489.
97. Brown, S. A.; Neish, A. C. *Can. J. Biochem. Physiol.* **1956**, *34*, 769-778.
98. Brown, S. A.; Neish, A. C. *J. Am. Chem. Soc.* **1959**, *81*, 2419-2424.
99. Neish, A. C. *Annu. Rev. Plant Physiol.* **1960**, *11*, 55-80.
100. Haworth, R. D. *Annu. Rept. Prog. Chem.* **1937**, *33*, 266-279.
101. Erdtman, H. *Research* **1950**, *3*, 63-67.
102. Lewis, N. G.; Newman, J.; Rej, R. N.; Just, G.; Ripmeister, J. *Macromolecules* **1987**, *20*, 1752-1756.
103. Lewis, N. G.; Razal, R. A.; Dhara, K. P.; Yamamoto, E.; Bokelman, G. H.; Wooten, J. B. *J. Chem. Soc., Chem. Commun.* **1988**, 1626-1628.
104. Lewis, N. G.; Razal, R. A.; Yamamoto, E.; Bokelman, G. H.; Wooten, J. B. In *Plant Cell Wall Polymers—Biogenesis and Biodegradation*; Lewis, N. G., Paice, M. G., Eds.; ACS Symposium Series: Washington, DC, 1989, Vol. 399, pp 169-181.
105. Eberhardt, T. L.; Bernards, M. A.; He, L.; Davin, L. B.; Wooten, J. B.; Lewis, N. G. *J. Biol. Chem.* **1993**, *268*, 21088-21096.
106. Karkunen, P.; Rummakko, P.; Sipilä, J.; Brunow, G. *Tetrahedron Lett.* **1995**, *36*, 4501-4504.
107. Karkunen, P.; Rummakko, P.; Pajunen, A.; Brunow, G. *J. Chem. Soc., Perkin Trans. I* **1996**, 2303-2308.
108. Harada, H.; Côté, W. A. J. In *Biosynthesis and Biodegradation of Wood Components*; Higuchi, T., Ed.; Academic Press: Orlando, Fl, 1985, pp 1-42 (and references therein).
109. Fry, S. C.; Miller, J. G. In *Plant Cell Wall Polymers—Biogenesis and Biodegradation*; Lewis, N. G., Paice, M. G., Eds.; ACS Symposium Series: Washington, DC, 1989, Vol. 399, pp 33-46.
110. Fengel, D.; Wegener, G. *Wood. Chemistry, Ultrastructure, Reactions*; Walter de Gruyter: Berlin, 1984, pp 613.
111. Wardrop, A. B. *TAPPI* **1957**, *40*, 225.
112. Wardrop, A. B. In *Lignins—Occurrence, Formation, Structure and Reactions*; Sarkanen, K. V., Ludwig, C. H., Eds.; Wiley Interscience: New York, NY, 1971, pp 19-41.
113. Wardrop, A. B. *Appl. Poly. Symp.* **1976**, *28*, 1041.
114. Takabe, K.; Fukazawa, K.; Harada, H. In *Plant Cell Wall Polymers—Biogenesis and Biodegradation*; Lewis, N. G., Paice, M. G., Eds.; ACS Symposium Series: Washington, DC, 1989, Vol. 399, pp 47-66.
115. Donaldson, L. A. *Wood Sci. Technol.* **1994**, *28*, 111-118.
116. Ryser, U.; Schorderet, M.; Zhao, G.-F.; Studer, D.; Ruel, K.; Hauf, G.; Keller, B. *Plant J.* **1997**, *12*, 97-111.
117. Gottlieb, O. R. *Phytochemistry* **1972**, *11*, 1537-1570.
118. Gottlieb, O. R. *Rev. Latinoamer. Quim.* **1974**, *5*, 1-10.

119. Gottlieb, O. R. *Progr. Chem. Org. Nat. Prod.* **1978**, *35*, 1-72.
120. Katayama, T.; Davin, L. B.; Lewis, N. G. *Phytochemistry* **1992**, *31*, 3875-3881.
121. Katayama, T.; Davin, L. B.; Chu, A.; Lewis, N. G. *Phytochemistry* **1993**, *32*, 581-591.
122. Paré, P. W.; Wang, H.-B.; Davin, L. B.; Lewis, N. G. *Tetrahedron Lett.* **1994**, *35*, 4731-4734.
123. Davin, L. B.; Lewis, N. G. In *Recent Advances in Phytochemistry*; Stafford, H. A., Ibrahim, R. K., Eds.; Plenum Press: New York, NY. (and references therein), 1992, Vol. 26, pp 325-375.
124. Koukol, J.; Conn, E. E. *J. Biol. Chem.* **1961**, *236*, 2692-2698.
125. Russell, D. W.; Conn, E. E. *Arch. Biochem. Biophys.* **1967**, *122*, 256-258.
126. Russell, D. W. *J. Biol. Chem.* **1971**, *246*, 3870-3878.
127. Grand, C. *FEBS Lett.* **1984**, *169*, 7-11.
128. Mansell, R. L.; Stöckigt, J.; Zenk, M. H. *Z. Pflanzenphysiol.* **1972**, *68S*, 286-288.
129. Zenk, M. H.; Gross, G. G. In *Recent Advances in Phytochemistry*; Runeckless, V. C., Watkin, J. E., Eds.; Appleton-Century-Crofts: New York, NY, 1972, Vol. 4, pp 87-106.
130. Gross, G. G.; Stöckigt, J.; Mansell, R. L.; Zenk, M. H. *FEBS Lett.* **1973**, *31*, 283-286.
131. Byerrum, R. U.; Flokstra, J. H. *Federation Proceedings* **1952**, *11*.
132. Byerrum, R. U.; Flokstra, J. H.; Dewey, L. J.; Ball, C. D. *J. Biol. Chem.* **1954**, *210*, 633-643.
133. Finkle, B. J.; Nelson, R. F. *Biochim. Biophys. Acta* **1963**, *78*, 747-749.
134. Finkle, B. J.; Masri, M. S. *Biochim. Biophys. Acta* **1964**, *85*, 167-169.
135. Teutsch, H.; Hasenfratz, M. P.; Lesot, A.; Stoltz, C.; Garnier, J.-M.; Jeltsch, J.-M.; Durst, F.; Werck-Reichhart, D. *Proc. Natl. Acad. Sci. USA* **1993**, *90*, 4102-4106.
136. Urban, P.; Werk-Reichhart, D.; Teutsch, G. H.; Durst, F.; Regnier, S.; Kazmaier, M.; Pompon, D. *Eur. J. Biochem.* **1994**, *222*, 843-850.
137. Meyer, K.; Cusumano, J. C.; Somerville, C.; Chapple, C. C. S. *Proc. Natl. Acad. Sci. USA* **1996**, *93*, 6869-6874.
138. Bugos, R. C.; Chiang, V. L. C.; Campbell, W. H. *Plant Mol. Biol.* **1991**, *17*, 1203-1215.
139. Ye, Z.-H.; Kneusel, R. E.; Matern, U.; Varner, J. E. *Plant Cell* **1994**, *6*, 1427-1439.
140. Ye, Z.-H.; Varner, J. E. *Plant Physiol.* **1995**, *108*, 459-467.
141. Meng, H.; Campbell, W. H. *Arch. Biochem. Biophys.* **1996**, *330*, 329-341.
142. Fell, D. A. *Biochem. J.* **1992**, *286*, 313-330.
143. ap Rees, T.; Hill, S. A. *Plant Cell Environ.* **1994**, *17*, 587-599.
144. Brand, M. D. *J. theor. Biol.* **1996**, *182*, 351-360.
145. Hawkins, S. W.; Boudet, A. M. *Plant Physiol.* **1994**, *104*, 75-84.
146. Lewis, N. G.; Davin, L. B. In *Isopentenoids and Other Natural Products: Evolution and Function*; Nes, W. D., Ed.; ACS Symposium Series: Washington, DC, 1994, Vol. 562, pp 202-246.
147. Davin, L. B.; Lewis, N. G. *An. Acad. bras. Ci.* **1995**, *67 (S3)*, 363-378.
148. Davin, L. B.; Wang, H.-B.; Dinkova-Kostova, A. T.; Gang, D. R.; Fujita, M.; Sarkanen, S.; Lewis, N. G. **1997** U.S. Patent Application Serial No. 60/033,298.
149. Timell, T. E. *Compression Wood in Gymnosperms*; Springer-Verlag: Berlin, 1986, pp 1338.
150. Guan, S.-y.; Mlynár, J.; Sarkanen, S. *Phytochemistry* **1997**, *45*, 911-918.
151. Marcinowski, S.; Grisebach, H. **1977**, *16*, 1665-1667.
152. Lewis, N. G.; Yamamoto, E. *Annu. Rev. Plant Physiol. Plant Mol. Biol.* **1990**, *41*, 455-496.
153. Ingold, E.; Sugiyama, M.; Komamine, A. *Physiol. Plant.* **1990**, *78*, 67-74.
154. Marcus, A.; Greenberg, J.; Averyhart-Fullard, V. *Physiol. Plant.* **1991**, *81*, 273-279.

Chapter 23

Stereochemistry in Lignan Biosynthesis and Neolignan Preparation Through Fungal Reduction

T. Katayama

Faculty of Agriculture, Kagawa University, Miki-cho, Kagawa 761-07, Japan

Cell-free extracts of *Zanthoxylum ailanthoides* have catalyzed the enantiospecific reduction of (+)-pinoresinol in the presence of NAD(P)H to afford (+)-lariciresinol, and intact cells of *Fusarium solani* M-13-1 have effected stereoselective carbonyl reduction in (±)-$\Delta^{8'}$-4-hydroxy-3,3′-dimethoxy-7-oxo-8-*O*-4′-neolignan to give $\Delta^{8'}$-4,7-dihydroxy-3,3′-dimethoxy-8-*O*-4′-neolignan in the presence of α-oxo-guaiacylglycerol-β-vanillyl ether. The mechanism of stereoselective and enantiospecific lignan and neolignan formation are interesting subjects of inquiry. However, except for some (8*R*,8′*R*)-lignans in *Forsythia* plants, little is known about the biosynthetic pathways involved. Interestingly (−)-pinoresinol and (−)-secoisolariciresinol isolated from *Z. ailanthoides* have opposite absolute configurations, (8*S*,8′*S*) and (8*R*,8′*R*), respectively. The first objective, therefore, was to clarify the biosynthetic route to the lignans in *Z. ailanthoides*. The absolute configuration of the (+)-lariciresinol, (8*R*,8′*R*), was identical with that of natural (−)-secoisolariciresinol. Following administration of [8-^{14}C]coniferyl alcohol and (±)-[8,8′-^{14}C]pinoresinols to an excised shoot of *Z. ailanthoides*, the preferential formation of (−)-[^{14}C]pinoresinol and (+)-[^{14}C]lariciresinol, respectively, was observed, but their enantiomeric purities were low. On the other hand, the [^{14}C]secoisolariciresinol isolated following [^{14}C]pinoresinol administration was almost racemic. It was suggested that the pathways to both (8*R*,8′*R*) and (8*S*,8′*S*)-lignans are present in *Z. ailanthoides*. Lastly, production of bioactive neolignans by stereochemically controlled bio-transformations is important, and enantiomerically pure neolignans are essential for studying their biosynthesis. The fungal reduction product of (±)-$\Delta^{8'}$-4-hydroxyl-3,3′-dimethoxy-7-oxo-8-*O*-4′-neolignan was a mixture of *erythro* and *threo* forms in the ratio 3:2, which would be derived from both enantiomers of the racemic substrate. Both diastereomers were enantiomerically pure. These observations indicate that the fungal reduction was enantiospecific but exhibited low substrate specificity.

Lignans are typically dimeric phenylpropanoid compounds that are C_8-$C_{8'}$ linked (*1*). They have diverse biological activities and are important pharmaceutically (*2-4*). Since 1990 enzymatic approaches to clarifying lignan biosynthesis have been initiated in *Forsythia* sp. by Lewis and coworkers (*5-7*), where the first biosynthetic pathway to a class of lignans was successfully established (*8-15*). Two molecules of coniferyl alcohol become coupled to afford (8*R*,8′*R*)-(+)-pinoresinol, the enantiospecific reduction of which first gives (8*R*,8′*R*)-(+)-lariciresinol and then (8*R*,8′*R*)-(−)-secoisolariciresinol (Figure 1, route A). Enantiospecific oxidation of (−)-secoisolariciresinol then affords (8*R*,8′*R*)-(−)-matairesinol (*5, 7*), which is methylated to give arctigenin (*15*).

Stereochemistry in Lignan Biosynthesis

Most naturally occurring lignans and neolignans are isolated in optically active form (*16, 17*). On the other hand, lignins, which are structurally related to lignans/neolignans, are optically inactive. Their dehydrogenatively formed dilignol/oligolignol structures are racemic (*18*), since lignins are biosynthesized *via* phenoxy radical formation from monolignols by the non-stereocontrolled action of peroxidase or laccase and subsequent non-enzymatic coupling of the formal phenoxy radical mesomers (discussed in references *19, 20*). Until the studies of Lewis and co-workers, there had been no experimental basis for explaining how the stereochemical differences between lignans/neolignans and lignins were effected. However, the recent studies on lignan biosynthesis in *Forsythia* plants have suggested, for the first time, that enzyme-catalyzed stereoselective coupling of coniferyl alcohol and stereospecific post-coupling transformations are responsible for the stereochemical differences.

To more fully establish the generality of the lignan biosynthetic pathway revealed in *Forsythia* plants, other biological systems also need to be examined. Interestingly, both the magnitude and sign of specific rotation of a particular lignan may vary with the plant species (*4, 21*). For example, *Forsythia suspensa* has (+)-pinoresinol (*22*), whereas *Z. ailanthoides* contains (−)-pinoresinol (*23*). The reason for this variation is unclear, but the presence of a pathway giving rise to (8*S*,8′*S*)-lignans that are the enantiomers of (8*R*,8′*R*)-lignans is one possibility (Figure 1, route B). More remarkable is the presence of (−)-secoisolariciresinol in *Z. ailanthoides* (*23*), the sign of its specific rotation being the same as that in *Forsythia* sp. (*7*). The analog of the above pathway (route A) indicates that (+)-secoisolariciresinol should be formed from (−)-pinoresinol.

In order to clarify this issue, we have begun investigating the enzymology and stereochemistry of lignan biosynthesis in *Zanthoxylum* sp. This section first describes the enzymatic reduction of (+)-pinoresinol by cell-free extracts from *Z. ailanthoides* (*24*), and then the enantiomeric composition of lignans obtained following administration of [8-^{14}C]coniferyl alcohol and (±)-[8,8′-^{14}C]pinoresinols to the excised shoot.

(+)-Lariciresinol Formation with *Zanthoxylum ailanthoides* Cell-free Extracts. Cell-free extracts from *Z. ailanthoides* were incubated with (±)-pinoresinols for 1 h in the presence of NADPH. The analysis of the products and their identification were accomplished by reversed-phase C_{18}-HPLC using two solvent systems and by chiral column HPLC using Chiralcel OC and OD. Previously (±)-pinoresinols, (±)-lariciresinols, and (±)-secoisolariciresinols had been synthesized (*11, 25*) and employed as substrates and authentic samples. This was because lariciresinol and secoisolariciresinol were expected to be products of the enzymatic reduction of (±)-pinoresinols in analogy with the processes occurring in *Forsythia* species (*9-12*).

Reversed-phase C_{18}-HPLC chromatograms of the reaction products after 1 h incubation gave a peak which was identical with lariciresinol. This peak was absent from the chromatograms of three control assays, without NADPH, without substrate, and with the boiled enzyme preparation, respectively. However, secoisolariciresinol was not detected in any of the enzyme assays.

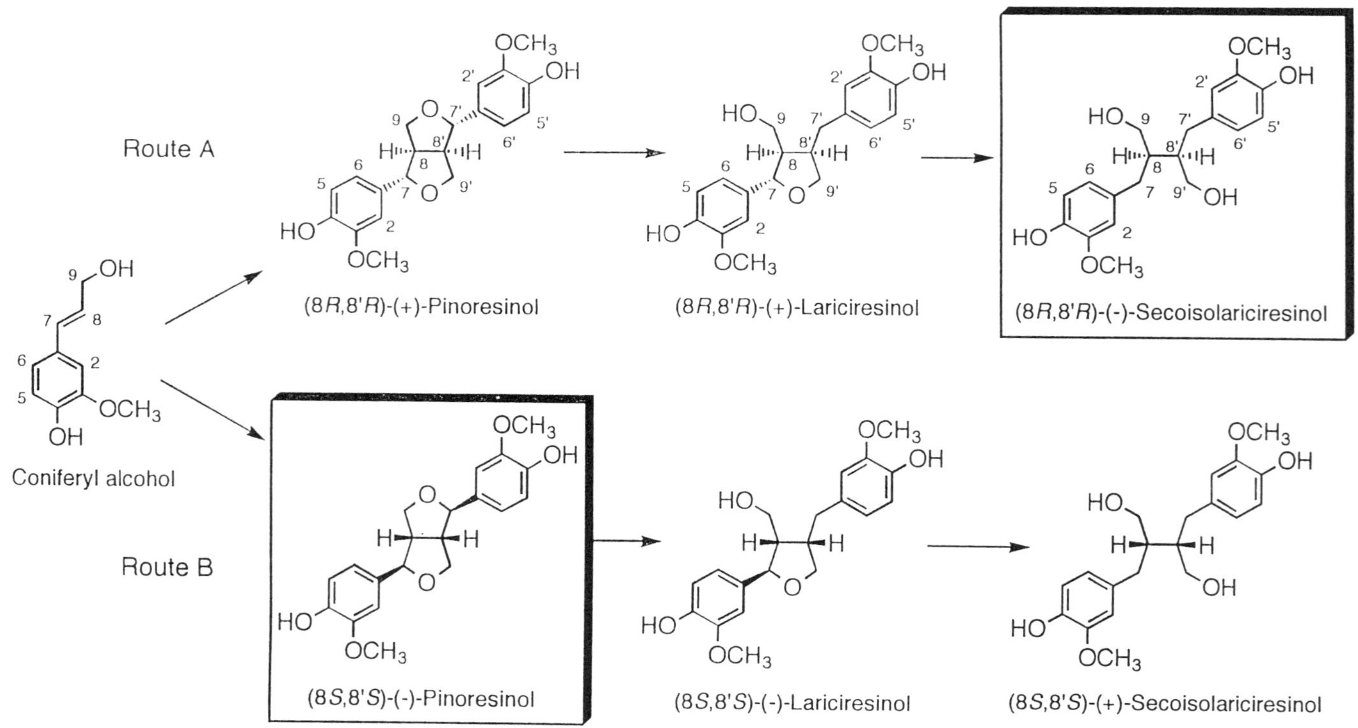

Figure 1. Possible biosynthetic routes from coniferyl alcohol to secoisolariciresinol *via* pinoresinol and lariciresinol. Route A was established using *Forsythia* plants. (−)-Pinoresinol and (−)-secoisolariciresinol were isolated from *Zanthoxylum ailanthoides* (*23*). Route B may account for formation of (8*S*,8′*S*)-lignans. (Redrawn from ref. 24).

Figures 2C and 2A show the chiral separation of the synthetic (±)-lariciresinols [(+)-form, V_R = 8.40 ml; (−)-form, V_R = 9.75 ml] and the lariciresinol fraction of the reaction products, respectively, under exactly the same conditions. The latter had a peak (V_R = 8.38 ml) corresponding to (+)-lariciresinol, whereas the (−)-lariciresinol amounted at most to a small shoulder in the chromatogram. Thus, the formation of (+)-lariciresinol from (±)-pinoresinols by the crude cell-free extracts was confirmed.

Next, in order to examine this enzymatic reaction quantitatively at higher sensitivity, radioactive (±)-[8,8′-^{14}C]pinoresinols were prepared by dehydrogenating [8-^{14}C]coniferyl alcohol. The cell-free extracts were treated with ammonium sulfate and the proteins precipitating between 10 and 80% saturation were used as the enzyme preparation.

When (±)-[8,8′-^{14}C]pinoresinols were incubated with this enzyme preparation in the presence of NAD(P)H at 30°C, the formation of (+)-[8,8′-^{14}C]lariciresinol was again confirmed (Figure 2B), but only a weak shoulder was observed at the elution volume corresponding to (−)-[8,8′-^{14}C]lariciresinol. Figure 3 shows the time course of the formation of (+)- and (−)-[8,8′-^{14}C]lariciresinols and (+)- and (−)-[8,8′-^{14}C] secoisolariciresinols. The formation of (+)-[8,8′-^{14}C]lariciresinol attained 5.21 nmol (1.09 nmol mg^{-1} protein; 1.2% level of incorporation from the substrate) after 1 h incubation, whereas the radioactivity corresponding to the (−)-antipode was almost at background level over the whole incubation period (0.09 nmol mg^{-1} protein). Neither (+) nor (−)-[8,8′-^{14}C]secoisolariciresinol formation was detected under these experimental conditions.

Then the cofactor requirements for the formation of (+)- and (−)-[8,8′-^{14}C] lariciresinols and (+)- and (−)-[8,8′-^{14}C]secoisolariciresinols were examined and the results are listed in Table I. When the incubation was carried out in the presence of NADPH and NADH, (+)-[8,8′-^{14}C]lariciresinol was produced. The activity with NADH was less (three quarters) than that with NADPH. If NADPH or NADH was omitted, [8,8′-^{14}C]lariciresinol formation was largely lost, although very small amounts of radioactivity remained corresponding to lariciresinol. If the enzyme preparation was denatured, [8,8′-^{14}C]lariciresinol formation was not detected either. In neither case was any [8,8′-^{14}C]secoisolariciresinol formation observed.

These results confirm the presence of the enzyme activity that converts (±)-pinoresinols to (+)-lariciresinol in *Z. ailanthoides* (Figure 4), where the (8*R*,8′*R*)-(+)-lariciresinol is derived from (8*R*,8′*R*)-(+)-pinoresinol.

Enantiomeric Composition of Radioactive Lignans Following Administration of [8-^{14}C]Coniferyl Alcohol and (±)-[8,8′-^{14}C]-Pinoresinols to *Z. ailanthoides*. When [8-^{14}C]coniferyl alcohol was administered to a shoot of *Z. ailanthoides*, the formation of radioactive [^{14}C]pinoresinol was observed. (8*S*,8′*S*)-(−)-[^{14}C]Pinoresinol, the naturally more abundant enantiomer, was somewhat more prevalent (54% stem; 59% leaves) than (8*R*,8′*R*)-(+)-[^{14}C]pinoresinol. Following administration of (±)-[8,8′-^{14}C]pinoresinols to a shoot of *Z. ailanthoides*, the formation of radioactive [^{14}C]lariciresinol and [^{14}C]secoisolariciresinol was observed, although they were only tentatively identified by retention volume in C_{18} reversed phase and chiral column HPLC. Table II shows the percentage incorporation and enantiomeric composition of the lignans obtained from both the stem and the leaves; the incorporation in the stem was much more than that in the leaves. As can be seen, (8*R*,8′*R*)-(+)-[^{14}C]lariciresinol was the predominant enantiomer in the stem, although its enantiomeric purity was low and the [^{14}C]secoisolariciresinol formed was almost racemic.

Summary. Enzymatic reduction of the benzyl ethers (+)-pinoresinol and (+)-lariciresinol had been observed *in vitro* for the first time using *Forsythia intermedia* extracts (*9-11*). Cell-free extracts from *Z. ailanthoides* also catalyzed (+)-lariciresinol formation from (+)-pinoresinol in the presence of NAD(P)H in the present work. The author and others also obtained the same enzyme activity from *Z. schinifolium* (unpublished results). These results indicate that pinoresinol reductase was not restricted to

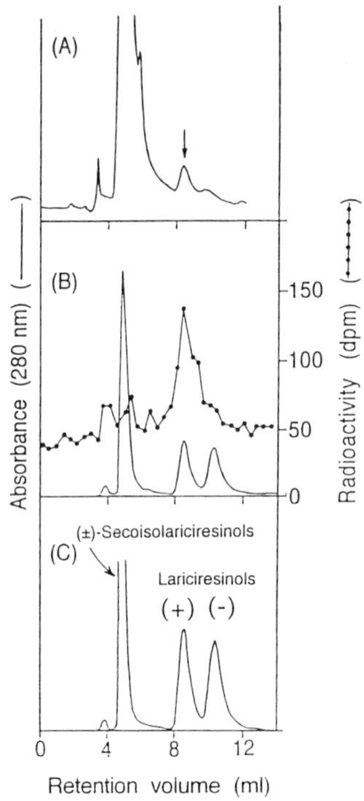

Figure 2. Chiral HPLC chromatograms of (A) unlabeled lariciresinol and (B) [8,8′-^{14}C]lariciresinol formed by the action of enzyme preparations 1 and 2, respectively (see Experimental Section), and of (C) synthetic (±)-lariciresinols. Antipode assignments were reported in ref. 11. In (B) unlabeled (±)-lariciresinols and (±)-secoisolariciresinols were added as cold carriers. (Redrawn from ref. 24).

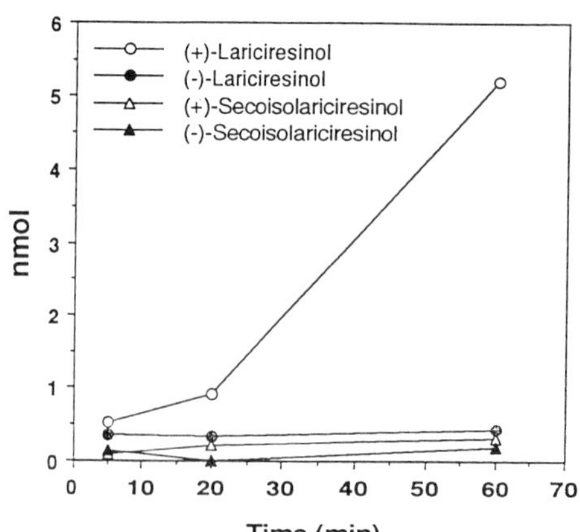

Figure 3. Time course for enzyme-catalyzed formation of (+)- and (−)-[8,8′-^{14}C]lariciresinols and (+)- and (−)-[8,8′-^{14}C]secoisolariciresinols. (Redrawn from ref. 24).

Table I. Cofactor Requirements for Formation of (+)- and (−)-[8,8′-^{14}C]Lariciresinols and (+)- and (−)-[8,8′-^{14}C]Secoisolariciresinols following Incubation of (±)-[8,8′-^{14}C]Pinoresinols with the Enzyme Preparation[a] from Cell-Free Extracts of *Zanthoxylum ailanthoides*.

	Lignans formed in 1 h (nmol mg^{-1} protein)			
Enzyme assay	Lariciresinols		Secoisolariciresinols	
	(+)	(−)	(+)	(−)
Standard				
NADPH	1.09	0.09	0.07	0.04
NADH	0.81	0.06	0.05	0.01
Controls				
Without NAD(P)H	0.16	0.05	0.03	0.02
Denatured enzyme/NADPH	0.00	0.00	0.03	0.03
Denatured enzyme/NADH	0.01	0.00	0.02	0.02

[a] Enzyme preparation 2 (see Experimental Section) with 4.78 g L^{-1} protein.

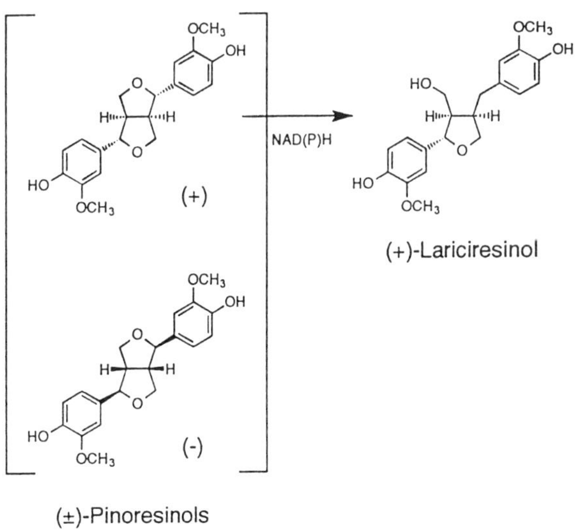

Figure 4. Enzymatic reduction of (+)-pinoresinol to (+)-lariciresinol in *Zanthoxylum ailanthoides*. (Redrawn from ref. 24).

Table II. Incorporation and Enantiomeric Composition of Radioactive [^{14}C]Lariciresinol and [^{14}C]Secoisolariciresinol Obtained from Stem and Leaves following Administration of [8,8′-^{14}C]Pinoresinol to Excised Shoot of *Zanthoxylum ailanthoides*

	Stem Incorporation (%) (8*R*,8′*R*) (8*S*,8′*S*)	Leaves Incorporation (%) (8*R*,8′*R*) (8*S*,8′*S*)
Lariciresinol	1.64	0.04
	(+) (−)	(+) (−)
	55.9 : 44.1	50.5 : 49.5
Secoisolariciresinol	1.01	0.09
	(−) (+)	(−) (+)
	50.2 : 49.8	42.2 : 57.8

Forsythia plants. When a plant contains a substituted 7-*O*-9′ tetrahydrofuran (*e.g.* lariciresinol), a dibenzylbutane (*e.g.* secoisolariciresinol), or a dibenzylbutyrolactone (*e.g.* matairesinol), the plant would be expected to exhibit pinoresinol reductase activity and these lignans would be formed by way of pinoresinol reduction.

Although secoisolariciresinol formation from (±)-pinoresinols has not been found under these experimental conditions yet, (+)-lariciresinol formation is a prerequisite for explaining the presence of (−)-secoisolariciresinol in *Z. ailanthoides*, because the absolute configuration of (+)-lariciresinol, (8*R*,8′*R*), is the same as that of the (−)-secoisolariciresinol isolated from this plant by Ishii and coworkers (*23*). On the other hand, when cell-free extracts from young stems of *Z. schinifolium*, which is closely related to *Z. ailanthoides*, were incubated with (±)-syringaresinols in the presence of NADPH, neither 5,5′-dimethoxylariciresinol nor 5,5′-dimethoxy-secoisolariciresinol, the corresponding reduction product, were obtained (*26*). Recently, the formation of (+)-secoisolariciresinol (*ca.* 20% enantiomeric excess, e.e.) from coniferyl alcohol by cell-free extracts of *Arctium lappa* in the presence of NADPH and H_2O_2 was reported (*27*).

Radioactivity from [8-^{14}C]coniferyl alcohol and [8,8′-^{14}C]pinoresinol was taken into pinoresinol, lariciresinol and secoisolariciresinol, respectively, in the feeding experiments that were performed. The cell-free extract of *Z. ailanthoides* catalyzed (+)-lariciresinol formation from (+)-pinoresinol in the presence of NAD(P)H. These facts suggest that this plant has the same pathway as *Forsythia* plants, if the *absolute* stereochemistry of the lignans is ignored; two molecules of coniferyl alcohol couple to give pinoresinol, which is transformed first to lariciresinol and then to secoisolariciresinol (Figures 1A and 1B).

Pinoresinol formation from coniferyl alcohol was not highly stereoselective in the feeding experiments. With *Z. ailanthoides,* the use of coniferyl alcohol as a precursor in intact shoots would be expected to result in competition between stereospecific and non-specific coupling. The latter could be due to the predominant action of non-specific peroxidases and/or laccases. Such results had already been observed in *Forsythia* sp. (*6*). (−)-Pinoresinol isolated from this plant by Ishii and coworkers (*23*) was considered to contain some of the (+)-antipode, because the absolute value of its specific rotation was lower than that of (+)-pinoresinol in *Forsythia*. The results of the feeding experiments indicated that transformation of (±)-pinoresinols to lariciresinol and secoisolariciresinol might conceivably occur without stereospecificity or stereoselectivity. It was suggested that pathways to both (8*R*,8′*R*) and (8*S*,8′*S*)-lignans are present in *Z. ailanthoides*.

Preparation of Neolignans through Fungal Reduction

Typical neolignans are dimeric phenylpropanoid compounds linked through bonds, such as C_8-C_3, C_8-*O*-$C_{4'}$, C_3-$C_{3'}$, etc., other than C_8-$C_{8'}$ (*1*), and most of them have been isolated in optically active forms (*16, 17*). It has been found that they have diverse biological activities (*2-4*). In order to produce bioactive lignans and neolignans, development of stereochemically controlled bioconversions to supplement organic synthesis is required. However, few studies on neolignan bioconversion have been reported.

8-*O*-4′-Neolignans containing *erythro*-$\Delta^{8'}$-7-hydroxy-3′,5′-methoxy-structures (*28-30*), and other types of 8-*O*-4′-neolignans (*31, 32*) had been isolated from the seeds and aril of the fruit of *Myristica fragrans*. Some of them had antifeedant effects on silkworm larvae (*29*) or exhibited bactericidal activity (*33*). Surinamensin (*threo*-$\Delta^{7'}$-7-hydroxy-3,4,5,3′-tetramethoxy-8-*O*-4′-neolignan) and virolin ($\Delta^{7'}$-7-hydroxy-3,4,3′-trimethoxy-8-*O*-4′-neolignan) had both been isolated from the leaves of *Virola surinamensis* and their strong activity against penetration of the cercaria of *Schistsoma mansoni* was reported (*34*). Production of these bioactive 8-*O*-4′-neolignans is thus very important.

On the other hand, there has been no research on the biosynthesis of 8-*O*-4′-neolignans, although these 8-*O*-4′-neolignans were considered to be formed by way of the oxidative coupling of eugenol, isoeugenol, and their derivatives (*35*). To clarify what steps are involved in the biosynthesis of optically active neolignans, their optical isomers are needed as authentic samples.

The 7-hydroxy-8-*O*-4′-neolignans have four stereoisomers (*erythro* and *threo* diastereomers and their enantiomers) as do the arylglycerol-β-aryl ethers, which are lignin substructure models. Stereoselective synthesis of a (±)-*threo* neolignan (*36*) and the asymmetric synthesis of both enantiomers of two kinds of *erythro*-7-hydroxy-8-*O*-4′-neolignans have been investigated by Zacchino and coworkers (*37, 38*). The absolute configuration of one *erythro*-7-hydroxy-8-*O*-4′-neolignan was determined as (7*R*,8*S*) by Mosher's approach (*38*). Zacchino synthesized *threo*-(−)-virolin through a combination of diastereoselective $NaBH_4$ reduction and an enantiospecific microbial hydrolysis (*39*). Recently, Kasahara and coworkers reported the bioconversion of a 7-hydroxy-8-*O*-4′-neolignan to the corresponding 7-methylene analog (*40*).

The preparation of 7-hydroxy-8-*O*-4′-neolignans by reduction of the 7-carbonyl group in 7-oxo-8-*O*-4′-neolignans with *Fusarium solani* M-13-1 is described below (*41*). This is the first example of neolignan synthesis by fungal reduction of a ketone.

The author and coworkers had studied the degradation of arylglycerol-β-aryl ethers, as lignin substructure model compounds, by *F. solani* M-13-1 (*42-44*). In the course of the study, it was found that the fungus reduced the α-carbonyl group of (±)-α-oxo-guaiacylglycerol-β-vanillin ether **1**, after oxidation of the α′-arylaldehyde to the corresponding carboxylic acid, to give guaiacylglycerol-β-(vanillic acid) ether **2** as a mixture of *erythro* and *threo* forms, both of which were enantiomerically pure (*43-45*) (Figure 5). This fungal reduction was applied to the preparation of 4,7-dihydroxy-3,3′-dimethoxy-8-*O*-4′-neolignan **4** and $\Delta^{8'}$-4,7-dihydroxy-3,3′-dimethoxy-8-*O*-4′-neolignan **6** from 4-hydroxy-3,3′-dimethoxy-7-oxo-8-*O*-4′-neolignan **3** and $\Delta^{8'}$-4-hydroxy-3,3′-dimethoxy-7-oxo-8-*O*-4′-neolignan **5**, respectively (*41*). Neolignan **6** is considered to be an initial coupling product of eugenol and isoeugenol in plants. Optical isomers of **6** will now be available for investigations of the enzymatic enantioselective coupling step.

Reduction of 4-Hydroxy-3,3′-dimethoxy-7-oxo-8-*O*-4′-neolignans by *F. solani* M-13-1. Preliminarily, the reduction of 4-hydroxy-3,3′-dimethoxy-7-oxo-8-*O*-4′-neolignans **3** by *F. solani* M-13-1 was investigated. According to our previous results (*42, 44*), when *F. solani* M-13-1 degraded guaiacylglycerol-β-coniferyl ether and α-oxo-guaiacylglycerol-β-vanillin ether **1**, the initial transformation of the former involved an oxidative degradation of the terminal allyl alcohol side chain ($C_{7'}$, $C_{8'}$, and $C_{9'}$) followed by cleavage of the Cα-C_1 bond, while that of the latter **1** involved oxidation of the α′-aldehyde to the carboxylic acid with the Cα(C_7)-ketone thereupon reduced to the alcohol. But our unpublished results suggested that a neolignan with a saturated 1-propyl side chain was stable with respect to degradation by the fungus. Therefore, the 7-oxo-8-*O*-4′-neolignan **3** was synthesized and used as a substrate for the fungal reduction.

7-Oxo-8-*O*-4′ neolignan **3** was incubated with intact cells of *F. solani* M-13-1 both in the absence and presence of ketone **1** (*41*). In the absence of **1**, the 7-oxo-8-*O*-4′-neolignan **3** was stable for the whole incubation (24, 98 and 184 h). On the other hand, in the presence of ketone **1**, when the transformation of **1** to guaiacylglycerol-β-(vanillic acid) ether **2** was observed, a new product was found to be formed concomitantly. The new product gave the same R_f value with TLC and the same positive color test with 2,6-dichloroquinone-4-chloroimide as the synthetic 7-hydroxy-8-*O*-4′-neolignan **4** which was obtained by $NaBH_4$ reduction of 7-oxo-8-*O*-4′-neolignan **3**. It was demonstrated that reduction of the carbonyl in **3** by *F. solani* M-13-1 gave the desired 7-hydroxy-8-*O*-4′-neolignan **4**. In this case, ketone **1** may act as an inducer for the reduction.

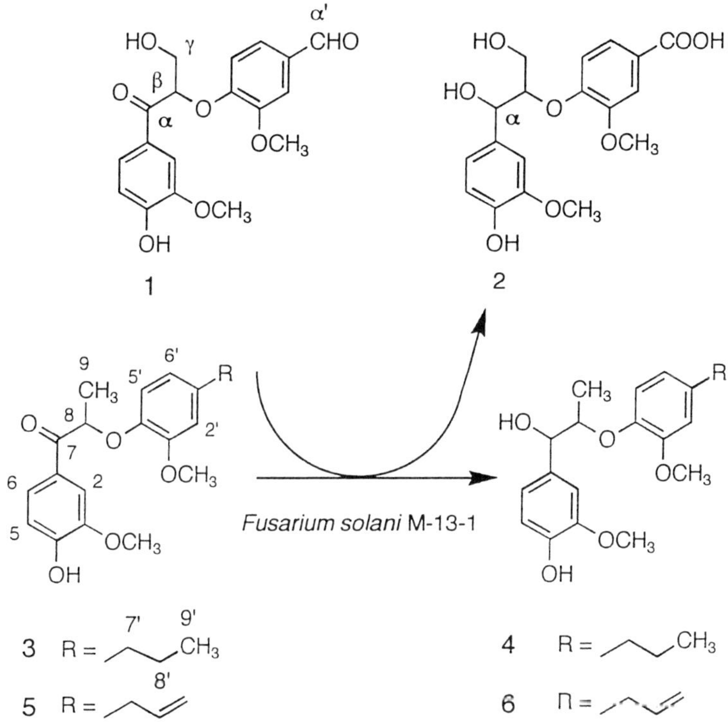

Figure 5. Formation of 4,7-dihydroxy- and $\Delta^{8'}$-4,7-dihydroxy-3,3′-dimethoxy-8-*O*-4′-neolignans (**4** and **6**) with concomitant reduction of (±)-4-hydroxy- and (±)-$\Delta^{8'}$-4-hydroxy-3,3′-dimethoxy-7-oxo-8-*O*-4′-neolignans (**3** and **5**) by *Fusarium solani* M-13-1 in the presence of α-oxo-guaiacylglycerol-β-vanillin ether (**1**).

Reduction of $\Delta^{8'}$-4-Hydroxy-3,3′-dimethoxy-7-oxo-8-*O*-4′-neolignan by *F. solani* M-13-1. $\Delta^{8'}$-4-Hydroxy-3,3′-dimethoxy-7-oxo-8-*O*-4′-neolignan **5** was next incubated with *F. solani* M-13-1 for 192 h in the presence of ketone **1** (*41*). When conversion of **1** to guaiacylglycerol-β-(vanillic acid) ether **2** occurred, formation of a new product was also observed in this case. This product gave the same R_f values with TLC and the same positive color test for 2,6-dichloroquinone-4-chloroimide as synthetic 7-hydroxy-8-*O*-4′-neolignan **6**. This compound was identified as a mixture of *erythro* and *threo* $\Delta^{8'}$-4,7-dihydroxy-3,3′-dimethoxy-8-*O*-4′-neolignans **6** by mass spectrometric and chromatographic comparisons with synthetic **6**. Each diastereomer was carefully separated by preparative TLC and each structure was confirmed individually by ^{1}H NMR and mass spectrometry. It was found that the carbonyl in **5** was reduced to a secondary alcohol. The yield was 57%, although this was not at this time optimized.

In the ^{1}H NMR spectra of **6**, characteristic differences between *erythro* and *threo* isomers were indicated in the chemical shifts and coupling constants for C_7-H: *erythro* δ 4.81, J = 3.2 Hz; *threo* δ 4.61, J = 8.6 Hz, in agreement with those reported in the literature (*31, 32*). The ratio of *erythro* and *threo* forms was determined to be 3:2 by integration of the two peaks. The *erythro* (*31*) and *threo* (*32*) isomers had been isolated from *M. fragrans*, but there was no information provided about their optical activity.

Then the enantiomeric purities of the fungal reduction products, *erythro*- and *threo*-**6** were investigated by chiral column HPLC. Previously, conditions for optical resolution were established, using synthetic *erythro* (±)-**6** and *threo* (±)-**6**, to be as follows: EtOH-hexanes (5:95) eluted at a flow rate 1.0 ml min^{-1} through a Daicel Chiralcel OD column with detection at 280 nm (*41*). It was found that both fungal reduction products, the *erythro* and *threo* 7-hydroxy-8-*O*-4′-neolignans **6**, respectively, gave a single peak which was identical with the peak for one enantiomer in the corresponding synthetic compound under the exactly same conditions while the peak for the other enantiomer was absent (*41*). Therefore, both the *erythro* and *threo* isomers of the fungal products were enantiomerically pure (e.e. 100%), indicating that the reduction had been highly enantioselective.

The *erythro* and *threo* products would be derived from both enantiomers of the racemic substrate. This was supported by the fact that their 57% yield was more than 50%. In this sense, the substrate specificity of the reduction would be low. Figure 5 shows a proposed pathway for the reduction of the 7-oxo-8-*O*-4′-neolignans by *F. solani* M-13-1 in the presence of compound **1** as an inducer. Determination of the absolute configuration of the products was under way at the time of writing. Zacchino and Badano obtained both enantiomers of *erythro* 8-*O*-4′-neolignans by asymmetric reduction of the corresponding 7-oxo-8-*O*-4′-neolignan first with LiAlH[(−)- and (+)-*N*-methyl-ephedrine]$(N\text{-ethylaniline})_2$ (e.e. 62.5 and 60%, respectively) (*37*) and second with $LiAlH_4$ modified with (*S*)- and (*R*)-binaphthol (e.e. 82 and 80%, respectively) (*38*).

Experimental Section

Plant. *Z. ailanthoides* from Mount Nyotai and Kamiyama Forest, Kagawa University, (Kagawa, Japan) were used in this study.

Enzyme Extraction. A young shoot (20 cm, 21.2 g) of *Z. ailanthoides* was defoliated, sectioned (less than 5 mm), frozen (liquid N_2) and ground to a powder. The powder was homogenized in PVPP (4.28 g), sea sand (25.9 g) and potassium phosphate (K-P_i) buffer (0.1 M, pH 7.0, 55 mL) containing DTT (10 mM). The homogenate was filtered through four layers of cheesecloth and the filtrate centrifuged (11 000 *g*, 20 min, 1°C). The supernatant was treated as follows to obtain two kinds of cell-free enzyme preparation: (a) an aliquot of the supernatant was passed through a PD-10 column (Sephadex G25M, equilibrated with K-P_i buffer) with the excluded fraction used as cell-free enzyme preparation 1; (b) the supernatant was fractionated with ammonium sulfate, with the proteins precipitating between 10 and 80%

ammonium sulfate saturation being recovered and redissolved in K-P_i buffer (0.1 M, pH 7.0), whereupon an aliquot of the solution was passed through a PD-10 column as before and used as cell-free enzyme preparation 2. Protein contents were determined by the method of Bradford (*47*).

Enzyme Assays. Each assay mixture (250 μL) consisted of 0.42 mM (±)-pinoresinols (5.2 mM, 20 μL), 4.0 mM NAD(P)H (50 mM, 20 μL), K-P_i buffer (10 μL) and the cell-free enzyme preparation (200 μL) (*24*). Assays were conducted in quintuplicate.

Enzyme preparation 1 (1 mL) was incubated with unlabelled (±)-pinoresinols for 1 h in the presence of NADPH. The products were extracted with ethyl acetate. Methanol soluble components of the extracts were analyzed by C_{18} reversed phase HPLC (Jasco Gulliver PU-980 HPLC system with a Chemco Pak Finesil C_{18}-5 column) using linear gradient elution with CH_3CN-3%AcOH in H_2O (from 10:90 at t = 0 to 40:60 at t = 30 min, which was then held for additional 10 min; flow rate 1 ml min^{-1}). Then the eluting components corresponding to lariciresinol and secoisolariciresinol were collected and evaporated. Methanol soluble components of the residue were further analyzed by C_{18} reversed phase HPLC with isocratic elution using CH_3OH-3%AcOH in H_2O (40:60, 0.5 ml min^{-1}) and chiral column HPLC (Daicel Chiralcel OC and OD columns) (*9, 11*). All UV detection was carried out at 280 nm.

Enzyme preparation 2 was incubated with (±)-[8,8′-^{14}C]pinoresinols (10.6 MBq $mmol^{-1}$, 0.42 mM) for 5, 20 and 60 min, in the presence of NADPH (0.4 mM) or NADH (0.4 mM) at 30°C. The products were extracted with ethyl acetate containing both (±)-lariciresinols and (±)-secoisolariciresinols as radiochemical carriers. Methanol soluble components of the extracts were subjected to C_{18} HPLC (Hitachi 6200 HPLC system with a Waters Nova-Pak C_{18}-5 column) with the same gradient elution using CH_3CN-3%AcOH in H_2O as before. The eluting components were collected at 30 sec intervals, and the radioactivity of the resulting fractions was counted with a liquid scintillation counter (LSC). The methanol soluble components were further subjected to C_{18} reversed phase HPLC repeatedly and the fractions containing lariciresinol and secoisolariciresinol were collected and evaporated. Methanol soluble components of the residues were further subjected to the Chiralcel OC (for resolution of lariciresinol) and the Chiralcel OD (secoisolariciresinol) columns. Eluting components were collected at 1 min intervals and their radioactivities were determined with LSC.

Precursor Administration Experiments. (±)-[8,8′-^{14}C]Pinoresinols (10.6 MBq $mmol^{-1}$, 1.30 mM) in a mixture of methanol and K-P_i buffer (1:1; 10.4 mL) was administered over a 5.3 h period to an excised young shoot of *Z. ailanthoides*. After 3 h of metabolism, the stem and leaves were separated, and each was extracted five times with hot methanol. The combined methanol soluble components were concentrated to 10 mL for the stem and 20 mL for the leaves *in vacuo*, to which was added water (50 and 100 mL, respectively), with the resulting mixture being individually filtered. Each filtrate was extracted four times with ether, and the ether soluble components were combined and evaporated *in vacuo*. As cold carriers, unlabeled (±)-lariciresinols and (±)-secoisolariciresinols were added to the ether extracts, and then the methanol soluble components of the extracts were analyzed with C_{18} reversed phase HPLC (Hitachi 6200 HPLC system with a Waters Nova-Pak C_{18}-5 column) using the same gradient elution with CH_3CN-3%AcOH in H_2O as before. The components were collected at 30 sec intervals with each fraction being subjected to LSC. The methanol soluble components were further subjected to C_{18} reversed phase HPLC repeatedly and the fractions containing both lignans were collected and evaporated. Methanol soluble components of the lignan fractions were separated with chiral column HPLC (Chiralcel OC and OD for lariciresinol and secoisolariciresinol, respectively), wherefrom the eluting components were collected at 1 min intervals with each fraction being subjected to LSC.

Microorganism and Culture Conditions. *Fusarium solani* M-13-1 was employed for the fungal reduction experiments (*48*). The composition of the basal inorganic medium and nutrition medium, as well as the preparation of the intact cells of *F. solani* M-13-1 have been described previously (*43-45*).

Fungal Reduction. Three cultures contained **3** (1 mg) in DMF (20 μL), α-oxo-guaiacylglycerol-β-vanillin ether **1** (1 mg) in DMF (20 μL) and mycelia (7 mg dry weight) in the inorganic medium (5 mL). Three other cultures were of the same composition except that ketone **1** was omitted. Another culture contained **5** (21.7 mg) in DMF (0.4 mL), the ketone **1** (20 mg) in DMF (0.2 mL) and mycelia (140 mg dry weight) in the inorganic medium (100 mL). All cultures were incubated with shaking at 28°C.

After the prescribed time intervals, the cultures were filtered, with the mycelia then being washed with water and methanol successively. The filtrate and the washings were combined and extracted three times with ether. The combined ether solution was washed with saturated brine, dried over anhydrous Na_2SO_4, and evaporated to dryness *in vacuo*. The residue was purified by preparative TLC and the compounds isolated were identified by NMR and mass spectrometry. Since the desired product was expected to be a mixture of *erythro* and *threo* isomers, half of the product was purified by preparative TLC at first without the separation of both diastereomers in order to determine its diastereomeric ratio by the integration of the appropriate peaks in the ^{1}H NMR spectrum, and the rest was subjected to preparative TLC to separate both diastereomers. The synthesis of **1**, **2**, **3**, **4**, **5**, and **6** has been described previously (*43, 46, 49*).

Conclusion

Cell-free extracts from the young stem of *Z. ailanthoides* have catalyzed the formation of (8*R*,8′*R*)-(+)-lariciresinol from (+)-pinoresinol in the presence of NAD(P)H; little (−)-lariciresinol was formed. The (8*R*,8′*R*) configuration of (+)-lariciresinol was identical to that of natural (−)-secoisolariciresinol. Following administration of [8-^{14}C]coniferyl alcohol and (±)-[8,8′-^{14}C]pinoresinols to excised young shoots of the plant, preferential formation of (−)-[^{14}C]pinoresinol, and (+)-[^{14}C]lariciresinol and [^{14}C]secoisolariciresinol, was found, respectively. But the enantiomeric purities of the lignans were low and the secoisolariciresinol was racemic. It was suggested that the pathways to both (8*R*,8′*R*) and (8*S*,8′*S*)-lignans are present in *Z. ailanthoides*.

$\Delta^{8'}$-4,7-Dihydroxy-3,3′-dimethoxy-8-*O*-4′-neolignan was prepared by incubation of intact *Fusarium solani* M-13-1 cells with the corresponding 7-carbonyl derivative in the presence of α-oxo-guaiacylglycerol-β-vanillin ether. The neolignan obtained was a mixture of *erythro* and *threo* isomers in a 3:2 ratio, and both were enantiomerically pure. The optically active 8-*O*-4′-neolignans will be useful for the investigation of 8-*O*-4′-neolignan biosynthesis in plants. This fungal reduction seems to be highly enantiospecific but of low substrate specificity, indicating that this reduction could be applied to preparation of other bioactive 8-*O*-4′-neolignans.

Acknowledgments

The author thanks T. Masaoka, H. Yamada, H. Sugimoto and S. Okamoto, graduates of the Faculty of Agriculture, Kagawa University, for their cooperation during this research, which was supported in part by a Grant-in-Aid for Scientific Research (No. 05660191) from the Ministry of Education, Science, Sports and Culture of Japan.

Literature Cited

1. Whiting, D. A. *Nat. Prod. Rep.* **1985**, *2*, 191-212.
2. MacRae, W. D.; Towers, G. H. N. *Phytochemistry* **1984**, *23*, 1207-1220.

3. Ayres, D. C.; Loike, J. D. *Lignans: Chemical, Biological and Clinical Properties*; Cambridge University Press: Cambridge, 1990, pp 85-112.
4. Davin, L. B.; Lewis, N. G. In *Phenolic Metabolism in Plants*; Stafford, H. A., Ibrahim, R. K., Eds.; *Recent Advances in Phytochemistry, Vol 26*; Plenum Press: New York, NY, 1992, pp 325-375.
5. Umezawa, T.; Davin, L. B.; Lewis, N. G. *Biochem. Biophys. Res. Commun.* **1990**, *171*, 1008-1014.
6. Umezawa, T.; Davin, L. B.; Yamamoto, E.; Kingston, D. G. I.; Lewis, N. G. *J. Chem. Soc., Chem. Commun.* **1990**, 1405-1408.
7. Umezawa, T.; Davin, L. B.; Lewis, N. G. *J. Biol. Chem.* **1991**, *266*, 10210-10217.
8. Davin, L. B.; Bedgar, D. L.; Katayama, T.; Lewis, N. G. *Phytochemistry* **1992**, *31*, 3869-3874.
9. Katayama, T.; Davin, L. B.; Lewis, N. G. *Phytochemistry* **1992**, *31*, 3875-3881.
10. Lewis, N. G.; Davin, L. B.; Katayama, T.; Bedgar, D. L. *Bull. Liaison Groupe Polyphénols* **1992**, *16*, 98-103.
11. Katayama, T.; Davin, L. B.; Chu, A.; Lewis, N. G. *Phytochemistry* **1993**, *33*, 581-591.
12. Katayama, T.; Davin, L. B.; Bedgar, D. L.; Lewis, N. G. *Proc. 7th Internatl. Symp. Wood Pulp. Chem.* **1993**, *2*, 984-992.
13. Chu, A.; Dinkova, A.; Davin, L. B.; Bedgar, D. L.; Lewis, N. G. *J. Biol. Chem.* **1993**, *268*, 27026-27033.
14. Paré, P. W.; Wang, H.-B.; Davin, L. B.; Lewis, N. G. *Tetrahedron Lett.* **1994**, *35*, 4731-4734.
15. Ozawa, S.; Davin, L. B.; Lewis, N. G. *Phytochemistry* **1993**, *32*, 643-652.
16. Gottlieb, O. R.; Yoshida, M. In *Natural Products of Woody Plants, I: Chemicals Extraneous to the Lignocellulosic Cell Wall*; Rowe, J. W., Ed.; Springer-Verlag: Berlin, 1989, pp 439-511.
17. Ayres, D. C.; Loike, J. D. *Lignans: Chemical, Biological and Clinical Properties*; Cambridge University Press: Cambridge, 1990, pp12-84.
18. Freudenberg, K. In *Constitution and Biosynthesis of Lignin;* Freudenberg, K., Neish, A. C., Eds.; Springer-Verlag: Berlin, 1968, pp 45-122.
19. Higuchi, T. *Wood Sci. Technol.* **1990**, *24*, 23-63.
20. Dean, J. F. D.; Eriksson, K.-E. L. *Holzforschung* **1994**, *48 Suppl.*, 21-33.
21. Yamaguchi, H.; Nakatsubo, F.; Katsura, Y.; Murakami, K. *Holzforshung* **1990**, *44*, 381-385
22. Chiba, M.; Tsukamoto, H.; Hisada, S.; Nishibe, S. *Shoyakugaku Zasshi* **1979**, *33*, 150-154.
23. Ishii, H.; Ishikawa, T.; Mihara, M.; Akaike, M. *Yakugaku Zasshi* **1983**, *103*, 279-292.
24. Katayama, T.; Masaoka, T.; Yamada, H. *Mokuzai Gakkaishi*, **1997**, *43*, 580-588.
25. Weinges, K. *Chem. Ber.* **1961**, *94*, 2522-2533.
26. Katayama, T.; Masaoka, T. *Tech. Bull. Fac. Agr. Kagawa Univ.* **1994**, *46*, 117-125.
27. Umezawa, T.; Shimada, M. *Biosci. Biotech. Biochem.* **1996**, *60*, 736-737.
28. Forrest, J. E.; Heacock, R. A.; Forrest, T. P. *J. Chem. Soc., Perkin Trans. I* **1974**, 205-209.
29. Isogai, A.; Murakoshi, S.; Suzuki, A.; Tamura, S. *Agric. Biol. Chem.* **1973**, *37*, 889-895.
30. Isogai, A.; Murakoshi, S.; Suzuki, A.; Tamura, S. *Agric. Biol. Chem.* **1973**, *37*, 1479-1486.
31. Hattori, M.; Hada, S.; Shu, Y.-Z.; Kakiuchi, N.; Namba, T. *Chem. Pharm. Bull.* **1987**, *35*, 668-674.
32. Hada, S.; Hattori, M.; Tezuka, Y.; Kikuchi, T.; Namba, T. *Phytochemistry* **1988**, *27*, 563-568.

33. Miyazawa, M.; Kasahara, H.; Kameoka, H. *Proc. 39th Lignin Symp. (Fukuoka)* **1994**, 45-48
34. Barata, L. E. S.; Baker, P. M.; Gottlieb, O. R.; Ruveda, E. A. *Phytochemistry* **1978**, *17*, 783-786.
35. Gottlieb, O. R. *Fortschr. Chem. Org. Naturstoffe* **1978**, *16*, 1-72.
36. Zacchino, S. A.; Badano, H. *J. Nat. Prod.* **1985**, *48*, 830-832.
37. Zacchino, S. A.; Badano, H. *J. Nat. Prod.* **1988**, *51*, 1261-1265.
38. Zacchino, S. A.; Badano, H. *J. Nat. Prod.* **1991**, *54*, 155-160.
39. Zacchino, S. A. *J. Nat. Prod.* **1994**, *57*, 446-451.
40. Kasahara, H.; Miyazawa, M.; Kameoka, H. *Phytochemistry* **1995**, *40*, 1515-1517.
41. Katayama, T.; Okamoto, S. *Abstr. 46th Ann. Mtg. Japan Wood Res. Soc. (Kumamoto)* **1996**, 408.
42. Katayama, T.; Nakatsubo, F.; Higuchi, T. *Arch. Microbiol.* **1980**, *126*, 127-132
43. Katayama, T.; Nakatsubo, F.; Higuchi, T. *Arch. Microbiol.* **1981**, *130*, 198-203
44. Katayama, T.; Sogo, M.; Higuchi, T. *Holzforschung* **1986**, *40*, 175-182.
45. Katayama, T.; Sogo, M. *Mokuzai Gakkaishi* **1989**, *35*, 1116-1124
46. Katayama, T. *Mem. Fac. Agric. Kagawa Univ.* **1989**, *53*, 1-97.
47. Bradford, M. M. *Anal. Chem.* **1976**, *72*, 248-252.
48. Iwahawa, S.; Kuwahara, M.; Higuchi, T. *Hakkokogaku Kaishi* **1977**, *55*, 325-329.
49. Katayama, T.; Nakatsubo, F.; Higuchi, T. *Mokuzai Gakkaishi* **1981**, *27*, 223-230.

Chapter 24

Stereochemical Differences in Lignan Biosynthesis Between *Arctium lappa, Wikstroemia sikokiana,* and *Forsythia* spp.

Toshiaki Umezawa, Tomoya Okunishi, and Mikio Shimada

Wood Research Institute, Kyoto University, Uji, Kyoto 611, Japan

During the past seven years, much has been learned about lignan biosynthesis in *Forsythia* spp., which produce (–)-secoisolariciresinol, (–)-matairesinol and many other lignans. Enzymatic formation of the naturally occurring enantiomers of *Forsythia* lignans has been achieved with enzyme preparations from the plant, and an outline of the stereochemical control mechanisms for *Forsythia* lignan biosynthesis has been established. On the other hand, there are many examples of naturally occurring lignans which are enantiomers of *Forsythia* lignans. Recently, (+)-secoisolariciresinol (78% enantiomeric excess, e.e.), which is the enantiomer of that occurring in *Forsythia* spp., was isolated from *Arctium lappa.* Moreover, cell-free extracts of this plant were found to catalyze the enantioselective formation of (+)-secoisolariciresinol (20% e.e.) from its achiral precursor, coniferyl alcohol. In addition, six lignans have been isolated from *Wikstroemia sikokiana*: (+)-matairesinol, (+)-kusunokinin, (+)-wikstromol, (–)-pinoresinol, (–)-lariciresinol, and (–)-secoisolariciresinol. The predominant enantiomers of pinoresinol, lariciresinol, and matairesinol from *W. sikokiana* are the opposite of those occurring in *Forsythia* spp. On the basis of these results, the diversity in stereochemical mechanisms for lignan biosynthesis in *A. lappa*, *W. sikokiana* and *Forsythia* spp. is discussed in this review.

The biosynthesis of lignans has been receiving widespread interest in many respects. This is mainly due to these compounds having unique stereochemical properties as well as a number of biological activities (*1-3*), *e.g.* antitumor, antimitotic, and antiviral properties.

Concerning their stereochemistry, most lignan molecules are chiral, although there are a small number of exceptions, such as mesosecoisolariciresinol and arylnaphthalenes which are achiral (*4*). The optical rotation of a particular lignan preparation can vary with the plant source (see structures **1-10**). For example, (–)-arctigenin **6** was isolated from *Forsythia intermedia* (*5-7*), whereas the

(+)-enantiomer **6** was obtained from *Wikstroemia indica* (*8*). Thus the stereochemical mechanisms involved in lignan biosynthesis may be different among different plants, and an overall elucidation of the stereochemical mechanisms involved may provide new insights into the field of asymmetric organic synthesis.

Enzymatic studies of lignan biosynthesis mostly have employed *Forsythia* plants as enzyme sources (*9-18*). In 1990 the first example of an enzymatic reaction to produce an optically pure lignan was demonstrated in the Lewis laboratory with cell-free extracts of *Forsythia intermedia* (*9*). The crude enzyme preparation catalyzed enantioselective formation of optically pure (–)-secoisolariciresinol **4** from the achiral coniferyl alcohol precursor **1** in the presence of NADPH and H_2O_2. Since then, the Lewis laboratory has defined the precise enzymology involved in lignan biosynthesis with *Forsythia* plants (*10-18*). On the other hand, little was known about the enzymes from plants other than *Forsythia* spp., especially those catalyzing selective formation of the enantiomers of *Forsythia* lignans.

Recently, we have reported the preferential enantioselective formation of (+)-secoisolariciresinol **4** in small enantiomeric excess (20%), which is the enantiomer of the one from *Forsythia* plants, from coniferyl alcohol **1** using cell-free extracts of *Arctium lappa* (*19*). Also, we have isolated several lignans from *Wikstroemia sikokiana*, and many were enantiomers of those isolated from *Forsythia* plants (*20, 21*). In this review, we discuss the differences in the stereochemical mechanisms of lignan biosynthesis in *A. lappa*, *W. sikokiana*, and *Forsythia* plants.

Enantiomeric Composition of Lignans Isolated from Various Plants

Since the enantiomeric composition of a particular lignan can vary with plant source, it is important to determine the precise values for any adopted for biosynthetic studies. They can be calculated from the optical rotation ([α]) which is usually reported in accounts of lignan isolation. However, the [α] values are often not enough to calculate the enantiomeric composition explicitly, because the $[\alpha]_{max}$ is not always known. Recently, the enantiomeric composition of several lignans has been determined precisely by chiral HPLC techniques.

***Forsythia* Lignans.** *Forsythia* plants contain large amounts of dibenzylbutyrolactone lignans as well as others based on furan, furofuran, and dibenzylbutane structures (*5-18*): (–)-matairesinol **5**, (–)-arctigenin **6**, and (–)-arctiin **11** (dibenzylbutyrolactones); (+)-pinoresinol **2**, (+)-epipinoresinol **9**, and (+)-phillygenin **10** (furofurans); (+)-lariciresinol **3** (furan); and (–)-secoisolariciresinol **4** (dibenzylbutane) (Figure 1).

It should be noted that the *Forsythia* lignans have the same absolute configuration at C8 and C8′. That is, they can be interconverted by chemical means with retention of configuration (*2, 22-24*).

Chiral HPLC analysis confirmed that the lignans isolated from *Forsythia intermedia* were optically pure, since no peaks corresponding to their respective enantiomers were detected on the chiral HPLC chromatograms (*9-16*). On the other hand, with *Forsythia koreana*, although **4**, **5**, **6**, **9**, and **10** were optically pure, the chromatograms for pinoresinol **2** and lariciresinol **3** were not, *i.e.* while the (+)-antipodes were the major enantiomer, the (–)-forms (based on HPLC) were also present—to a smaller amount (*7*). However, the identities of the peaks remain to be confirmed.

***Arctium lappa* Lignans.** In 1929 Shinoda and Kawagoye isolated arctiin **11** from seeds of *A. lappa*, the content of which was relatively high (*ca.* 2% of seeds) (*25*). Omaki showed that the aglycone, arctigenin **6**, was levorotatory (*26*). At present the seeds of *Arctium* plants are known to contain the following dibenzylbutyrolactone lignans: **11** in *A. lappa* (*25, 27-29*), *Arctium minus* (*27*), *Arctium tomentosum* (*27*), and *Arctium leiospermum* (*30*); matairesinol **5** (*28*) and (–)-arctigenin **6** (*28, 29*) from *A. lappa* (Figure 1). In addition, dibenzylbutyrolactone type sesqui- and dilignans were isolated from the seeds of *A. lappa* (*28, 29, 31-33*).

In order to determine the enantiomeric compositions of the *Arctium* dibenzylbutyrolactone lignans, we isolated matairesinol **5** and arctigenin **6** from MeOH extracts of *A. lappa* cv. Kobarutogokuwase seeds after β-glucosidase treatment, and submitted them to chiral HPLC analysis (Umezawa, T., Shimada, M., unpublished data). Both lignans were found to be the optically pure (–) antipodes, as were those from *Forsythia* spp. (*9-16*). In contrast to the dibenzylbutyrolactone lignans, however, secoisolariciresinol **4,** which was isolated from MeOH extracts of *A. lappa* cv. Kobarutogokuwase petioles after β-glucosidase treatment (0.0016% yield based on dried petioles), was not optically pure, and whose enantiomeric composition was 78% enantiomeric excess (e.e.) in favor of the (+)-enantiomer **4** (*19*). The result was quite interesting, because **4** and the levorotatory dibenzylbutyrolactone lignans, **5** and **6**, have opposite configurations with respect to each other at C8 and C8′ *i.e*. reduction of **5** with retention of configuration yields (–)-secoisolariciresinol **4** but not the (+)-enantiomer **4**. This contrasts sharply with the occurrence of the levorotatory and optically pure **4** as well as **5** and **6** in *Forsythia* plants (*9-16*).

Thymelaeaceae Lignans. As mentioned in the previous sections, levorotatory dibenzylbutyrolactone lignans have been isolated, sometimes as aglycones after hydrolysis of the corresponding glycosides, from many plants (*1, 2*) including *Forsythia* spp. (*5-7, 10*), *Arctium lappa* (*26, 29, 34*), and *Trachelospermum asiaticum* var. *intermedium* (*35*); these include (–)-matairesinol **5**, (–)-arctigenin **6**, and (–)-nortrachelogenin (= wikstromol) **7**. In contrast, *Wikstroemia*, *Passerina* and *Daphne* plants (Thymelaeaceae) produce dextrorotatory dibenzylbutyrolactone lignans: (+)-wikstromol **7** from *W. indica* (*8, 36, 37*), *Wikstroemia viridiflora* (= *indica*) (*38*), *Wikstroemia foetida* var. *oahuensis* (*39*), *Wikstroemia uva-ursi* (*39*), *Passerina vulgaris* (*40*), and *Daphne odora* (*41*); (+)-arctigenin **6** from *W. indica* (*8*); and (+)-matairesinol **5** from *D. odora* (*41*).

Furofuran and furan lignans have also been isolated from *Wikstroemia*, *Daphne*, *Passerina*, and *Dirca* plants (Thymelaeaceae) (*38, 40, 42-47*). Again many of them have the opposite configuration at C8 and C8′ to those of the (+)-pinoresinol **2** and (+)-lariciresinol **3** isolated from *Forsythia* plants (*5, 7, 11-15, 17*) as in the following examples: (–)-pinoresinol **2** (*42*), (–)-lariciresinol **3** (*42*) and (–)-dihydrosesamin **12** (*42, 43*) from *Daphne tangutica*; and **3** (*45*) from *Dirca occidentalis*. However, two instances of the isolation of dextrarotatory furofuran lignans, (+)-pinoresinol **2** from *W. viridiflora* (*38*) and (+)-syringaresinol **13** from *P. vulgaris* (*40*) and *D. occidentalis* (*45*), have been reported, and even racemic furan lignans, (±)-lariciresinols **3** and its monomethoxy analog (±)-5-methoxylariciresinols, have been isolated from *Wikstroemia elliptica* (*46*).

Since no precise determination of enantiomeric compositions based on chiral HPLC techniques had been reported for the Thymelaeaceae lignans, a survey of lignans from *W. sikokiana* was conducted, and the enantiomeric compositions of the isolated lignans were deduced (*20, 21*). Pinoresinol **2**, lariciresinol **3**, secoisolariciresinol **4**, matairesinol **5**, wikstromol **7** and kusunokinin **8** were isolated from MeOH extracts of stems of the plant. Chiral HPLC analysis showed that the dibenzylbutyrolactone lignans **5**, **7** and **8** thus obtained were dextrorotatory and optically pure. On the other hand, **2**, **3** and **4** were not optically pure, exhibiting the following enantiomeric excess values: **2**, 74% e.e., (–)>(+); **3**, 39% e.e., (–)>(+); and **4**, 53% e.e., (–)>(+).

The isolation of the dextrorotatory dibenzylbutyrolactone lignans, (+)-matairesinol **5**, (+)-wikstromol **7** and (+)-kusunokinin **8,** from *W. sikokiana* accords well with the previous isolation of **5**, **6** and **7** from other Thymelaeaceae plants (*8, 36-41*). These results suggest that the occurrence of the dextrorotatory dibenzylbutyrolactone lignans (**5**, **6**, **7** and/or **8**] is characteristic of these plants. In contrast, the optical rotation of a furofuran lignan can vary even within plants of the same genus: (–)-pinoresinol **2** (74% e.e.) was isolated from *W. sikokiana* (*20*), but the (+)-antipode **2** was obtained from *W. viridiflora* (*38*). A comparable difference in enantiomeric composition was observed for a furan lignan isolated from *Wikstroemia* plants: **(–)-3**

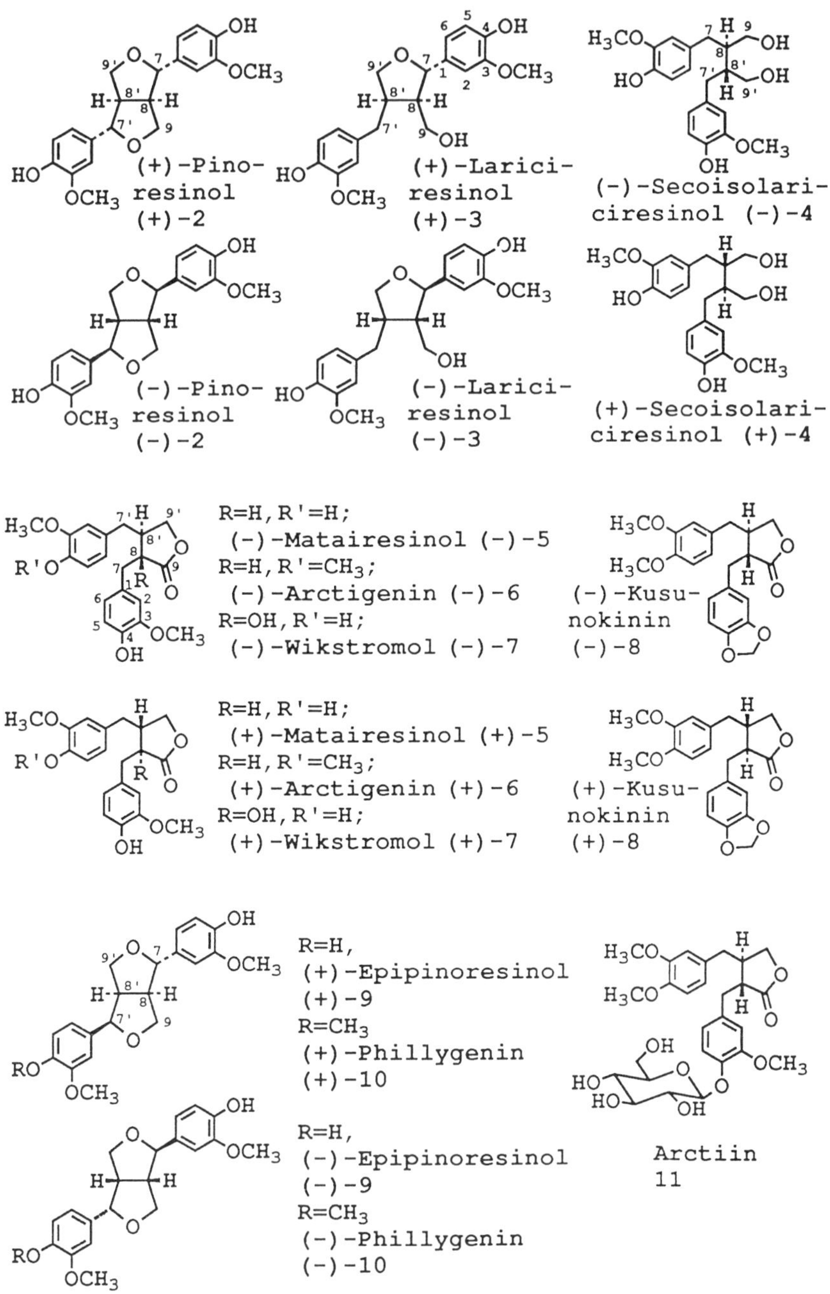

Figure 1. Structures of Lignans.

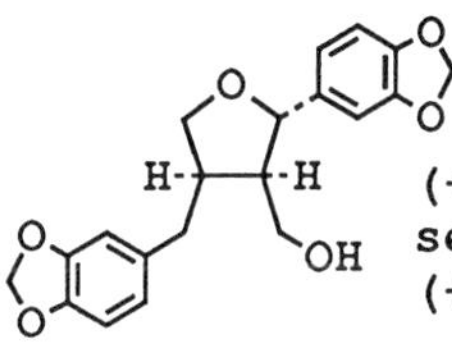

(+)-Dihydro-sesamin
(+)-12

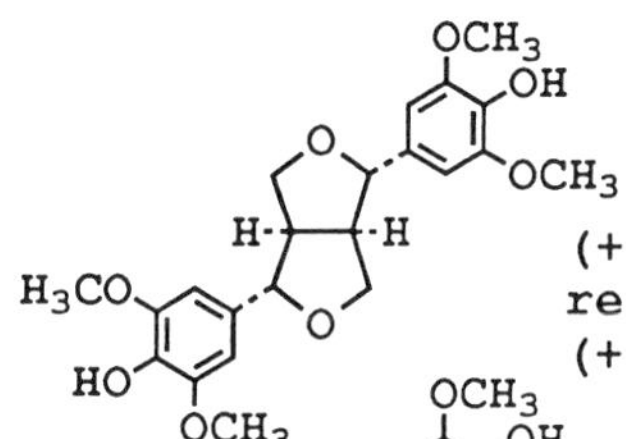

(+)-Syringa-resinol
(+)-13

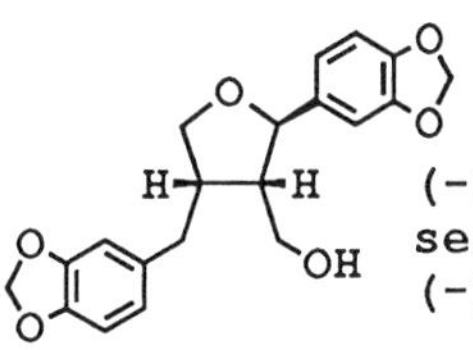

(-)-Dihydro-sesamin
(-)-12

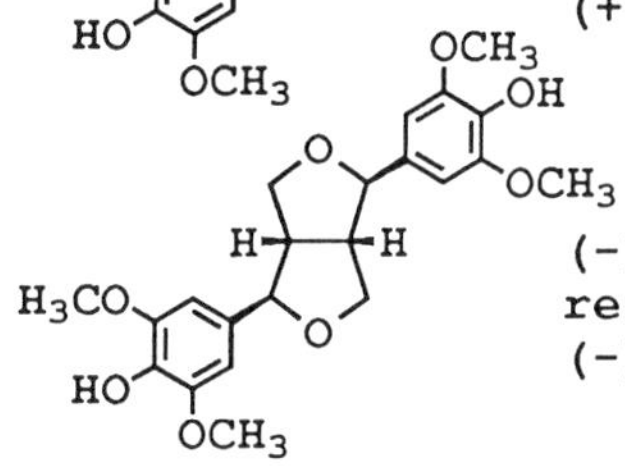

(-)-Syringa-resinol
(-)-13

Figure 1. *Continued.*

(39% e.e.) was extracted from *W. sikokiana* (*21*), while racemic **(±)-3** was obtained from *W. elliptica* (*46*).

Stereochemistry of Lignan Biosynthesis

Stereochemistry of Lignan Formation with *Forsythia* Enzymes. Most of our knowledge about the enzymatic basis of lignan biosynthesis has been obtained with *Forsythia* spp. Since 1990 many studies of lignan formation by *Forsythia* enzymes have been reported (*9-18, 48, 49*), mostly by Lewis and co-workers, and the overall conversion of coniferyl alcohol **1** to the natural enantiomers of *Forsythia* lignans by *Forsythia* enzyme preparations was established to occur as shown in Figure 2. Recently, the gene of pinoresinol/lariciresinol reductase was cloned (*48*), this being the first example of the cloning of a gene for an enzyme involved in lignan biosynthesis.

Each conversion, except for the final methylation, is well-controlled stereochemically; (+)-pinoresinol **2** is enantioselectively formed from achiral **1**, and then **2**, (+)-lariciresinol **3**, and **4** are transformed preferentially over their antipodes into almost optically pure **3**, **4**, (–)-matairesinol **5**, respectively. Very recently, a 78-kilodalton protein lacking a catalytically active (oxidative) center has been found to be involved in the enantioselective formation of (+)-pinoresinol **2** from coniferyl alcohol **1** (*49*).

These enzymatic studies indicate that the complete biosynthetic pathway from **1** to matairesinol **5** *via* pinoresinol **2**, lariciresinol **3**, and secoisolariciresinol **4** (Figure 2) is present in *Forsythia* plants. Regarding the intermediacy of **2** and **3** in the *in vivo* transformation, however, an alternative but remote possibility could not be ruled out. As shown in Figure 2, quinone methides **14-16** were proposed as conceivable intermediates in each step of the enzymatic conversion of **1** to **4** (*14, 15, 17, 18*). If the enzymes involved were located closely enough together in the plants so that the quinone methides **14** and **16** were trapped by the active site of the next reductase before furan ring formation could occur, **4** might be formed directly through quinone methides **14** and **16** but not *via* **2** and **3**. Studies with purified enzymes would be required before final conclusions can be drawn about which intermediates are ultimately involved.

Enantioselective Formation of (+)-Secoisolariciresinol with Cell-free Extracts of *Arctium lappa*. Recently enantioselective formation of (+)-secoisolariciresinol **4** was demonstrated with cell-free extracts from petioles of *Arctium lappa* cv. Kobarutogokuwase (*19*) (Figure 3a). When [9-$^2H_2$]coniferyl alcohol **1-*d*$_2$** was incubated with the cell-free extracts in the presence of NADPH and H_2O_2, [$^2H_4$]secoisolariciresinol **4-*d*$_4$** was formed. Chiral LC-MS analysis indicated that the (+)-enantiomer **4-*d*$_4$** predominated in the enzymatically formed **4-*d*$_4$** but only to the extent of 20% enantiomeric excess (e.e.) (*19*). Nevertheless, this is the first report of the preferential enzymatic formation of (+)-secoisolariciresinol **4**.

The result with *A. lappa* cv. Kobarutogokuwase (Figure 3a) (*19*) is in marked contrast to those obtained with *Forsythia* plants (Figure 3b) (*9, 17*). The (+)-enantiomer of secoisolariciresinol **4** was also the predominant one (78% e.e.) isolated from *A. lappa* cv. Kobarutogokuwase petioles. On the other hand, optically pure (–)-secoisolariciresinol **4** was isolated from *Forsythia* plants (*7, 10*), and this was formed regio- and sterospecifically from coniferyl alcohol **1** by *Forsythia* enzymes (Figure 3b) (*9, 17*).

Besides **(+)-4**, optically pure (–)-matairesinol **5** and (–)-arctigenin **6** were isolated from the MeOH extracts of *A. lappa* cv. Kobarutogokuwase seeds after β-glucosidase treatment. Since **(–)-4**, but not **(+)-4**, has the same configuration at C8 and C8′ as does **(–)-5** and **(–)-6**, it is unlikely that the *A. lappa* plant converts **(+)-4** or its glycoside directly into **(–)-5** with inversion of configuration. This is supported by the finding that *Forsythia* enzymes convert only (–)-secoisolariciresinol **4**, but not **the** corresponding enantiomer, to (–)-matairesinol **5** (*10, 11, 14, 15*). In any event, the

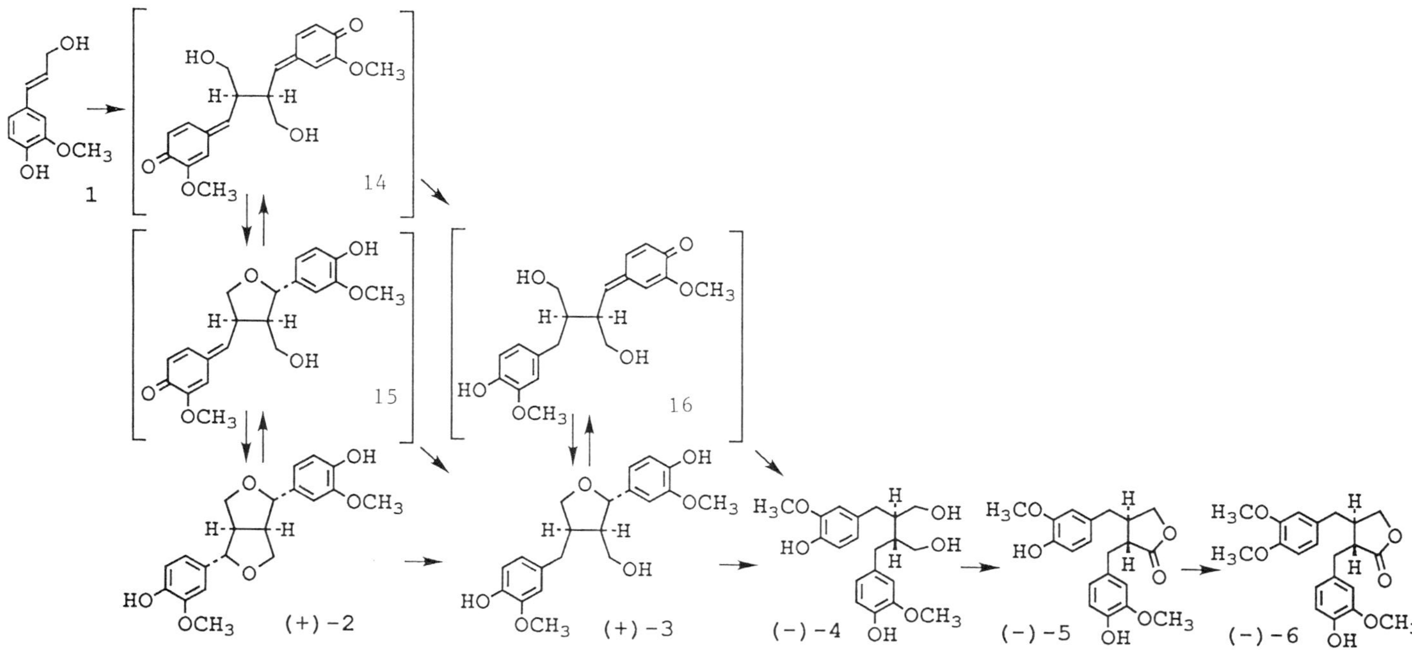

Figure 2. Enzymatic Formation of *Forsythia* Lignans.

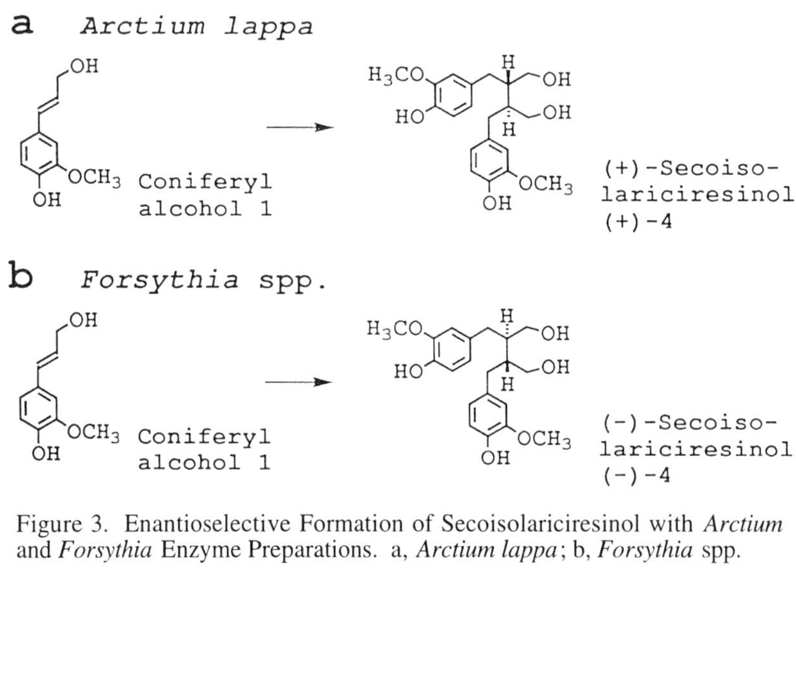

Figure 3. Enantioselective Formation of Secoisolariciresinol with *Arctium* and *Forsythia* Enzyme Preparations. a, *Arctium lappa*; b, *Forsythia* spp.

present result clearly indicates that the stereochemical mechanism for formation (or accumulation) of the lignans in *A. lappa* differs from those in *Forsythia* plants, although the precise mechanism that accounts for the observed enantiomeric composition of **4** and other lignans in *A. lappa* remains unknown.

Proposed Biosynthetic Pathway to Lignans in *Wikstroemia sikokiana*. Analysis of the enantiomeric compositions of the *W. sikokiana* lignans has indicated that the stereochemical control mechanisms involved in lignan biosynthesis in *W. sikokiana* are different from those in *Arctium lappa* and *Forsythia* spp.

Wikstroemia sikokiana and *Arctium lappa* produce (or accumulate) different enantiomers of dibenzylbutyrolactone lignans (*20, 21,* Umezawa, T., Shimada, M., unpublished data). Secoisolariciresinols **4** isolated from these plants were mixtures of both enantiomers, and the predominant ones were the opposite to each other (*19, 21*). Hence, it is obvious that the stereochemical mechanisms of lignan biosynthesis in both plants are different.

As for the differences between *Wikstroemia sikokiana* and *Forsythia* spp., two aspects can be pointed out.

First, although the enantiomeric composition of **4** from *W. sikokiana* apparently favored formation of the (–)-enantiomer **4** (*21*) as did that from *Forsythia* plants (*7, 10, 13-15*), the predominant enantiomers of pinoresinol **2**, lariciresinol **3**, and matairesinol **5** isolated from *W. sikokiana* were opposite to those obtained from *Forsythia* plants (*5-7, 10-15, 17*). In addition, the optically pure dibenzylbutyrolactone lignans from *W. sikokiana*, (+)-matairesinol **5**, (+)-wikstromol **7**, and (+) kusunokinin **8** (*20, 21*), have the same configurations at C8 and C8′ as each other (Figure 1), which are opposite to those of the *Forsythia* dibenzylbutyrolactone lignans, (–)-matairesinol **5**, (–)-arctigenin **6**, and arctiin **11** (*5-7, 10*). Thus it is obvious that different stereochemical mechanisms are operating in plants that produce (or accumulate) this different series of enantiomeric lignans.

Second, the other aspect pertains to the metabolic steps that produce optically pure lignans. The three lignans from *W. sikokiana,* (–)-pinoresinol **2** (74% e.e.), (–)-lariciresinol **3** (39% e.e.), and (–)-secoisolariciresinol **4** (53% e.e.), are not optically pure (*20, 21*). The enantiomeric purities are much lower than that of the optically pure **4** isolated from *Forsythia* plants (*7, 10*), and those of the lignans obtained from *in vitro* reactions with *Forsythia* enzymes [*viz.* (+)-pinoresinol **2**, more than 97% e.e. (*18*); (+)-lariciresinol **3**, almost optically pure (*14*); and (–)-secoisolariciresinol **4**, optically pure (*9, 14, 17*)]. The findings indicated that the formation of these lignans in *W. sikokiana* was less enantioselective than that in *Forsythia* plants. Catalysis by less enantioselective enzymes or contributions from two types of enzymes, highly enantioselective enzymes to produce optically pure lignans and non enantioselective enzymes to afford racemic lignans, could account for the enantiomeric composition of the *Wikstroemia* lignans.

In contrast to **2**, **3** and **4,** which are not optically pure, the dibenzylbutyrolactone lignans isolated from *W. sikokiana*, **(+)-5**, **(+)-7**, and **(+)-8**, were found to be optically pure (*20, 21*). Since feeding experiments with deuterium labeled coniferyl alcohol and lignans strongly suggested that the conversion of coniferyl alcohol **1** to **5** takes place *via* **4** (Figure 4) in *W. sikokiana* (*50*), not only the initial formation of **2** (or the corresponding bis(quinone methide)) from **1** but also the post-coupling processes to afford the dibenzylbutyrolactone lignans, especially the oxidation of **4** to **5**, may involve the selective formation of only one enantiomer of these lignans, *i.e.* the optically pure lignans in the plant. This is in marked contrast to lignan biosynthesis in *Forsythia* plants; the coupling to give **(+)-2** by the *Forsythia* enzyme system is highly enantioselective (Figure 2) (*18, 49*). The precise stereochemical mechanisms involved in formation of the optically pure dibenzylbutyrolactone lignans in *W. sikokiana* will be the subject of a future study with isolated enzymes.

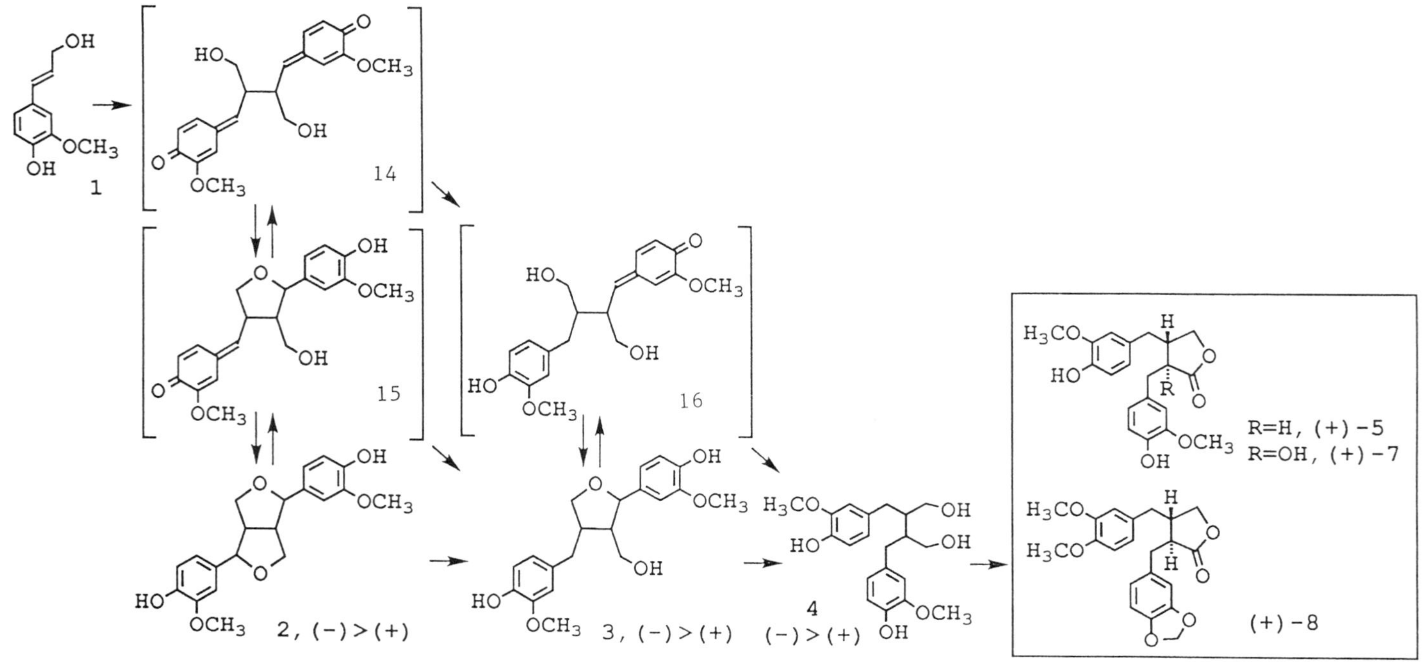

Figure 4. Proposed Pathway for the Biosynthesis of *Wikstroemia sikokiana* Lignans.

Conclusions

An analysis of the enantiomeric compositions of lignans isolated from *Forsythia* spp., *Arctium lappa*, and *Wikstroemia sikokiana* has indicated that these plants produce (or accumulate) different enantiomers of these lignans with various enantiomeric compositions. Some are optically pure while others are mixtures of both enantiomers but not racemic. Enantioselective formation of the naturally predominating enantiomers of some of the lignans has been demonstrated with enzyme preparations from the plants in question.

These results indicate that different stereochemical mechanisms are operating which give rise to the different enantiomers in these plants, and that the metabolic steps which produce the optically pure lignans are probably different in the plants. Thus there is a great diversity in the stereochemical mechanisms for lignan biosynthesis in *Forsythia*, *Arctium* and *Wikstroemia* plants.

Literature Cited

1. *Chemistry of Lignans*; Rao, C. B. S., Ed.; Andhra University Press: Andhra Paradesh, India, 1978.
2. *Lignans: Chemical, Biological and Clinical Properties;* Ayres, D. C., Loike, J. D., Eds.; Cambridge University Press: Cambridge, 1990.
3. Umezawa, T. *Mokuzai Gakkaishi* **1996**, *42*, 911-920.
4. Umezawa, T. In *Biochemistry and Molecular Biology of Wood*; Higuchi, T., Ed.; Springer-Verlag: Berlin, 1997; pp 181-194.
5. Kitagawa, S.; Nishibe, S.; Benecke, R.; Thieme, H. *Chem. Pharm. Bull.* **1988**, *36*, 3667-3670.
6. Rahman, M. M. A.; Dewick, P. M.; Jackson, D. E.; Lucas, J. A. *Phytochemistry* **1990**, *29*, 1971-1980.
7. Umezawa, T.; Isohata, T.; Kuroda, H.; Higuchi, T.; Shimada, M. In *Biotechnology in Pulp and Paper Industry*; Kuwahara, M., Shimada, M., Eds.; Uni Publishers: Tokyo, 1992; pp 507-512.
8. Suzuki, H.; Lee, K.-H.; Haruna, M.; Iida, T.; Ito, K.; Huang, H.-C. *Phytochemistry* **1982**, *21*, 1824-1825.
9. Umezawa, T.; Davin, L. B.; Lewis, N. G. *Biochem. Biophys. Res. Commun.* **1990**, *171*, 1008-1014.
10. Umezawa, T.; Davin, L. B.; Lewis, N. G. *J. Biol. Chem.* **1991**, *266*, 10210-10217.
11. Umezawa, T.; Davin, L. B.; Yamamoto, E.; Kingston, D. G. I.; Lewis, N. G. *J. Chem. Soc. Chem. Commun.* **1990**, 1405-1408.
12. Davin, L. B.; Bedgar, D. L.; Katayama, T.; Lewis, N. G. *Phytochemistry* **1992**, *31*, 3869-3874.
13. Katayama, T.; Davin, L. B.; Lewis, N. G. *Phytochemistry* **1992**, *31*, 3875-3881.
14. Katayama, T.; Davin, L. B.; Chu, A.; Lewis, N. G. *Phytochemistry* **1993**, *33*, 581-591.
15. Chu, A.; Dinkova, A.; Davin, L. B.; Bedgar, D. L.; Lewis, N. G. *J. Biol. Chem.* **1993**, *268*, 27026-27033.
16. Ozawa, S.; Davin, L. B.; Lewis, N. G. *Phytochemistry* **1993**, *32*, 643-652.
17. Umezawa, T.; Kuroda, H.; Isohata, T.; Higuchi, T.; Shimada, M. *Biosci. Biotech. Biochem.* **1994**, *58*, 230-234.
18. Paré, P. W.; Wang, H.-B.; Davin, L. B.; Lewis, N. G. *Tetrahedran Lett.* **1994**, *35*, 4731-4734.
19. Umezawa, T.; Shimada, M. *Biosci. Biotech. Biochem.* **1996**, *60*, 736-737.
20. Umezawa, T.; Shimada, M. *Mokuzai Gakkaishi* **1996**, *42*, 180-185.
21. Okunishi, T.; Umezawa, T.; Shimada, M. *Abstr. 46th Annu. Mtg. Japan Wood Res. Soc.* **1996**, 410.

22. Weinges, K.; Spänig, R. In *Oxidative Coupling of Phenols*; Taylor, W. I., Battersby, A. R., Eds.; Marcel Dekker: New York, NY, 1967; pp 323-355.
23. Stevenson, R. In *Chemistry of Lignans*; Rao, C. B. S., Ed.; Andhra University Press: Andhra Paradesh, India, 1978; pp 65-94.
24. Kato, Y.; Munakata, K. In *Chemistry of Lignans*; Rao, C. B. S., Ed.; Andhra University Press: Andhra Paradesh, India, 1978; pp 95-122.
25. Shinoda, J.; Kawagoye, M. *Yakugaku Zasshi* **1929**, *49*, 565-575.
26. Omaki, T. *Yakugaku Zasshi* **1935**, *55*, 816-827.
27. Hänsel, R.; Schulz, H.; Leuckert, C. *Z. Naturforsch.* **1964**, *19*, 727-734.
28. Yamanouchi, S.; Takido, M.; Sankawa, U.; Shibata, S. *Yakugaku Zasshi* **1976**, *96*, 1492-1493.
29. Han, B. H.; Kang, Y. H.; Yang, H. O.; Park, M. K. *Phytochemistry* **1994**, *37*, 1161-1163.
30. Yakhontova, L. D.; Kibalchich, P. N. *Kim. Prir. Soedin.* **1971**, *7*, 299-301.
31. Ichihara, A.; Oda, K.; Numata, Y.; Sakamura, S. *Tetrahedron Lett.* **1976**, 3961-3964.
32. Ichihara, A.; Numata, Y.; Kanai, S.; Sakamura, S. *Agric. Biol. Chem.* **1977**, *41*, 1813-1814.
33. Ichihara, A.; Kanai, S.; Nakamura, Y.; Sakamura, S. *Tetrahedron Lett.* **1978**, 3035-3038.
34. Földeák, S.; Hegyes, P.; Dombi, G. Y. *Acta Phys. Chem. Nova Ser.* **1974**, *20*, 459-463.
35. Nishibe, S.; Hisada, S.; Inagaki, I. *Phytochemistry* **1971**, *10*, 2231-2232.
36. Kato, A.; Hashimoto, Y.; Kidokoro, M. *J. Nat. Prod.* **1979**, *42*, 159-162.
37. Lee, K.-H.; Tagahara, K.; Suzuki, H.; Wu, R.-Y.; Haruna, M.; Hall, I. H.; Huang, H.-C.; Ito, K.; Iida, T.; Lai, J. S. *J. Nat. Prod.* **1981**, *44*, 530-535.
38. Tandon, S.; Rastogi, R. P. *Phytochemistry* **1976**, *15*, 1789-1791.
39. Torrance, S. J.; Hoffmann, J. J.; Cole, J. R. *J. Pharm. Sci.* **1979**, *68*, 664-665.
40. Guo, J.-X.; Handa, S. S.; Pezzuto, J. M.; Kinghorn, A. D.; Farnsworth, N. R. *Planta Med.* **1984**, *50*, 264-265.
41. Kogiso, S.; Wada, K.; Munakata, K. *Rept. 26th Internatl. Congr. Pure Appl. Chem.* (Tokyo) **1977,** 291.
42. Zhuang, L.-g.; Seligmann, O.; Jurcic, K.; Wagner, H. *Planta Med.* **1982**, *45*, 172-176.
43. Zhuang, L.-g.; Seligmann, O.; Lotter, H.; Wagner, H. *Phytochemistry* **1983**, *22*, 265-267.
44. Thusoo, A.; Raina, N.; Minhaj, N.; Ahmed, S. R.; Zaman, A. *Indian J. Chem.* **1981**, *20B*, 937-938.
45. Badawi, M. M.; Handa, S. S.; Kinghorn, A. D.; Cordell, G. A.; Farnsworth, N. R. *J. Pharm. Sci.* **1983**, *72*, 1285-1287.
46. Duh, C.-Y.; Phoebe, C. H., Jr.; Pezzuto, J. M.; Kinghorn, A. D.; Farnsworth, N. R. *J. Nat. Prod.* **1986**, *49*, 706-709.
47. Kreher, B.; Neszmélyi, A.; Wagner, H. *Phytochemistry* **1990**, *29*, 3633-3637.
48. Dinkova-Kostova, A. T.; Gang, D. R.; Davin, L. B.; Bedgar, D. L.; Chu, A.; Lewis, N. G. *J. Biol. Chem.*, **1996**, *271*, 29473-29482.
49. Davin, L. B.; Wang, H.-B.; Crowell, A. L.; Bedgar, D. L.; Martin, D. M.; Sarkanen, S.; Lewis, N. G. *Science* **1997**, *275*, 362-366.
50. Okunishi, T.; Umezawa, T.; Shimada, M. *Proc. 41st Lignin Symp* (Nagoya) **1996,** 29-32.

Chapter 25

The 'Abnormal Lignins': Mapping Heartwood Formation Through the Lignan Biosynthetic Pathway

David R. Gang, Masayuki Fujita, Laurence B. Davin, and Norman G. Lewis

Institute of Biological Chemistry, Washington State University, Pullman, WA 99164–6340

A significant portion of '*non-lignin*' phenolic extractives in heartwood tissues requires solubilization conditions normally used for lignin dissolution. Although they have incorrectly been characterized as 'abnormal' or 'secondary' lignins, their formation and accumulation differs profoundly from that leading to the lignins in three *primary* ways: first, they are transported through specialized cells (such as ray parenchyma) and are infused into surrounding pre-lignified cells; second, they are formed *via distinct* biochemical pathways, affording lignan or (iso)flavonoid-derived substances of various molecular sizes; third, their accumulation begins in the pith and over time extends out centrifugally towards the cambial regions, unless formed locally elsewhere as an inducible response. By contrast, lignification results *via* direct monomer transport from the cytoplasm of a lignifying cell into its polysaccharide-rich cell wall with subsequent polymerization; this process represents the first and final committed step to lignification, being primarily *initiated* and *completed* in maturing cell walls not far from the cambial zone. In this study, western red cedar, western hemlock and loblolly pine were examined to establish how their species-specific, *non-lignin* heartwood substances are biosynthesized. Specific enzymes and genes were obtained for pivotal steps leading to plicatic acid (western red cedar), α-conidendrin (western hemlock) and dihydrodehydrodiconiferyl alcohol and related structures (loblolly pine). In each case, coniferyl alcohol served as the initial precursor, being subsequently metabolized with precise regio- and stereochemical control to first give species-specific lignans from which the 'secondary', or 'abnormal lignins' derive. The results again underscore the need to systematically determine the precise temporal and spatial biochemical processes involved in phenolic coupling, as well as any subsequent metabolic events *in planta*. These findings pave the way to a full delineation of the temporal, spatial, tissue- and cell-specific events involved in heartwood formation. Ultimately this will offer new strategies for biotechnological manipulation. It will also lead to the adoption of a coherent terminology to define these *non-lignin* phenolics.

The durability, longevity and resistance to rot of different tree species are, in large part, dictated by their particular heartwood properties (*1*). It is also this portion of trees that either ends up as lumber, solid wood products, building materials and fine furniture or becomes used for pulp and paper production. Heartwood, which can account for more than 95% of the merchantable bole of mature wood, is often distinguished from neighboring sapwood by the amounts of so-called 'extractive' components. Depending on the species, these primarily consist of significant levels of colored phenolic compounds derived from lignans, tannins and (iso)flavonoids, although certain species have additional substances such as isoprenoids and alkaloids. The 'extractives' can often make up a considerable proportion of heartwood (*e.g.* the lignans in western red cedar [*Thuja plicata*]) (*2-10*). They also significantly help define the distinguishing features of particular woods, such as durability, color, quality, odor and texture. On the other hand, the presence of heartwood compounds can often adversely affect wood-pulping processes, leading to increased production costs associated with their removal. The term 'extractive' is a misnomer, however, since many heartwoods contain colored phenolic substances which cannot be fully removed by simple aqueous/solvent extraction; rather, they require more rigorous procedures normally used for lignin removal. Thus, many of these lignan or (iso)flavonoid-derived phenolics have been erroneously described as 'abnormal lignins' or 'secondary lignins', even though they are produced *via* distinct biochemical pathways.

Determining how heartwood formation occurs represents an important biotechnological goal, given that the specific characteristics of many woods (and hence biodiversity itself) are defined in large part by the nature of their particular 'extractive' constituents. Put in another way, any biotechnological endeavor to modify wood quality must also target heartwood, whose formation represents one of the largest unsolved questions in plant science. Heartwood's significance becomes even more apparent when its commercial value for all purposes, prior to mill processing, is taken into account: this value exceeds $135 *billion* annually in the U.S. alone (1990 figures). That is, heartwood is the largest contributor to all plant (agricultural and forestry) derived income.

The purpose of this chapter is to summarize our recent advances in the definition of lignan biosynthetic pathways, and how this is shedding light on the *general processes* involved in heartwood formation. Selected woody plants, *i.e.* western red cedar (*Thuja plicata*), western hemlock (*Tsuga heterophylla*) and loblolly pine (*Pinus taeda*), were chosen for investigation, each generating different lignan-derived heartwood metabolic products.

Theories of Heartwood Formation

Within living, growing trees exists the remnant of years, and in some cases, centuries past: the heartwood. An excellent and striking example of this can be seen from the cross-section of a mature tamarack larch (*Larix laricina*) stem that primarily contains darker heartwood (Figure 1). This contrasts markedly to the pale outer few centimeters of sapwood and the even paler thin ring (the so-called 'transition' zone) between both tissue types. Not all woody species, however, biosynthesize heartwood components to the same extent, or even use the same range of metabolic products. Consequently, woody plants can often visually differ dramatically depending on their particular heartwood constituents, such as western red cedar (red-brown), ebony (black) and southern pines (yellow-orange).

Regardless of the amount of metabolites actively deposited in heartwood, this tissue no longer participates in any of the growth processes associated with the living part of the tree. That is, it does not conduct photosynthate to the roots, which is transported in the phloem of inner bark. Nor does it conduct minerals and water upward to the leaves, which is the main function of sapwood, lying just inside the growing, cambial region. Nor does it directly inhibit growth of disease-causing organisms which primarily attack living, nutrient-rich, sapwood, since the latter

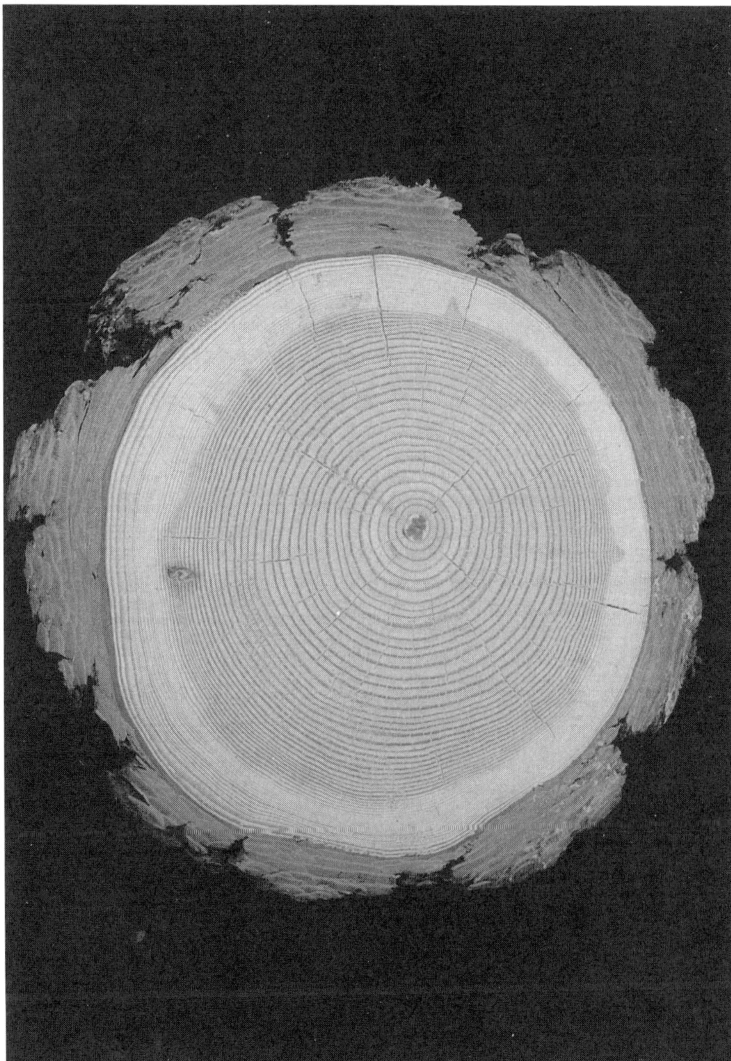

Figure 1. Cross-section of a tamarack larch (*Larix laricina*) log revealing both (dark, inner) heartwood and the (lighter, outer) sapwood zones.

effects are minimized by the compartmentalizing ability of trees, which wall off attacking organisms from living sapwood, cambium and phloem (*11-15*).

Heartwood does, however, play a significant role in the lives of trees. In addition to sapwood, it helps provide the mechanical strength necessary to stand upright in a world ruled by gravity, without which trees could not successfully compete in the struggle for light within the forest canopy. Accordingly, many woody plant species also have mechanisms to protect heartwood from rotting fungi and bacteria. Yet, how is this achieved when, being dead (*i.e.* possessing no living, metabolizing parenchyma cells), heartwood cannot actively respond to attack? As discussed in detail below, this protection involves the infusion of protective compounds (the so-called 'extractives') into pre-lignified sapwood.

Conflicting theories have arisen about the initiation and process of heartwood formation (*16*), which is formed in a manner highly variable between species. It has been proposed, for example, that an increasingly less aerobic environment exists toward the interior of trees, and as this condition progresses, cells die and heartwood forms. A second theory suggests that over time, air enters the tree trunks and that the resulting embolisms cause heartwood formation to begin. A third theory proposes that some heartwood-initiating hormone/substance (*17*) moves centripetally along the rays from the cambium, and once it reaches a minimal threshold level, heartwood formation occurs. A fourth theory states that it is simply an aging process, where organelles in parenchyma cells become progressively degraded as the distance from the cambium increases (*18-21*). This particular view is apparently contradicted, however, in species such as maple (*Acer*), oak (*Quercus*) and walnut (*Juglans*), which have larger nuclei in cells near the heartwood/sapwood boundary than in cells in the middle of the sapwood, suggesting that higher metabolic activity actually occurs toward the heartwood/sapwood boundary than towards the periphery ((*22*), in (*16*)). Other species such as *Pinus radiata* show an increase in metabolism at the sapwood/ heartwood boundary that is *seasonal* (*23, 24*). The fact that this enhanced metabolism in the transition zone occurs only during a small portion of the year may explain why investigators have not always observed the same phenomenon. Work on other species, such as *Nothofagus cunninghamii*, *Sloanea woollsii* and *Diospyros pentamera* (*25*), have demonstrated that active parenchyma cells continue to live for years right next to dead, colored cells, that could be mistaken for heartwood cells. In these species, the transition zone is very wide, although it does surround a core of heartwood where all of the cells are dead. More recent work on *Garuga pinnata* and *Ougeinia oojeinensis* (*26*) has also supported this finding. Indeed, it has been cautioned that initiation and mode of heartwood formation may actually differ between given species (*i.e.* simultaneous initiation has been observed in the ray and axial parenchyma cells in *Ougeinia* and initiation only in the ray parenchyma cells in *Garuga*), a point that has rarely been emphasized by the investigators in this field. This would perhaps begin to provide some explanation for the conflicting hypotheses that have plagued heartwood formation investigations for the last half century.

Perhaps the most controversial hypothesis for heartwood formation holds that the cells die because they are filled with lethal 'waste' products from external cells (*27, 28*). That is, heartwood metabolites are putatively transported in sub-toxic concentrations along the rays toward the interior of the plant, where they are then 'disposed of', thus performing a 'detoxication' process. This hypothesis proposes that compounds laid down in the heartwood, such as tannins, (iso)flavonoids, lignans and stilbenes, are simply by-products or 'wastes' of normal cellular metabolism. The authors even went so far as to state that lignin formation itself may be simply a 'detoxication' process to remove the 'excess' phenolic compounds derived through normal cellular respiration (*27, 28*) and that the role of lignin as a structural component of terrestrial plants may only be of *secondary evolutionary significance*!

Such an hypothesis has, however, many inconsistencies. For example, careful examination of the heartwood 'waste' products (see next section), reveal that they are most often optically active products of specific biosynthetic pathways (*e.g.* such as to (–)-plicatic acid **1** in *Thuja plicata*, see Figure 2) and that the actual metabolites can

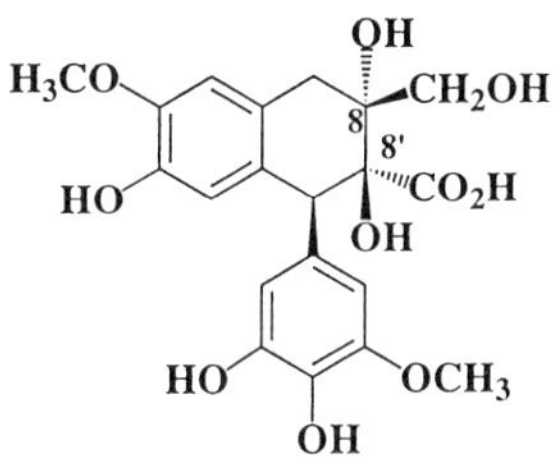

1: (–)-Plicatic acid

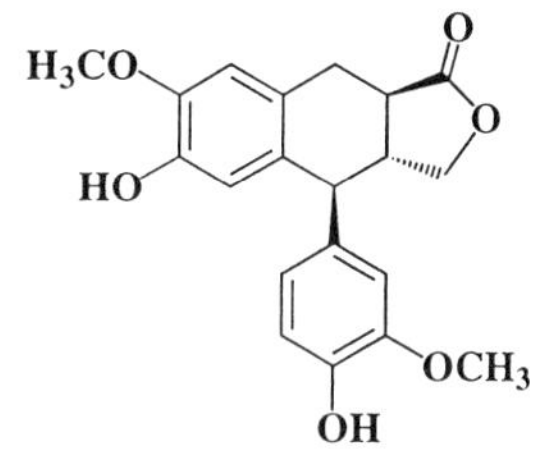

2: (–)-α-Conidendrin

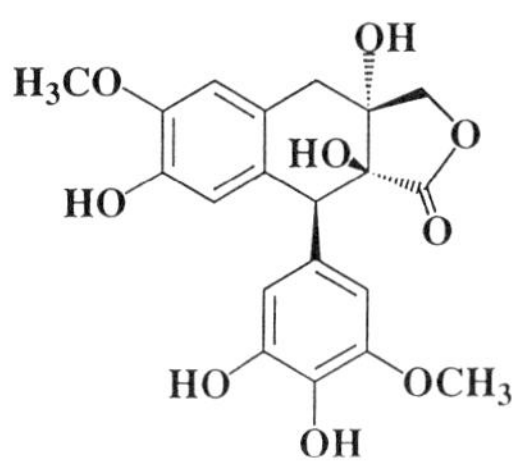

3: (–)-Plicatin

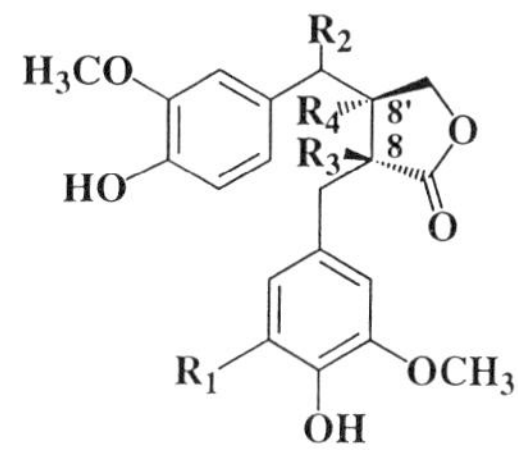

4 $R_1=R_2=R_3=R_4=H$: (–)-Matairesinol
8 $R_1=OH$, $R_2=R_3=R_4=H$: (–)-Thujaplicatin a
9 $R_1=R_3=R_4=OH$, $R_2=H$: (–)-Thujaplicatin b
10 $R_1=OCH_3$, $R_2=R_3=R_4=H$: (–)-Thujaplicatin methyl ether a (T.M.E. a)
11 $R_1=OCH_3$, $R_3=OH$, $R_2=R_4=H$: (–)-T.M.E. b
12 $R_1=OCH_3$, $R_3=R_4=OH$, $R_2=H$: (–)-T.M.E. c
13 $R_1=R_3=R_4=H$, $R_2=OH$: (–)-7′-Hydroxymatairesinol

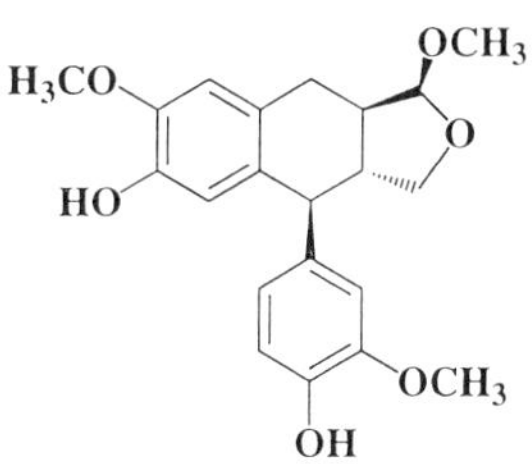

14: (+)-Tsugacetal

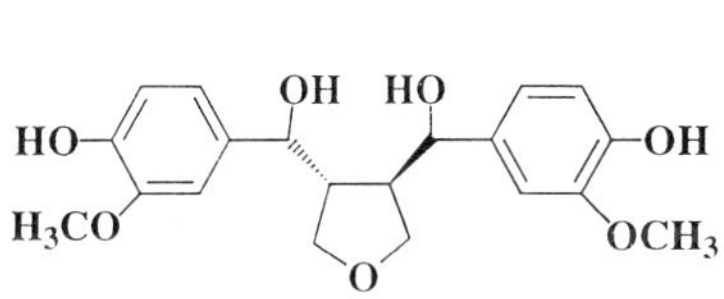

15: Liovil

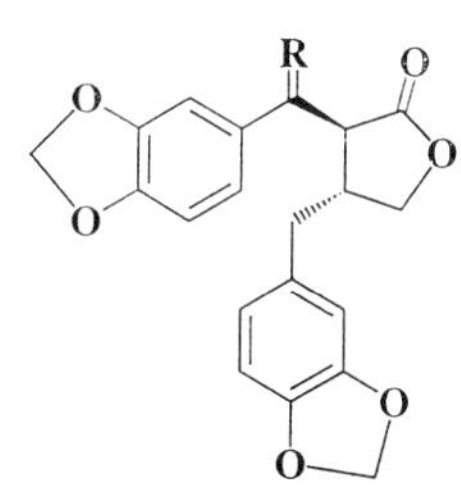

16 R=H,H: (–)-Hinokinin
17 R=O: (–)-7-Oxohinokinin

Figure 2a. Selected lignans listed in Table I.

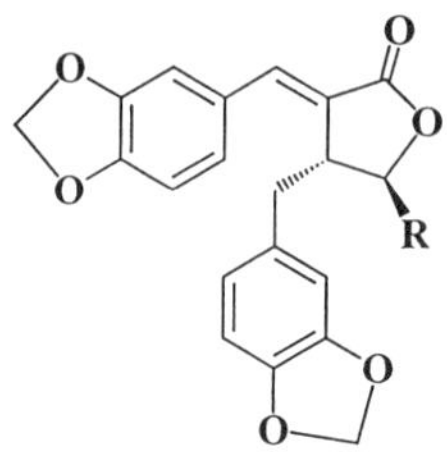

18 R=H: (–)-Savinin
19 R=OH: (+)-Calocedrin

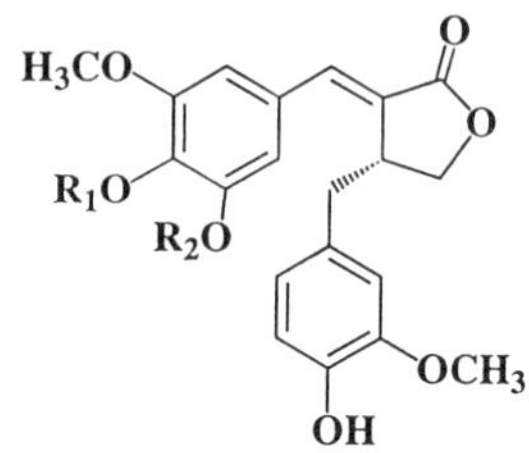

20 R_1=CH_3, R_2=H: (–)-Savininolide a
21 R_1=H, R_2=CH_3: (–)-Savininolide b

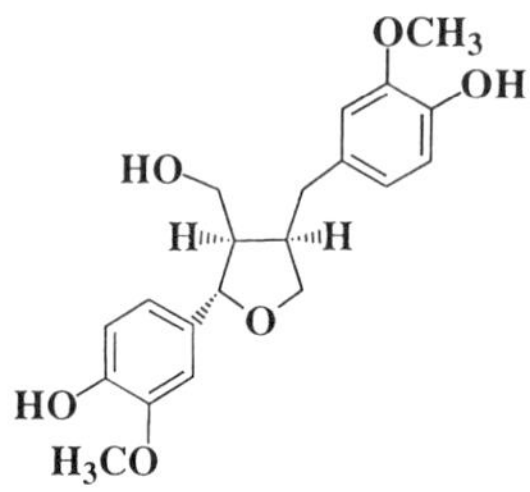

22a (+)-Lariciresinol

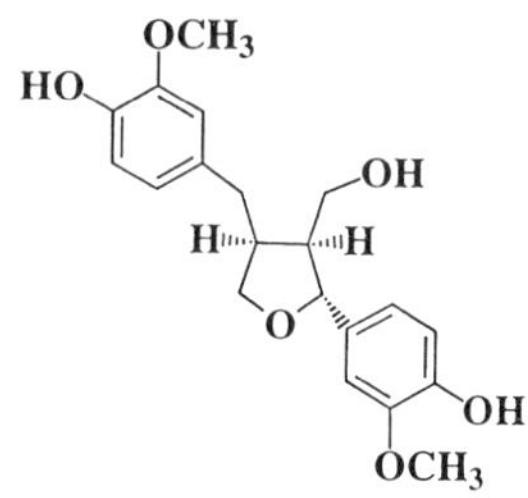

22b (–)-Lariciresinol

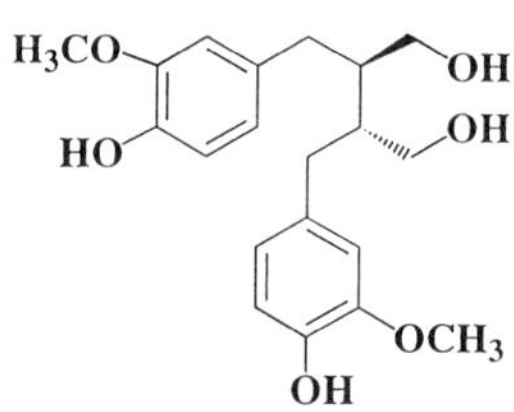

23a (–)-Secoisolariciresinol

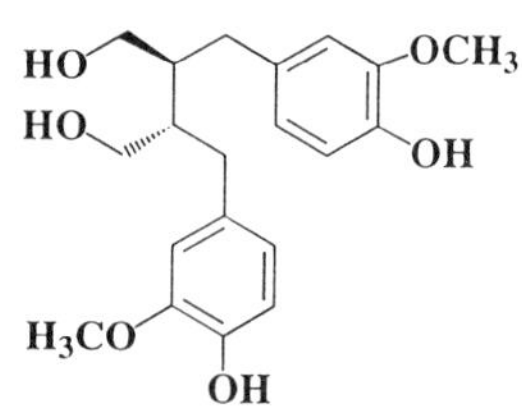

23b (+)-Secoisolariciresinol

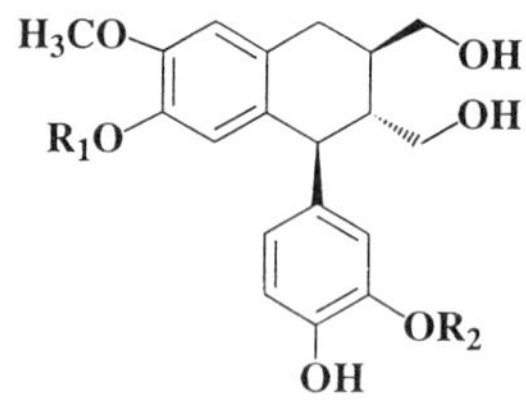

24 R_1=H, R_2=CH_3: (+)-Isolariciresinol
25 R_1=R_2=H: (+)-Isotaxiresinol
26 R_1=CH_3, R_2=H:
(+)-Isotaxiresinol-4-methyl ether

Figure 2b. Selected lignans listed in Table I (cont.).

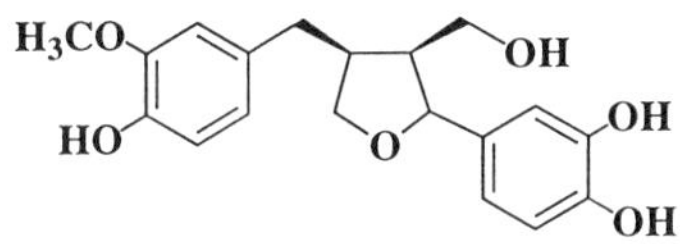

27: (+)-Taxiresinol

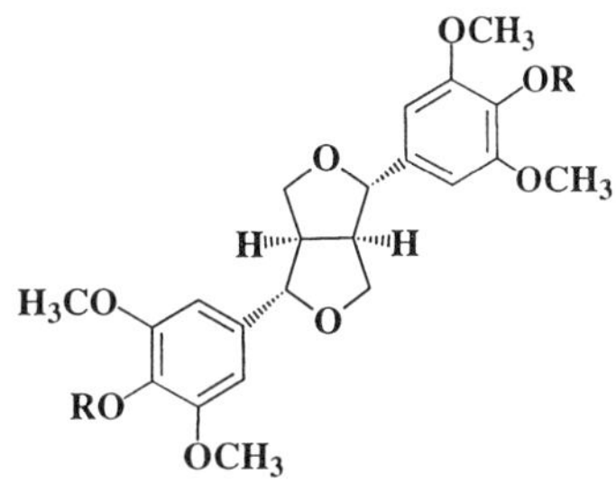

28 R=H: (+)-Syringaresinol
29 R=CH_3: (+)-Yangambin

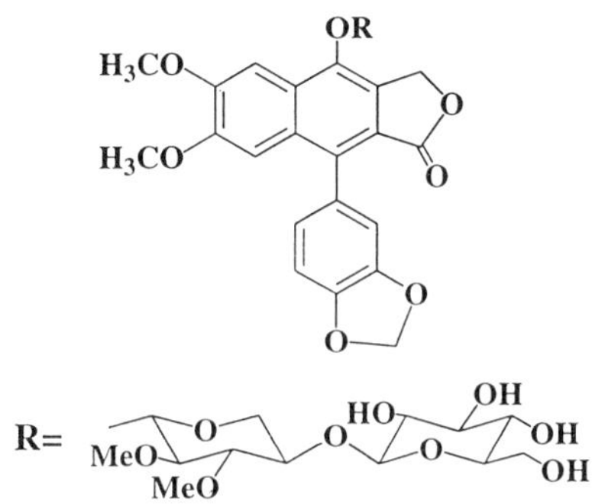

30: Ramontoside

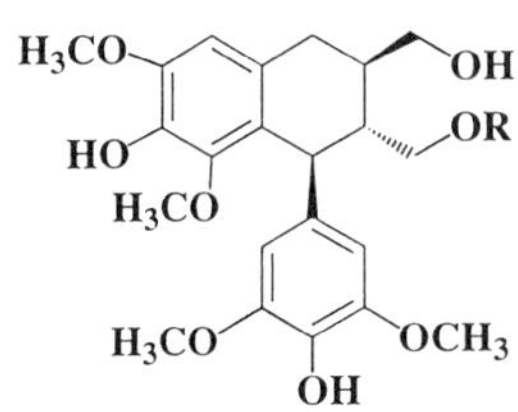

31 R=Xyl: Lyoniside =(+)-Lyoniresinol 2a-*O*-ß-D-xyloside
33 R=H: (+)-Lyoniresinol
35 R=Glc: (+)-Lyoniresinol 2a-*O*-ß-D-glucoside
36 R=Rhamnose: (+)-Lyoniresinol 2a-*O*-ß-D-rhamnoside

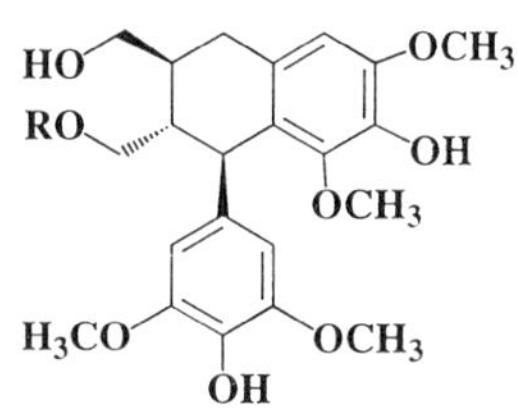

32 R=Xyl: (–)-Lyoniresinol 2a-*O*-ß-D-xyloside
34 R=H: (–)-Lyoniresinol

Figure 2c. Selected lignans listed in Table I (cont.).

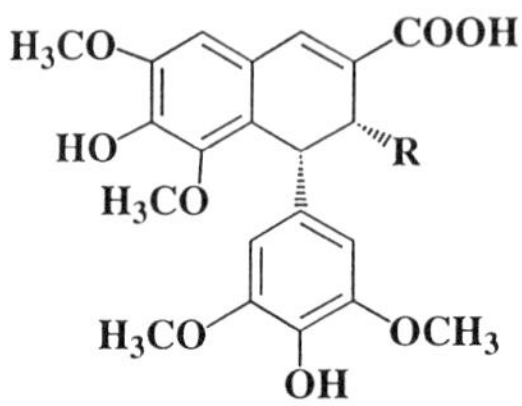

37 R=CH_2OH: *rac*-Thomasic acid
38 R=COOH: *rac*-Thomasidioic acid

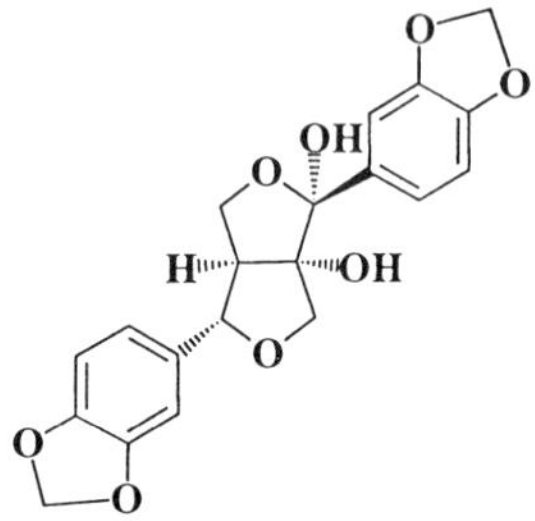

39: Arboreol

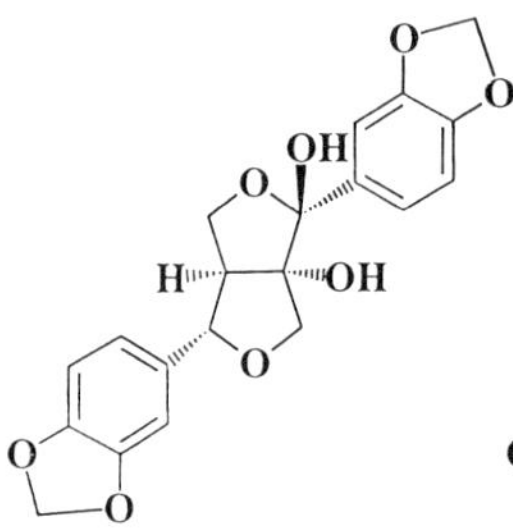

40: Isoarboreol

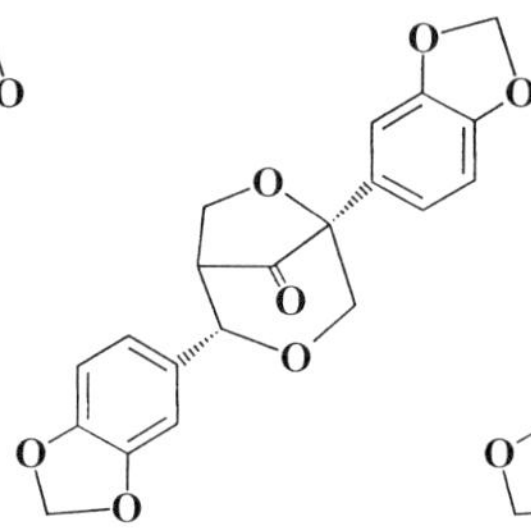

41: Gmelanone

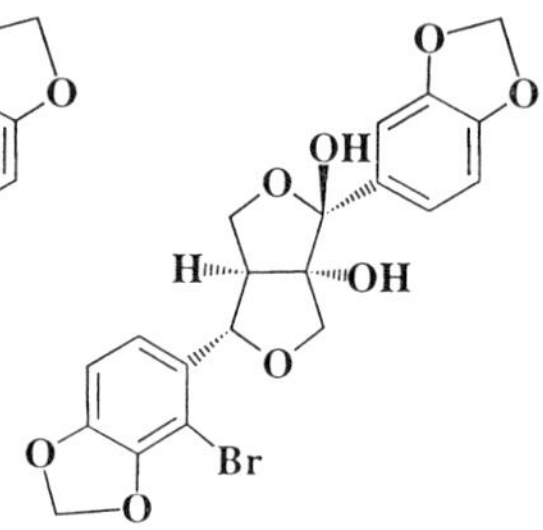

42: (–)-2′-Bromo-isoarboreol

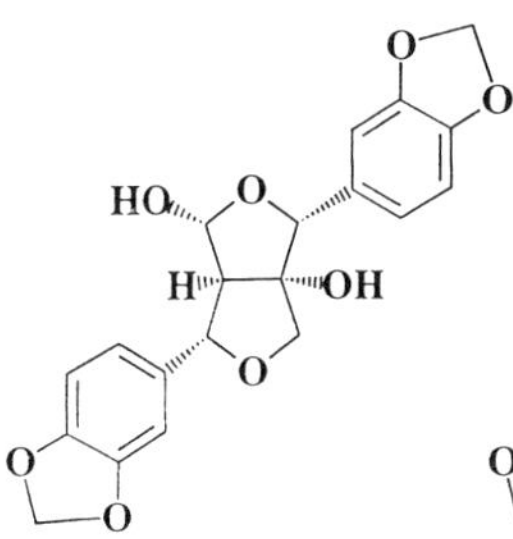

43: (+)-Gummadiol

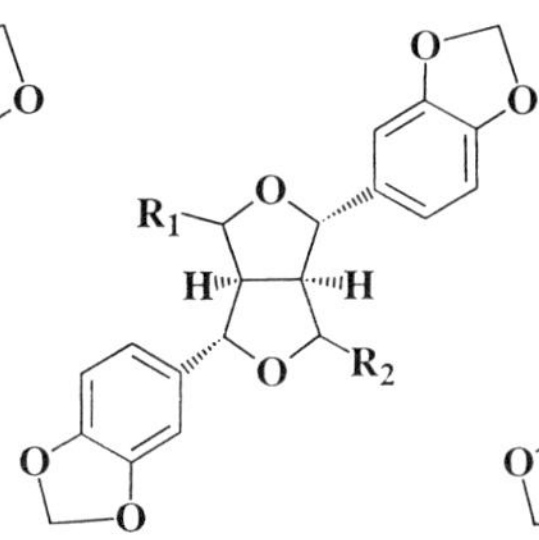

44 R_1=OH, R_2=H: (+)-9-Hydroxysesamin
45 R_1=R_2=OH: (+)-9,9′-Dihydroxysesamin

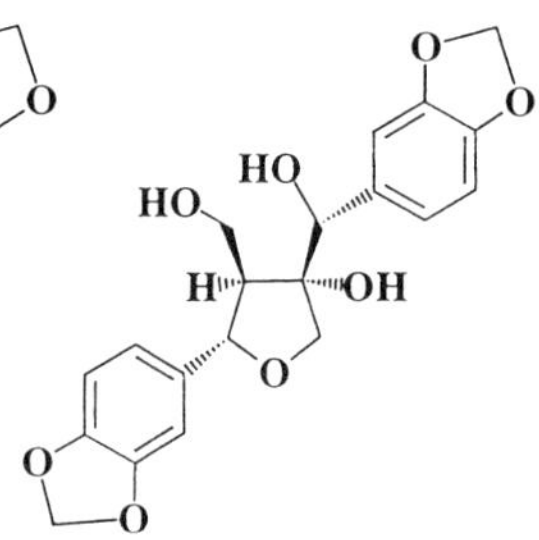

46 Dihydro-gummadiol

Figure 2d. Selected lignans listed in Table I (cont.).

vary substantially between species, thus establishing that no common 'excretion' products are being formed. That is, this particular hypothesis was based on an wholly erroneous and overly simplistic assumption. Instead, heartwood metabolites are not 'wastes' at all, but are instead distinct components of an elaborate defensive arsenal that has evolved to protect these diverse heartwoods from invasion by rotting fungi and bacteria, as well as from other biological challenges.

Heartwood metabolites are only significantly laid down, albeit in differing amounts depending on the species, following secondary wall formation and lignification, *i.e.* their deposition is a *post-lignification* non-structural mechanism. Moreover, according to Hergert (*29*), the heartwood 'extractive' precursors are formed and transported along the ray parenchyma cells. They are, perhaps, eventually converted further in the transition zone, at the sapwood/heartwood boundary, into specific heartwood constituents. Thereafter, these metabolites are employed to further seal off bordered pits in the heartwood zone (*11, 30*) and infuse into adjacent pre-lignified cells. Interestingly, although the precise biochemical events need to be defined, a significant portion of these 'extractive' substances, but not all, become insoluble deposits and can only be removed using 'lignin' dissolution procedures, such as from milled wood lignins (*29*). It must be emphasized, however, that these heartwood substances, even if partially polymerized in some manner, are *not* 'abnormal', 'native' or 'secondary' lignins, as previously and erroneously concluded (*31*). Instead, they are partially polymerized extractives such as polylignans and polymeric (iso)flavonoids. That is, they represent a distinct and unique set of metabolic *non-lignin* products which are deposited into the lignified secondary xylem tracheids *via* infusion through the neighboring specialized cells. As Hergert showed many years ago with infrared spectroscopy (*29, 32, 33*), such heartwood forming substances (then called Braun's native lignins) in western red cedar and western hemlock contain functional groups such as 5-membered lactone rings (*32*). These cannot directly result from the coupling of monolignols, but instead are derived in these species from lignans, such as α-conidendrin **2**, plicatin **3** and matairesinol **4** (see Figure 2), a fact that has been further confirmed by NMR spectroscopic analyses (Hergert, personal communication).

Thus, regardless of their chemical composition, the deposition of heartwood characteristic metabolites occurs in a controlled and regulated manner after lignification has ceased and coincides with the death of parenchyma cells in the outer heartwood layers and in the transition zone. In fact, heartwood formation may give an additional perspective to the type of programmed cell death that has become the focus of so much research in recent years. It is also evident that comparable localized 'heartwood-like' forming processes can also occur in regions, for example, surrounding insect attack (and other stresses) where the plant has attempted to seal off the affected zone.

Heartwood Formation and the 'Abnormal Lignins'

Our knowledge of heartwood formation has not substantially changed over the last 45 years. It should be clear, however, that there is an urgent need to understand both the formation and the precise roles that heartwood constituents play in the physiology of plants, whether solely as defense elements or for additional purposes. Indeed, until their precise biochemical formation and regulation are determined, a definitive understanding of how and why heartwood is formed will not be possible. As Chattaway stated in 1952, 'although various workers have studied heartwood formation intensively, and have even offered explanations for its development in individual genera and species, no one has yet formulated an answer to the fundamental question of what lies behind this uniformity of pattern in the life of so many and such varied trees, so that trees from a very wide variety of families growing in a great range of different habitats produce a stem which contains two types of wood, one an outer layer of relatively pale-colored sapwood, and the other a central core of darker, more durable heartwood' (*25*).

She went on to state that 'although the physical changes that give heartwood its characteristics are dependent on the death of the living cells of the sapwood, *the cause of heartwood formation must be sought further back in the life of the ray cells*, in the agent that stimulates the cells to produce increased quantities of extractives' (emphasis added) (*25*). Both this summation and its implications hold to this day. It is therefore curious that some scientists would, even to this present time, describe such substances as 'abnormal lignins' in particular woods. Not only do they more closely resemble heartwood metabolites formed by distinct and unique biochemical pathways (discussed below), but they are generated through a distinctive infusion system. Indeed, wood cannot be treated as if it were somehow an homogeneous material with no cell types other than lignified xylem, and with all phenolics being lignin-derived! This is not the case.

This chapter will, therefore, not revolve around the initiation or causes of heartwood formation, as such a discussion is premature at this point. This is because we still do not know the precise regulatory control of the biosynthesis of these heartwood compounds. Rather, this discussion will center around what is currently known about the biosynthesis of an important class of heartwood constituents, the lignans, with the belief that an understanding of this biochemical pathway will eventually help lead to the uncovering of the underlying causes of heartwood formation itself.

Prior to describing progress to date in defining the unique biochemistry involved in lignan heartwood metabolite accumulation, some commentary on the fundamental differences between heartwood metabolite deposition and lignification is required. Indeed, an appreciation of these fundamental differences is necessary in order to begin to bring a coherent terminology into the field and to eliminate the laxness used in current nomenclature.

This is because lignification and heartwood metabolite generating processes (*i.e.* affording certain 'extractives', 'abnormal lignins' and 'secondary lignins') differ profoundly in several principal ways, namely in their biochemical mechanisms and in their temporal and spatial deposition:

1. Heartwood metabolite formation, in contrast to lignin synthesis, is first initiated in the center (pith) region of mature woody tissues at some undetermined time. Apparently, comparable substances can also be formed in sapwood in response to biological challenges, such as insect attack. Heartwood metabolites are released from specialized ray parenchyma cells and then further infuse into neighboring cells. Some 48 years ago, Chattaway attempted to graphically illustrate this phenomenon as redrawn in Figure 3 (*25*). That is, the various heartwood-characteristic substances are released by parenchyma cells through the pit apertures into the lumen of adjacent, dead (lignified) cells and then infuse into neighboring, pre-lignified cells (tracheids). The overall biochemistry involved in their formation and transport is now coming to light, and it is fundamentally distinct from the lignin forming processes.

2. Heartwood metabolites can vary extensively between species in terms of amount, molecular size and chemical composition, since they can include lignan, (iso)flavonoid, isoprenoid and even alkaloid-derived substances. It is important to note that such compounds can also constitutively accumulate in sapwood, albeit to a much lower extent, presumably due to their lower concentrations in the 'conducting' sapwood ray cells or as a phytoalexin response.

3. Imperfect characterization of these heartwood/sapwood metabolites, particularly when lignan-derived, has led to vague definitions, such as 'abnormal lignins', 'secondary lignins' and 'Braun's native lignins'. They are not lignins, but instead represent a distinct biochemical class. Indeed, the fact that they can be produced in a range of sizes and can form insoluble deposits, which can only be removed under conditions normally used for lignin removal (*e.g.* milled wood lignins), in no way permits their classification as lignins.

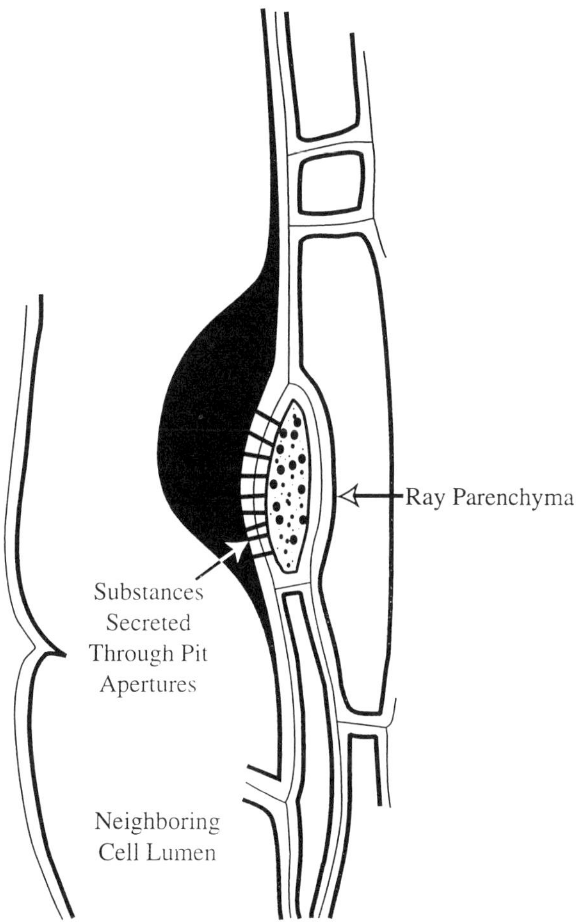

Figure 3. Secretion of heartwood constituents by ray parenchyma cells into the lumen of neighboring cells appears to occur through pit apertures. Redrawn from reference *(25)*.

4. In contrast, lignification is a process of individual cell wall maturation. The monolignols, *E-p*-coumaryl **5**, *E*-coniferyl **6** and *E*-sinapyl **7** alcohols, are transported across the plasma membrane in a lignifying cell where they are then polymerized at specific sites (or regions) within the maturing cell wall (see Chapters 15 and 22).

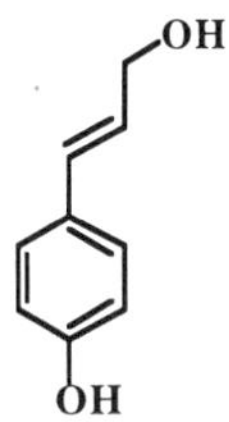

5: *E-p*-Coumaryl alcohol

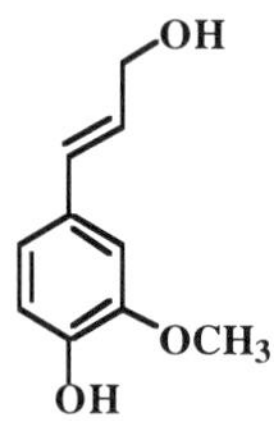

6: *E*-Coniferyl alcohol

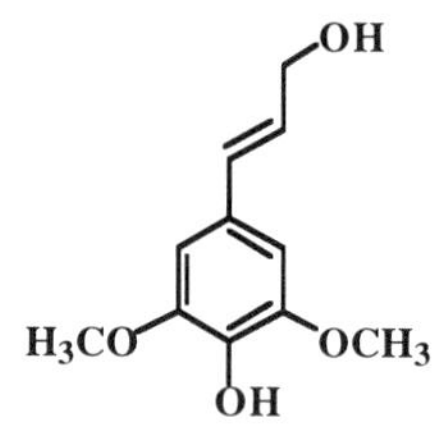

7: *E*-Sinapyl alcohol

5. Lignification occurs in distinct and predictable phases of cell wall development, and is the first major transportation of phenolics into the cell wall of the maturing xylem. It occurs in a progressive manner in cells near to the cambial region, and results in secondary wall (xylem) formation during sapwood generation.
6. Lignins in mature wood tissues that contain little or no heartwood metabolites are essentially colorless; discoloration of both sap- and heartwood is most often due to formation and deposition of non-lignin phenolic metabolites.
7. Polymerization of monolignols represents the first committed, as well as the final, step in the lignin-forming process in woody plants. On the other hand, heartwood metabolite accumulation, including the generation of appropriate precursors in sapwood parenchymal ray cells, often uses monolignols as the entry point into a distinct biochemical pathway to generate the various lignans.

Selected Elements of the Heartwood Arsenal

As mentioned above, various heartwood species contain large and (bio)chemically distinct groups of so-called 'extractives', *i.e. non-lignin* phenolic compounds that help confer particular properties to specific heartwoods. In this chapter, western red cedar, western hemlock and loblolly pine have been chosen as representatives of heartwood-forming species whose main phenolic heartwood constituents are the lignans. It must be emphasized, however, that the underlying principles of deposition described here are viewed as being applicable to all other classes of heartwood substances in other species, such as those also leading to the tannins, (iso)flavonoids, isoprenoids, tropolones, stilbenes and alkaloids, to name only a few compound classes.

Table I and Figure 2 list some of the lignans (compounds **1-4**, **8-46**) (*3-10, 34-45*) known to date as constituents of the substances deposited in the heartwoods of various gymnosperms and angiosperms. The ones shown represent several subclasses of abundant 8,8′-linked lignans, including the furofuran (**28**), tetrahydrofuran (**22**), dibenzylbutane (**23**), dibenzylbutyrolactone (**4, 8**), aryltetrahydronaphthalene (**1**), aryldihydronaphthalene (**37**), arylnaphthalene (**30**) and divanillyltetrahydrofuran (**15**) skeletal types. Four important points about their formation, in terms of the heartwood metabolic arsenal, are discussed below: optical activity, biological activities, magnitude of deposition and biosynthesis.

Table I. Lignans found in heartwood

Compound	Species	Common Name	Family	Reductase Derived?	Reference
1 (–)-plicatic acid	*Thuja plicata*	Western Red Cedar	Cupressaceae	Y	*3-5, 10*
2 (–)-α-conidendrin	*Tsuga heterophylla*	Western Hemlock	Cupressaceae	Y	*34*
	Picea abies	Norway Spruce	Pinaceae	Y	*35*
3 (–)-plicatin	*Thuja plicata*	Western Red Cedar	Cupressaceae	Y	*10*
4 (-)-matairesinol	*Tsuga heterophylla*	Western Hemlock	Cupressaceae	Y	*34, 36*
	Tsuga mertensiana	Mountain Hemlock	Cupressaceae		*36*
	Abies amabilis	Balsam Fir	Pinaceae		*36*
	Abies grandis	Grand Fir	Pinaceae		*36*
	Picea abies	Norway Spruce	Pinaceae	Y	*35*
	Calocedrus formosana		Cupressaceae	Y	*37*
8,9 (–)-thujaplicatin a & b	*Thuja plicata*	Western Red Cedar	Cupressaceae	Y	*6, 9, 10*
10-12 (–)-thujaplicatin methyl ethers	*Thuja plicata*	Western Red Cedar	Cupressaceae	Y	*6-8, 10*
13 (–)-7′-hydroxymatairesinol	*Tsuga heterophylla*	Western Hemlock	Cupressaceae	Y	*34*
	Picea abies	Norway Spruce	Pinaceae	Y	*35*
15 liovil	*Picea abies*	Norway Spruce	Pinaceae	Y	*35*
14 (+)-tsugacetal	*Tsuga chinensis var. formosana*	Taiwan hemlock	Cupressaceae	Y	*38*
16 (–)-hinokinin	*Calocedrus formosana*		Cupressaceae	Y	*37*
17 (–)-7-oxohinokinin	*Calocedrus formosana*		Cupressaceae	Y	*37*
18 (–)-savinin	*Calocedrus formosana*		Cupressaceae	Y	*37*

Continued on next page.

Table I. (cont.) Lignans found in heartwood

Compound	Species	Common Name	Family	Reductase Derived?	Reference
19 (+)-calocedrin	*Calocedrus formosana*	Taiwan hemlock	Cupressaceae	Y	*37*
20,21 (–)-"savininolide a & b"	*Calocedrus formosana*		Cupressaceae	Y	*37*
22a (+)-lariciresinol	*Taxus cuspidata*	Japanese yew	Taxaceae	Y	*39*
23a (–)-secoisolariciresinol	*Taxus baccata* *Taxus cuspidata*	Common yew Japanese yew	Taxaceae	Y	*40* *39*
24 (+)-isolariciresinol	*Taxus baccata* *Taxus cuspidata*	Common yew Japanese yew	Taxaceae	Y	*39*
25 (+)-isotaxiresinol	*Taxus baccata* *Taxus cuspidata*	Common yew Japanese yew	Taxaceae	Y	*40* *39*
26 (+)-isotaxiresinol-4-methyl ether	*Taxus baccata* *Taxus cuspidata*	Common yew Japanese yew	Taxaceae	Y	*39*
27 (+)-taxiresinol	*Taxus baccata*	Common yew	Taxaceae	Y	*40*
28 (+)-syringaresinol	*Liriodendron tulipifera*	Yellow poplar	Magnoliaceae	N	*41*
29 (+)-yangambin	*Liriodendron tulipifera*	Yellow poplar	Magnoliaceae	N	*41*
30 ramontoside	*Flacourtia ramontchi*		Flacourtiaceae Violales	Y	*42*
31 lyoniside =(+)-lyoniresinol 2a-*O*-ß-D-xyloside	*Quercus petraea*	European oak	Fabaceae	Y	*43*

32 (–)-lyoniresinol 2a-*O*-ß-D-xyloside	*Quercus petraea*	European oak	Fabaceae	Y	*43*
33 (+)-lyoniresinol major	*Quercus petraea* *Ulmus thomasii*	European oak	Fabaceae Ulmaceae	Y	*43* *44*
34 (–)-lyoniresinol minor	*Quercus petraea* *Ulmus thomasii*	European oak	Fabaceae Ulmaceae	Y	*43* *44*
35 (+)-lyoniresinol 2a-*O*-ß-D-glucoside	*Quercus petraea*	European oak	Ulmaceae	Y	*43*
36 (+)-lyoniresinol-2a-*O*-rhamnoside	*Ulmus thomasii*		Ulmaceae	Y	*44*
37 *rac*-thomasic acid	*Ulmus thomasii*		Ulmaceae	Y	*44*
38 *rac*-thomasidioic acid	*Ulmus thomasii*		Ulmaceae	Y	*44*
39 arboreol	*Gmelina arborea*	Yemane/Gumhar	Verbenaceae	N	*45*
40 isoarboreol	*Gmelina arborea*	Yemane/Gumhar	Verbenaceae	N	*45*
41 gmelanone	*Gmelina arborea*	Yemane/Gumhar	Verbenaceae	N	*45*
42 (-)-2′-bromo-isoarboreol	*Gmelina arborea*	Yemane/Gumhar	Verbenaceae	N	*45*
43 (+)-gummadiol	*Gmelina arborea*	Yemane/Gumhar	Verbenaceae	N	*45*
44 (+)-9-hydroxysesamin	*Gmelina arborea*	Yemane/Gumhar	Verbenaceae	N	*45*
45 (+)-9,9′-dihydroxysesamin	*Gmelina arborea*	Yemane/Gumhar	Verbenaceae	N	*45*
46 dihydrogummadiol	*Gmelina arborea*	Yemane/Gumhar	Verbenaceae	Y	*45*

First, with the exception of thomasic acid **37** and related substances, lignans in heartwood (and other tissues as well) are normally optically active, although the particular enantiomer present may differ with the plant species (see Table I). For example, (+)-pinoresinol **47a** and (+)-lariciresinol **22a** are found in *Forsythia suspensa* (*46*), but (–)-pinoresinol **47b** occurs in *Xanthoxylum ailanthoides* (*47*), and (–)-lariciresinol **22b** is present in *Daphne tangutica* (*48*) (see below and Figure 2). This optical activity may have important ramifications regarding biological activity. Using a mammalian treatment, as an example, (–)-trachelogenin **48a** inhibits the *in vitro* replication of HIV-I, whereas its (+)-enantiomer **48b** has much reduced activity (*49*).

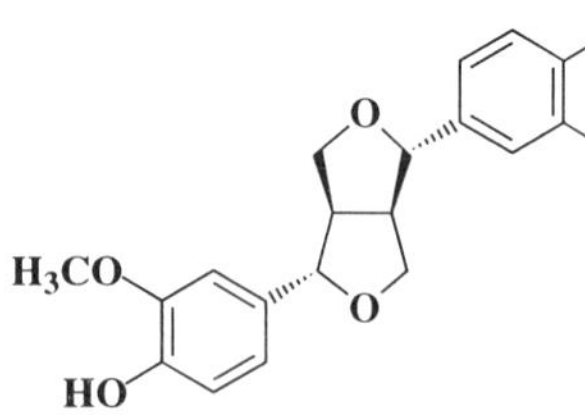

47a: (+)-Pinoresinol

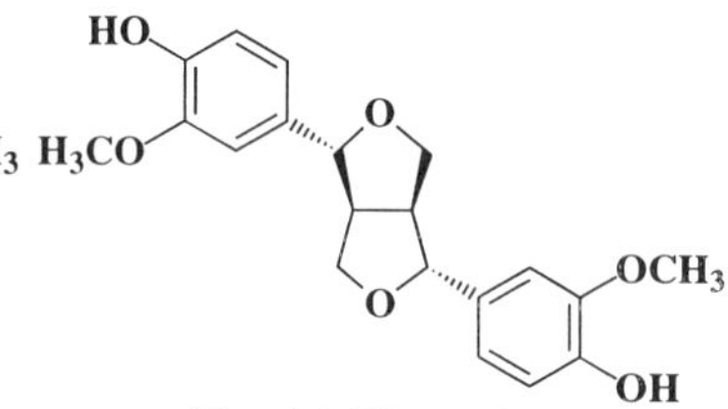

47a: (–)-Pinoresinol

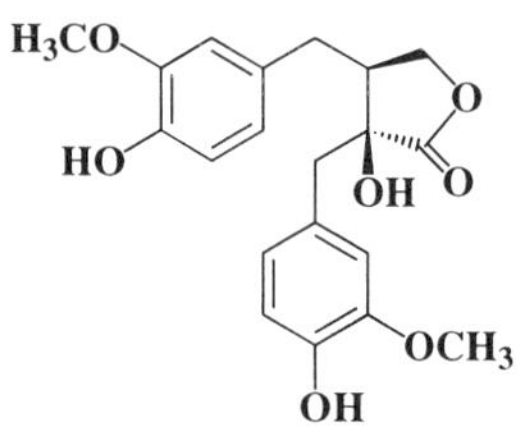

48a: (–)-Trachelogenin

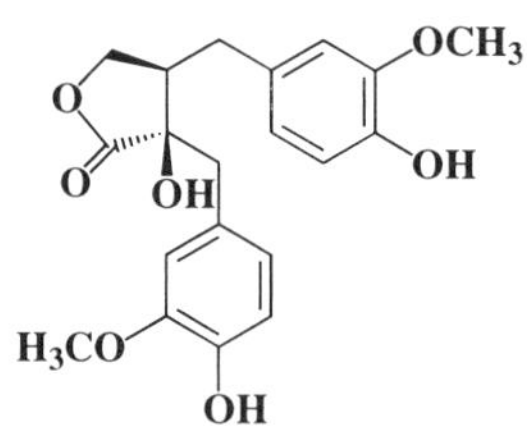

48b: (+)-Trachelogenin

Second, lignans in heartwood can possess important biological activities that also help confer resistance to microbial rotting of heartwood. (–)-Matairesinol **4** and (–)-7′-hydroxymatairesinol **13** are produced in *Picea abies* sapwood after infection by the pathogenic fungus *Fomes annosus* and are thought to limit further fungal growth (*35*). In this example, a compartmentalization response is implicated, thereby preventing death of the living tree (*15*), rather than protection of the dead heartwood *per se*. The fact that matairesinol **4** and 7′-hydroxymatairesinol **13** are found, however, in the heartwoods of several Cupressaceae and Pinaceae species indicates that they may also be involved in protection against microorganisms that attack the dead heartwood and weaken the mechanical support of the tree. In fact, (–)-matairesinol **4** has been reported to inhibit the growth of the wood rotting fungus *Lentinus lepideus* at a concentration of 1% (*50*).

Third, lignans such as (–)-plicatic acid **1** and its derivatives in *Thuja plicata* can accumulate in heartwood in large amount (*2-10*). Indeed, when taken together with the tropolones (which are also thought to contribute to the high quality of this wood), these readily solubilized compounds can make up more than 20% of the mass of the heartwood, based on dry weight. In fact, (–)-plicatic acid **1** by itself accounts for 3.5% of the dry mass of a western red cedar trunk, as does its lactone, (–)-plicatin **3** (*10*). Note, however, that these lignan-derived components can also range from monomers to oligomers and that a portion of them are only removed under conditions required for lignin removal (*29, 32, 33*). Moreover, since the heartwood of gymnosperms is typically *circa* 60% water (*51*), the actual concentration of each lignan alone, (–)-plicatic acid **1** or (–)-plicatin **3**, exceeds 1% of the total heartwood weight. These values compare well to the inhibition concentration of (–)-matairesinol

4 against *L. lepideus* given above (*50*). Thus, they appear to be major contributors in the defense against wood rotting fungi.

Fourth, 8,8′-linked lignans are derived through the lignan biosynthetic pathway as previously established in our laboratory using *Forsythia sp.* (*52-55*). Their formation includes the involvement of dirigent proteins, which help engender formation of (+)-pinoresinol **47a** from *E*-coniferyl alcohol **6**, stereospecific reductases, such as (+)-pinoresinol/(+)-lariciresinol reductase, and stereospecific dehydrogenases, such as (–)-secoisolariciresinol dehydrogenase (outlined in Figure 4).

Salient features of the dirigent proteins, reductases and dehydrogenases are discussed below, since all species examined to date form 8,8′-linked lignans *via* this pathway.

Biosynthesis

Heartwood forming vascular woody plants are not readily amenable to detailed biochemical and molecular analyses, due to their slow growth as well as high levels of phenolic constituents which cause difficulties during DNA and protein purification. It was, therefore, deemed more prudent to first delineate the lignan biosynthetic pathway(s) in *Forsythia intermedia* stems, and then to apply that knowledge to investigate heartwood formation in important wood species such as western red cedar, western hemlock and loblolly pine. This strategy proved remarkably successful, since much of the overall pathway established in *F. intermedia* was also responsible for heartwood lignan biosynthesis in western red cedar and western hemlock. A somewhat analogous system is also present in loblolly pine.

The ‘Abnormal Lignins’: Lignan Biosynthesis in Western Red Cedar and Western Hemlock. Figure 4 shows the proposed biosynthetic pathway to various heartwood lignans based on studies using *Forsythia* species, western red cedar and western hemlock. Initially, with *Forsythia intermedia*, it was established that two monolignols (*E*-coniferyl alcohols **6**) are first coupled to form the furofuran lignan (+)-pinoresinol **47a**, this representing the entry point into the pathway. This is then sequentially reduced, first to the tetrahydrofuran, (+)-lariciresinol **22a**, and then to the dibenzylbutane lignan, (–)-secoisolariciresinol **23a**, which subsequently undergoes a two-step dehydrogenation to form the dibenzylbutyrolactone lignan, (–)-matairesinol **4**.

It was envisaged that this pathway must also operate in western red cedar and western hemlock heartwood metabolite forming processes, and that these precursors can then undergo further metabolism to give the aryltetrahydronaphthalene lignans, (–)-plicatic acid **1** (western red cedar) and (–)-α-conidendrin **2** (western hemlock). Further modifications beyond matairesinol **4** in the western red cedar lignans, such as methylenedioxy bridge and/or aryltetrahydronaphthalene ring formation, hydroxylations and/or methylations, can also occur to produce the variety of lignan structures that are observed. In addition, matairesinol **4** is also the putative precursor of (–)-α-conidendrin **2** in western hemlock.

Taken together, it was, therefore, considered (Figure 4) that the furofuran/tetrahydrofuran reductases (*e.g.* (+)-pinoresinol/(+)-lariciresinol reductase) play a pivotal role in controlling the branching of this pathway to the various subclasses of lignans that are involved in western red cedar and western hemlock heartwood formation. Interestingly, in western hemlock, the lumen of different tracheid cells were examined (*34*) for their lignan contents, where it was found that some contained 7′-hydroxymatairesinol **13**, whereas others contained α-conidendrin **2**, again reflecting the beautifully orchestrated control that is being exercised at the cellular level and that ultimately leads to the ‘abnormal lignins’ in the heartwoods. This is a particularly interesting observation, since it strongly suggests that certain parenchyma cells may be involved in formation of specific metabolites, regardless of the fact that

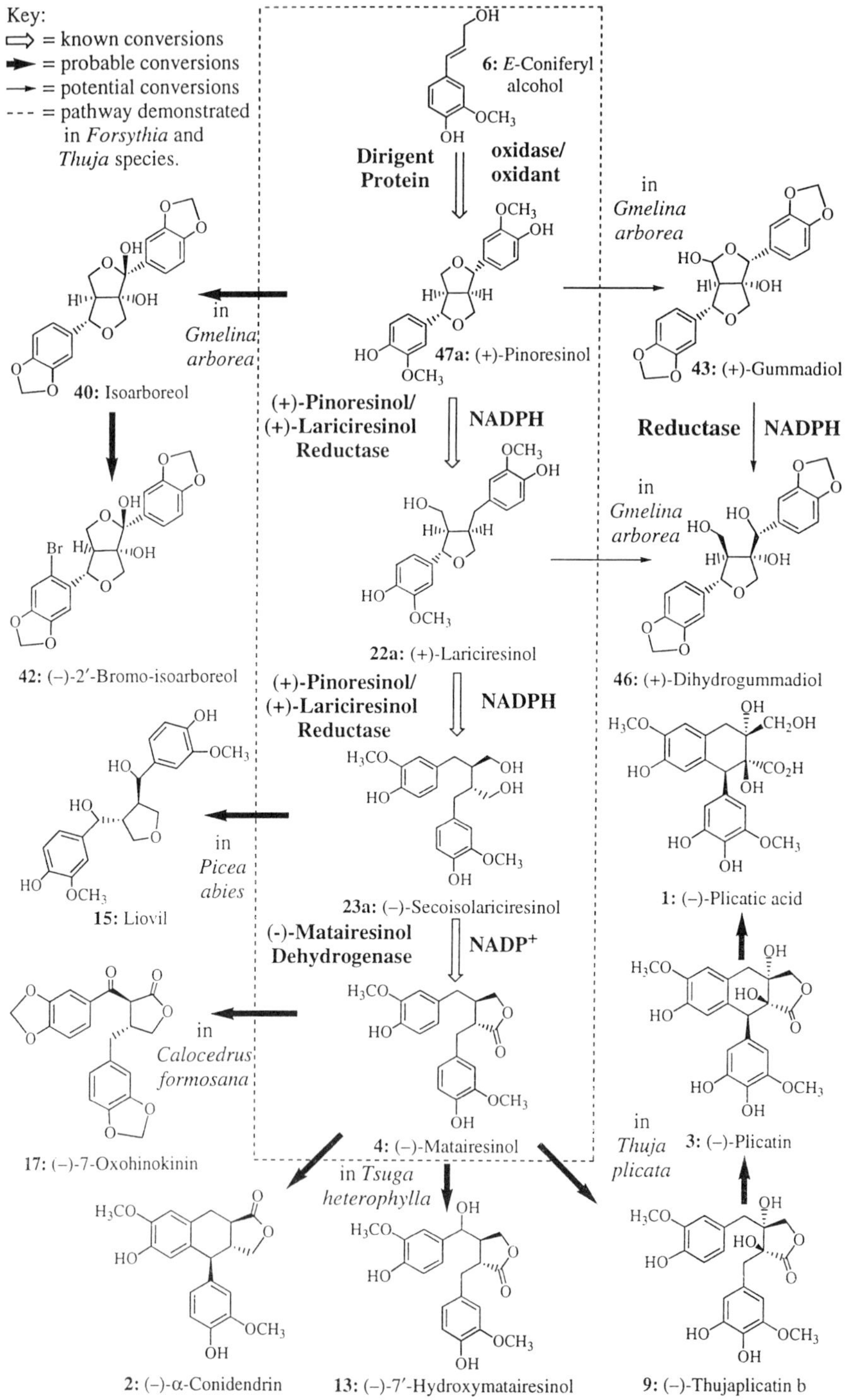

Figure 4. Proposed biosynthetic pathway to various heartwood lignans.

complex mixtures may result in the developed heartwood *via* the infusion process previously discussed.

In any event, as far as the lignans are concerned, pinoresinol **47**, lariciresinol **22**, secoisolariciresinol **23** and matairesinol **4** appear to be general precursors for a wide array of 8,8′-linked lignans (Figure 4). Indeed, as described below, all of the data obtained to date with western red cedar and western hemlock point to the fact that the biochemical transformations leading to heartwood formation are carried out by very specific enzymes, with strict stereochemical specificities for their substrates.

Stereoselective Entry into the Heartwood Lignan Pathway: Dirigent Proteins. In initial investigations in lignan biosynthesis using *Forsythia intermedia*, it was found that the entry point into lignan formation occurred *via* coupling of two *E*-coniferyl alcohol **6** molecules to give (+)-pinoresinol **47a**. From this observation, the discovery was made that in order to control this regio- and stereospecificity, a coupling agent, identified as a 78 kDa dirigent protein, was necessary (*52, 53*). This finding has since been extended to various plant systems, including western red cedar and western hemlock. The 78 kDa dirigent protein apparently only serves to bind and orient the substrate molecules and has no detectable active (oxidative capacity) site (*53*). As discussed in Chapter 22, stereoselective coupling occurs only when auxiliary oxidative capacity is provided through addition of either a non-specific oxidase or a free radical initiator. Thus, the means to achieve stereoselective coupling leading to optical activity differs markedly from that currently viewed to produce lignins, where only non-specific coupling has *to this* point been implicated. The effect of the 78 kDa protein on stereoselective coupling of *E*-coniferyl alcohol **6** is shown in Figure 5.

The 78 kDa protein, which apparently exists as a homotrimer of 27 kDa subunits, was first cloned from a *Forsythia intermedia* stem cDNA library. The resulting cDNA encoded an 18 kDa secreted protein containing several potential N-glycosylation sites, with sugar moieties accounting for the difference in native versus cloned polypeptide size. This cDNA was transferred to an eukaryotic expression system, which employs protein production by a *Spodoptera frugiperda* (fall army worm) cell line following infection with the baculovirus strain containing the cDNA of interest, and subsequently shown to produce functional recombinant dirigent protein. Thus, reactions of *Forsythia* laccase with *E*-coniferyl alcohol **6** in the presence of the recombinant dirigent protein afforded only (+)-pinoresinol **47a** formation (see Figure 5). Therefore, the protein conferring regio- and stereospecificity in the first step of 8,8′-linked lignan biosynthesis has unambiguously been cloned.

This cDNA was then used as a probe to screen western red cedar and western hemlock cDNA libraries, resulting in the cloning of several homologs (see Chapter 22). Current work is underway to heterologously express these and verify that the isolated putative dirigent protein clones are functional. However, their presence strongly suggests that the pathway is also present in western red cedar and western hemlock. Significantly, sequence analysis of all of the cloned dirigent protein cDNAs revealed no homology to any other protein of known function (manuscript in preparation).

The identification and cloning of the dirigent protein is an unprecedented finding, since not only does it represent the *first* example of stereoselective bimolecular free-radical coupling in nature, but it is manifested in a manner that could not have been predicted based on existing chemical, biochemical or molecular precedents. Moreover, since different plant species clearly catalyze distinct coupling modes (*i.e.* to give lignans with linkages other than 8,8′), it is envisaged that this is also directed by comparable dirigent proteins. Put in another way, a new *class* of proteins has apparently been discovered.

Stereospecific Reduction of Heartwood Lignan Precursors. The first enzyme identified in the lignan pathway was initially isolated from *F. intermedia*, *i.e.* the bi-functional reductase that converts (+)-pinoresinol **47a** into (+)-lariciresinol **22a**, and (+)-lariciresinol **22a** into (−)-secoisolariciresinol **23a**, and trivially called

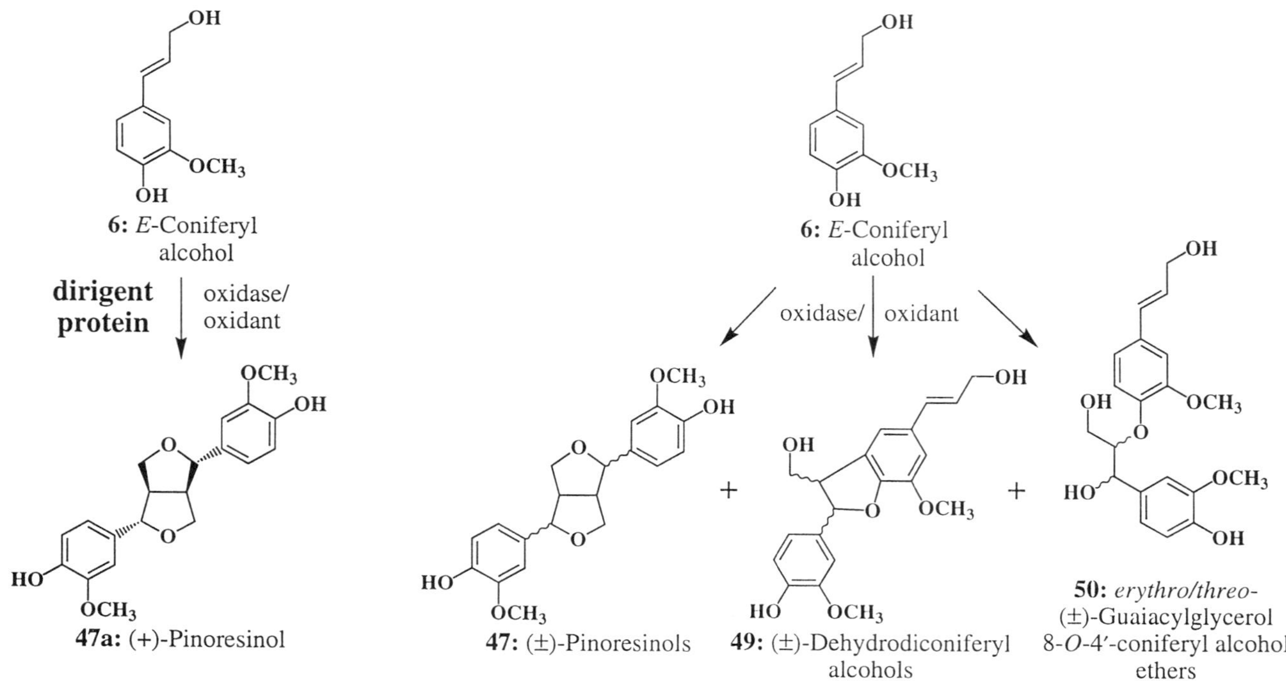

Figure 5. Bimolecular phenoxy radical coupling products from *E*-coniferyl alcohol **6** incubated with an oxidase/oxidant in the presence (left) and absence (right) of the dirigent protein.

(+)-pinoresinol/(+)-lariciresinol reductase (54). This 35 kDa enzyme is a 4A-NADPH-dependent reductase that catalyzes the stereospecific reductions of (+)-pinoresinol **47a** and (+)-lariciresinol **22a**, resulting in the inversion of configuration about C_7 and $C_{7'}$. The gene (called plr-Fi1) encoding this enzyme was subsequently cloned, and its recombinant protein over-expressed in *Escherichia coli* using a pET-based expression system (pSBETa vector) as described in reference (*56*). After purification, the recombinant protein was obtained in catalytically active form. That is, the only products formed from (+)-pinoresinol **47a** are (+)-lariciresinol **22a** and (–)-secoisolariciresinol **23a** (see Figure 6). Figure 7 shows a typical chiral HPLC separation of the products of the NADPH-dependent enzymatic reaction of (±)-pinoresinols **47a/b** catalyzed by *E. coli*-expressed *F. intermedia* (+)-pinoresinol/(+)-lariciresinol reductase (plr-Fi1). As can be seen, only (+)-pinoresinol **47a** and (+)-lariciresinol **22a,** and not (–)-pinoresinol **47b** nor (–)-lariciresinol **22b**, serve as substrates (see Figure 6), *i.e.* the expressed protein functions exactly as the plant-isolated protein does.

After having cloned the gene encoding pinoresinol/lariciresinol reductase from *F. intermedia* and purifying the functional recombinant protein, attention was next directed to obtaining the putative comparable reductases from both western red cedar and western hemlock. The results obtained thus far are very interesting. In western red cedar, *two* reductase cDNAs were identified whose recombinant proteins, expressed in *E. coli*, displayed different pinoresinol/lariciresinol reductase activities. The first, called plr-Tp1, catalyzed the conversion of (–)-pinoresinol **47b** into (–)-lariciresinol **22b**, which was subsequently reduced to give (+)-secoisolariciresinol **23b** (see Figure 6). Figure 7 shows a typical chiral HPLC separation of the products of the conversion of (±)-pinoresinols **47a/b** catalyzed by *E. coli*-expressed plr-Tp1 in the presence of NADPH. That is, it catalyzed the same sequence of reactions as the *Forsythia* reductase but used the opposite enantiomers as substrates. The second, by contrast, called plr-Tp2 (and isolated from the same tree), had the same stereochemical specificity as plr-Fi1, *i.e.* (+)-pinoresinol **47a** was again first converted to (+)-lariciresinol **22a**, and then into (–)-secoisolariciresinol **23a** (see Figures 6 and 7). These radiochemical data were also further substantiated using deuterated substrates (data not shown). Thus, in western red cedar, parallel pathways to lignan formation have been found in the same tree, but with *different* enantiospecificities. This is an important result, since it further underscores the need for caution in interpretation of crude enzymatic data, as well as in the analysis of lignan enantiomeric compositions from either whole plants or particular tissues. That is, there is a much higher degree of control exerted biochemically than previously thought for the lignan-derived metabolites. Moreover, this data may begin to partially explain the enantiomeric compositions of lignans from various plants observed by Umezawa *et al.* in Chapter 24.

Further characterization of the two cedar lignan reductases, plr-Tp1 and plr-Tp2, led to another interesting discovery. In contrast to the *Forsythia* reductase, both are 'leaky' in their reduction of (±)-pinoresinols **47a/b**: plr-Tp1 preferentially converts (–)-pinoresinol **47b** into (–)-lariciresinol **22b** but can also reduce (+)-pinoresinol **47a** to give (+)-lariciresinol **22a**, although at a much slower rate (see Figure 6). The same is true for plr-Tp2: it preferentially converts (+)-pinoresinol **47a** to (+)-lariciresinol **22a**, but can also reduce (–)-pinoresinol **47b** into (–)-lariciresinol **22b** albeit at a much slower rate. Significantly, neither enzyme is able to reduce the opposite antipode of lariciresinol **22**, *i.e.* plr-Tp1 can only convert (–)-lariciresinol **22b** into (+)-secoisolariciresinol **23b**, but does not act on (+)-lariciresinol **22a**. On the other hand, plr-Tp2 can only convert (+)-lariciresinol **22a** into (–)-secoisolariciresinol **23a**, but does not act on (–)-lariciresinol **22b**. The biological significance of this interesting discovery needs to be determined.

What is remarkable about both western red cedar lignan reductases is that plr-Tp2, which has the same major catalytic function as plr-Fi1, has a higher level of amino acid sequence similarity to plr-Tp1 than it has to plr-Fi1 (71% identity, 81% similarity to plr-Tp1 *vs.* 66% identity, 75% similarity to plr-Fi1). Figure 8 is a

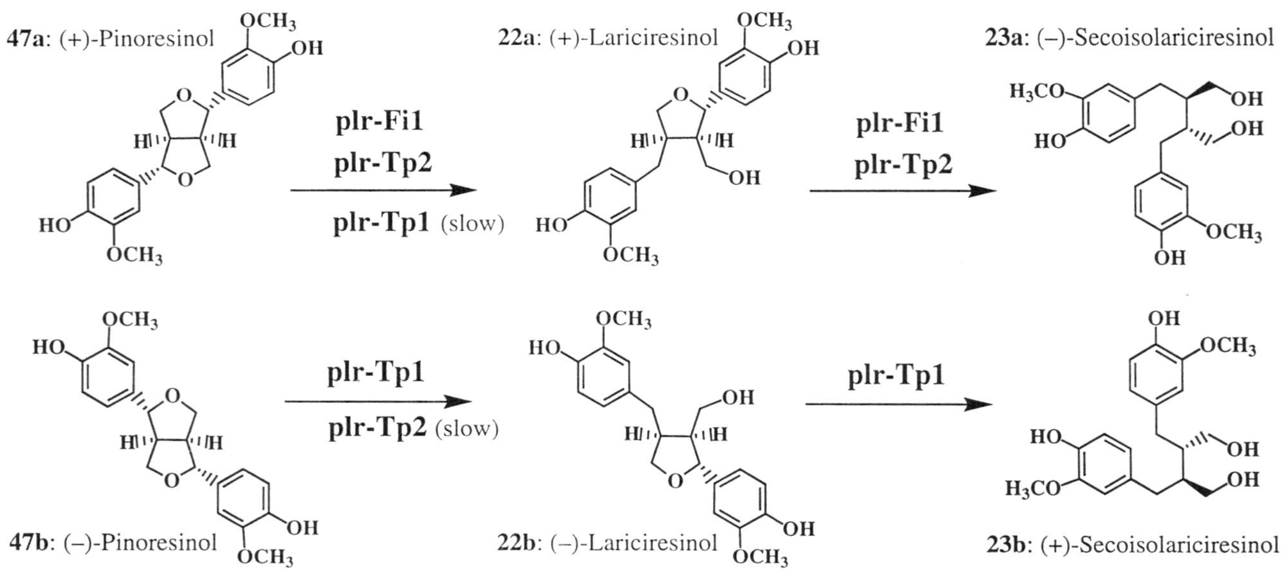

Figure 6. Reactions catalyzed by three recombinant pinoresinol/lariciresinol reductases encoded by genes from *Forsythia* (plr-Fi1) and western red cedar (plr-Tp1 and plr-Tp2). Gene abbreviation over the reaction arrow indicates that the reaction proceeds via that particular protein. All three enzymes catalyze the transfer of the 4A-hydride from the required NADPH cofactor. See text for further details.

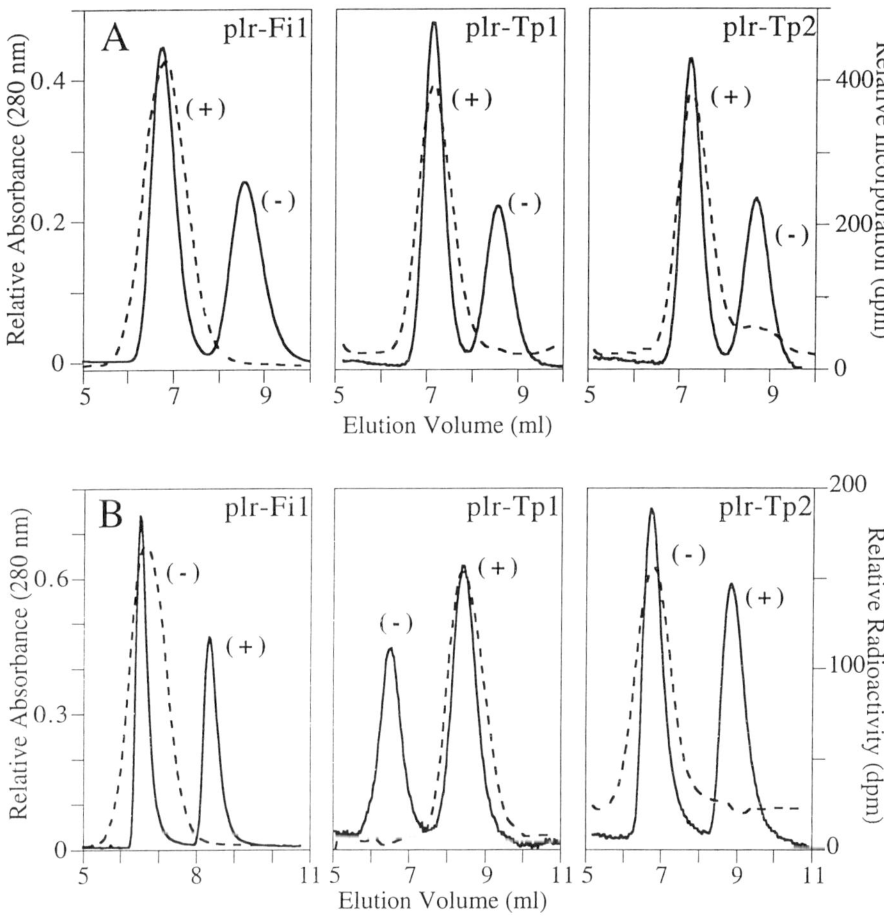

Figure 7. Chiral HPLC separation of lariciresinol **22** (top, **A**) and secoisolariciresinol **23** (bottom, **B**) products from reactions of (±)-pinoresinols **47a/b** with 4A-^{3}H-NADPH in the presence of pinoresinol/lariciresinol reductases from *Forsythia intermedia* [(+)-reductase, plr-Fi1] and *Thuja plicata* [(–)-reductase, plr-Tp1; (+)-reductase, plr-Tp2]. Unlabeled (±)-lariciresinols **22a/b** and (±)-secoisolariciresinols **23a/b** were added as radiochemical carriers after the reactions and are indicated by the solid lines (absorbance at 280 nm). The dashed lines indicate incorporation of the ^{3}H from the NADPH cofactor into the lignan product.

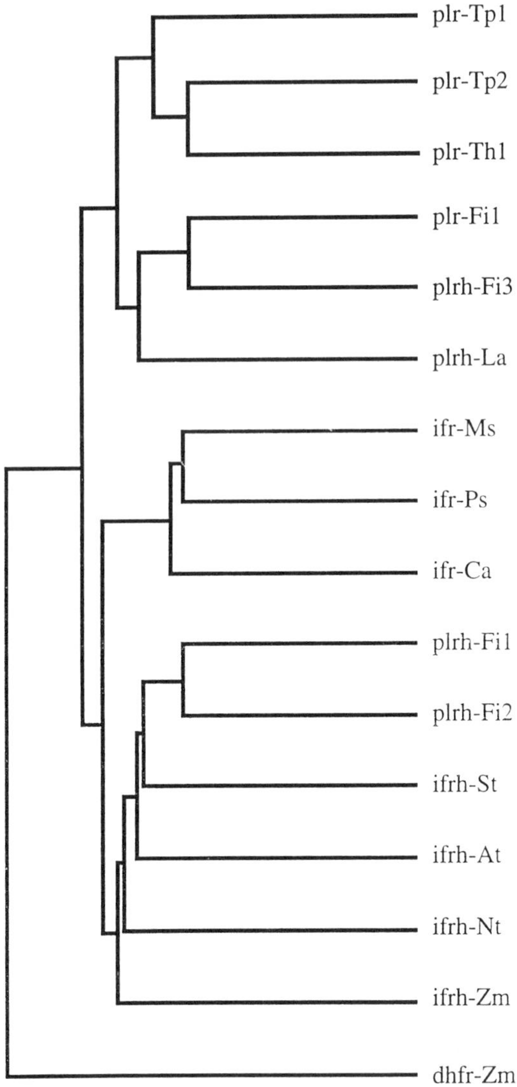

Figure 8. Dendogram showing amino acid sequence homology between pinoresinol reductases (plr-Tp1 and plr-Tp2 from *Thuja plicata,* plr-Fi1 from *Forsythia intermedia*), putative pinoresinol reductase (plr-Th1 from *Tsuga heterophylla*), isoflavone reductases (ifr-Ms from *Medicago sativa*, ifr-Ps from *Pisum sativum*, ifr-Ca from *Cicer arietinum*), reductase 'homologs' (plrh-Fi1, plrh-Fi2 and plrh-Fi3 from *Forsythia intermedia*, plrh-La from *Lupinus alba*, ifrh-St from *Solanum tuberosum*, ifrh-At from *Arabidopsis thaliana*, ifrh-Nt from *Nicotiana tabacum*, ifrh-Zm from *Zea mays*) and dihydroflavonol-4-reductase (dhfr-Zm from *Zea mays*).

dendogram showing the relative similarity of the lignan reductases to each other, as well as to a putative pinoresinol/lariciresinol reductase from western hemlock, to isoflavone reductases and to several other similar genes of unknown function (so-called 'homologs'). Note that plr-Tp2 groups closer to plr-Tp1 than it does to plr-Fi1. Thus, many of the differences between the sequences of these proteins may be catalytically neutral, with only a few critical changes necessary to invert stereospecificity of catalysis. On the other hand, the sequence similarity between plr-Tp1 and plr-Tp2 may reflect the 'leakiness' of these two reductases. Clarifying this matter is the subject of ongoing studies. In addition, a cDNA, most similar to plr-Tp2, from western hemlock has also recently been cloned (called plr-Th1), and its catalytic activity will soon be determined.

Perhaps the most significant finding is, again, that in all of the enzymatic conversions with the lignans observed to date, there is a strict control of stereochemical outcome.

Further Putative Steps in the Western Red Cedar and Western Hemlock Heartwood 8,8′-linked Lignan Biosynthetic Pathway. In addition to the first three steps of the lignan biosynthetic pathway, a fourth has also been identified, namely, the dehydrogenation of secoisolariciresinol **23** to matairesinol **4**. This two-step reaction is catalyzed by (–)-secoisolariciresinol dehydrogenase (see Figure 4), a 37 kDa protein that has been purified from *F. intermedia* and partially characterized in our laboratory (Xia *et al.*, manuscript in preparation). This enzyme is specific for (–)-secoisolariciresinol **23a** as substrate, with only (–)-matairesinol **4** formed as product; $NADP^+$ serves as cofactor for the reaction. (+)-Secoisolariciresinol **23b** does not serve as a substrate; (+)-matairesinol is not formed as product. Thus, the stereospecificity of the lignan pathway continues to be supported. It is anticipated that this enzyme will be found in western red cedar and western hemlock. Additionally, the biochemical steps beyond (–)-matairesinol **4** to (–)-plicatic acid **1** and (–)-α-conidendrin **2** are under investigation, as well as the biochemistry involved in the synthesis of the higher oligomeric forms (so-called 'abnormal lignins' and 'secondary lignins') in these species.

Summary of Known Features of the Heartwood Forming Lignin Pathway in Western Red Cedar and Western Hemlock. To summarize, the screening of cDNA libraries from western red cedar and western hemlock, whose heartwoods are rich in *non-lignin* phenolic, lignan-derived constituents, has been successfully used to delineate the early biochemical steps involved in their heartwood metabolite formation. It was concluded that the entry point in their biosynthesis also occurred *via* stereoselective coupling of two molecules of *E*-coniferyl alcohol **6** to give pinoresinol **47**, the subsequent enantiospecific reduction of which afforded lariciresinol **22** and secoisolariciresinol **23**. Dehydrogenation of the latter gave matairesinol **4**, the presumed precursor of plicatic acid **1**, α-conidendrin **4** and related substances.

Although the initial steps in the biosynthetic pathway (and the corresponding cDNAs) were determined first using *Forsythia* species, the cDNAs putatively encoding dirigent proteins were found for both western red cedar and western hemlock, as were those encoding the reductive steps to give lariciresinol **22** and secoisolariciresinol **23**. Importantly, analysis of the catalytic properties of two recombinant pinoresinol/lariciresinol reductases from western red cedar revealed that this species contains *two* distinct forms, each of which reduces a different enantiomeric form of pinoresinol.

This again underscores the fact that phenolic coupling and subsequent transformations are under exquisite control and not left to chance. It will be of considerable interest in the future to establish whether these two reductases are functionally expressed in the same or different tissues and/or stages of development. Accordingly, future work will be directed towards defining the tissue-specificity (in both a temporal and spatial manner) of these important reactions, as a means to probe and dissect the systems involved in the general process of heartwood formation in these woody plants.

It is now critical that the precise cell types and tissues involved in the formation of the heartwood-bound metabolites be fully established and unambiguously clarified. It would be folly to ignore the fact that they can accumulate in tracheids as distinct, individual constituents, as shown by the Hillis laboratory (*34*), and then ultimately as more complex oligomeric/polymeric components, as demonstrated by Hergert (*29, 32, 33*). This would suggest that some metabolic individualization of transport/biosynthetic processes is associated with the 'conducting' cell types, perhaps particularly at the sapwood-heartwood boundary zone. In any event, the infusion that ultimately occurs to give a complex mixture of these lignans in the heartwood, which can range from simple molecules to higher oligomers, needs to be systematically studied in order to unravel the subtleties involved. Moreover, given that these cell types (*i.e.* ray parenchyma) appear to represent the conduit for transport of different classes of metabolites, this perhaps might also explain how formation of mixed (iso)flavonoid-lignan substances can also occur, *e.g.* in *Distemonanthus benthamianus* (*57*).

The 'Abnormal Lignins': Lignan Formation in Loblolly Pine. It has long been known that various members of the Pinaceae accumulate lignans, such as dihydrodehydrodiconiferyl alcohol **51** (*58, 59*), guaiacylglycerol 8–*O*–4′-dihydroconiferyl alcohol **54**, and their corresponding demethylated analogues **53** (*58, 60, 61*) and **55** (see Figures 9 and 10). Although they are typically present only at low constitutive levels, they can accumulate quite markedly during, for example, gall formation in *Picea glauca* resulting from attack by the gall-forming aphid *Adelges abietis* (*59*). These constituents, as well as formation of other well known metabolites such as dihydroconiferyl alcohol **56** [see reference (*59*)], are thus part of the *normal 'inducible defense' response*. It must be emphasized, however, that all of the factors influencing their induction need to be identified, such as effects of disease, nutrient stress, pathogen attack, and so forth.

Additionally, a relatively low molecular-weight polymeric *non-lignin* fraction, originally erroneously described as 'Brauns native lignins' (*31*), can also be extracted from the Pinaceae. Significantly, depending upon the species in question, the constitutive levels of these *non-lignin* components can be as much as ten times higher in heartwood than in sapwood. Hergert had shown earlier that they result from substances accumulating in ray parenchyma cells. They are often also found as insoluble deposits in tracheid cells, which can only be solubilized under conditions normally used for lignin removal (*29, 32, 33*). (To our knowledge, it is not yet known how their levels change in localized containment areas, such as in response to insect attack.) Accordingly, it is for these reasons that Hergert cautioned that analytical results obtained from the analysis of wood samples *must* take into account the physiological condition, *i.e.* whether the sample comes from sapwood, heartwood, compression wood, diseased wood, *etc.*

In spite of these cautionary admonitions about specifying the physiological condition of the wood sample, various scientific studies to the present day have summarily ignored this aspect in their research undertakings. For example, acidolysis of spruce (Pinaceae) milled wood lignins resulted in the isolation of so-called 'reduced substructures' such as the divanillyl furan, shonanin **57** (see Figure 10) (*62, 63*). Indeed, it was these findings coupled with related studies that suggested that lignin contains ~2% of these 'reduced substructures'. The results were interpreted further as implying that, in addition to oxidative polymerization, reductive steps were also involved in lignin biosynthesis. An alternative and much more likely explanation is, however, that such metabolites are infused into the pre-lignified xylem *via* the ray parenchymal pit apertures, as previously described for the other lignan heartwood constituents.

Given our considerable interest in *systematically* establishing the actual biochemical events controlling lignin and lignan assembly *in vivo*, it was deemed

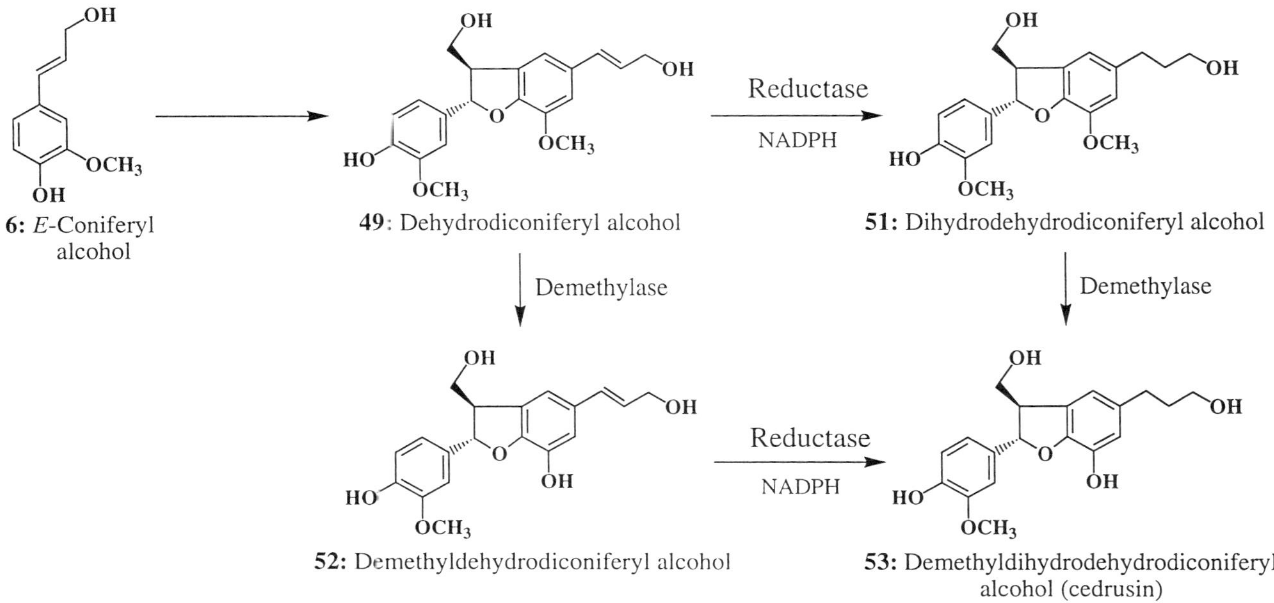

Figure 9. Conversion of dehydrodiconiferyl alcohol **49** to its reduced **51**, demethylated **52**, and reduced/demethylated **53** forms in *Pinus taeda*. Note: dehydrodiconiferyl alcohol **49** is viewed to be produced from *E*-coniferyl alcohol **6** via direct coupling.

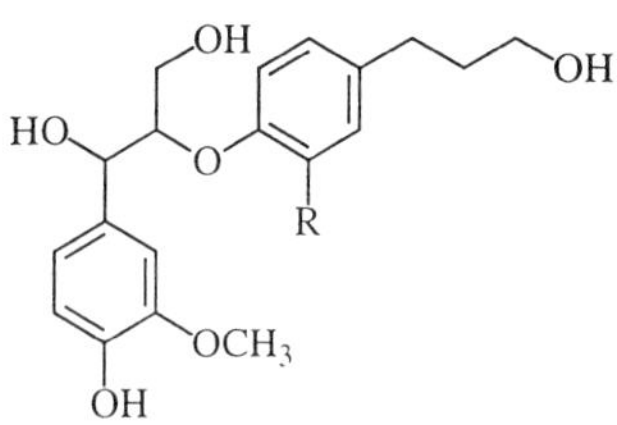

54: R = OCH_3, Guaiacylglycerol 8-*O*-4′-dihydroconiferyl alcohol ethers

55: R = OH, Guaiacylglycerol 8-*O*-4′-demethyl-dihydroconiferyl alcohol ethers

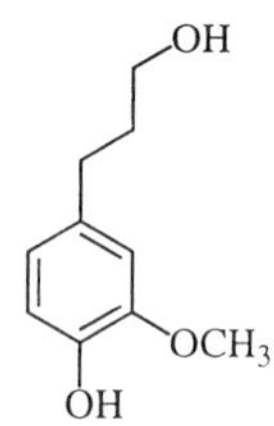

56: Dihydroconiferyl alcohol

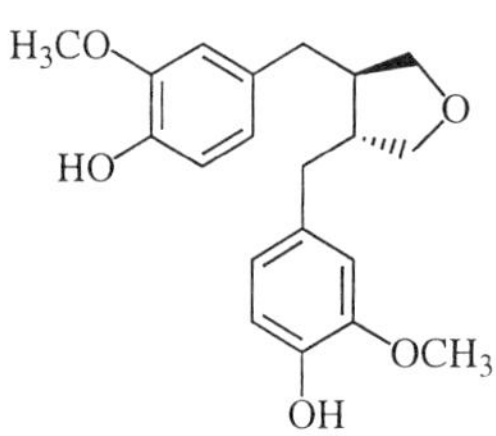

57: Shonanin

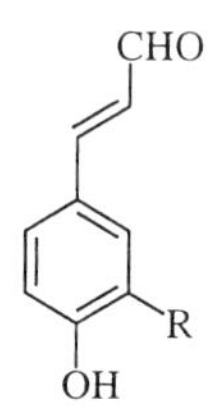

58: R = OCH_3, *E*-Coniferyl aldehyde

59: R = H, *E*-*p*-Coumaryl aldehyde

Figure 10. Selected lignans and monolignols mentioned in text.

instructive some years ago (*64*) to investigate how formation of a so-called 'extracellular lignin' precipitate was induced in loblolly pine (*Pinus taeda*, Pinaceae) cell suspension cultures. This occurs when unlignified *P. taeda* cell cultures are grown in a medium containing 8% sucrose. It is manifested as both a 'cell-wall lignification' response and by secretion of a complex mixture of lignans into the extracellular medium, which are subsequently 'polymerized' to give the 'extracellular lignin precipitate'. Careful and systematic identification of the dimeric substances initially released into the medium revealed that at least part of the so-called 'reduced lignin' substructures were derived from dihydrodehydrodiconiferyl alcohol **51** and shonanin **57** (*64*). Further, over a 96 hour time period, these substances undergo a series of conversions, including demethylation reactions in the extracellular medium, to afford the so-called 'extracellular lignin' precipitate. Such isolates, however, as shown by both the Lewis (*64*) and Brunow (*65*) laboratories, using both loblolly pine and spruce cell suspension cultures, respectively, are *not* lignins proper. Nor are they generated in a biochemical manner that can in any satisfactory way duplicate lignification *in vivo*.

Indeed, formation of the 'extracellular lignin precipitate' appears to have more similarity to either the heartwood phenolic metabolite infusion process or to an inducible (phytoalexin-like) response.

In spite of the rather extensive literature on *non-lignin* phenolics and their infusion processes, and the need for caution in interpreting crude analytical data, a recent claim (*66, 67*) has stipulated that, if monolignols are not available for lignin biosynthesis, then 'abnormal' lignification results. According to these authors, the plant(s) in question will utilize any other available phenolic precursors to make 'abnormal lignins'. This claim originated with the analysis of phenolic constituents released from a 'milled wood lignin' preparation from discolored loblolly pine wood, which was considered by the authors to be deficient in cinnamyl alcohol dehydrogenase, the enzyme catalyzing the final step in monolignol biosynthesis.

Acceptance of such a forthright claim would constitute nothing less than a shift to a paradigm that has no precedence elsewhere in biochemistry. Accordingly, some detailed consideration and discussion of this recent hypothesis, and its full implications, is required, since it has a direct bearing on the subject of this chapter.

The hypothesis arose following the analysis of a 12-year old, putative CAD^- loblolly pine mutant, which contained discolored woody tissue, and was described as harboring an 'abnormal lignin' derived from dihydroconiferyl alcohol **56**. It was based solely on the analysis of a soluble 'milled wood lignin' preparation, released by dioxane-H_2O, and its comparison to another preparation from loblolly pine plants that did not duplicate this particular phenotype, *even though they had the identical genotype*. The 'mutant' lignin preparation was reported to contain some 30% of dihydroconiferyl alcohol **56** substructures (*66*), even though only a very small percentage of the putative lignin present (~17%) was actually solubilized. Perhaps most bewildering of all, it was claimed to result from the oxidative coupling of dihydroconiferyl alcohol **56** rather than *E*-coniferyl alcohol **6**.

Calculation of the contribution of 'dihydroconiferyl alcohol' **56** to that of the presumed 'lignin' content of the sample revealed that it only accounted for some 5.1% of the total. This value does not stray very far from the original reports by Adler *et al.* (*63*) of *circa* 2% reduced substructures in lignin for *typical* Pinaceae wood specimens, particularly when one considers that significant losses are incurred and expected during the acidolysis procedure.

The primary distinguishing feature of this recent hypothesis, however, was that the monomer was claimed to have resulted *directly via* a two-step reduction from *E*-coniferyl aldehyde **58** to give *E*-dihydroconiferyl alcohol **56**, since the pathway to *E*-coniferyl alcohol **6** was *putatively* blocked. The authors did not explain, however, how *p*-coumaryl alcohol **5** formation still occurred since this would presumably involve a reduction analogous to that catalyzed by CAD. Nor did they explain how the putative lignin substructure, dihydrodehydrodiconiferyl alcohol **51** was being formed, if coniferyl alcohol **6** biosynthesis was blocked as claimed. Indeed, as can be seen by simple inspection of substructure **51**, it is built up from *at least* 50% of coniferyl alcohol **6** moieties. Clearly, *E*-coniferyl alcohol **6** is in fact being formed, in contrast to the claim that its formation is being severely repressed.

It seems remarkable that a new paradigm would, therefore, be implicated when existing explanations are already in place for each and every observation made. Put in another way, the results obtained match that already known for infusion of non-lignin metabolites, *e.g.* from specialized cells (such as ray parenchyma cells) into neighboring pre-lignified cells, whether involved as an inducible response or for constitutive heartwood formation. Yet, this was not in any way discussed or referenced as a possibility by these authors, in spite of the extensive literature on this subject spanning more than five decades. [It was also most unsettling that only one of the putative CAD$^-$ loblolly pine plants, in a homozygous population, had this particular discolored wood phenotype (J. Vernon, Clemson University, personal communication), and from a plantation that had experienced significant hurricane damage some years earlier (L. Pearson, Westvaco, personal communication).]

On the other hand, as already documented from previous systematic research in our laboratory ((*64, 68*) and Jiao *et al.*, manuscript in preparation), using crude cell-free extracts from *P. taeda* cell suspension cultures, dehydrodiconiferyl alcohol **49** undergoes allylic bond reduction, sometimes with concomitant demethylation, to afford dihydrodehydrodiconiferyl alcohol **51** and its demethylated analog cedrusin **53** (see Figure 9). That is, this lignan can undergo two regiospecific transformations, involving double bond reduction and demethylation. Although the precise enzymology and co-factor dependence of *both* transformations must await purification of the enzymes involved to apparent homogeneity, it nevertheless provides a plausible explanation as to how their formation can occur without any necessity for imposing a new paradigm. Put in another way, it is not necessary to implicate any biochemical process that obviates the involvement of cinnamyl alcohol dehydrogenase at this juncture.

Taken together, these results demonstrate that loblolly pine, like western red cedar and western hemlock, is able to generate non-lignin substances *via* a lignan/monolignol biosynthetic pathway as before. Perhaps the most important lesson to be learned from the reports of 'abnormal lignins' is that future work in this area must be more *circumspect* in differentiating between distinct areas of plant metabolism.

In short, in order to propose that a particular plant phenolic substance is somehow conscripted into the lignification process, specific conditions must first be met:

1. Monomers must be formed in the cytoplasm of the lignifying cell prior to their transportation across the plasma membrane and into the cell wall, where they are subsequently polymerized. If they do not, they are not part of the lignification response.

2. It must be established that the resulting structures in the cell wall are indeed polymeric, and are not merely insoluble deposits of oligomeric components.

3. It should be shown that such substances are not, in fact, part of the *non-lignin* phenolic metabolic system that is associated with the biosynthesis/transport/infusion process released from specialized cells, such as ray parenchyma, into the surrounding pre-lignified cells and tissues.

4. That putative lignin mutants are not, in fact, plants that are diseased, infected or challenged in some other manner, or whose metabolite accumulation is part of the normal defense process.

Acknowledgments

The authors thank the United States Department of Agriculture (9603622), the National Science Foundation (MCB-9631980), the United States Department of Energy (DEFG0397ER20259) and McIntire-Stennis for generous support of this study, as well as the Organisation for Economic Co-operation and Development for support to Dr. Fujita (partial stipend and travel award), and the Loyal and Helen Davis Trust for a partial fellowship to D. R. Gang.

Literature Cited

1. Panshin, A. J.; de Zeeuw, C. *Textbook of Wood Technology*; McGraw-Hill: New York, NY, 1980, pp 722.
2. MacLean, H.; Gardner, J. A. F. *For. Prod. J.* **1956**, *6*, 510-516.
3. Gardner, J. A. F.; Barton, G. M.; MacLean, H. *Can. J. Chem.* **1959**, *37*, 1703-1709.
4. Gardner, J. A. F.; MacDonald, B. F.; MacLean, H. *Can. J. Chem.* **1960**, *38*, 2387-2394.
5. Gardner, J. A. F.; Swan, E. P.; Sutherland, S. A.; MacLean, H. *Can. J. Chem.* **1966**, *44*, 52-58.
6. MacLean, H.; Murakami, K. *Can. J. Chem.* **1966**, *44*, 1541-1545.
7. MacLean, H.; Murakami, K. *Can. J. Chem.* **1966**, *44*, 1827-1830.
8. MacLean, H.; Murakami, K. *Can. J. Chem.* **1967**, *45*, 305-309.
9. MacLean, H.; MacDonald, B. F. *Can. J. Chem.* **1967**, *45*, 739-740.
10. Swan, E. P.; Jiang, K. S.; Gardner, J. A. F. *Phytochemistry* **1969**, *8*, 345-351.
11. Liese, W.; Bauch, J. *J. Inst. Wood Sci.* **1967**, *19*, 3-14.
12. Shigo, A. L.; Hillis, W. E. *Annu. Rev. Phytopathol.* **1973**, *11*, 197-222.
13. Shigo, A. L. *Northern Logger and Timber Processing* **1975**, *24*, 28-29.
14. Shigo, A. L.; Shortle, W. C. *Phytopathology* **1979**, *69*, 710-711.
15. Shigo, A. L. *Scientific American* **1985**, *252*, 96-103.
16. Hillis, W. E. In *The Structure, Biogenesis and Degradation of Wood*; Loewus, F. A.; Runeckles, V. C., Ed.; Recent Advances in Phytochemistry; Plenum Press: New York, NY, 1977, Vol. 11, pp 247-309.
17. Bamber, R. K. *Wood Sci. Technol.* **1976**, *10*, 1-8.
18. Frey-Wyssling, A.; Bosshard, H. H. *Holzforschung* **1959**, *13*, 129-137.
19. Bosshard, H. H. *News Bull. Int. Assoc. Wood Anat.* **1966**, 11-14.
20. Bosshard, H. H. *Holzals Roh-und Werkstoff* **1967**, *25*, 409-416.
21. Bosshard, H. H. *Wood Sci. Technol.* **1968**, *2*, 1-12.
22. Hugentobler, U. H. *Vierteljahrsschrift der Naturforschenden Gesellschaft in Zurich* **1965**, *110*, 321.
23. Shain, L.; Hillis, W. E. *Can. J. Bot.* **1973**, *51*, 1331-1335.
24. Shain, L.; MacKay, J. F. G. *Can. J. Bot.* **1973**, *51*, 737-741.
25. Chattaway, M. M. *Aust. For.* **1952**, *16*, 25-34.
26. Bhat, K. V.; Patel, J. D. *Flora* **1980**, *170*, 144-157.
27. Stewart, C. M. *Nature* **1960**, *186*, 374-375.
28. Stewart, C. M. *Science* **1966**, *153*, 1068-1074.
29. Hergert, H. L. In *Cellulose Chemistry and Technology*; Arthur, J. C., Jr., Ed.; ACS Symposium Series; American Chemical Society: Washington, DC, 1977, Vol. 48, pp 227-243.
30. Bauch, J.; Schweers, W.; Berndt, H. *Holzforschung* **1974**, *28*, 86-91.
31. Brauns, F. E. *J. Am. Chem. Soc.* **1939**, *61*, 2120-2127.
32. Hergert, H. L. *J. Org. Chem.* **1960**, *25*, 405-413.
33. Hergert, H. L. In *Lignins, Occurrence, Formation, Structure and Reactions*; Sarkanen, K. V.; Ludwig, C. H., Ed.; Wiley-Interscience: New York, NY, 1971, pp 267-297.

34. Krahmer, R. L.; Hemingway, R. W.; Hillis, W. E. *Wood. Sci. Technol.* **1970**, *4*, 122-139.
35. Shain, L.; Hillis, W. E. *Phytopathology* **1971**, *61*, 841-845.
36. Barton, G. M.; Gardner, J. A. F. *J. Org. Chem.* **1962**, *27*, 322-323.
37. Fang, J.-M.; Liu, M.-Y.; Cheng, Y.-S. *Phytochemistry* **1990**, *29*, 3048-3049.
38. Fang, J.-M.; Wei, K.-M.; Cheng, Y.-S.; Cheng, M.-C.; Wang, Y. *Phytochemistry* **1985**, *24*, 1363-1365.
39. Erdtman, H.; Tsuno, K. *Phytochemistry* **1969**, *8*, 931-932.
40. Mujumdar, R. B.; Srinivasan, R.; Venkataraman, K. *Indian J. Chem.* **1972**, *10*, 677-680.
41. Chen, C. L.; Chang, H. M.; Cowling, E. B. *Phytochemistry* **1976**, *15*, 547-550.
42. Satyanarayana, V.; Krupadanam, G. L. D.; Srimannarayana, G. *Phytochemistry* **1991**, *30*, 1026-1029.
43. Dada, G.; Corbani, A.; Manitto, P.; Speranza, G.; Lunazzi, L. *J. Nat. Prod.* **1989**, *52*, 1327-1330.
44. Hostettler, F. D.; Seikel, M. K. *Tetrahedron* **1969**, *25*, 2325-2337.
45. Anjaneyulu, A. S. R.; Madhusudhana Rao, A.; Kameswara Rao, V.; Ramachandra Row, L.; Pelter, A.; Ward, R. S. *Tetrahedron* **1977**, *33*, 133-143.
46. Kitagawa, S.; Nishibe, S.; Benecke, R.; Thieme, H. *Chem. Pharm. Bull.* **1988**, *36*, 3667-3670.
47. Ishii, H.; Ishikawa, T.; Mihara, M.; Akaike, M. *Yakugaku Zasshi* **1983**, *103*, 279-292.
48. Lin-gen, Z.; Seligmann, O.; Jurcic, K.; Wagner, H. *Planta Med.* **1982**, *45*, 172-176.
49. Schröder, H. C.; Merz, H.; Steffen, R.; Müller, W. E. G.; Sarin, P. S.; Trumm, S.; Schulz, J.; Eich, E. *Z. Naturforsch* **1990**, *45c*, 1215-1221.
50. Rudman, P. *Holzforschung* **1965**, *19*, 57-58.
51. Hillis, W. E. *Heartwood and Tree Exudates*; Springer-Verlag: Berlin, 1987, pp 268.
52. Davin, L. B.; Lewis, N. G. *An. Acad. bras. Ci.* **1995**, *67 (Supl. 3)*, 363-378.
53. Davin, L. B.; Wang, H.-B.; Crowell, A. L.; Bedgar, D. L.; Martin, D. M.; Sarkanen, S.; Lewis, N. G. *Science* **1997**, *275*, 362-366.
54. Dinkova-Kostova, A. T.; Gang, D. R.; Davin, L. B.; Bedgar, D. L.; Chu, A.; Lewis, N. G. *J. Biol. Chem.* **1996**, *271*, 29473-29482.
55. Gang, D. R.; Dinkova-Kostova, A. T.; Davin, L. B.; Lewis, N. G. In *Phytochemicals for Pest Control*; Hedin, P. A.; Hollingworth, R. M.; Masler, E. P.; Miyamoto, J.; Thompson, D. G., Ed.; ACS Symposium Series; Washington, DC, 1997, Vol. 658, pp 58-89.
56. Shenk, P. M.; Baumann, S.; Mattes, R.; Steinbiß, H.-H. *BioTechniques* **1995**, *19*, 196-200.
57. Malan, E.; Swinny, E.; Ferreira, D. *Phytochemistry* **1994**, *37*, 1771-1772.
58. Popoff, T.; Theander, O. *Phytochemistry* **1975**, *14*, 2065-2066.
59. Kraus, C.; Spiteller, G. *Phytochemistry* **1997**, *44*, 59-67.
60. Higuchi, R.; Aritomi, M.; Donnelly, D. M. X. *Phytochemistry* **1977**, *16*, 1007-1011.
61. Agrawal, P. K.; Agarwal, S. K.; Rastogi, R. P. *Phytochemistry* **1980**, *19*, 1260-1261.
62. Adler, E.; Lundquist, K.; Miksche, G. E. *Adv. Chemn. Ser.* **1966**, *59*, 22.
63. Adler, E. *Wood Sci. Technol.* **1977**, *11*, 169-218.
64. Nose, M.; Bernards, M. A.; Furlan, M.; Zajicek, J.; Eberhardt, T. L.; Lewis, N. G. *Phytochemistry* **1995**, *39*, 71-79.
65. Brunow, G.; Kilpeläinen, I.; Lapierre, C.; Lundquist, K.; Simola, L. K.; Lemmetyinen, J. *Phytochemistry* **1993**, *32*, 845-850.
66. Ralph, J.; MacKay, J. J.; Hatfield, R. D.; O'Malley, D. M.; Whetten, R. W.; Sederoff, R. R. *Science* **1997**, *277*, 235-239.

67. MacKay, J. J.; O'Malley, D. M.; Presnell, T.; Booker, F. L.; Campbell, M. M.; Whetten, R. W.; Sederoff, R. R. *Proc. Natl. Acad. Sci., USA* **1997**, *94*, 8255-8260.
68. Davin, L. B.; Gang, D. R.; Fujita, M.; Anterola, A.; Lewis, N. G. 9th International Symposium on Wood and Pulping Chemistry; Montréal, Québec, Canada; Technical Section CPPA: 1997, Vol. 1, pp H3-1 - H3-4.

Author Index

Subject Index

C

D

E

F

M

N

O

P

Q

R

S

T

V

W

X

Z